Lecture Notes in Mathematics

Edited by A. Dold and B. Eckmann

564

Ordinary and Partial Differential Equations

Proceedings of the Fourth Conference
Held at Dundee, Scotland
March 30 – April 2, 1976

Edited by
W. N. Everitt and B. D. Sleeman

Springer-Verlag
Berlin · Heidelberg · New York 1976

Editors
William N. Everitt
Brian D. Sleeman
Department of Mathematics
University of Dundee
Dundee DD1 4HN/Scotland

Library of Congress Cataloging in Publication Data

Ordinary and Partial Differential Equations, Dundee 1976
Proceedings of the Fourth Conference held at
Dundee, Scotland, March 30 - April 2, 1976

(Lecture notes in mathematics ; 564)
1. Differential equations--Congresses. 2. Differential equations, Partial--Congresses. I. Everitt, William Norrie. II. Sleeman, B. D. III. Title.
QA3.L28 no. 564 [QA371] 510'.8s [515'.35] 76-51380

AMS Subject Classifications (1970): 26A84, 33A40, 34A10, 34A15, 34A30, 34A40, 34B05, 34B15, 34B20, 34B25, 34C10, 34C15, 34D10, 34D15, 34D20, 34E05, 34E10, 34E15, 34G05, 34H05, 34J05, 34J10, 34K05, 34K10, 35A05, 35A15, 35A20, 35A25, 35A30, 35A35, 35B05, 35B10, 35B20, 35B25, 35B30, 35B40, 35B45, 35C05, 35C15, 35D05, 35D10, 35E05, 35E15, 35F05, 35F15, 35F25, 35F30, 35G05, 35G15, 35G20, 35G30, 35J05, 35J10, 35J25, 35J60, 35K05, 35K20, 35K45, 35L05, 35L20, 35P05, 35P15, 35P25, 35Q10, 35R05, 35R15, 35R30, 39A05, 39A10, 65N30, 73A05, 73G05, 73G10, 78A30, 92A05, 92A10, 92A15, 92A20.

ISBN 3-540-08058-9 Springer-Verlag Berlin · Heidelberg · New York
ISBN 0-387-08058-9 Springer-Verlag New York · Heidelberg · Berlin

Printed in Germany

Printing and binding: Beltz Offsetdruck, Hemsbach/Bergstr.

PREFACE

These Proceedings form a record of the lectures delivered at the fourth Conference on Ordinary and Partial Differential Equations which was held at the University of Dundee, Scotland during the period of four days Tuesday 30 March to Friday 2 April 1976.

The Conference was attended by 140 mathematicians from the following countries: Belgium, BRD Germany, Canada, France, Italy, The Netherlands, Norway, Poland, South Africa, Sweden, Switzerland, the United Kingdom and the United States of America.

The Conference was organised by the following Committee: W N Everitt (Chairman); M W Green, I M Michael and B D Sleeman (Organising Secretaries).

Following the tradition set by the earlier Dundee Conferences the Organising Committee named as Honorary Presidents of the 1976 Conference:

Professor Dr Lothar Collatz (BRD Germany)

Professor Gaetano Fichera (Italy).

On behalf of the Committee we thank all mathematicians who took part in the work of the Conference; many travelled long distances to be in Dundee at the time of the Conference.

The Committee thanks: the University of Dundee for generously supporting the Conference in providing all necessary facilities; the Warden and Staff of Belmont Hall for their help in providing accommodation for participants; colleagues and research students in the Department of Mathematics for help during the week of the Conference; the Finance Office of the University of Dundee.

The Committee wishes to record special appreciation of a grant from the European Research Office of the United States of America which made available travel grants to participants from North America and Europe, and also covered the cost of some secretarial support for the Conference organisation.

As in previous years the Committee places on record its indebtedness to Mrs Norah Thompson, Secretary in the Department of Mathematics, for her invaluable help in seeing through the work of the Conference from early days to completion.

W N Everitt B D Sleeman

Dundee, Scotland.

CONTENTS

A. Acker and W. Walter:

The quenching problem for nonlinear parabolic differential equations 1

J. M. Ball:

On the calculus of variations and sequentially weakly continuous maps ... 3

M. F. Barnsley and P. D. Robinson:

Bivariational bounds on <g,ϕ> for non-linear problems $F\phi = 0$ 26

H. E. Benzinger:

The spectral analysis of a Stone regular differential operator 34

L. Collatz:

Bifurcation diagrams 41

D. Colton:

Walsh's theorem for the heat equation 54

C. Conley:

A new statement of Wazewski's theorem and an example 61

M. S. P. Eastham:

On the absence of square-integrable solutions of the Sturm-Liouville equation 72

W. D. Evans:

On limit-point and Dirichlet-type results for second-order differential expressions 78

W. N. Everitt:

Spectral theory of the Wirtinger inequality 93

W. E. Fitzgibbon:

Nonlinear evolution operators and delay equations 106

J. A. Gatica and P. Waltman:

A singular functional differential equation arising in an immunological model 114

H. Grabmüller:

On linear partial integro differential equations with a small parameter .. 125

P. Grisvard:

Smoothness of the solution of a monotonic boundary value problem for a second order elliptic equation in a general convex domain 135

P. Habets:

On the method of strained coordinates 152

K. P. Hadeler:

Nonlinear diffusion equations in Biology 163

J. K. Hale:

Discrete dissipative processes 207

W. S. Hall:

Integrating a differential equation with a weak* continuous vector field .. 225

W. A. Harris Jr.:

On asymptotic integration 23

G. Hecquet:

Existence globale des solutions de quelques problemes aux limites 239

G. C. Hsiao and R. Weinacht:

Singular perturbation problems for a class of singular partial differential equations 249

R. M. Kauffman:

On the limit-n classification of ordinary differential operators with positive coefficients 259

M. Kisielewicz:

On the non-convergence of successive approximations in the theory of ordinary differential equations 267

R. J. Knops:

Comments on nonlinear elasticity and stability 271

E. L. Koh:

A Mikusinski Calculus for the Bessel operator B_μ 291

K. Kreith:

An oscillation theory for fourth order differential equations 301

T. Küpper:

Pointwise error bounds for the eigenfunctions of one-dimensional Schrödinger operators 309

L. Lara-Carrero:

Stability of shock waves 316

N. G. Lloyd:

Remarks on L_2 solutions 329

V. Marić and M. Tomić:

Regularly varying functions and differential equations 333

J. W. Neuberger:

Projection methods for linear and nonlinear systems of partial differential equations 361

K. Nickel:

New results on strongly coupled systems of parabolic differential equations 350

L. A. Peletier:

On a nonlinear diffusion equation arising in population genetics 365

A. T. Plant:

A generalization of the Flaschka-Leitman theorem 372

A. J. B. Potter:

Hilbert's projective metric applied to a class of positive operators 377

T. T. Read:

A limit-point criterion for -(py')' + qy 383

E. E. Rosinger:

Nonsymmetric Dirac distributions in scattering theory 391

P. W. Schaefer and R. P. Sperb:

A maximum principle for a class of functionals in non-linear Dirichlet problems 400

G. Scheu:

Global methods for the construction of convergent sequences of bounds for systems of ordinary initial value problems 407

M. Shearer:

Bifurcation from a multiple eigenvalue 417

C. G. Simader:

Another approach to the Dirichlet problem for very strongly nonlinear elliptic equations 425

B. D. Sleeman:

Global estimates for non-linear wave equations and linear wave equations with non-linear boundary constraints 438

E. Stephan:

Difference approximations for some pseudifferential equations in R^n 452

E. Stephan and W. Wendland:

Remarks to Galerkin and least squares methods with finite elements for general elliptic problems 461

C. A. Stuart:

Boundary-value problems with discontinuous non-linearities 472

A. van Harten:

On an elliptic singular perturbation problem 485

P. Volkmann:

Über die Existenz von Lösungen der Differentialgleichung u' = f(u) in einer abgeschlossenen Menge, wenn f eine k-Mengenkontraktion ist 496

J. Walter: Bemerkungen zur Verwendung der Pfaffschen Formen bei der Definition der absoluten Temperatur nach Carathéodory 504

H. O. Walther:

On the eigenvalues of linear autonomous differential delay equations 513

N. Weck:

An explicit St. Venant's principle in three-dimensional elasticity 518

J. Weidmann:

Trace class methods for scattering in a homogeneous electro-static field 527

A. D. Wood and F. D. Zaman:

Eigenvalue problems for free vibrations of rectangular elastic plates with random inhomogeneities 533

A. Zettl:

Limit point conditions for powers 540

W.D.Evans and A.Zettl

Addendum to "Limit Point Conditions for Powers" 550

Lectures given at the Conference which are not represented by contributions to these Proceedings.

A. Borzymowski:

A boundary value problem for an infinite system of integro-differential equations in a non-cylindrical domain

B. L. J. Braaksma:

Asymptotic expansions for a linear difference equation

K. J. Brown:

An existence and uniqueness theorem for non-linear boundary value problems

H. Budin:

Existence and multiplicity of positive solutions of a non-linear boundary value problem

J. Carr:

Decay to zero in critical cases of second-order ordinary differential equations of Duffing type

W. D. Collins:

Dual extremum principles for the heat equation

E. R. Dawson and W. N. Everitt:

On recent results in integral inequalities involving derivatives

J. Dyson and R. Villella Bressan:

On an abstract functional differential equation with infinite delay

E M de Jager:

Singular perturbations of hyperbolic type

M. Essén:

A singular integral inequality associated with a problem in estimating harmonic measure

G. Fichera:

Uniqueness, existence and estimate of the solution in the dynamical problem of thermodiffusion in an elastic solid

K. P. Hadeler:

Travelling and standing waves in semi-linear parabolic equations

J. K. Hale:

Restricted generic bifurcation

G. S. Halvorsen:

Sharp bounds and L^p stability for solutions of second order linear ordinary differential equations

H. Kalf:

Krylov's embedding theorem, Kato's inequality and the adjoint of Schrödinger operators

U. Kirchgraber:

On systems with an unstable equilibrium

H-W. Knobloch:

Integral manifolds and stability in ordinary differential equations

D. A. Lutz:

Recent results on invariance for meromorphic differential equations

T. K. Puttaswamy:

A two-point connection problem for a certain nth order differential equation

R. M. Redheffer:

Invariance properties of matrix differential equations

M. Reeken:

The equations of motion of a chain

R. Reissig:

Periodic solutions of some second order differential equations

J. Serrin:

The foundations of thermodynamics: what is a reversible process?

J. Serrin:

Isoperimetric inequalities and the torsion problem. Global geometry of Weingarten surfaces

R. A. Smith:

Resonance sets of the periodic feedback control equation

M. A. Sneider:

Theory of Volterra integral equations of the first kind not reducible to the second kind

W. von Wahl:

Regularity of solutions of semi-linear partial differential equations

Address list of authors and speakers

A. Acker: Mathematisches Institut I, Universität Karlsruhe, 75 KARLSRUHE, West Germany

J. M. Ball: Department of Mathematics, Heriot-Watt University, Riccarton, Midlothian, Scotland

M. F. Barnsley: Department of Mathematics, University of Bradford, BRADFORD, England

H. E. Benzinger: Department of Mathematics, University of Illinois, URBANA, Illinois 61801, USA

A. Borzymowski: Institute of Mathematics, Warsaw Technical University, WARSAW, Poland

B. L. J. Braaksma: Mathematisch Instituut, University of Groningen, PO Box 800, GRONINGEN, The Netherlands

K. J. Brown: Department of Mathematics, Heriot-Watt University, Riccarton, Midlothian, Scotland

H. Budin: Department of Mathematics, Heriot-Watt University, Riccarton, Midlothian, Scotland

J. Carr: Department of Mathematics, Heriot-Watt University, Riccarton, Midlothian, Scotland

L. Collatz: Institut für Angewandte Mathematik, 2 HAMBURG 13, Rothenbaumchaussee 41, West Germany

W. D. Collins: Department of Applied Mathematics and Computing Science, University of Sheffield, SHEFFIELD, England

D. Colton: Department of Mathematics, University of Strathclyde, Livingstone Tower, Richmond Street, GLASGOW, Scotland

C. Conley: Department of Mathematics, University of Wisconsin, MADISON, Wisconsin 53706, USA

E. R. Dawson: Department of Mathematics, The University, DUNDEE, Scotland

E. M. de Jager:	Mathematisch Instituut, Universiteit Amsterdam, Roetersstrasse 15, AMSTERDAM, The Netherlands
J. Dyson:	The Mathematical Institute, University of Oxford, 24-29 St Giles, OXFORD, England
M. S. P. Eastham:	Department of Mathematics, Chelsea College, Manresa Road, LONDON
M. Essén:	Institutionen för Matematik, Kungliga Tekniska Högskolan, 10044 STOCKHOLM 70, Sweden
W. D. Evans:	Department of Pure Mathematics, University College, CARDIFF, Wales
W. N. Everitt:	Department of Mathematics, University of Dundee, DUNDEE, Scotland
G. Fichera:	Via Pietro Mascagni 7, 00199 ROMA, Italy
W. E. Fitzgibbon	Department of Mathematics, University of Houston, HOUSTON, Texas 77009, USA
J. A. Gatica:	Department of Mathematics, University of Iowa, IOWA CITY, Iowa 52240, USA
H. Grabmüller:	Fachbereich Mathematik, Technische Hochschule 61 DARMSTADT, West Germany
P. Grisvard:	I.M.S.P., Parc Valrose, 06034 NICE CEDEX, France
P. Habets:	Institut Mathematique, Université Catholique de Louvain, Chemin du Cyclotron 2, 1348 LOUVAIN LA NEUVE, Belgium
K. P. Hadeler:	Universität Tübingen, Lehrstuhl für Biomathematik, D 7400 TUBINGEN 1, Auf der Morganstelle 28, West Germany
J. K. Hale:	Division of Applied Mathematics, Brown University, PROVIDENCE, Rhode Island 02912, USA
W. S. Hall:	Department of Mathematics, The University of Pittsburgh, PITTSBURGH, Pennsylvania 15260, USA
G. S. Halvorsen:	Institute of Mathematics, University of Trondheim, NTH, 7034 TRONDHEIM-NTH, Norway

W. A. Harris Jr:	Department of Mathematics, University of Southern California, LOS ANGELES, California 90007, USA
G. Hecquet:	Department of Mathematics, University of Lille, 82 Rue Meurem, LILLE, France
G. C. Hsiao:	Fachbereich Mathematik, Technische Hochschule Darmstadt, 61 DARMSTADT, West Germany
H. Kalf:	Institut für Mathematik, RWTH Aachen, D51 AACHEN, Templergraben 55, West Germany
R. M. Kauffman:	Department of Mathematics, Western Washington State College, BELLINGHAM, Washington 98225, USA
U. Kirchgraber:	Seminar für Angewandte Mathematik der ETH, Clausiusstr. 55, 8006 ZURICH, Switzerland
M. Kisielewicz:	Ul. Akademicka 10, 65-240 Zielona Góra, Poland
H-W. Knobloch:	Mathem. Institut der Universität, 87 WURZBURG, Am Hubland, West Germany
R. J. Knops:	Department of Mathematics, Heriot-Watt University, Riccarton, Midlothian, Scotland
E. L. Koh:	Fachbereich Mathematik, Technische Hochschule Darmstadt, 6100 DARMSTADT, Kantplatz 1, West Germany
K. Kreith:	Department of Mathematics, University of California, DAVIS, California 95616, USA
T. Küpper:	Mathematisches Institut der Universität zu Köln, D-5000 KÖLN 41, Weyertal 86, West Germany
L. Lara-Carrero:	Departamento de Matemáticas, I.V.I.C., Apartado 1827, CARACAS 101, Venezuela
N. G. Lloyd:	Department of Pure Mathematics, University College of Wales, Penalais, ABERYSTWYTH, Dyfed, Wales
D. A. Lutz:	Department of Mathematics, University of Winconsin - Milwaukee, MILWAUKEE, WI 53211, USA

V. Marić Faculty of Technical Science, University of Novi Sad NOVI SAD, Yugoslavia

J. W. Neuberger: Department of Mathematics, Emory University, ATLANTA, Georgia 30322, USA

K. Nickel: Mathematisches Institut, 78 Freiburg i. Br., Hebelstrasse, 40, West Germany

L. A. Peletier: Department of Mathematics, Delft University of Technology, DELFT, The Netherlands

A. T. Plant Fluid Mechanics Research Institute, University of Essex, COLCHESTER, England

A. J. B. Potter Department of Mathematics, Edward Wright Building, King's College, University of Aberdeen, ABERDEEN, Scotland

T. K. Puttaswamy: Department of Mathematical Sciences, Ball University, MUNCIE, Indiana 47304, USA

T. T. Read: Department of Mathematics, Western Washington State College, BELLINGHAM, Washington 98225, USA

R. M. Redheffer: Department of Mathematics, University of California, LOS ANGELES, California 90024, USA

M. Reeken: Mathematische Abteilung, Universität Bochum, D463 BOCHUM, West Germany

R. Reissig: Institut für Mathematik, Ruhr-Universität, D-463 BOCHUM, Universitätsstrasse 150, West Germany

P. D. Robinson: School of Mathematics, Bradford University, BRADFORD, West Yorkshire, England

E. E. Rosinger: Department of Computer Science, Haifa Technion, HAIFA, Israel

P. W. Schaefer: Department of Mathematics, University of Tennessee, KNOXVILLE, Tennessee 37916, USA

G. Scheu Institut für Angewandte Mathematik, Universität Karlsruhe, D-7500 KARLSRUHE, Kaiserstrasse 12, West Germany

J. Serrin: Department of Mathematics, University of Minnesota, MINNEAPOLIS, Minnesota 55455, USA

M. Shearer: Fluid Mechanics Research Institute, University of Essex, Wivenhoe Park, COLCHESTER, England

C. G. Simader: Mathematisches Institut der Universität München, D 8 MÜNCHEN 2, West Germany

B. D. Sleeman: Department of Mathematics, The University, DUNDEE, Scotland

R. A. Smith: Department of Mathematics, University of Durham, Science Laboratories, South Road, DURHAM, England

M. A. Sneider: Via A. Torlonia N.12, 00161 ROMA, Italy

R. P. Sperb: Department of Mathematics, University of Tennessee, KNOXVILLE, Tennessee 37916, USA

E. Stephan: Fachbereich Mathematik, Technische Hochschule Darmstadt, 61 DARMSTADT, West Germany

C. A. Stuart: Department of Mathematics, Edward Wright Building, King's College, University of Aberdeen, ABERDEEN, Scotland

M. Tomic: Faculty of Technical Science, University of Novi Sad, NOVI SAD, Yugoslavia

A. van Harten: Wiskundig Seminarium der Vrij Universiteit, de Boelelaan 1081, AMSTERDAM, Holland

R. Villella Bressan: Istituto Matematico "G Castelnuovo", Città Universitaria, Università di Roma, ROMA, Italy

P. Volkmann: Mathematisches Institut I, Universität (TH), 75 KARLSRUHE 1, West Germany

W. von Wahl: Institut für Mathematik, Ruhr-Universität Bochum, 463 BOCHUM, West Germany

J. Walter: Lehrstuhl für Mathematik, Universität Aachen, 51 AACHEN, Templergraben 55, West Germany

W. Walter: Mathematisches Institut I, Universität Karlsruhe, 75 KARLSRUHE, West Germany

H. O. Walther: Mathematisches Institut der Universität München, D 8000 MÜNCHEN 2, Theresienstrasse 39, West Germany

P. Waltman: Department of Mathematics, University of Iowa, IOWA CITY, Iowa 52240, USA

N. Weck: Fachbereich Mathematik, Technische Hochschule, Universität Darmstadt, 61 DARMSTADT, Kantplatz 1, West Germany

J. Weidmann: Fachbereich Mathematik der Universität Frankfurt, 6000 FRANKFURT AM MAIN, Robert-Mayer-Strasse 10, West Germany

R. Weinacht: Fachbereich Mathematik, Technische Hochschule Darmstadt, 61 DARMSTADT, Kantplatz 1, West Germany

W. Wendland: Fachbereich Mathematik, Technische Hochschule Darmstadt, 61 DARMSTADT, Kantplatz 1, West Germany

A. D. Wood: Department of Mathematics, Cranfield Institute of Technology, CRANFIELD, Bedford, England

F. D. Zaman: Department of Mathematics, Cranfield Institute of Technology, CRANFIELD, Bedford, England

A. Zettl: Department of Mathematics, University of Northern Illinois, DEKALB, Illinois 60115, USA

The Quenching Problem for Nonlinear Parabolic Differential Equations

Andrew Acker and Wolfgang Walter

1. Introduction. The following problem was recently posed by Kawarada [1]. Let $u(t,x)$ be the solution of the boundary value problem

$$u_t = u_{xx} + \frac{1}{1-u} \qquad \text{in } (0,T) \times (-a,a) \tag{1}$$

$$u(t,a) = u(t,-a) = 0 \quad \text{in } [0,T],\ u(0,x) = 0 \text{ in } [-a,a] \tag{2}$$

when $0 < T \leq \infty$ and $a > 0$. If u exists for all $t > 0$ ($T = \infty$), we have the case of global existence. If, on the contrary, $T < \infty$ and $\max\{u(t,x): |x| \leq a\} \to 1$ as $t \to T-0$, then u cannot be continued beyond $t = T$, i.e., the solution "quenches". Blow-up phenomena for parabolic differential equations, when solutions become infinite in finite time, have been extensively studied. Quenching (this terminology was introduced by Kawarada) is a related phenomenon, where the solution remains finite, but derivatives blow up. Naturally, the reason for such a behavior is the fact that the nonlinear term in (1) has a singularity for $u = 1$.

The problem consists of determining those values of $a > 0$ such that the solution of (1), (2) exists globally. It is easily seen that there exists a positive number a^* such that for $0 < a < a^*$ we have global existence, and for $a > a^*$ quenching occurs.

Kawarada has given the estimate $a^* \leq \sqrt{2}$. The second author found the bounds $0.765 < a^* \leq \frac{\pi}{4}$, using differential inequality methods [5]. He also obtained bounds for the maximum value of T in the case where $a > \frac{\pi}{4}$.

In this article, we consider the more general problem

$$u_t = Lu + f(u) \quad \text{in } (0,T) \times D \tag{3}$$

$$u = 0 \quad \text{for } t = 0 \text{ and for } x \in \partial D . \tag{4}$$

Here, D is a bounded open set in $\mathbb{R}^n$, $x = (x_1,\ldots,x_n) \in \mathbb{R}^n$, and L is an elliptic differential operator independent of t,

$$Lu := \sum_{i,j=1}^{n} a_{ij}(x)u_{x_i x_j} + \sum_{i=1}^{n} b_i(x)u_{x_i} + c(x)u \tag{5}$$

for short

$$Lu = au_{xx} + bu_x + cu . \tag{5'}$$

It is assumed throughout that L is locally uniformly elliptic, i.e., that there exists a function $\delta(x)$ continuous and positive in D such that $a(x) \geq \delta(x)I$, or explicitly

$$\sum_{i,j=1}^{n} a_{ij}(x)\xi_i\xi_j \geq \delta(x) \sum_{i=1}^{n} \xi_i^2 \quad \text{for } \xi \in \mathbb{R}^n . \tag{6}$$

Of course, the matrix a is symmetric.

Using differential inequality methods, we prove several theorems on the behavior of solutions of (3),(4). The connection with the "stationary problem" for $w = w(x)$

$$Lw + f(w) = 0 \quad \text{in } D, \quad w = 0 \quad \text{on } \partial D \tag{7}$$

is crucial. The question of global existence for (3),(4) can be reduced to the existence problem for the boundary value problem (7). In particular, the number a^* in Kawarada's problem (1),(2) can be given explicitly.

Let us note, in concluding, that the problems studied here were initiated from physical considerations; see [1].

<u>2. Parabolic Differential Inequalities.</u> Let D be a bounded open set in $\mathbb{R}^n$, let $G := (0,T) \times D$ $(0 < T \leq \infty)$ and $\Gamma :=$ $(0,T) \times \partial D \cup \{0\} \times \bar{D}$; Γ is the parabolic boundary of G. For a function $u(t,x)$, u_x denotes the gradient vector, u_{xx} the $n \times n$ - matrix of second order derivatives with respect to all the x_i. For functions $u,v,\ldots$, which are solutions of parabolic differential equations or inequalities, we assume that they are continuous in $G \cup \Gamma$ and that the derivatives which appear in (3) are continous in G. The operator L is defined by (5). There are no regularity assumptions on the coefficients of L unless explicitly stated. Yet (6) will be assumed, although several of our conclusions hold under the weaker condition $a \geq 0$.

<u>Nagumo's Lemma.</u> If $F(t,x,z)$ is Lipschitz continuous in z and $c(x)$ is bounded, then

$$u_t - Lu - F(t,x,u) \leq v_t - Lv - F(t,x,u) \quad \text{in } G$$

$$u \leq v \quad \text{on } \Gamma$$

implies

$$u \leq v \quad \text{in } G .$$

<u>Strong Minimum Principle.</u> Let a,b,c be bounded. If

$$u_t \geq Lu \quad \text{in } G \quad \text{and} \quad u \geq 0 \quad \text{on } \Gamma ,$$

then $u \geq 0$ in G. Furthermore, if $u(t_o,x_o) = 0$, where $(t_o,x_o) \in G$, then $u \equiv 0$ in $[0,t_o] \times \bar{D}$.

Both theorems are stated here in a special form sufficient for our purposes. Proofs are found, e.g., in [4; 24 VI and 26 III] and [3].

If u is a solution of the equation $u_t = Lu + F(t,x,u)$ and v satisfies $v_t \leq Lv + F(t,x,v)$, and if $v \leq u$ on Γ, then $v \leq u$ in G by Nagumo's lemma. In this case, v is said to be a subfunction (or lower function) for u. A superfunction (upper function) is defined correspondingly.

3. One Space Variable. We consider the special problem

$$u_t = u_{xx} + f(u) \quad \text{in } G_a := (0,T_a) \times (-a,a) \tag{8}$$

$$u = 0 \quad \text{on } \Gamma_a \tag{9}$$

where Γ_a is the parabolic boundary corresponding to G_a (see (2)) and $0 < T_a \leq \infty$. Throughout this no. it is assumed that f(z) is locally Lipschitz continuous in the range of u and that $f(0) > 0$. Let us remark that in the case $f(0) = 0$ the function $u \equiv 0$ is the only solution of (8),(9). Monotonicity is understood to be weak monotonicity. If equality is to be excluded, we use the notion of strict monotonicity.

Theorem 1. Let u be a solution of (8),(9).

(a) u is positive in G and strictly increasing in t for $|x| < a$.

(b) $u(t,x) = u(t,-x)$ and hence $u_x(t,0) = 0$. Furthermore, u(t,x) is, for fixed $t > 0$, strictly increasing in x for $-a \leq x \leq 0$.

(c) If $T_a = \infty$, $0 \leq u(t,x) \leq B < \infty$ and if f is Lipschitz continuous in [0,B], then the limit

$$w(x) := \lim_{t\to\infty} u(t,x)$$

exists uniformly in $[-a,a]$. The function w is a solution of the boundary value problem

$$w'' + f(w) = 0 \quad \text{in } [-a,a], \quad w(x) > 0 \quad \text{in } (-a,a)$$
$$w(a) = w(-a) = 0 \; . \tag{10}$$

Proof. (a) The solution of (8),(9) is unique, according to Nagumo's lemma. If $f(z_o) = 0$, $z_o > 0$, then $v(t,x) := z_o$ is a superfunction for u. This reasoning shows that $f(u) \geq 0$. Hence the strong minimum principle implies that $u > 0$ in G.

For $h > 0$, the function $v(t,x) := u(t+h,x)$ is a superfunction for u. Hence u is increasing in t. The strong minimum principle, applied to $w := v-u$, shows that $w > 0$ in G. Note that w satisfies an equation $w_t = w_{xx} + d(t,x)w$ with bounded d.

(b) The first part is implied by uniqueness. Now we consider the function $w(t,x) := u(t,x+h) - u(t,x-h)$ in $H := (0,T) \times (-a+h,0)$, where $0 < h < a$. Again, w satisfies an equation

$$w_t = w_{xx} + d(t,x)w \, , \quad d \text{ bounded.}$$

Since $w \geq 0$ on the parabolic boundary of H, the inequality $w > 0$ in H follows from the strong minimum principle.

(c) Let

$$U(t,x) = \int_t^{t+1} u(s,x)\, ds \; .$$

Upon integration of (8) we get

$$u(t+1,x) - u(t,x) = U_{xx} + \int_t^{t+1} f(u(s,x))\, ds \; ;$$

moreover

$$0 \leq u(t,x) \leq U(t,x) \leq u(t+1,x) \leq w(x) \; .$$

Obviously, if in Taylor's theorem

$$U(t,x) = U(t,0) + \int_0^x (x-\xi) U_{xx}(t,\xi)\, d\xi \, ,$$

we let $t \to \infty$, we obtain by using the theorem on majorized convergence

$$w(x) = w(0) - \int_0^x (x-\xi) f(w(\xi))\, d\xi \; .$$

It follows that w is in C^2 and satisfies the differential equations $w'' + f(w) = 0$ in $(-a,a)$.

In order to obtain the boundary condition for w, we construct an upper function ϕ for u. Let $f(z) \le 2C$ in $0 \le z \le B$ and $\phi := C(a^2-x^2)$. Since f may be continued by $f(z) = f(B)$ for $z \ge B$, ϕ is indeed an upper function for u. Hence $0 \le u(t,x) \le \phi(x)$ and $0 \le w(x) \le \phi(x)$, which proves the rest of (c).

Now we prove some relations between solutions for different values of a. We use the notation $u = u(t,x;a)$ for a solution of (8),(9).

<u>Theorem 2.</u> (a) Let $\alpha > 0$ and $T_a \le T_{a+\alpha}$. Then $u(t,x;a) < u(t,x;a+\alpha)$ and even

$$u(t,x;a) < \min \{u(t,x+h;a+\alpha) : |h| \le \alpha\} \quad \text{in } G_a .$$

(b) Let $w(x;a)$ be a solution of the boundary value problem (10). Then

$$u(t,x;a) < w(x;a) \quad \text{in } G_a .$$

(c) Let $T_a = \infty$, $u(t,0;a) \to b$ as $t \to \infty$ and $f(z) \to \infty$ as $z \to b-0$. Then $T_{a+\alpha} < \infty$ for $\alpha > 0$, i.e., the solution $u(t,x;a+\alpha)$ quenches.

<u>Proof.</u> (a) Let $v(t,x) := u(t,x+h;a+\alpha) - u(t,x;a)$, where $|h| \le \alpha$. It is easily seen that $v_t = v_{xx} + d(t,x)v$ in G_a and $v \ge 0$ on Γ_a, where d is a bounded function. Then (a) follows from the

strong minimum principle.

Proposition (b) is proved in a similar fashion.

(c) Let $u := u(t,x;a)$ and $U := u(t,x;a+\alpha)$. We choose positive numbers ε and t_o in such a way that

$$\alpha^2 + 2\varepsilon/\alpha^2 \leq f(z) \text{ for } b-\varepsilon \leq t < b \quad \text{and} \quad u(t_o;0) \geq b-\varepsilon.$$

Let S be the strip $(t_o,\infty) \times (-\alpha,\alpha)$. If U exists for all $t > 0$, then, according to (a) and Theorem 1 (a), $U \geq b-\varepsilon$ on the parabolic boundary of S. On the other hand, the function

$$v(t,x) := b - \varepsilon + (t-t_o)(\alpha^2-x^2)$$

is equal to $b-\varepsilon$ on the parabolic boundary of S. Furthermore,

$$v_t \leq v_{xx} + f(v) \quad \text{in } R\text{: } t_o < t < t_o + \varepsilon/\alpha^2, \quad |x| < \alpha.$$

Hence, v is subfunction for U in the rectangle R. In particular, $U(t_o + \varepsilon/\alpha^2,0) > v(t_o + \varepsilon/\alpha^2,0) = b$. This contradiction shows that $T_{a+\alpha} \leq t_o + \varepsilon/\alpha^2$.

Conclusions. We consider the quenching phenomenon, assuming that $f(z) \to \infty$ as $z \to b-0$ $(b > 0)$. Since $f(0) > 0$, it is easily seen that the boundary value problem (10) has a solution w for small positive values of a. Let a^* be the supremum of all values of $a > 0$ such that a solution $w(x;a)$ to (10) exists. Then $T_a = \infty$ (global existence) for $a < a^*$ and $T_a < \infty$ (quenching) for $a > a^*$. The first part follows from Theorem 2 (b), the second part from Theorem 2 (c) in conjunction with Theorem 1 (c).

In the limiting case $a = a^*$ the answer is not complete. If $w(x;a^*)$ exists, it is clear that global existence prevails.

If (10) has no solution for $a = a^*$, then we have either global existence with $u(t,0;a^*) \to b$ as $t \to \infty$ or quenching.

In the special problem (1),(2) studied by Kawarada, the number a^* is easily found to be $\sqrt{2}\,M \approx 0.765$, where M is the maximum of the function $e^{-s^2} \int_0^s e^{t^2}$ for $s > 0$. In this case, $w(x;a^*)$ exists, i.e., $T_a = \infty$ iff $a \leq a^*$.

4. Several Space Variables. The problem under consideration is given by

$$u_t = Lu + f(u) \quad \text{in } G := (0,T) \times D \tag{3}$$

$$u = 0 \quad \text{on } \Gamma \tag{4}$$

where L is the operator defined by (5),(6). We assume that the coefficients a,b,c are bounded, that f is locally Lipschitz continuous in the range of u and that $f(0) > 0$ (again, $u \equiv 0$ is the unique solution if $f(0) = 0$).

Theorem 3. Let u be a solution of (3),(4).

(a) u is positive in G and strictly increasing in t for $x \in D$.

(b) Let $T = \infty$, $0 \leq u \leq B$, f Lipschitz continuous in $[0,B]$ and $w(x) := \lim_{t\to\infty} u(t,x)$. If Green's function corresponding to the Dirichlet problem for L and, say, for small balls contained in D exists, then w satisfies

$$Lw + f(w) = 0 \quad \text{in } D, \quad w \in C^2(D), \quad w > 0 \quad \text{in } D\ . \tag{11}$$

(c) In addition to the assumptions in (b) we assume that to each $y \in \partial D$ there exists a "barrier function" $v(x) = v(x;y)$ in $C^2(D) \cap C^0(\bar{D})$ with the properties

$$v \geq 0 \quad \text{in } D, \quad v(y) = 0, \quad Lv \leq -1 \quad \text{in } D\ .$$

Then the function w defined in (b) is continuous in $\bar{D}$ and vanishes on ∂D.

Proof. The proof of (a) is similar to the one given in Theorem 1. For the proof of (b), we assume that K is a closed ball in D and that functions G, G_n exist with the property that ($d\sigma$ surface element)

$$(*) \qquad \phi(x) = \int_K G(x,y)(L\phi)(y)dy + \int_{\partial K} G_n(x,y)\phi(y)d\sigma(y) \text{ in } K$$

for all $\phi \in C^{2+\alpha}(\bar{K})$ $(\alpha > 0)$. G is Green's function, and G_n is a normal derivative of G, with respect to the Dirichlet problem for L in K; see [2; (10.4)]. As in Theorem 1, we introduce

$$U(t,x) = \int_t^{t+1} u(s,x)\, ds .$$

Integrating (9) from t to t+1 we obtain

$$u(t+1,x) - u(t,x) = LU + \int_t^{t+1} f(u(s,x))\, ds .$$

Using (*) for $\phi(x) = U(t,x)$ (t fixed) and letting $t \to \infty$, the equation

$$w(x) = -\int_K G(x,y)f(w(y))dy + \int_{\partial K} G_n(x,y)w(y)d\sigma(y)$$

is obtained. It follows that w is of class C^1 and hence $f(w(x))$ is Lipschitz continuous. Therefore, $f \in C^2$ and (11) holds in K.

(c) Let $K \geq f(z)$ for $0 \leq z \leq B$. Then $Kv(x;y)$ is a superfunction for u, i.e., $w(x) \leq K \inf\{v(x;y)\colon y \in \partial D\}$. Now (c) easily follows.

Remark. It is not difficult to give sufficient conditions for the existence of barrier functions. For example, if L is uniformly elliptic ($\delta(x) = \delta > 0$) and if there exists a closed ball K such that $D \cap K = \{y\}$ (i.e., D satisfies an outer sphere condition at $y \in \partial D$), then a barrier of the form $v(x;y) = \phi(|x-y|)$ is easily constructed.

The next theorem corresponds to Theorem 2. It gives relations between solutions of (3),(4) for different regions D.

Theorem 4. Let D and D_1 be bounded open subsets of $\mathbb{R}^n$ such that $D \subset D_1$ and $D \neq D_1$. Let u and U be solutions of (3),(4) in $(0,T) \times D$, $(0,T_1) \times D_1$ respectively, and let $T_o = \min(T,T_1)$. We assume that the coefficients of L are defined in D_1.

(a) $u < U$ for $x \in D$, $0 < t < T_o$.

(b) If D_1 contains an α-neighbourhood of D ($\alpha > 0$), and if the coefficients of L are constant, then

$$u(t,x) < \min\{U(t,x+h): |x| \leq \alpha\} \text{ in } (0,T_o) \times D.$$

(c) If f is increasing and convex, then U-u is strictly increasing in t (for $x \in D$, $0 < t < T_o$).

(d) Let $T = \infty$, $\max\{u(t,x): x \in \bar{D}\} \to b$ as $t \to \infty$ and $f(z) \to \infty$ as $z \to b-0$. Then, in each one of the following two cases, $T_1 < \infty$:

(d_1) D_1 contains an α-neighbourhood of D ($\alpha > 0$), the coefficients of L are constant.

(d_2) f(z) is increasing and convex for $0 \leq z < b$.

Proof. (a) The function $v = U-u$ satisfies an equation $v_t = Lv + d(t,x)v$, $d(t,x)$ bounded. The assertion follows from the strong minimum principle ($u \equiv U$ is impossible in view of Theorem 3 (a)).

(b) The proof is similar to the proof of Theorem 2 (a).

(c) Let $v := U-u$, let $\bar{u}(t,x) := u(t+h,x)$ ($h > 0$ fixed), and let $\bar{U},\bar{v}$ be defined similarly. The function v satisfies

$$v_t = Lv + f(U) - f(u) = Lv + F(t,x,v),$$

where $F(t,x,z) := f(u(t,x)+z) - f(u(t,x))$. On the other hand, $\bar{v}$ satisfies

$$\bar{v}_t = L\bar{v} + f(\bar{U}) - f(\bar{u}) \geq L\bar{v} + F(t,x,\bar{v}),$$

because $f(\bar{u}+\bar{v}) - f(u+\bar{v}) \geq f(\bar{u}) - f(u)$. Furthermore, $\bar{v} \geq v$ on Γ by Theorem 3 (a). Hence, $\bar{v}$ is an upper function for v, $v \leq \bar{v}$. The equality sign is excluded by the strong minimum principle since $(\bar{v}-v)_t \geq L(\bar{v}-v)$.

(d_1) There exists a closed ball $K \subset \bar{D}$ of radius $\alpha/2$ such that $\phi(t) := \max \{u(t,x): x \in K\} \to b$ as $t \to \infty$. It follows from (b) that

(*) $\quad U(t,x) \geq \phi(t) \quad$ for $x \in K$.

The rest of the proof runs along the same lines as in Theorem 2 (c), S being the cylinder $(t_o,\infty) \times K$.

(d_2) The proof follows immediately from (*) and the fact that U-u is positive and increasing.

<u>Concluding Remarks.</u> We again consider the case where $f(u) \to \infty$ as $u \to b-0$. Theorems 3 and 4 show that in the case of several space dimensions the problem of quenching versus global existence yields to an anlysis very similar to that in the one dimensional case studied in No. 3.

Assume $\{D(\alpha) | \alpha > 0\}$ is a monotone increasing class of domains.

Let L and each domain $D(\alpha)$ satisfy conditions under which a local existence theorem holds for equations (3),(4) (with $D = D(\alpha)$). We consider two sets of additional conditions. In case I, we assume that $\overline{D(\alpha)} \subset D(\beta)$ whenever $\alpha < \beta$ and that the coefficients of L are constant. In case II, we assume $D(\alpha) \subsetneq D(\beta)$ whenever $\alpha < \beta$ and that $f(u)$ is a convex increasing function on $[0,b)$. Now, let $\alpha_0 > 0$ be the supremum of the values of $\alpha > 0$ for which a solution w of (11) on $D(\alpha)$ exists with $w = 0$ on $\partial D(\alpha)$. Then in both cases I and II, the solution of (3),(4) for $D = D(\alpha)$ exists globally for all $0 < \alpha < \alpha_0$, but quenches at some finite T for all $\alpha > \alpha_0$. The solution of (3),(4) for $D = D(\alpha_0)$ also exists globally if a solution of (11) on $D(\alpha_0)$ satisfying $w = 0$ on $\partial D(\alpha_0)$ exists.

A further consequence of our results is that the solution w of (11) on $D(\alpha)$ with $w = 0$ on $\partial D(\alpha)$ exists for all $0 < \alpha < \alpha_0$, but for no $\alpha > \alpha_0$.

Literature.

[1] H. Kawarada, On Solutions of Initial-Boundary Problem for $u_t = u_{xx} + (1/(1-u))$. Publ. RIMS, Kyoto Univ. 10 (1975), pp. 729-736.

[2] C. Miranda, Partial differential equations of elliptic type, Ergebnisse der Mathematik und ihrer Grenzgebiete, Vol. 2, Springer-Verlag 1970.

[3] M.H. Protter and H.F. Weinberger, Maximum principles in Differential Equations. Prentice Hall, Englewood Cliffs, N.J., 1967.

[4] W. Walter, Differential and Integral Inequalities, Ergebnisse der Mathematik und ihrer Grenzgebiete, Vol. 55, Springer-Verlag 1970.

[5] W. Walter, Parabolic Differential Equations with a Singular Nonlinear Term. Funkcialaj Ekvacio (to appear).

ON THE CALCULUS OF VARIATIONS AND SEQUENTIALLY WEAKLY CONTINUOUS MAPS

J. M. BALL

1. Introduction

Consider the problem of finding a function $\underset{\sim}{u}: \Omega \to \mathbb{R}^n$ minimizing

$$I(\underset{\sim}{u},\Omega) = \int_\Omega f(\underset{\sim}{x},\underset{\sim}{u}(\underset{\sim}{x}),\nabla\underset{\sim}{u}(\underset{\sim}{x}))d\underset{\sim}{x} \tag{1}$$

subject to certain constraints, such as boundary conditions. In (1) Ω is a bounded open subset of $\mathbb{R}^m$, $f:\Omega \times \mathbb{R}^n \times M^{n\times m} \to \mathbb{R}$ (where $M^{n\times m}$ denotes the linear space of real $n\times m$ matrices), $\underset{\sim}{x} = (x_1,\ldots,x_m)$ and $d\underset{\sim}{x} = dx_1\ldots dx_m$.

In the direct method of the calculus of variations it is customary to seek conditions on f such that $I(\underset{\sim}{u},\Omega)$ is sequentially weakly lower semicontinuous on a subset K of a suitable Banach space X (i.e. $\underset{\sim}{u}_r \rightharpoonup \underset{\sim}{u}$ in K implies $I(\underset{\sim}{u},\Omega) \leqslant \liminf_{r\to\infty} I(\underset{\sim}{u}_r,\Omega)$). X is usually a space of Sobolev type. If I is bounded below on K and certain growth conditions are satisfied then the existence of a minimizer is assured.

The purpose of this paper is to show that the study of sequentially weakly <u>continuous</u> maps leads quickly to conditions on f guaranteeing lower <u>semi</u>continuity of $I(\underset{\sim}{u},\Omega)$, and thus to new existence theorems for nonlinear elliptic systems such as those arising in nonlinear elasticity.

<u>Notation</u>: The spaces $L^p(\Omega)$, $W^{k,p}(\Omega)$ are defined in the usual

way (cf Adams [1]). We deal throughout with vector and matrix functions $\underset{\sim}{w} = (w_i)_{1 \leqslant i \leqslant r}$. If Y is a Banach space and r a positive integer we define Y_r to be the Cartesian product $\prod_{i=1}^{r} Y$ equipped with the norm $\|\underset{\sim}{w}\|_{Y_r} = \sum_{i=1}^{r} \|w_i\|_Y$. $\overline{\mathbb{R}}$ denotes $\mathbb{R} \cup \{+\infty\}$. We employ the summation convention throughout.

2. The L^∞ case

To gain intuition we first consider maps between L^p spaces which arise from pointwise evaluation by a function. Corollary 1.1 characterizes maps of this type which are sequentially weakly continuous.

Theorem 1

Let $\phi: \mathbb{R}^n \to \mathbb{R}$ satisfy $\phi(\underset{\sim}{u}(\cdot)) \in L^1(\Omega)$ whenever $\underset{\sim}{u} \in L_n^\infty(\Omega)$. Then

$$J(\underset{\sim}{u}) \overset{\mathrm{def}}{=} \int_\Omega \phi(\underset{\sim}{u}(\underset{\sim}{x}))\,d\underset{\sim}{x}$$

is sequentially weak * lower semicontinuous on $L_n^\infty(\Omega)$ if and only if ϕ is convex.

Proof

Suppose J is sequentially weak * lower semicontinuous. Let $\underset{\sim}{a}, \underset{\sim}{b} \in \mathbb{R}^n$ and $\lambda \in [0,1]$. Let Q be the unit cube $\{\underset{\sim}{x} \in \mathbb{R}^m : 0 \leqslant |x_i| < \tfrac{1}{2}\}$ and define $\underset{\sim}{v} \in L_n^\infty(Q)$ by $\underset{\sim}{v}(\underset{\sim}{x}) = \underset{\sim}{a}$ if $\underset{\sim}{x} \in A_1$, $\underset{\sim}{v}(\underset{\sim}{x}) = \underset{\sim}{b}$ if $\underset{\sim}{x} \in A_2$, where $Q = A_1 \cup A_2$, $\mu(A_1) = \lambda$, $\mu(A_2) = 1-\lambda$, and μ denotes m-dimensional Lebesgue measure. Tesselate $\mathbb{R}^m$ by disjoint congruent open cubes Q_j with centre $\underset{\sim}{x}_j$ and side $1/k$. For i = 1,2 let $E_{k,i} = \bigcup_j (\underset{\sim}{x}_j + \frac{1}{k} A_i)$. Define a sequence $\underset{\sim}{u}_k \in L_n^\infty(\Omega)$ (k = 1,2...) by $\underset{\sim}{u}_k(\underset{\sim}{x}) = \underset{\sim}{v}(k(\underset{\sim}{x}-\underset{\sim}{x}_j))$ if $\underset{\sim}{x} \in Q_j \cap \Omega$. If $E \subseteq \Omega$ is measurable and $\underset{\sim}{c} \in \mathbb{R}^n$ then

$$\int_\Omega \underset{\sim}{u}_k \cdot \underset{\sim}{c}\, \chi_E(\underset{\sim}{x})\,d\underset{\sim}{x} = \int_E \underset{\sim}{u}_k \cdot \underset{\sim}{c}\, d\underset{\sim}{x} = \mu(E \cap E_{k,1})\underset{\sim}{a}\cdot\underset{\sim}{c} + \mu(E \cap E_{k,2})\underset{\sim}{b}\cdot\underset{\sim}{c},$$

which as $k \to \infty$ tends to

$$\mu(E)[\lambda \underset{\sim}{a} + (1-\lambda)\underset{\sim}{b}]\cdot\underset{\sim}{c} = \int_\Omega [\lambda\underset{\sim}{a} + (1-\lambda)\underset{\sim}{b}]\cdot\underset{\sim}{c}\, \chi_E(\underset{\sim}{x})\,d\underset{\sim}{x}.$$

Since finite linear combinations of functions of the form $\underset{\sim}{c}\,\chi_E$ are dense in $L^1_n(\Omega)$, and since $\|\underset{\sim}{u}\|_{L^\infty_n(\Omega)}$ is bounded, it follows that $\underset{\sim}{u}_k \overset{*}{\longrightarrow} \lambda\underset{\sim}{a} + (1-\lambda)\,\underset{\sim}{b}$ in $L^\infty_n(\Omega)$. Hence

$$\phi(\lambda\underset{\sim}{a} + (1-\lambda)\underset{\sim}{b}) \leq \liminf_{k\to\infty} \frac{1}{\mu(\Omega)} \int_\Omega \phi(\underset{\sim}{u}_k(\underset{\sim}{x}))\,d\underset{\sim}{x}$$

$$= \lim_{k\to\infty} \left[\frac{\mu(\Omega\cap E_{k,1})}{\mu(\Omega)}\,\phi(\underset{\sim}{a}) + \frac{\mu(\Omega\cap E_{k,2})}{\mu(\Omega)}\,\phi(\underset{\sim}{b})\right]$$

$$= \lambda\phi(\underset{\sim}{a}) + (1-\lambda)\,\phi(\underset{\sim}{b}),$$

so that ϕ is convex.

Conversely, let ϕ be convex, so that in particular ϕ is continuous. For $c,d\in\mathbb{R}$ the set $K(c,d) = \{\underset{\sim}{u}\in L^1_n(\Omega) : \|\underset{\sim}{u}\|_{L^\infty_n(\Omega)} \leq c,\ J(\underset{\sim}{u}) \leq d\}$ is closed in $L^1_n(\Omega)$ (by the bounded convergence theorem) and convex, hence weakly closed. Thus J is sequentially weak * lower semicontinuous. □

Corollary 1.1

Let ϕ be as above. Then $\phi: (L^\infty_n(\Omega),\ \text{weak}\ *) \longrightarrow (L^1(\Omega),\ \text{weak})$ is sequentially continuous if and only if ϕ is affine i.e. $\phi(\underset{\sim}{u}) = \alpha + \underset{\sim}{k}.\underset{\sim}{u}$ for constant $\alpha,\underset{\sim}{k}$.

Proof

If ϕ is affine the stated continuity property holds trivially. The converse follows by applying Theorem 1 to ϕ and $-\phi$. □

Remark: Theorem 1 is closely related to many known lower semicontinuity results. Note, however, that no assumption is made about continuity of ϕ.

3. The $W^{1,\infty}$ case

Consider now a function $\phi: M^{n\times m} \to \mathbb{R}$ satisfying $\phi(F(\cdot)) \in L^1(\Omega)$ whenever $F\in L^\infty_{mn}(\Omega)$. For $\underset{\sim}{u}: \mathbb{R}^m \to \mathbb{R}^n$ we pose the question: For which ϕ is the map $\underset{\sim}{u} \longmapsto \phi(\nabla\underset{\sim}{u}(\cdot))$ sequentially continuous from

$(W_n^{1,\infty}(\Omega)$, weak $*) \to (L^1(\Omega)$, weak)? (By the weak * topology on $W_n^{1,\infty}(\Omega)$ we mean the topology induced by the canonical embedding of $W_n^{1,\infty}(\Omega)$ into a finite product of $L^\infty(\Omega)$ spaces, each being endowed with the weak * topology). Bearing Corollary 1.1 in mind one might think that only affine ϕ are possible. However this is not the case unless $m = 1$ or $n = 1$. The actual situation is characterized by the following result of Morrey [6].

Theorem 2

Let $\psi: \Omega \times \mathbb{R}^n \times M^{n\times m} \to \mathbb{R}$ be continuous. Define

$$J(\underset{\sim}{u}) = \int_\Omega \psi(\underset{\sim}{x}, \underset{\sim}{u}(\underset{\sim}{x}), \nabla \underset{\sim}{u}(\underset{\sim}{x}))\,d\underset{\sim}{x}.$$

Then J is sequentially weak * lower semicontinuous on $W_n^{1,\infty}(\Omega)$ if and only if ψ is quasiconvex i.e. for each fixed $\underset{\sim}{x}_0 \in \Omega, \underset{\sim}{u}_0 \in \mathbb{R}^n, F_0 \in M^{n\times m}$, and for every bounded open subset D of $\mathbb{R}^m$ the inequality

$$\int_D \psi(\underset{\sim}{x}_0, \underset{\sim}{u}_0, F_0 + \nabla \underset{\sim}{\zeta}(\underset{\sim}{x}))\,d\underset{\sim}{x} \geqslant \int_D \psi(\underset{\sim}{x}_0, \underset{\sim}{u}_0, F_0)\,d\underset{\sim}{x} = \mu(D)\psi(\underset{\sim}{x}_0, \underset{\sim}{u}_0, F_0) \quad (2)$$

holds for all $\underset{\sim}{\zeta} \in C_0^\infty(D)$.

Corollary 2.1

Let $\psi: \Omega \times \mathbb{R}^n \times M^{n\times m} \to \mathbb{R}$ be continuous. The map $\underset{\sim}{u} \mapsto \psi(\cdot, \underset{\sim}{u}(\cdot), \nabla \underset{\sim}{u}(\cdot))$ is sequentially continuous from $(W_n^{1,\infty}(\Omega)$, weak $*) \to (L^1(\Omega)$, weak) if and only if for each fixed $\underset{\sim}{x}_0 \in \Omega, \underset{\sim}{u}_0 \in \mathbb{R}^n$, $F_0 \in M^{n\times m}$, and for every bounded open subset D of $\mathbb{R}^m$,

$$\int_D \psi(\underset{\sim}{x}_0, \underset{\sim}{u}_0, F_0 + \nabla \underset{\sim}{\zeta}(\underset{\sim}{x}))\,d\underset{\sim}{x} = \mu(D)\psi(\underset{\sim}{x}_0, \underset{\sim}{u}_0, F_0) \quad (3)$$

for all $\underset{\sim}{\zeta} \in C_0^\infty(D)$.

Proof of Corollary

Suppose $\underset{\sim}{u} \mapsto \psi(\cdot, \underset{\sim}{u}(\cdot), \nabla \underset{\sim}{u}(\cdot))$ has the stated continuity property. Applying Theorem 2 to $\pm \psi$ we obtain (3). Conversely let ψ satisfy (3) and let $\underset{\sim}{u}_r \overset{*}{\to} \underset{\sim}{u}$ in $W_n^{1,\infty}(\Omega)$. Then the sequence $\psi(\cdot, \underset{\sim}{u}_r(\cdot), \nabla \underset{\sim}{u}_r(\cdot))$ is bounded in $L^\infty(\Omega)$, so that in particular there exists a subsequence $\underset{\sim}{u}_\mu$ of $\underset{\sim}{u}_r$ such that $\psi(\cdot,\underset{\sim}{u}_\mu(\cdot), \nabla\underset{\sim}{u}_\mu(\cdot)) \xrightarrow{*} \Theta$ in $L^\infty(\Omega)$. Let α: $\mathbb{R}^m \to \mathbb{R}$ be continuous, and define $\psi_1(\underset{\sim}{x},\underset{\sim}{a},F) = \pm\, \psi(\underset{\sim}{x},\underset{\sim}{a},F)\alpha(\underset{\sim}{x})$. Then ψ_1 is quasiconvex, so that by Theorem 2

$$\int_\Omega \psi(\underset{\sim}{x},\underset{\sim}{u}_\mu(\underset{\sim}{x}),\nabla\underset{\sim}{u}_\mu(\underset{\sim}{x}))\alpha(\underset{\sim}{x})d\underset{\sim}{x} \quad\to\quad \int_\Omega \psi(\underset{\sim}{x},\underset{\sim}{u}(\underset{\sim}{x}),\nabla\underset{\sim}{u}(\underset{\sim}{x}))\alpha(\underset{\sim}{x})d\underset{\sim}{x}.$$

The arbitrariness of α implies that $\Theta = \psi(\cdot,\underset{\sim}{u}(\cdot),\nabla\underset{\sim}{u}(\cdot))$, and hence

$$\psi(\cdot,\underset{\sim}{u}_r(\cdot),\ \nabla\underset{\sim}{u}_r(\cdot)) \xrightarrow{*} \psi(\cdot,\underset{\sim}{u}(\cdot),\nabla\underset{\sim}{u}(\cdot)) \text{ in } L^\infty(\Omega)$$

which is stronger than the required conclusion. □

For the relationship of quasiconvexity to ellipticity see [2,6].

4. The null-space of the Euler-Lagrange operator

Let ψ: $\mathbb{R}^m \times \mathbb{R}^n \times M^{n\times m}$ be C^1. We say that ψ belongs to the null-space N of the Euler-Lagrange operator if and only if

$$\int_D \left(\frac{\partial\psi}{\partial u^i}\,\zeta^i + \frac{\partial\psi}{\partial u^i_{,\alpha}}\,\zeta^i_{,\alpha}\right) d\underset{\sim}{x} = 0 \tag{4}$$

for every bounded open set $D \subseteq \mathbb{R}^m$ and for all $\underset{\sim}{u}\epsilon C^1(\bar{D}),\underset{\sim}{\zeta}\epsilon C_0^\infty(D)$.

Theorem 3

Let ψ: $\Omega \times \mathbb{R}^n \times M^{n\times m} \to \mathbb{R}$ be continuous, and suppose that for each fixed $\underset{\sim}{x}_0\epsilon\Omega$, $\underset{\sim}{u}_0\epsilon\mathbb{R}^n$, $\psi(\underset{\sim}{x}_0,\underset{\sim}{u}_0,\ \cdot)$ is C^1. Then the map $\underset{\sim}{u} \longmapsto \psi(\cdot,\underset{\sim}{u}(\cdot),\ \nabla\underset{\sim}{u}(\cdot))$ is sequentially continuous from $(W^{1,\infty}_n(\Omega)$, weak *) $\longrightarrow$ $(L^1(\Omega)$, weak) if and only if for each fixed $\underset{\sim}{x}_0\epsilon\Omega,\underset{\sim}{u}_0\epsilon\mathbb{R}^n$, $\psi(\underset{\sim}{x}_0,\underset{\sim}{u}_0,\cdot)\ \epsilon$ N.

Proof

Let $\underset{\sim}{u} \longmapsto \psi(\cdot,\underset{\sim}{u}(\cdot),\ \nabla\underset{\sim}{u}(\cdot))$ have the stated continuity property. Let $\underset{\sim}{x}_0\epsilon\Omega,\underset{\sim}{u}_0\epsilon\mathbb{R}^n$ and define $\phi(F) = \psi(\underset{\sim}{x}_0,\underset{\sim}{u}_0,F)$. By Corollary 2.1 we have that

$$\int_D \phi(F_0 + \nabla \underset{\sim}{\zeta}(\underset{\sim}{x}))d\underset{\sim}{x} = \mu(D)\ \phi(F_0) \tag{5}$$

for all bounded open subsets $D \subseteq \mathbb{R}^m$, $F_0 \varepsilon M^{n\times m}$, $\underset{\sim}{\zeta}\varepsilon C_0^\infty(D)$. Let $\rho\varepsilon C_0^\infty(M^{n\times m})$ satisfy $\rho \geqslant 0$, $\rho(F) = 0$ if $|F| \geqslant 1$, $\int_{M^{n\times m}} \rho(F)dF = 1$. For $\varepsilon > 0$ let $\rho_\varepsilon(F) = \varepsilon^{-mn}\ \rho(F/\varepsilon)$. Then

$\phi_\varepsilon \overset{\text{def}}{=} \rho_\varepsilon * \phi$ is C^∞ and satisfies (5). Hence (cf for example Morrey [7 p 11])

$$\frac{\partial^2\phi_\varepsilon(F)}{\partial F^i_\alpha \partial F^j_\beta}\ \lambda^i\lambda^j\mu_\alpha\mu_\beta = 0$$

for all $F\varepsilon M^{n\times m}$, $\underset{\sim}{\lambda}\varepsilon\mathbb{R}^n$, $\underset{\sim}{\mu}\varepsilon\mathbb{R}^m$. Thus $\dfrac{\partial^2\phi_\varepsilon(F)}{\partial F^i_\alpha \partial F^j_\beta} = -\dfrac{\partial^2\phi_\varepsilon(F)}{\partial F^i_\beta \partial F^j_\alpha}$,

so that (4) holds for ϕ_ε. Letting $\varepsilon \longrightarrow 0$ we see that $\phi\varepsilon N$.

The converse follows by noting that if $\phi\varepsilon N$ then (5) holds for all $F_0, \underset{\sim}{\zeta}$. □

The null-space N has been characterized for arbitrary m,n by Edelen [3] (see also Ericksen [4]). Edelen assumes that the functions $\underset{\sim}{u}$ in (4) are C^2, but his results hold for $\underset{\sim}{u}$ that are C^1 by approximation. By Theorem 3 we are interested only in elements $\phi(F)$ of N which do not depend on $\underset{\sim}{x}, \underset{\sim}{u}$. These are given by linear combinations of 1 and all $r \times r$ subdeterminants of F for $1 \leqslant r \leqslant \min(m,n)$. Thus, for example, if $m = n = 1$, 2 or 3 then $\phi(F)\ \varepsilon\ N$ if and only if ϕ has the form

$(n = 1)$ $\quad \phi(F) = a + bF$

$(n = 2)$ $\quad \phi(F) = a + A^\alpha_i F^i_\alpha + B\det F$

$(n = 3)$ $\quad \phi(F) = a + A^\alpha_i F^i_\alpha + B^\alpha_i(\operatorname{adj}F)^i_\alpha + C\det F$,

where a, b, A^α_i, B^α_i, B, C are constants.

5. Sequentially weakly continuous functionals on $W^{1,p}(\Omega)$

Corollary 2.1 and Theorem 3 show in particular that if ψ is continuous and such that for some $1 \leq p < \infty$ the map $\theta : \underset{\sim}{u} \longmapsto \psi(\cdot, \underset{\sim}{u}(\cdot), \nabla \underset{\sim}{u}(\cdot))$ is sequentially continuous from $(W_n^{1,p}(\Omega), \text{weak}) \longrightarrow (L^1(\Omega), \text{weak})$, then $\psi(\underset{\sim}{x}_0, \underset{\sim}{u}_0, \cdot) \in N$ for all $\underset{\sim}{x}_0 \in \mathbb{R}^m$, $\underset{\sim}{u}_0 \in \mathbb{R}^n$. In this section we investigate to what extent the converse holds.

Lemma 1

Let $K \geq 2$, $m \geq 2$, $n \geq 2$, and suppose that $y^i \in W^{1,p}(\Omega)$ for $1 \leq i \leq K$, where $p \geq p_0 = \min(m,n)$. Then the formula

$$\frac{\partial(y^1,\ldots,y^K)}{\partial(x_1,\ldots,x_K)} = \sum_{s=1}^{K} (-1)^{s+1} \frac{\partial}{\partial x_s}\left(y^1 \frac{\partial(y^2,\ldots\ldots\ldots\ldots,y^K)}{\partial(x_1,\ldots,x_{s-1},x_{s+1},\ldots x_K)}\right) \quad (6)$$

holds in the sense of distributions, where

$$\frac{\partial(y^1,\ldots,y^K)}{\partial(x_1,\ldots,x_K)} \stackrel{\text{def}}{=} \det\left(\frac{\partial y^i}{\partial x_j}\right).$$

Proof

First suppose that each $y^i \in C^2(\Omega)$. Then the right hand side of (6) equals

$$\sum_{s=1}^{K} (-1)^{s+1} y^1_{,s} \frac{\partial(y^2,\ldots\ldots\ldots\ldots,y^K)}{\partial(x_1,\ldots,x_{s-1},x_{s+1},\ldots,x_K)} +$$

$$+ y^1 \sum_{s=1}^{K} (-1)^{s+1} \frac{\partial}{\partial x_s} \frac{\partial(u^2,\ldots\ldots\ldots\ldots,u^K)}{\partial(x_1,\ldots,x_{s-1},x_{s+1},\ldots,x_K)} .$$

The second term is zero by Morrey [7 Lemma 4.4.6], while the first equals $\frac{\partial(y^1,\ldots,y^K)}{\partial(x_1,\ldots,x_K)}$ as required.

Now suppose $y^i \in W_n^{1,p}(\Omega)$ for $1 \leq i \leq K$ and let $\alpha \in C_0^\infty(\Omega)$. There exists a sequence y_r^i of $C^2(\Omega)$ functions such that $y_r^i \longrightarrow y^i$ in $W^{1,p}(\Omega')$ for some $\Omega' \supset \text{supp}\,\alpha$.

Then

$$\int_{\Omega} \frac{\partial(y_r^1,\ldots,y_r^K)}{\partial(x_1,\ldots,x_K)} \alpha(\underset{\sim}{x})\, d\underset{\sim}{x} = \sum_{s=1}^{K} (-1)^s \int_{\Omega} y_r^1 \frac{\partial(y_r^2,\ldots\ldots\ldots\ldots\ldots,y_r^K)}{\partial(x_1,\ldots,x_{s-1},x_{s+1},\ldots,x_K)} \frac{\partial\alpha}{\partial x_s}\, d\underset{\sim}{x}.$$

Note that $y_r^1 \longrightarrow y^1$ in $L^q(\Omega')$ for $q \geq 1$, $\frac{1}{q} > \frac{1}{p_0} - \frac{1}{m}$, and that $\frac{2}{p_0} - \frac{1}{m} < 1$. Using the Hölder inequality we obtain (6). □

Theorem 4

Let $1 \leq K \leq p_0 = \min(m,n)$, $1 \leq i_1 < i_2 < \ldots < i_K \leq n$, $1 \leq j_1 < j_2 < \ldots < j_K \leq m$, and define

$$\phi(\nabla \underset{\sim}{u}) = \frac{\partial(u^{i_1},\ldots,u^{i_K})}{\partial(x_{j_1},\ldots,x_{j_K})}.$$

Let $p \geq p_0$ and let $\underset{\sim}{u}_r \longrightarrow \underset{\sim}{u}$ in $W_n^{1,p}(\Omega)$. Then $\phi(\nabla \underset{\sim}{u}_r) \longrightarrow \phi(\nabla \underset{\sim}{u})$ in the sense of distributions.

If $p > p_0$ then $\phi(\nabla \underset{\sim}{u}_r) \rightharpoonup \phi(\nabla \underset{\sim}{u})$ in $L^{p/p_0}(\Omega)$.

Proof

Suppose $K \geq 2$, $m \geq 2$, $n \geq 2$, since the other cases are trivial. Let $\alpha \in C_0^\infty(\Omega)$. Then $\underset{\sim}{u}_r \longrightarrow \underset{\sim}{u}$ in $L_n^q(\Omega')$ for some $\Omega' \supset \operatorname{supp}\alpha$ and for $q \geq 1$, $\frac{1}{q} > \frac{1}{p_0} - \frac{1}{n}$. Hence

$$u_r^1 \frac{\partial(u_r^{i_2},\ldots\ldots\ldots\ldots\ldots,u_r^{i_K})}{\partial(x_{j_1},\ldots,x_{j_{s-1}},x_{j_{s+1}},\ldots,x_{j_K})} \longrightarrow u^1 \frac{\partial(u^{i_2},\ldots\ldots\ldots\ldots\ldots,u^{i_K})}{\partial(x_{j_1},\ldots,x_{j_{s-1}},x_{j_{s+1}},\ldots,x_{j_K})}$$

in $L^1(\Omega')$ as $r \to \infty$. The result follows from Lemma 1.

If $p > p_0$ then $\phi(\nabla \underset{\sim}{u}_r)$ is bounded in $L^{p/p_0}(\Omega)$, so that a subsequence $\phi(\nabla \underset{\sim}{u}_\mu) \rightharpoonup \Theta$ in $L^{p/p_0}(\Omega)$. By the first part

$\Theta = \phi(\nabla \underset{\sim}{u})$ and thus the whole sequence converges to $\phi(\nabla \underset{\sim}{u})$. □

Note that the right hand side of (6) may have meaning as a distribution when $p < p_0$. In fact we just need that the products

$$y^1 \frac{\partial(y^2, \ldots\ldots\ldots\ldots, y^K)}{\partial(x_1, \ldots, x_{s-1}, x_{s+1}, \ldots, x_K)} \tag{7}$$

are in $L^1(\Omega)$ when $y^i \varepsilon W^{1,p}(\Omega)$ for $1 \leqslant i \leqslant K$, and conditions on p,m for this to hold are easily derivable from the imbedding theorems. In fact, one may go further and define the Jacobians in (7) under correspondingly weaker conditions. Rather than give a complete inductive definition of these generalized Jacobians we here restrict ourselves to an illustrative example.

Let $m = n = 3$. Define the distributions

$$(\text{Adj } \nabla \underset{\sim}{u})^{\alpha}_{i} = (u^{i+2} u^{i+1}_{,\alpha+1})_{,\alpha+2} - (u^{i+2} u^{i+1}_{,\alpha+2})_{,\alpha+1} ,$$

where the indices are taken modulo 3, and

$$\text{Det } \nabla \underset{\sim}{u} = [u^1 (\text{Adj } \nabla \underset{\sim}{u})^j_1]_{,j} .$$

Note that if $\underset{\sim}{u} \varepsilon W^{1,p}_3(\Omega)$, $p \geqslant 2$ then Adj $\nabla \underset{\sim}{u}$ = adj $\nabla \underset{\sim}{u}$, where $(\text{adj } \nabla \underset{\sim}{u})^T$ is the matrix of cofactors of $\nabla \underset{\sim}{u}$, and that if $\underset{\sim}{u} \varepsilon W^{1,p}_3(\Omega)$, $p \geqslant 2$, and Adj $\nabla \underset{\sim}{u} \ \varepsilon\ L^{p'}_q(\Omega)$ then Det $\nabla \underset{\sim}{u}$ = det $\nabla \underset{\sim}{u}$. In general, however, Adj $\nabla \underset{\sim}{u} \neq$ adj $\nabla \underset{\sim}{u}$, Det $\nabla \underset{\sim}{u} \neq$ det $\nabla \underset{\sim}{u}$. The following theorem may be proved by similar methods to Theorem 4 (cf [2] for details).

<u>Theorem 5</u>

(i) Let $p > 3/2$. If $\underset{\sim}{u}_r \longrightarrow \underset{\sim}{u}$ in $W^{1,p}_3(\Omega)$ then Adj $\nabla \underset{\sim}{u}_r \longrightarrow$ Adj $\nabla \underset{\sim}{u}$ in the sense of distributions.

(ii) Let $1 < p < \infty$, $1 < q < \infty$, $\frac{1}{p} + \frac{1}{q} < \frac{4}{3}$. If $\underset{\sim}{u}_r \longrightarrow \underset{\sim}{u}$ in $W^{1,p}_3(\Omega)$ and if Adj $\nabla \underset{\sim}{u}_r \longrightarrow$ Adj $\nabla \underset{\sim}{u}$ in $L^q_9(\Omega)$ then

Det $\nabla \underset{\sim}{u}_r \longrightarrow$ Det $\nabla \underset{\sim}{u}$ in the sense of distributions.

Remark: Results analogous to Theorems 4,5 can be proved in an Orlicz-Sobolev space setting (see [2]).

6. Lower semicontinuity theorems

Let $\phi_1(F), \ldots, \phi_K(F)$ belong to N and let $g: \Omega \times \mathbb{R}^n \times \mathbb{R}^K \to \overline{\mathbb{R}}$ satisfy the conditions

(a) for almost all $\underset{\sim}{x} \in \Omega$, $g(\underset{\sim}{x}, \cdot, \cdot)$ is continuous on $\mathbb{R}^n \times \mathbb{R}^K$,

(b) for all $\underset{\sim}{u} \in \mathbb{R}^m, \underset{\sim}{a} \in \mathbb{R}^K, g(\cdot, \underset{\sim}{u}, \underset{\sim}{a})$ is measurable,

(c) for almost all $\underset{\sim}{x} \in \Omega$ and for all $\underset{\sim}{u} \in \mathbb{R}^n, g(\underset{\sim}{x}, \underset{\sim}{u}, \cdot)$ is convex,

(d) $g(\underset{\sim}{x}, \underset{\sim}{u}, \underset{\sim}{a}) \geqslant \alpha(\underset{\sim}{x}) + \eta(|\underset{\sim}{a}|)$,

where $\alpha \in L^1(\Omega)$ and $\eta(t)$ is a real-valued, continuous, even, convex function of $t \in \mathbb{R}$ satisfying $\eta(t) > 0$ for $t > 0$, $\eta(t)/t \longrightarrow 0$ as $t \longrightarrow 0$, $\eta(t)/t \longrightarrow \infty$ as $t \longrightarrow \infty$.

Define $f: \Omega \times \mathbb{R}^n \times M^{n \times m} \longrightarrow \overline{\mathbb{R}}$ by

$$f(\underset{\sim}{x}, \underset{\sim}{u}, F) = g(\underset{\sim}{x}, \underset{\sim}{u}, \phi_1(F), \ldots, \phi_K(F)) \tag{8}$$

and let $I(\underset{\sim}{u}, \Omega)$ be given by (1).

Theorem 6

Let $\underset{\sim}{u}_r \rightharpoonup \underset{\sim}{u}$ in $W_n^{1,p}(\Omega)$, where $p > p_0 = \min(m,n)$. Then

$$I(\underset{\sim}{u}, \Omega) \leqslant \liminf_{r \to \infty} I(\underset{\sim}{u}_r, \Omega).$$

Proof

For $i = 1,2,\ldots$ let Ω_i be the union of all open balls contained in Ω of radius less than $1/i$. Each Ω_i satisfies the cone condition, so that by the imbedding theorems a subsequence $\underset{\sim}{u}_\mu \longrightarrow \underset{\sim}{u}$ almost everywhere on Ω_i. A standard diagonal argument shows that we may assume that $\underset{\sim}{u}_\mu \longrightarrow \underset{\sim}{u}$ almost everywhere on each Ω_i and thus almost everywhere on Ω. Since each $\phi_i(F)$ is a finite linear combination of subdeterminants of F of order less than or equal to p_0, we may suppose without loss of generality that $\phi_i(\nabla \underset{\sim}{u}_\mu) \rightharpoonup \Theta_i$ in $L^1(\Omega)$, and hence, by Theorem 4, $\Theta_i = \phi_i(\nabla \underset{\sim}{u})$. By a known theorem [5 p 226]

$$I(\underset{\sim}{u},\Omega) \leq \liminf_{\mu \to \infty} I(\underset{\sim}{u}_\mu,\Omega),$$

and the result follows. □

Remarks: 1. If $g: \Omega \times \mathbb{R}^n \times \mathbb{R}^K \longrightarrow \mathbb{R}$ is continuous, then by Theorems 2 and 6 it follows that f is quasiconvex.

2. Other lower semicontinuity theorems can be proved using Theorem 5 and analogous results.

Integrands of the form (8) occur in nonlinear elasticity. An example is the Mooney-Rivlin strain-energy function.

$$W(F) = A(I-3) + B(II-3),$$

where $A > 0$, $B > 0$ are constants, and where $I = \mathrm{tr}(FF^T)$, $II = \mathrm{tr}[(\mathrm{adj}F)(\mathrm{adj}F)^T]$. Theorems 4 and 6 can be applied simply to prove the existence of equilibrium solutions for various boundary value problems for the Mooney-Rivlin material subject to the pointwise constraint of incompressibility

$$\det \nabla \underset{\sim}{u} = 1 \quad \text{almost everywhere in } \Omega. \tag{9}$$

More general existence theorems are proved in [2].

7. Conclusion

The method in this paper would seem to have the following advantages.

(i) It enables the existence of minimizers for integrands of the form (8) to be established under significantly weaker continuity and growth conditions than those of Morrey [5 Thm 4.4.5], and the proofs are much simpler.

(ii) It can treat 'weakly continuous' pointwise constraints such as (9).

(iii) It can be extended to equations which do not arise from the calculus of variations.

(iv) It can be extended to higher order equations.

On the other hand, there are examples of quasiconvex integrands which cannot be written in the form (8), so that Morrey's theorem applies but Theorem 6 does not.

We end with a few examples illustrating (iv)†. We consider integrands depending only on second derivatives of $\underset{\sim}{u}$. In the case m = 2, n = 1, the only nonlinear element of the null-space of the corresponding Euler-Lagrange operator is

$$u_{,11}\, u_{,22} - u^2_{,12} = (u_{,1}u_{,22})_{,1} - (u_{,1}u_{,12})_{,2} \qquad (10)$$

while if m = n = 2 the basic nonlinear elements of the null-space are

$$u^1_{,11}\, u^2_{,12} - u^1_{,12}\, u^2_{,11}\ ,\ u^1_{,22}\, u^2_{,21} - u^1_{,12}\, u^2_{,22}\ ,\ u^1_{,11}u^1_{,22}-(u^1_{,12})^2$$

$$u^1_{,11}\, u^2_{,22} - u^1_{,12}\, u^2_{,12}\ ,\ u^2_{,11}\, u^1_{,22} - u^2_{,12}\, u^1_{,12}\ ,\ u^2_{,11}u^2_{,22}-(u^2_{,12})^2$$

Results analogous to Theorems 4, 5 and 6 may be proved. We remark that expressions of the form (10) occur, for example, in connection with the Monge-Ampère equation and the von Karman plate equations.

REFERENCES

[1] R. A. Adams, "Sobolev spaces", Academic Press, New York, 1975.

[2] J. M. Ball, Convexity conditions and existence theorems in nonlinear elasticity, to appear.

† see forthcoming work with J. C. Currie.

[3] D. G. B. Edelen, The null set of the Euler-Lagrange operator, Arch. Rat. Mech. Anal. 11 (1962) 117-121.

[4] J. L. Ericksen, Nilpotent energies in liquid crystal theory, Arch. Rat. Mech. Anal., 10 (1962) 189-196.

[5] I. Ekeland and R. Temam, "Analyse convexe et problèmes variationnels", Dunod, Paris, 1974.

[6] C. B. Morrey, Quasiconvexity and the lower semicontinuity of multiple integrals, Pac. J. Math. 2(1952) 25-53.

[7] ___________ , "Multiple integrals in the calculus of variations", Springer, Berlin, 1966.

BIVARIATIONAL BOUNDS ON $\langle g,\phi\rangle$ FOR NON-LINEAR PROBLEMS $F\phi = 0$

M.F. BARNSLEY and P.D. ROBINSON

Abstract

Complementary (upper and lower) bivariational bounds are presented on the inner product $\langle g,\phi\rangle$ associated with the solution ϕ of an arbitrary non-linear problem $F\phi = 0$ in a real Hilbert space $\mathcal{H}$. They take the form $J(\Psi,\Phi) \pm C(\Psi,\Phi)$, where $C(\Psi,\Phi)$ is a positive bivariational approximation to zero and

$$J(\Psi,\Phi) = -\langle\Psi,F\Phi\rangle + \langle g,\Phi\rangle$$

is a bivariational approximation to $\langle g,\phi\rangle$. The applicability of the bounds is briefly discussed.

1. Introduction

Let $\mathcal{H}$ be a real Hilbert space with symmetric inner product $\langle\, ,\, \rangle$. Let F be a non-linear mapping in $\mathcal{H}$, which for simplicity of presentation we assume maps the whole of $\mathcal{H}$ into itself. Denote the image of $\Phi \in \mathcal{H}$ by $F\Phi$, and suppose that the non-linear problem

$$F\phi = 0, \quad \phi \in \mathcal{H}, \tag{1.1}$$

has a solution ϕ. We seek upper and lower bounds on the real number $\langle g,\phi\rangle$ for a given $g \in \mathcal{H}$. Such quantities $\langle g,\phi\rangle$ are often of direct physical interest, and can also lead to information about the solution ϕ itself.

We make the following further simplifying assumptions:

(i) a positive constant c is known such that

$$\|F\Phi_1 - F\Phi_2\| \geqslant c\|\Phi_1 - \Phi_2\| \quad \text{for all} \quad \Phi_1, \Phi_2 \in S \subset \mathcal{H}; \tag{1.2}$$

(ii) the Fréchet derivative $F'(\Phi)$ exists for all $\Phi \in \mathcal{H}$; (1.3)

(iii) a non-negative functional $K(\Psi)$ is known such that

$$|\langle\Psi, F\Phi_1 - F\Phi_2 - F'(\Phi_1)(\Phi_1-\Phi_2)\rangle| \leq \tfrac{1}{2}K(\Psi)\|\Phi_1-\Phi_2\|^2 \text{ for all } \Phi_1,\Phi_2 \in S \subset \mathcal{H}. \tag{1.4}$$

It follows from assumption (1.2) that a solution $\phi \in S$ of (1.1) must be unique. It follows from assumption (1.3) that $F'(\Phi)$ is a bounded linear mapping with domain $\mathcal{H}$; further the adjoint $F'(\Phi)^*$ exists, and is also a bounded linear mapping with domain $\mathcal{H}$ [1].

Under these conditions, we establish the results

$$J(\Psi,\Phi) - C(\Psi,\Phi) \leq \langle g,\phi\rangle \leq J(\Psi,\Phi) + C(\Psi,\Phi), \quad \Psi \in \mathcal{H},\ \Phi \in S, \tag{1.5}$$

where the functional

$$J(\Psi,\Phi) = -\langle\Psi,F\Phi\rangle + \langle g,\Phi\rangle \tag{1.6}$$

is a bivariational approximation to $\langle g,\phi\rangle$, and where the functional

$$C(\Psi,\Phi) = \frac{1}{c}\|F'(\Phi)^*\Psi - g\|\|F\Phi\| + \frac{1}{2c^2}K(\Psi)\|F\Phi\|^2 \tag{1.7}$$

is a positive bivariational approximation to zero.

2. The functional $J(\Psi,\Phi)$

The functional $J(\Psi,\Phi)$ is associated with equation (1.1) and also with the auxiliary equation

$$F'(\phi)^*\psi = g \tag{2.1}$$

where g is such that this equation does have a solution ψ. If J is regarded as a mapping

$$J : D(J) = \mathcal{H} \times S \to \mathcal{R}, \tag{2.2}$$

and a typical element in $\mathcal{H} \times S$ is denoted by $\begin{bmatrix}\Psi\\ \Phi\end{bmatrix}$, then $J\begin{bmatrix}\Psi\\ \Phi\end{bmatrix}$ is in fact a <u>variational</u> approximation to

$$J\begin{bmatrix}\psi\\ \phi\end{bmatrix} = \langle g,\phi\rangle \tag{2.3}$$

because its Fréchet derivative $J'\left(\begin{bmatrix}\Psi\\ \Phi\end{bmatrix}\right)$ exists and is the null operator when $\Psi = \psi$, $\Phi = \phi$. This result can be established from consideration of the identity

$$J\begin{bmatrix}\Psi\\ \Phi\end{bmatrix} = J\begin{bmatrix}\psi\\ \phi\end{bmatrix} + \left\langle\begin{bmatrix}-F\phi\\ g-F'(\phi)^*\psi\end{bmatrix}, \begin{bmatrix}\Psi-\psi\\ \Phi-\phi\end{bmatrix}\right\rangle - \left\langle\begin{bmatrix}\Psi\\ \Phi-\phi\end{bmatrix}, \begin{bmatrix}F\Phi-F\phi-F'(\phi)(\Phi-\phi)\\ F'(\phi)^*(\Psi-\psi)\end{bmatrix}\right\rangle, \tag{2.4}$$

where the inner product in $\mathcal{H}\times\mathcal{H}$ is

$$\left\langle\begin{bmatrix}\Psi_1\\ \Phi_1\end{bmatrix}, \begin{bmatrix}\Psi_2\\ \Phi_2\end{bmatrix}\right\rangle = \langle\Psi_1,\Psi_2\rangle + \langle\Phi_1,\Phi_2\rangle. \tag{2.5}$$

However, it is sensible to revert to the notation (1.6) and call $J(\Psi,\Phi)$ a bivariational approximation to $\langle g,\phi\rangle$.

In the same manner, the functional C in (1.7) is actually a variational approximation to zero if the underlying space is $\mathcal{H}\times S$, but again it is sensible to regard it as bivariational.

It is interesting to note that the functional J is closely related to Newton's approximation for ϕ, which is [2]

$$N\Phi = \Phi - F'(\Phi)^{-1}F\Phi. \tag{2.6}$$

To see this, suppose that the vector Ψ can be chosen so that

$$\Psi = F'(\Phi)^{*-1}g, \tag{2.7}$$

in line with (2.1). Then it follows from (1.6) and (2.6) that

$$J\left(F'(\Phi)^{*-1}g,\Phi\right) = \langle g,N\Phi\rangle. \tag{2.8}$$

3. Derivation of the bounds

To prove the inequalities in (1.5), we first set $\Phi_1 = \Phi$ and $\Phi_2 = \phi$ in (1.2) and (1.4). Since $F\phi = 0$, we obtain

$$\|F\Phi\| \geqslant c\|\Phi-\phi\|, \quad \Phi \in S, \tag{3.1}$$

and

$$|\langle\Psi,F\Phi-F'(\Phi)(\Phi-\phi)\rangle| \leqslant \tfrac{1}{2}K(\Psi)\|\Phi-\phi\|^2, \quad \Phi \in S. \tag{3.2}$$

Using (3.1) and Schwarz's inequality, we see that

$$\frac{1}{c}\|F'(\Phi)^*\Psi-g\|\,\|F\Phi\| \geqslant \|F'(\Phi)^*\Psi-g\|\,\|\Phi-\phi\| \geqslant |\langle F'(\Phi)^*\Psi-g,\Phi-\phi\rangle|$$
$$= |\langle\Psi,F'(\Phi)(\Phi-\phi)\rangle - \langle g,\Phi\rangle + \langle g,\phi\rangle|. \tag{3.3}$$

With the further help of (3.2), we have

$$\frac{1}{2c^2}\ K(\Psi)\|F\Phi\|^2 \geqslant \tfrac{1}{2}K(\Psi)\|\Phi-\phi\|^2 \geqslant |\langle\Psi,F\Phi-F'(\Phi)(\Phi-\phi)\rangle| \ . \tag{3.4}$$

Finally from (1.7), (3.3), (3.4) and the triangle inequality for moduli it follows that

$$C(\Psi,\Phi) \geqslant |\langle\Psi,F\Phi\rangle - \langle g,\Phi\rangle + \langle g,\phi\rangle| \ = \ |\langle g,\phi\rangle - J(\Psi,\Phi)|, \tag{3.5}$$

which can be rearranged to give the inequalities (1.5).

4. The quasilinear case

The bivariational bounds in (1.5) take a simple form when

$$F\Phi \ = \ A\Phi - f, \tag{4.1}$$

where A is a bounded linear mapping from $\mathcal{H}$ into itself. In this case we have

$$J(\Psi,\Phi) \ = \ -\langle\Psi,A\Phi\rangle + \langle\Psi,f\rangle + \langle g,\Phi\rangle, \tag{4.2}$$

which is the inhomogeneous Rayleigh-Ritz stationary approximation to $\langle g,\phi\rangle$.

Also

$$\begin{aligned}\|F\Phi_1-F\Phi_2\|^2 &= \langle A(\Phi_1-\Phi_2),A(\Phi_1-\Phi_2)\rangle\\ &= \langle\Phi_1-\Phi_2,A^*A(\Phi_1-\Phi_2)\rangle\\ &\geqslant a^2\ \|\Phi_1-\Phi_2\|^2\end{aligned} \tag{4.3}$$

where $a^2 \geqslant 0$ is the smallest eigenvalue of the non-negative self-adjoint mapping A^*A. If $a > 0$, a suitable choice in (1.2) is $c = a$.

Further, since

$$F'(\Phi) = A, \tag{4.4}$$

it follows that we can choose $K(\Psi) \equiv 0$ in (1.4). Here we may take $S = \mathcal{H}$, and the functional C becomes simply

$$C(\Psi,\Phi) = \frac{1}{a} \|A^*\Psi - g\| \|A\Phi - f\| . \tag{4.5}$$

The resulting bounds are not new [3], but their realization as a particular case of non-linear theory is.

5. Applicability

We first emphasize that the domain of F need not be the whole of $\mathcal{H}$; for example it is possible to modify the theory to embrace the typical differential-equation situation where the domain of F is a dense linear subspace of $\mathcal{H}$ [4]. It may also happen that the Fréchet derivative $F'(\Phi)$ does not exist, as for example in the quasilinear situation (4.1) if A is unbounded. In such cases it does seem possible to employ instead a linear Gâteaux derivative, which is allowed to be unbounded.

The conditions (1.2) and (1.4) are not unduly restrictive. For example, consider a problem of the type

$$F\phi = A\phi + f(\phi) = 0, \tag{5.1}$$

where here $f(\Phi)$ is a straightforward function of the real variable Φ with $d^2f/d\Phi^2$ continuous, and A is a positive, self-adjoint, linear mapping which is bounded below by $a > 0$. With the aid of monotonicity theorems one can often establish a suitable closed, bounded region S wherein a solution of (5.1) is known to exist, and an appropriate value of the constant c can then be inferred.

There are two situations in which one readily finds a suitable functional $K(\Psi)$ for (1.4). Firstly, in cases where Kantorovich's Theorem on the convergence of Newton's method holds, the inequality

$$\|F\Phi_1 - F\Phi_2 - F'(\Phi_1)(\Phi_1 - \Phi_2)\| \leqslant \tfrac{1}{2}k\|\Phi_1 - \Phi_2\|^2, \qquad \Phi_1, \Phi_2 \in S, \tag{5.2}$$

pertains [5]. Accordingly

$$\begin{aligned} |\langle\Psi, F\Phi_1 - F\Phi_2 - F'(\Phi_1)(\Phi_1 - \Phi_2)\rangle| &\leqslant \|\Psi\|\|F\Phi_1 - F\Phi_2 - F'(\Phi_1)(\Phi_1 - \Phi_2)\| \\ &\leqslant \tfrac{1}{2}k\|\Psi\|\|\Phi_1 - \Phi_2\|^2, \end{aligned} \tag{5.3}$$

so that we may take in (1.4)

$$K(\Psi) = k\|\Psi\| . \tag{5.4}$$

Here the domain of K is the whole of $\mathcal{H}$.

Secondly, for a differential-equation problem of type (5.1) with A represented by a differential operator adjoined to appropriate boundary conditions, we have

$$\begin{aligned} |F\Phi_1 - F\Phi_2 - F'(\Phi_1)(\Phi_1 - \Phi_2)| &= |f(\Phi_1) - f(\Phi_2) - f'(\Phi_1)(\Phi_1 - \Phi_2)| \\ &\leqslant \tfrac{1}{2}\tilde{k}|\Phi_1 - \Phi_2|^2 \end{aligned} \tag{5.5}$$

for some finite $\tilde{k}$, since $d^2f/d\Phi^2$ is continuous. Now suppose that the inner product is an integral of the type

$$\langle\Psi_1, \Psi_2\rangle = \int \Psi_1(\underline{r})\Psi_2(\underline{r})\, \rho(\underline{r})\, d\underline{r}, \qquad \rho(\underline{r}) \geqslant 0. \tag{5.6}$$

Then

$$\begin{aligned} |\langle\Psi, F\Phi_1 - F\Phi_2 - F'(\Phi_1)(\Phi_1 - \Phi_2)\rangle| &\leqslant \langle|\Psi|, |F\Phi_1 - F\Phi_2 - F'(\Phi_1)(\Phi_1 - \Phi_2)|\rangle \\ &\leqslant \tfrac{1}{2}\tilde{k}\; |\Psi|_{max} \langle 1, |\Phi_1 - \Phi_2|^2\rangle \\ &= \tfrac{1}{2}\tilde{k}\; |\Psi|_{max} \|\Phi_1 - \Phi_2\|^2, \end{aligned} \tag{5.7}$$

so that we may take in (1.4)

$$K(\Psi) = \tilde{k}\; |\Psi|_{max} . \tag{5.8}$$

As an example, consider the problem

$$\left.\begin{aligned} -\frac{d^2\phi}{dx^2} + (\phi+1) + \tfrac{1}{4}(\phi+1)^2 &= 0, \qquad -1 \leq x \leq 1, \\ \phi(-1) = 0, \qquad \phi(1) &= 0, \end{aligned}\right\} \tag{5.9}$$

which might describe the temperature distribution $\phi(x)$ in a rod with surroundings at temperature -1 and ends at temperature 0. The heat in the bar would be proportional to

$$Q = \langle 1,\phi\rangle = \int_{-1}^{1} \phi(x)\, dx . \tag{5.10}$$

We find that $\bar{\phi}(x) \equiv 0$ and $\underline{\phi}(x) \equiv -1$ constitute an upper and lower solution pair [6], thereby establishing the existence of a regular solution $\phi(x)$ satisfying

$$-1 \leq \phi(x) \leq 0 \quad \text{for all} \quad -1 \leq x \leq 1. \tag{5.11}$$

Using (5.11) as a prescription for S, we obtain

$$c = 1 + \frac{\pi^2}{4}, \quad \tilde{k} = \frac{1}{2} . \tag{5.12}$$

Trial vectors

$$\Phi = -0.4(1-x^2) \in S, \qquad \Psi = 0.34(1-x^2), \tag{5.13}$$

give the result

$$-0.55 \leq Q \leq -0.52 . \tag{5.14}$$

References

[1] F. Riesz and B. Sz-Nagy, Functional Analysis, Ungar, New York, 1955.

[2] L.B. Rall, Computational Solution of Non-linear Operator Equations, Wiley, New York, 1969.

[3] M.F. Barnsley and P.D. Robinson, Bivariational bounds associated with non-self-adjoint linear operators, Proc. Roy. Soc. Edinburgh, A, to appear, 1976.

[4] M.M. Vainberg, Variational Method and Method of Monotone Operators, Wiley, New York, 1973.

[5] R.A. Tapia, The Differentiation and Integration of Non-linear Operators, in 'Nonlinear Functional Analysis and Applications', ed. L.B. Rall, Academic Press, New York, 1971, 45-101.

[6] D.H. Sattinger, Monotone methods in nonlinear elliptic and parabolic boundary value problems, Indiana University Mathematics Journal, 21, 979-1000, 1972.

The authors would like to acknowledge support from the Science Research Council.

The Spectral Analysis of a Stone Regular Differential Operator

by

Harold E. Benzinger

1. Introduction. This paper is a continuation of earlier work of the author [4,5] on the spectral theory in $L^p = L^p(0,1)$ of ordinary differential operators generated by two-point boundary value problems. To define such an operator, let $n \geq 1$ be a given integer, and let τ denote an n-th order differential expression defined for suitable functions $u(x)$, $o \leq x \leq 1$, by

$$(1.1) \qquad \tau(u) = u^{(n)} + a_{n-2}(x)u^{(n-2)} + \cdots + a_o(x)u \ ,$$

where $a_j(x)$ is in L^∞ for $j = o,\cdots,n-2$. Let M,N denote $n \times n$ matrices of complex constants, and let

$$(1.2) \qquad U(u) = M\hat{u}(o) + N\hat{u}(1) \ ,$$

where $\hat{u}(x)$ is the transpose of the vector $(u(x),\cdots,u^{(n-1)}(x))$. For fixed p, $1 \leq p < \infty$, the subspace $\mathcal{D}(T)$ of L^p consisting of all functions u such that $u^{(n-1)}$ is absolutely continuous, $u^{(n)}$ is in L^p, and $U(u) = o$, is dense in L^p, and the linear operator $T : L^p \to L^p$ defined by

$$(1.3) \qquad Tu = \tau(u) \ , \quad u \text{ in } \mathcal{D}(T) \ ,$$

is a closed, densely defined operator. In [4,5], we restricted consideration to operators T which are Birkhoff regular [6] and which have simple spectrum: all (but finitely many) eigenvalues have algebraic multiplicity one, and the distance between distinct eigenvalues cannot become arbitrarily small.

Let $\{u_k(x)\}$, $k = o, \pm 1, \cdots$ denote the root functions of T, and let $\{v_k(x)\}$ denote the root functions of the adjoint T^*, normalized so that $(u_k, v_j) = \delta_{kj}$. If T is Birkhoff regular with simple spectrum, we can further require that for some $M > o$, $|u_k(x)| \leq M$, $|v_k(x)| \leq M$. Let D denote the operator defined by the special case $n = 1$, $\tau(u) = u^{(1)}$, $U(u) = u(o) - u(1)$, and with eigenfunctions $\varphi_k(x) = \exp(2k\pi i x)$, $k = o, \pm 1, \cdots$ With this notation, the results of [4,5] can be summarized as follows:

Theorem 1. Let p be fixed, $1 < p < \infty$. Let T be an n-th order Birkhoff regular operator with simple spectrum. Then there exists a bicontinuous linear map $A : L^p \to L^p$, and a Fourier series multiplier transform $M : L^p \to L^p$, such that

1) $T = A(D+M)^n A^{-1}$;

2) $u_k = A\varphi_k$, $v_k = A^{-1^*} \varphi_k$;

3) the map $f \to S^*(x,f) = \sup_N \left| \sum_{-N}^{N} (f, v_k) u_k(x) \right|$ is of weak type (p,p).

Statement (2) implies that with respect to convergence in the norm of L^p, the eigenfunction expansion $\Sigma(f, v_k)u_k$ has all of the essential features of the Fourier series expansion of $A^{-1}f$, and 3) implies that $f(x) = \Sigma(f, v_k)u_k(x)$ a.e. if f is in L^p, $p > 1$.

In this paper we consider an explicit 2-nd order operator which is not Birkhoff regular, but instead is <u>Stone regular</u> [1], still with simple spectrum. We show how theorem 1 must be modified in this case, and thus illustrate the abstract ideas which will be useful in discussing the general class of Stone regular problems. Stone regular problems have been discussed previously in [9,1,2,3]. For the special case considered here, we obtain more precise results on the behavior of eigenfunction expansions.

2. Preliminary Considerations. For the remainder of this paper, T will correspond to $\tau(u) = u^{(2)}$, with boundary conditions $u^{(1)}(o) + u(1) - u^{(1)}(1) = o$, $u(o) + u(1) = o$. This problem has eigenvalues $\lambda_k = -\rho_k^2$, $k = o,\pm 1,\cdots$, where

$$\rho_k = 2k\pi,\ k = -1,-2,\cdots,\quad \rho_k = (2k+1)\pi,\ k = o,1,\cdots .$$

The adjoint T^* corresponds to $\tau(v) = v^{(2)}$, with boundary conditions $v(o) - v^{(1)}(o) - v^{(1)}(1) = o$, $v(o) + v(1) = o$, and has the same spectrum as T. To specify the eigenfunctions, let

$$s_k(x) = \sqrt{2}\sin\rho_k x\ ,\quad c_k(x) = \sqrt{2}\cos\rho_k x\ ,\quad k = o,\pm 1,\cdots .$$

Then the eigenfunctions $\{u_k\}$ of T and $\{v_k\}$ of T^* are

$$u_k(x) = s_k(x) + \delta_{k+}[2\pi(2k+1)c_k(x)] \qquad k = o,\pm 1,\cdots,$$

$$v_k(x) = s_k(x) + \delta_{k-}[4\pi k\ c_k(x)] \qquad k = o,\pm 1,\cdots,$$

where $\delta_{k+} = 1$ if $k \geq o$, $\delta_{k+} = o$ if $k < o$, and $\delta_{k-} = 1-\delta_{k+}$.

It is easy to verify that $(u_k,v_j) = \delta_{kj}$.

For a given function f, let

$$\mathcal{S}_N(x,f) = \sum_{-N}^{N}(f,v_k)u_k(x),\quad S_N(x,f) = \sum_{-N}^{N}(f,\varphi_k)\varphi_k(x)\ ,$$

$$\mathcal{S}^*(x,f) = \sup_N|\mathcal{S}_N(x,f)|,\quad S^*(x,f) = \sup_N|S_N(x,f)|\ ,$$

$$Af = \sum_{-\infty}^{\infty}(f,\varphi_k)u_k\ ,\qquad A^{-1}f = \sum_{-\infty}^{\infty}(f,v_k)\varphi_k\ ,$$

$$L_+f = \sum_{1}^{\infty}k(f,\varphi_k)\varphi_k\ ,\qquad L_-f = \sum_{-\infty}^{-1}k(f,c_k)\varphi_k\ .$$

It is easily seen that $\mathcal{D}(A)$ consists of those functions f in L^p such that L_+f is defined, and similarly for $\mathcal{D}(A^{-1})$ and L_-. Also, L_+ and L_- are closed operators, $1 \leq p < \infty$, so $\mathcal{D}(A)$, $\mathcal{D}(A^{-1})$ are Banach spaces with respective norms $\|f\|_+ = \|f\|_p + \|L_+f\|_p$, $\|f\|_- = \|f\|_p + \|L_-f\|_p$, and A, A^{-1} are bounded operators of their respective domains into L^p. For given f, let $\hat{f}(x) = f(1-x)$, $F = f + \hat{f}$. Let C denote the collection of those f in $\mathcal{D}(A^{-1})$ such that $A^{-1} f$ is in $\mathcal{D}(A)$. Then for f in C, $AA^{-1} f = f$. Also, it is possible to prove that f is in C if and only if $F^{(1)}$ exists and is in L^p, and $F(o) = o$. With this characterization of C, it is possible to prove that C is a Banach space with norm $\|f\|_C = \|f\|_-$

Let $\alpha_k = \pi\delta_{k+}$, and let $M : L^p \to L^p$ be defined by

$$Mf = \sum_{-\infty}^{\infty} \alpha_k (f,\varphi_k)\varphi_k \ .$$

Since L^p admits conjugation for $1 < p < \infty$ [7;p.48], we see that M defines a bounded linear operator for $1 < p < \infty$.

For a given sequence $\{\beta_k\}$, $k = o, \pm 1, \cdots$ of complex numbers, let

$$\sigma_N f = \sum_{-N}^{N} \beta_k (f,v_k)u_k \quad , \quad \eta_N f = \sum_{-N}^{N} \beta_k (f,\varphi_k)\varphi_k$$

3. The Main Theorem. If T_o is an operator on L^p, let $\overline{T}_o$ denote its closure. Let m denote Lebesgue measure on $[0,1]$.

Theorem 2. Assume $1 < p < \infty$, and f is in C. Let T_o be defined by

$$T_o u = A(D+M)^2 A^{-1} u \tag{3.1}$$

for all u in $\mathcal{D}(T)$ such that Tu is in C. Then

1) $\overline{T}_o = T$;

2) for all f in C $\lim_{N\to\infty} \sigma_N f$ converges in the norm of C if and only if $\lim_{N\to\infty} \eta_N A^{-1} f$ converges in the norm of L^p for all f in C ;

3) for all $y \geq o$ and some $K > o$,

(3.2) $\qquad m\,\{x : S^*(x,f) > y\} \leq K\,[\,\|f\|_C\, y^{-1}]^p$.

Remarks. From Theorem 2, we can show the following conclusions for f in C;

i) $S_N f \to f$ in the norm of C and also in the norm of L^p;

ii) $\{\beta_k\}$ is a multiplier sequence for $\{u_k\}$ in the norm of C if and only if $\{\beta_k\}$ is a multiplier sequence for $\{\varphi_k\}$ in the norm of L^p;

iii) the expansion of f in eigenfunctions $\{u_k\}$ converges unconditionally to f in the norm of C when $p = 2$;

iv) $f(x) = \Sigma(f,v_k)u_k(x)$, a.e.

4. Proof of Theorem 2. 1) Let T_{oo} be defined by $T_{oo}\,u = u^{(2)}$ for u in $\mathcal{D}(T)$ such that $u^{(3)}$ exists and is in L^p, and $u^{(2)}(o) = o = u^{(2)}(1)$. Then $T_{oo} \subset T_o$, so it suffices to prove that $\overline{T}_{oo} = T$. Thus given u in $\mathcal{D}(T)$ and $\varepsilon > o$, we must construct v in $\mathcal{D}(T_{oo})$ such that $\|u-v\|_p < \varepsilon$ and $\|u^{(2)} - v^{(2)}\|_p < \varepsilon$. To this end, let λ_o not be an eigenvalue of T, and let $G(x,t,\lambda_o)$ be the Green's function of T. Let $U(t) = u^{(2)}(t) - \lambda_o u(t)$. Then $u(x) = \int_o^1 G(x,t,\lambda_o)\,U(t)\,dt$. Select w in C^∞ such that $\|U-w\|_p < \varepsilon$. Let h be a C^∞ mollifier, i.e., given $\delta > o$, h is in C^∞, $|h(x)| \leq 1$, $h(x) = 1$ for $\delta \leq x \leq 1-\delta$, $h(x) = o$ and $h^{(1)}(x) = o$, for $x = o,1$. Then for δ sufficiently small, $\|U-hw\|_p < 2\varepsilon$, and $(hw)^{(1)}$ is zero for $x = o$, $x = 1$. Let $v(x) = \int_o^1 G(x,t,\lambda_o)h(t)w(t)\,dt$. Then v is in $\mathcal{D}(T_{oo})$ and $\|u-v\|_p < K\varepsilon$ for some constant $K > o$. Finally, we see that $\|u^{(2)} - v^{(2)}\|_p < 2\varepsilon$.

2) It is easily verified that

$$(4.1) \qquad \sigma_N f = \sum_{-N}^{N}\beta_k(f,s_k)s_k + 2\pi\sum_{o}^{N}\beta_k(f,s_k)c_k + 4\pi\sum_{o}^{N}\beta_k k(f,s_k)c_k + 4\pi\sum_{-N}^{-1}\beta_k k(f,c_k)s_k$$

For f in C, all of these terms can be expressed as Fourier series. Also,

$$(4.2) \qquad \left[\sigma_N f + (\sigma_N f)^{\sim}\right]^{(1)} = \sum_{-N}^{N}\beta_k(F^{(1)}, c_k)c_k ,$$

which can also be expressed as a Fourier series. Thus if for f in C, the right sides of (4.1), (4.2) converge in L^p, then $\sigma_N f$ converges in the norm of C. The converse follows from the formula $\sigma_N f = A\, n_N\, A^{-1} f$. If $\sigma_N f$ converges in the norm of C, then $A^{-1}\sigma_N f$ converges in L^p.

3) This also follows from (4.1) and the fact [8;p.8] that for $1 < p < \infty$, and f in L^p,

$$m\{x : S^*(x,f) > y\} \le K\,[\|f\|_p\, y^{-1}]^p .$$

Since the first two terms of (4.1) can be expressed as partial sums of Fourier series in f, and the last two terms as partial sums of Fourier series in $F^{(1)}$, the result follows.

References

1. H.E. Benzinger, Green's function for ordinary differential operators, J. Differential Equations 7(1970), 478-496.

2. ————————, The L^p behavior of eigenfunction expansions, Trans. Amer. Math. Soc. 174(1972), 333-344.

3. ————————, Completeness of eigenvectors in Banach spaces, Proc. Amer. Math. Soc. 38(1973), 319-324.

4. ————————, Pointwise and norm convergence of a class of biorthogonal expansions, submitted.

5. ————————, A canonical form for a class of ordinary differential operators, submitted.

6. G.D. Birkhoff, Boundary value and expansion problems of ordinary linear differential equations, Trans. Amer. Math. Soc. 9(1908), 373-395.

7. Y. Katznelson, An introduction to harmonic analysis, Wiley, New York, 1968.

8. C.J. Mozzochi, On the pointwise convergence of Fourier series, Springer Lecture Notes, vol. 199, Springer-Verlag, New York, 1970.

9. M.H. Stone, Irregular differential systems of order two and related expansion problems, Trans. Amer. Math. Soc. 29(1927), 23-53.

BIFURCATION DIAGRAMS

L. Collatz

Content: Different bifurcation phenomena and occurence in boundary value problems of ordinary and partial nonlinear differential equations and integral equations, geometry and algebra, physics, mechanics, hydrodynamics, stability, diffusion and other applications.
Topological definition of bifurcation points under certain smoothness conditions. Topological classification of bifurcation diagrams, trees, circles, webs, cocoons a.o. .
Different kinds of bifurcation diagrams, directed and planar diagrams.

Bifurcation phenomena have been studied extensively in the field of differential equations, compare for instance Keller-Amtman [69], Stakgold [71], Kirchgässner [71], Sattinger [71], [73], Wainberg-Trenogin [73] and many others, but they occur in so many different areas in integral equations, algebra, geometry, mechanics, physics and other applications (see for instance Velte [66], Bazley-McLeod [72], Crandall-Rabinowitz [73], Chow-Hale-Mallet Paret [75] and many more) that we consider bifurcations from a more general point of view and with the purpose of a general classification.

I. Different bifurcation phenomena

Let us begin with certain typical examples for showing the variety of phenomena.

§1 Differential equations

Here and in the following we consider simple examples in the real field, for which we know the exact explicit solutions so that we can be sure, to get the whole bifurcation diagram.
The equation for the unknown function y(x)

$$(1.1)\qquad y'' = \lambda\left(y'^2 + 1\right)^{3/2}$$

has as general solution $(x - c_1)^2 + (y - c_2)^2 = \lambda^{-2}$ with c_1, c_2 as real constants; these are circles with centre (c_1, c_2) and λ as curvature. λ may be > 0 or $= 0$ or < 0.

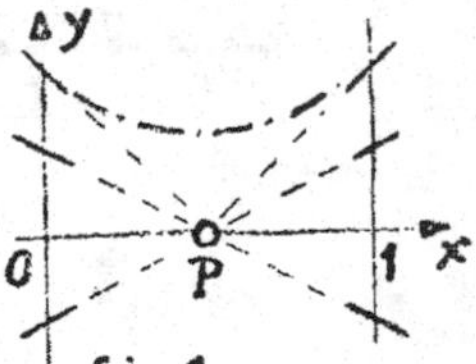

fig.1

We consider different boundary conditions:

a) $2y(0) + y'(0) = 2y(1) - y'(1) = 0$

Solutions are circles the tangents of which at $x = 0$ and $x = 1$ go through the point $P(x = \frac{1}{2}, y = 0)$ fig.1 .
There are two families of solutions, fig.2 ; the bifurcation element is $y \equiv 0$, fig.3 shows the bifurcation diagram with $y(1)$ depending on λ. The straight lines in fig.2 correspond to $\lambda = 0$, $y(1)$ arbitrary.

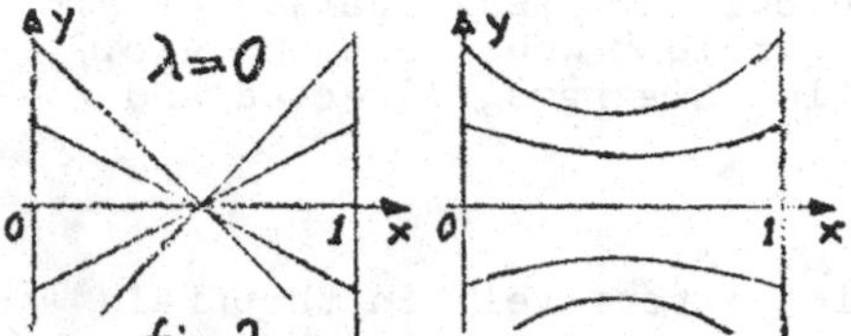

fig.2

b) $3y(0)+y'(0)= 3y(1)-y'(1)=0$.

The tangent to the circles at $x = 0$ and $x = 1$, fig.4 , go through the points $P_1(x=\frac{1}{3},y=0)$ and $P_2(x=\frac{2}{3},y=0)$. Now one has two bifurcation points and the diagram is in fig.5 .

c) Boundary conditions $y(0) = 0$, $y(1) = 1$. This is for (1.1) no bifurcation problem, but shows, that for a classical nonlinear boundary value problem it can happen, that the curve in the corresponding diagram ends suddenly, fig.6, because the arc C of the circle in fig.7 is situated partly outside the interval $[0,1]$ and therefore gives no solution.

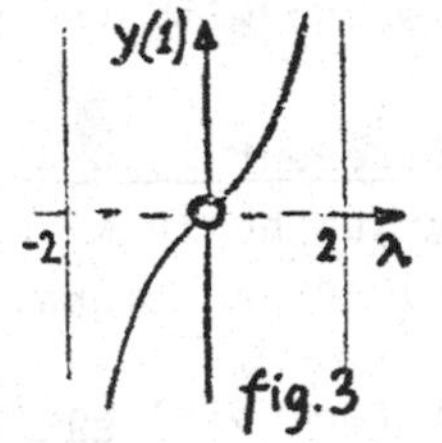

fig.3

d) The same phenomenon causes two endpoints P_1, P_2 in the diagram fig.8 , for the boundary conditions for (1.1)

$$y(0) = 1, \quad 2y(1) = y'(1)$$

e) Stability in mechanical systems in which some of the mechanical quantities may depend on a parameter λ, compare fig.9 . Vibrations of the system may be described by a system of ordinary differential equation with constant coefficients or a single equation for a function $y(t)$ of time t of the form

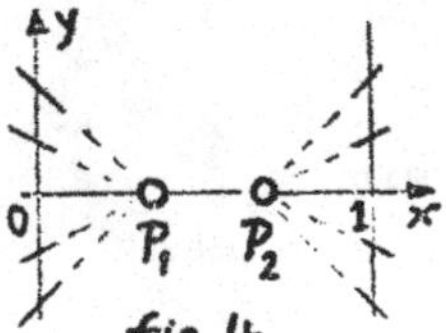

fig.4

$$\sum_{\nu=0}^{M} p_\nu(\lambda) y^{(\nu)}(t) = 0 \quad (1.2)$$

with $y(t) = e^{kt}$ one gets the characteristic equations

$$\phi(k) = \sum_{\nu=0}^{M} p_\nu(\lambda)k^\nu = 0$$

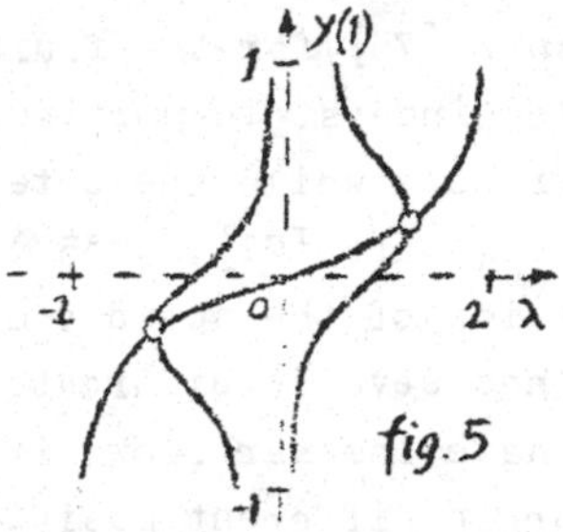

fig.5

The real parts p of the roots $k = p + iq$ (with p,q real) decide stability of the system. Consider for example the equation

(1.3) $y''' + \lambda(y'' - y) = 0$ with $k^3 + \lambda(k^2 - 1) = 0$

fig.10 gives the dependence of the real roots k on λ. But the complex roots are as important as the real roots. Fig.11 shows the dependence of the real part p of k on λ, where the dotted line corresponds to two different roots $p \pm iq$. These can be separated: fig. 12 shows $p + \frac{1}{2}(q + |q|)$ as a function of λ with a web as bifurcation diagram.

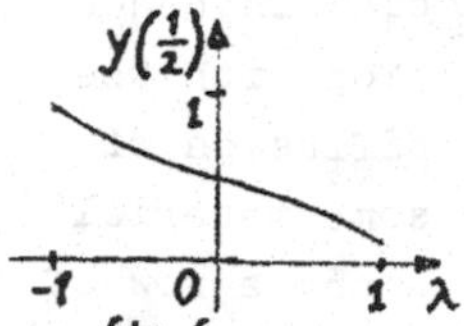

fig.6.

f) Partial differential equation for a function u(x,y) of two independent variables x,y

$$-\Delta u = -\left[\frac{\partial^2 u}{\partial x^2} + \frac{\partial^2 u}{\partial y^2}\right] = \left\{\lambda + [u(0,0)]^2\right\} u(x,y)$$

$$\text{in } B = \left\{ |x|<1, |y|<1 \right\}$$

with u = 0 on the boundary ∂B. We have then the bifurcation diagram fig.13 for u(a,b) as a function of λ. We suppose a,b, $^a/b$ as irrational numbers with $|a|<1, |b|<1$ for instance $a = \frac{1}{2}\sqrt{2}$, $b = \frac{1}{2}\sqrt{3}$.

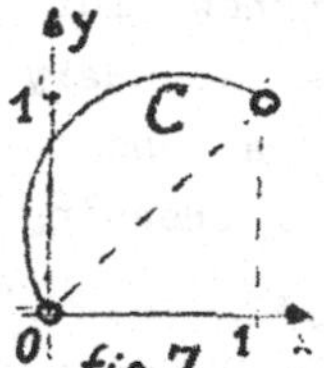

fig.7

§ 2 Physics

a) A mass m with weight G is movable along a vertical circle of radius 1 and subjected by a rope to a force K, fig.14 . We look on possible positions (described by the angle φ , fig.14) of equilibrium. φ ≡ 0 is such a position for all values of G and K. Consider K as fixed. Then fig.15 gives the

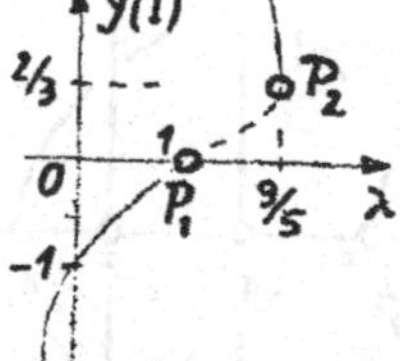

fig.8

bifurcation diagram with φ depending on G and $G = \frac{1}{2}K$ as bifurcation point.

b) A classical example is nonlinear stability theory (Kirchgässner [75]) for the fluid flow between concentric rotating circular cylinders. In particular the inner cylinder may rotate while the outer is at rest. Fig.16 (J.T. Stuart [71], p.357) shows the torque t as a function of the speed v of the inner cylinder; one has several bifurcation points. For small v we have laminar flow; if v increases vortices occur in different positions and for great values of v we have turbulence. In fig.16 stable branches are lined and unstable branches are dotted.

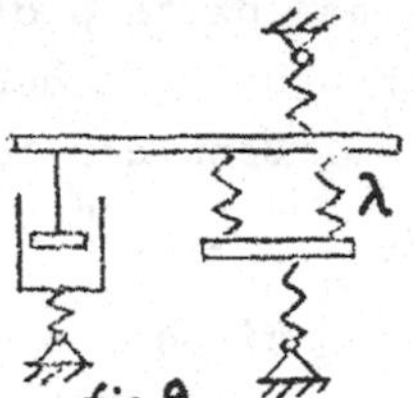

fig.9.

c) A model for Reactor Theory was considered by Poore [73]. A similar problem in connection with the diffusion of some material among a row of cells was studied by Rheinboldt [76].

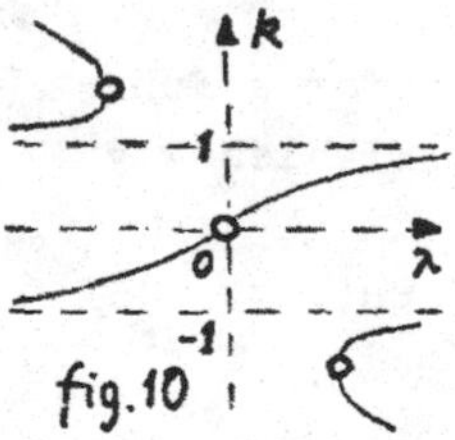

fig.10

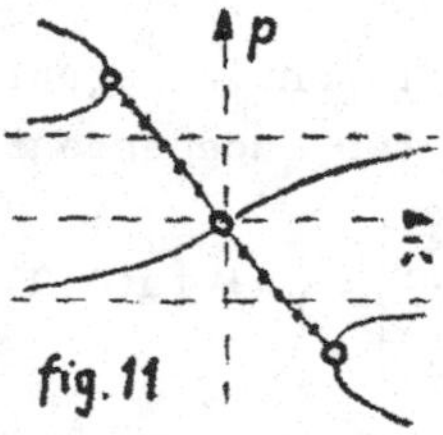

fig.11

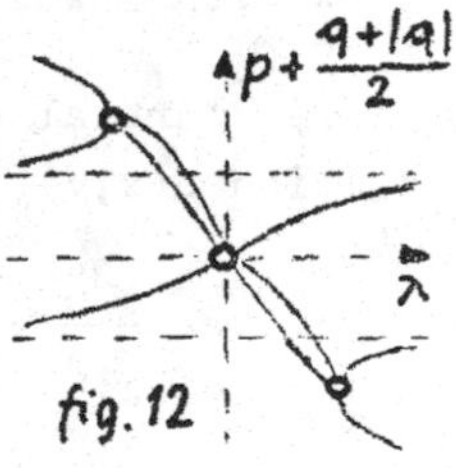

fig.12

He replaces a differential equation of the form

$$y'' + \phi(y)\cdot y = \lambda \quad \text{with} \quad \phi(y) = \frac{a+by}{1+y+y^2} \quad \text{and} \quad y(0) = y(1) = 0$$

by a system of finite difference equations with meshsize $h = \frac{1}{n}$. The bifurcation diagram (the dependence of one of the unknown function values on λ, fig.17) shows secondary and tertiary complicated bifurcations, the number of bifurcation points increasing rapidly with n.

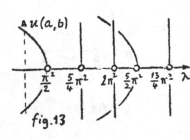

fig.13

§ 3 Algebraic equations

a) A matrix-eigenvalue problem may be described by

$$Ax - \lambda x = 0, \quad A = \begin{pmatrix} 1 & 0 & 1 \\ 0 & 1 & 1 \\ 0 & 0 & 2 \end{pmatrix}, \quad x = \begin{pmatrix} x_1 \\ x_2 \\ x_3 \end{pmatrix}.$$

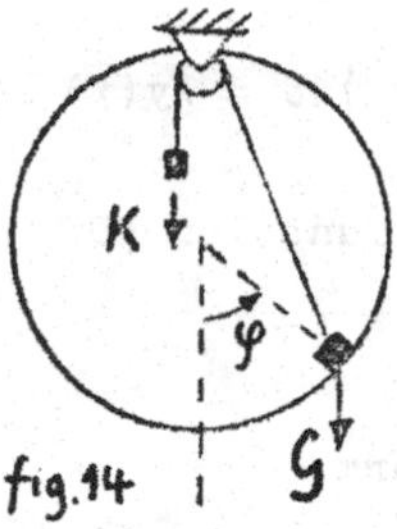

fig.14

Solutions are:
Line g_1: $x=o$, λ arbitrary (the trivial solution $x=o$ is usually counted in bifurcation problems) $\lambda=1$, $x_3=o$, x_1 and x_2 arbitrary, plan E,

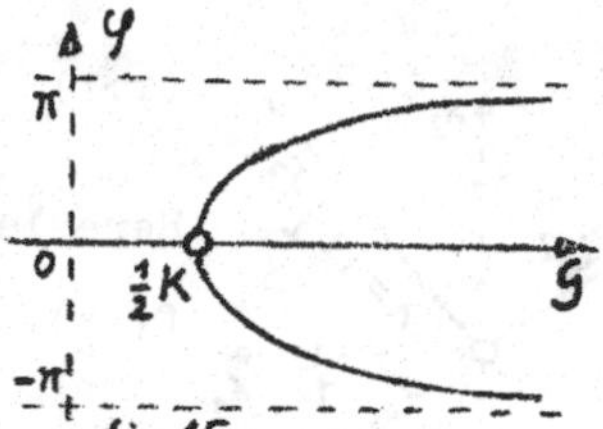

fig.15

Line g_2: $\lambda=2$, $x_1=x_2=x_3$; fig.18 shows the bifurcation diagram containing a cocoon and a web.

b) For the problem $Tx = \lambda x$ with $x=\begin{pmatrix}x_1\\x_2\end{pmatrix}$, $Tx = \begin{pmatrix}|x_1 + \frac{3}{4}x_2|\\ x_1+x_2\end{pmatrix}$

we have the bifurcation diagram of fig.19.

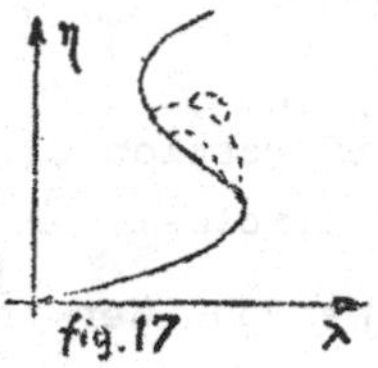

fig.17

§4 Nonlinear integral equations

a) For the equation

$$(1.4)\quad u(x) = \lambda \int_{-1}^{1} x\cdot|u(t)|\,dt$$

we have the bifurcation diagram of fig.20.

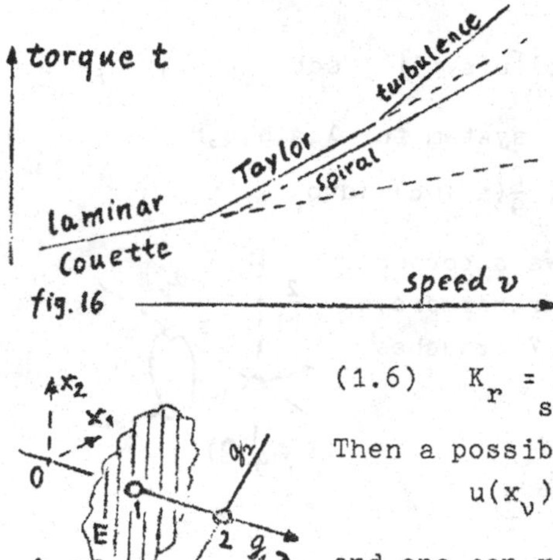

fig.16

b) The equation (1.5) $Tu(x_\nu) = \int_B \left\{\sum_{r=o}^{p} K_r(x_\nu,t_\nu)\,|u(t_\nu|^r\right\} dt_\nu = \lambda u(x_\nu)$ for a function $u(x_\nu)$ in the domain B of the n-dimensional point space R^n may have only given degenerate kernels K_r

$$(1.6)\quad K_r = \sum_{s=1}^{q} g_{rs}(x_\nu)\cdot h_{rs}(t_\nu).$$

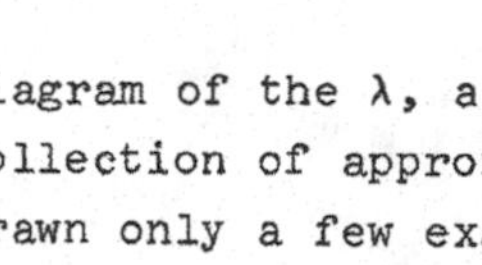

fig.18

Then a possible solution u has the form

$$u(x_\nu) = \sum_{r,s} a_{rs}\, g_{rs}(x\nu)$$

and one can reduce (1.5) to a nonlinear system for the unknowns λ, a_{rs}. Discussing this system one can draw a projection of the bifurcation diagram of the λ, a_{rs}-space into a twodimensional plane. From the collection of approximate 80 bifurcation diagrams the author has drawn only a few examples.

c) One of the most simple types for a function y(x) is

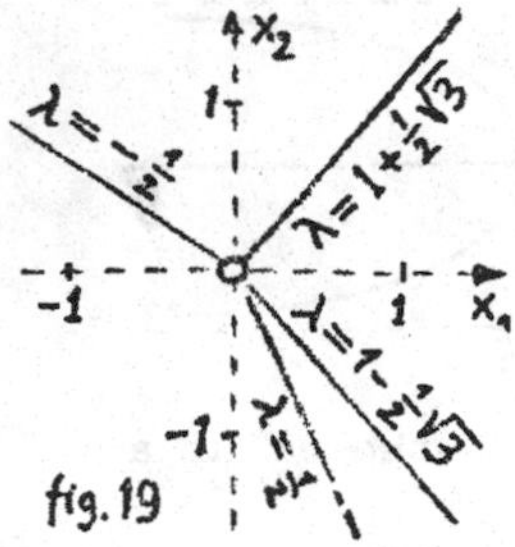

fig.19

$$(1.7)\quad Ty(x) = \int_{-1}^{1}\left\{y(t) + f(x)\,|y(t)|^2\right\}dt = \lambda y(x)$$

Here let $f(s) \in L_2(-1,1)$ be an odd function $\not\equiv 0$:

$$f(-x) = -f(x), \quad M = \int_{-1}^{1}|f(t)|^2 dt > 0$$

A solution $y(x)$ of (1.7) has the form $y(x) = a + bf(x)$ and one gets the equations for λ, a, b: $(\lambda-2)a = 0$, $2a^2+b^2M-\lambda b = 0$

with the solutions: $a = b = o$, λ arbitrary

$a = o$, $\lambda = bM$

$\lambda = 2$, $o \leq b \leq \frac{2}{M}$, $a^2 = b - \frac{b^2}{2}M$.

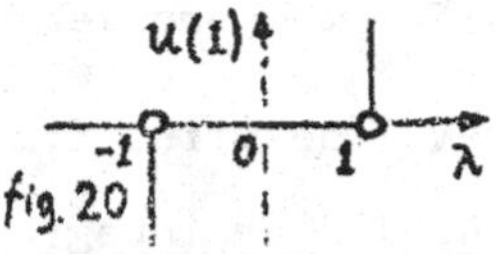

fig.20

The bifurcation diagram Fig.21 is a web with 3 bifurcation points and 2 circles.

d) We consider the equation for a function $u(x,y)$ in

$$B = \left\{|x| \leq 1, |y| \leq 1\right\}:$$

$$(1.8)\quad Tu(x,y) = \iint_B \left\{(1+xt)u(s,t) + ys(1+xt)[u(s,t)]^2\right\}dsdt = \lambda u(x,y) ;$$

with $u(x,y) = a+bx+cy+dxy$ we get the system for λ,a,b,c,d:

$$(4-\lambda)\,a=o,\quad \frac{4}{3}c - \lambda b=o,\quad \frac{8}{9}(3ab+cd)-\lambda c=o,\quad \frac{8}{9}(ad+bc)-\lambda d=o .$$

The bifurcation diagram, fig.22, shows a cocoon for $\lambda=a=c=o$, b and d arbitrary and furthermore a web with 5 bifurcation points P_ν and 7 branches. The coordinates of the points P_ν are

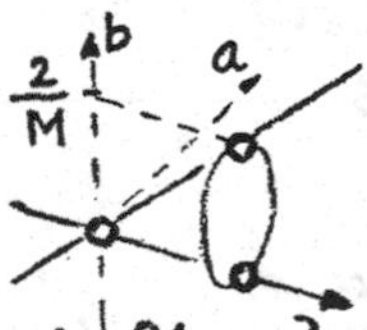

fig.21

	λ	a	b	c	d
P_1	0	0	0	0	0
P_2	4	0	0	0	0
P_3	4	0	β	3β	$9/2$
P_4	4	$9/2$	0	0	0
P_5	4	0	$-\beta$	-3β	$9/2$

with $\beta = \frac{3}{2}\sqrt{3}$

Perhaps a symbolic scetch as in fig.23 gives a better impression of the connections of the branches.

e) For $Tu(x,y,z)=\iint_B\left\{(s+3x)u(s,t\ )+y(1+2xs)[u(s,t)]^2\right\}dsdt=\lambda u(x,y)$

with the domain B and a,b,c,d as before one gets the bifurcation diagram of fig.24 with 9 bifurcation points. It is easy to get bifurcation diagrams with many more bifurcation points and circles, even with infinitely many circles.

f) For $Tu(x,y)=\int_B(1+x)\left(tu(s,t)+\ y\ s[u(s,t)]^2\right)dsdt\ =\ \lambda u(x,y)$

with the domain B as in d) fig.25 shows two cocoons and a web.

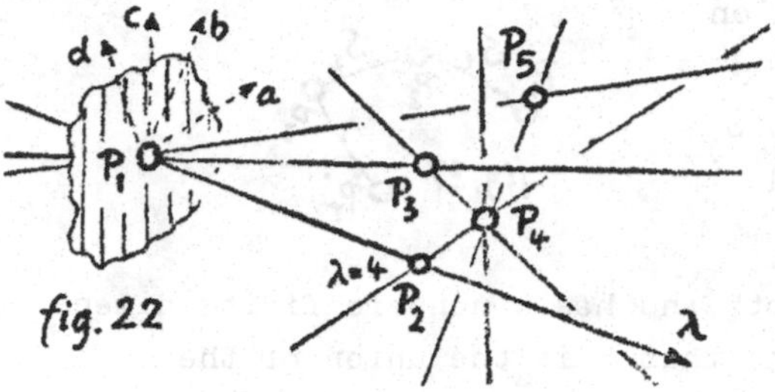

fig.22

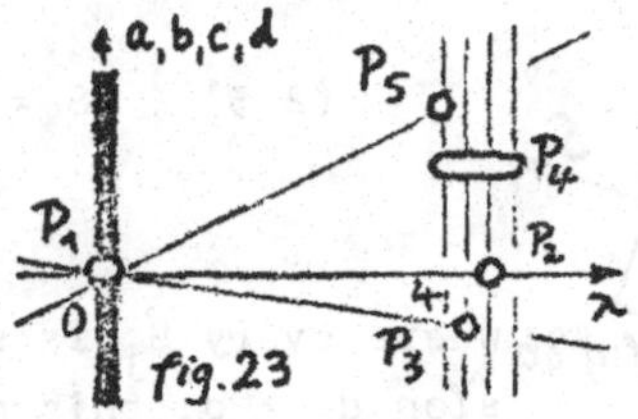

fig.23

II. Topological classification of bifurcations with smooth branches

There are many different ways to define bifurcation points. For the case of smooth branches we use the following geometrical introduction:

Let M be a set of elements u and R the set of real numbers; let F be a mapping

$$(2.1)\qquad F:\ M\times R\longrightarrow N$$

where the set N may be the same set as M or not. The elements of M may be for instance functions, systems of functions, geometrical or physical quantities etc.. We suppose N contains a "zeroelement" Θ, but N is not supposed to be a linear space.

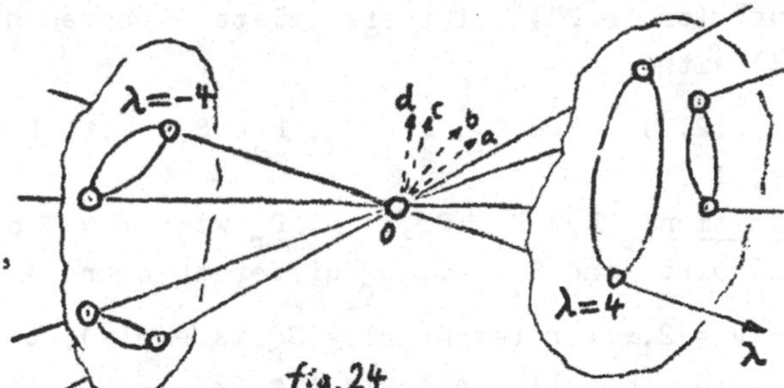

fig.24

Let Q be the set of all ordered pairs (u,λ) with $u\epsilon M$, $\lambda\epsilon R$ and S the set

$$(2.2)\qquad S=\left\{(u,\lambda)\ \text{with}\ F(u,\lambda)=\Theta,\ u\epsilon M,\ \lambda\epsilon R\right\}$$

We call the elements of S"solutions". We make the following assumptions

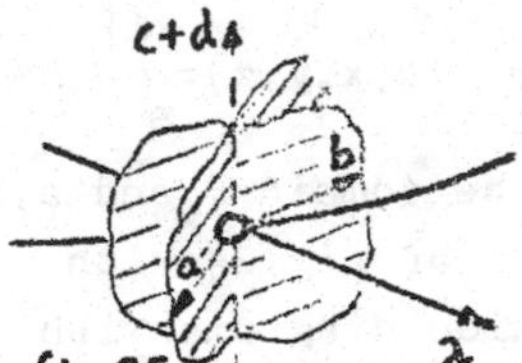

fig.25

1) We have for certain subsets $\overline{Q}$ of Q a "dimension" d $(\overline{Q})$, where d is a nonnegative integer; d = ∞ is allowed. Furthermore there exist subsets of Q which we call "smooth". In the case that the elements u of M are functions one can often define u as smooth if u is "piecewise smooth" in the classical sense, compare f.i. Courant Hilbert [62] p.245 .

2) There exists a representation

$$(2.3) \qquad S = \bigcup_{j=1}^{p} S_j$$

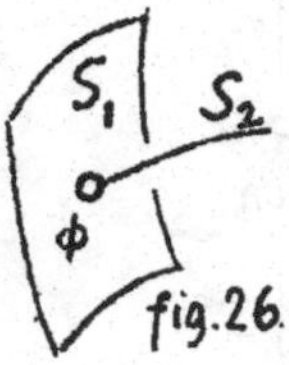

fig.26

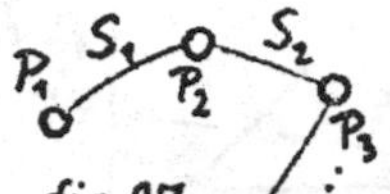

fig.27

where every S_j is smooth and has a nonzero finite dimension $d_j > 0$; this means that S is the union of the "branches" S_j; in (2.3) p = ∞ is allowed. For given S the representation (2.3) is usually not unique.

3) For the decomposition (2.3) we have

$$(2.4) \qquad \dim \bigcap(S_j,S_k) < \mathrm{Min}(d_j,d_k) \text{ for all } j \neq k .$$

Example: S_1 may be a twodimensional surface, S_2 a onedimensional curve and the intersection $\phi = \bigcap(S_1,S_2)$ a point of dimension zero, fig. 26

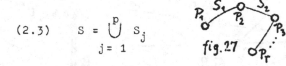

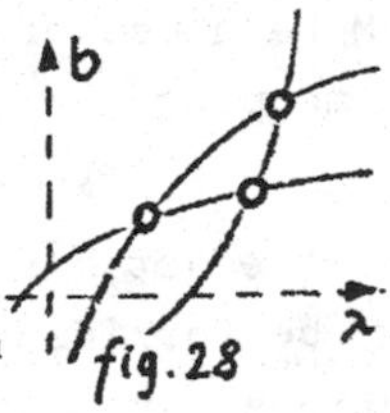

fig.28

<u>Definition:</u> A point P ∊ S is called a"bifurcation point" (or "B.P"), if there exists a representation (2.3) with

(2.) $\quad P \in S_j,\ P \in S_k,\ P \in S_l$ with three different indices j, k, l .

<u>Definition:</u> Let $P_1, P_2, \ldots, P_r$ with $r \geq 3$ be different bifurcation points and $S_1, \ldots, S_r$ different branches with $P_\nu \in S_{\nu-1} \cup S_\nu$ for $\nu = 2,\ldots,r$ (eventually S_r is empty), then the union of $S_1,\ldots,S_r$ is called a chain fig. 27. If furthermore $P_1 \in S_1 \cup S_r$, then $S_1,\ldots,S_r$ is called a circle.

Now we can state the following classification:

<u>Case I</u>: There exists a branch S_j with $\dim S_j = d_j > 1$. Then S is

called a cocoon (in German: "Gespinst").

Case II: All branches S_j have dim $S_j = d_j = 1$

Case IIa: S contains at least one circle; S is called a web (interweaving, wickerwork, in German: "Geflecht").

Case IIb: S contains no circle. In this case we get the scheme:

	S is connected	S is not connected
S contains a bifurcation point	S is a tree	S cons ists of several trees
S " no " "	S is a thread ("Faden")	S is a thread-bundle ("Bündel")

III. Different kinds of bifurcation diagrams

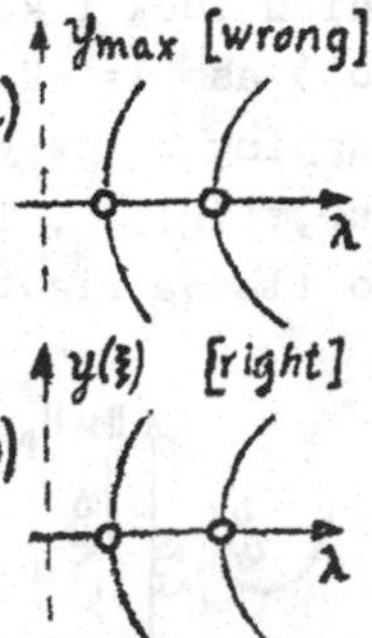

fig.29

For the applications one wishes very often to have a clear picture of the branches which show immediately the bifurcation points. This can cause great difficulties.

We introduce the subset $\hat{M}$ of all elements u of M for which there exists a λ with $F(u,\lambda) = \theta$ and we make the following

assumption: 4) There exists a one to one mapping T of $\hat{M}$ into a subset B of a q-dimensional real point space R^q with finite q:

$$(3.1) \qquad T: \ \hat{M} \longrightarrow B \subset \mathbb{R}^q .$$

Case I: $q = 1$. Then $T: u \to b$, with b a real number fig.28. Many of the classical diagrams are of this type ("directed bifurcation diagrams"). The mapping T is often obvious, but one can sometimes observe errors in the literature. For instance in the classical case of Euler's critical load for a beam

$$(3.2) \qquad \frac{dy}{dx} = \sin z, \ \frac{dz}{dx} = -\lambda y, \ y(o) = y(a) = o$$

one can find the picture, fig.29a, but the right mapping is the following: Let ξ be a point of the interval J = (o,a) with $|y(\xi)| = \max_{x \in J} |y(x)|$; then one draws the mapping $\lambda \longrightarrow y(\xi)$, fig.29b.

Many examples are given in de Mottoni-Tesei [76], Meyer-Spasche [76] a.o.

<u>Example</u>: $u = \begin{pmatrix} x \\ y \end{pmatrix}$, $F(u,\lambda) = \begin{pmatrix} y^2 \\ y(|x|-y^2) \end{pmatrix} - \lambda \begin{pmatrix} x \\ y \end{pmatrix} = \theta$

Fig.30 shows the different branches A,B,C, D,E,F of solutions. We consider different mappings $u \mapsto b$ (b real). Possibilities to be tried for b are $\| u \|_p$ (vector norm, usually $p \equiv 1$ or 2 or ∞), y or x, Max (x,y) or Min (x,y). All these trials, fig. 31 a) - e), are unsatifactory. Either different branches coincide or some branches are reduced to a single point (as indicated in fig.31e)) or there appear points, which look like bifurcation points (but are not) as $\lambda = -2$ for b = Min (x,y); but the mapping b = x-y is correct and shows the right bifurcation behaviour, fig.32 . This example illustrates that it may not be trivial to find a right mapping.

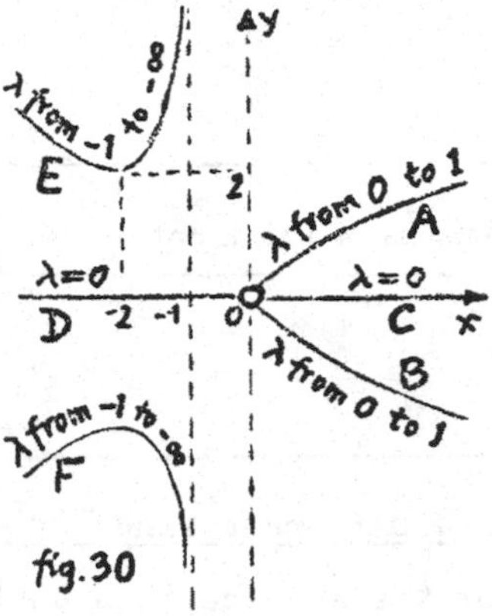

fig.30

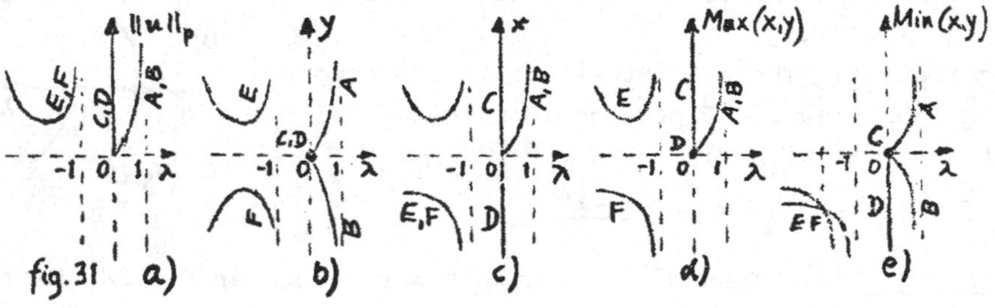

fig.31 a) b) c) d) e)

Of Course one can sometimes avoid this difficulty in drawing by using interrupted lines as in fig. 13 .

<u>Case II</u>, <u>$q > 1$</u>, Case IIa, all $d_j = 1$

In (3.1) let b be the vector $b = (b_1,\ldots,b_q)$. One chooses one function or several functions $\varphi_\mu(b_\nu)$. Every branch S_j is then represented in a λ-φ_μ space of dimension m + 1 by a curve $\mathcal{C}_j$ and can be drawn in a diagram D , fig.33 , for instance by parallel projection into a plane. One tries to choose the functions φ_u , if possible, in such a way, that in the diagram D intersections of the curves C_j are avoided, if they donot correspond to bifurcation points. If this is possible the diagram is called "planar". The choice of suitable functions φ_μ may be difficult.

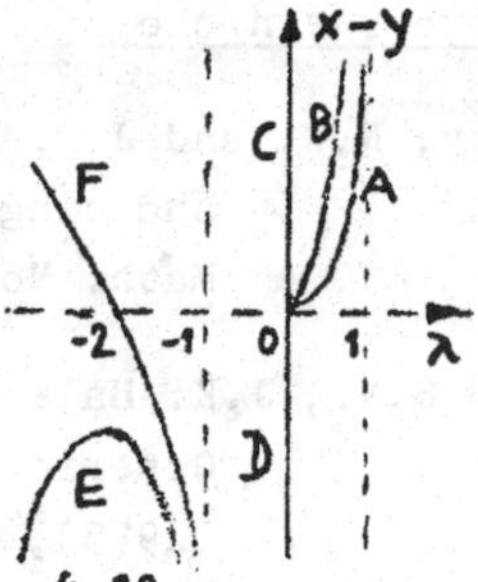

fig.32

In a slightly different way one chooses only one function $\varphi_1(b_\nu)$, m = 1, and uses diagrams as in fig.23 .

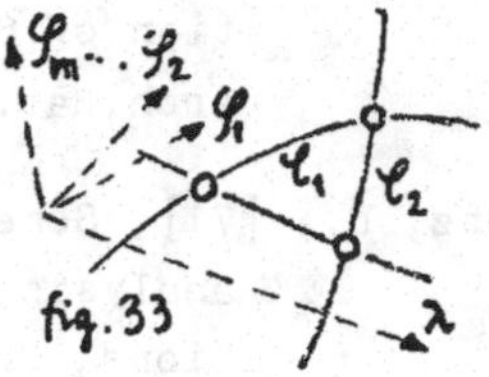

fig.33

I thank Mrs. Susanne Böttger and Mr. Rolf Wildhack for numerical calculations on a computer.

References

Bazley, N.W. and J.B. McLeod [72], Bifurcation from infinity and singular Eigenvalue Problems, Battelle Report Math. No 70, (1972), 26 p.

Chow, S.N., J.K. Hale and J. Mallet-Paret [75], Applications of Generic Bifurcation I, Arch.-Rat. Mech.Anal. 59 (1975), 159-188

Crandall, M.G. and P.H. Rabinowitz [73], Bifurcation, Perturbation of Simple Eigenvalues and Linearized Stability, Arch.Rat.Mech. Anal. 52, (1973), 161-180

Collatz, L., [71], Some applications of functional analysis to analysis, particularly to nonlinear integral equations, Proc. Nonlinear funct. Anal. and Applications, edited by L.B. Rall, Academic Press 1971, 1-43

Courant, R. and D. Hilbert [62] Methods of mathematical physics, Vol.II Interscience Publishers (1962), 830 p.

Keller, J.B. and S. Antman [69](editors) Bifurcation Theory and Nonlinear Eigenvalue Problems, W.A. Benjamin, New York, 1969

Kirchgässner, K. [71] Multiple Eigenvalue Bifurcation for Holomorphic Mappings, in Contributions to Nonlinear Funct. Anal., ed. Zarantonello, Academic Press (1971), 69-99

Kirchgässner, K. [75], Bifurcation in nonlinear Hydrodynamic Stability, SIAM REVIEW 17 (1975), 652-683

Meyer-Spasche, R. [76] Numerische Behandlung von elliptischen Randwertproblemen mit mehreren Lösungen u. von MHD Gleichgewichtsproblemen, Diss. Hamburg, 1976

de Mottoni, P. and A. Tesei [76] On the solutions of a class of nonlinear Sturm-Liouville-Problems, to appear

Poore, A.B. [73] A Model Equation Arising from Chemical Reactor Theory, Arch.Rat.Mech. Anal. 52 (1973) 358-388

Rheinboldt, W. [76] Mündliche Mitteilung über Verzweigungen bei einem Diffusionsproblem

Sattinger, D.H. [71] Stability of bifurcating solutions by Leray-Schauder degree, Arch.Rat.Mech. Anal.43 (1971) 154-166

Sattinger, D.H. [73] Topics in stability and bifurcation theory, Springer Lecture Notes in Math. 309 (1973)

Stakgold, I. [71] Branching of Solutions of nonlinear Equations, SIAM REVIEW 13 (1971) 289-332

Stuart, J.T. [71] Nonlinear Stability Theory, Annual Review of Fluid Mechanics 3 (1971) 347-370

Toland, J.F. [75] Bifurcation and Asymptotic bifurcation for non-compact nonsymmetric gradient operators, Proc. Royal Soc. Edinburgh, A 73 (1975) 137-147

Velte, W. [66] Stabilität und Verzweigung stationärer Lösungen der Navier-Stokes - schen Gleichungen beim Taylor-Problem, Arch.Rat.Mech.Anal. 22 (1966) 1-14

Wainberg, M.M und W.A. Trenogin [73] Theorie der Lösungsverzweigung bei nichtlinearen Gleichungen, Berlin 1973, 408p.

Walsh's Theorem for the Heat Equation

David Colton*

I. Introduction

Let D be a bounded simply connected domain in the complex plane with smooth boundary ∂D. Walsh's theorem states that if f(z) is an analytic function defined on D and continuous on $D \cup \partial D$ then f(z) can be uniformly approximated in the maximum norm over $D \cup \partial D$ by a polynomial ([5]). There is an almost immediate generalization of this theorem to analytic functions of n complex variables defined on polycylindrical domains. As pointed out by Widder ([6]) there is a close analogue between the set $\{z^n\}$ for analytic functions of one complex variable and the set $\{h_n(x,t)\}$ of heat polynomials for the heat equation

$$u_{xx} = u_t \tag{1.1}$$

where $h_n(x,t)$ is defined by

$$h_n(x,t) = [\tfrac{n}{2}]! \sum_{k=0}^{[\frac{n}{2}]} \frac{x^{n-2k} t^k}{(n-2k)!k!} \quad . \tag{1.2}$$

This similarity can also be extended to n dimensions by observing that the polynomials

$$\begin{aligned} h_m(x,t) &= h_{m_1}(x_1,t) h_{m_2}(x_2,t) \ldots h_{m_n}(x_n,t) \\ x &= (x_1, x_2, \ldots, x_n) \in \mathbb{R}^n \\ m &= (m_1, m_2, \ldots, m_n) \in \mathbb{N}^n \end{aligned} \tag{1.3}$$

are solutions of the n dimensional heat equation

$$\Delta_n u = u_t \quad . \tag{1.4}$$

*This research was supported in part by AFOSR Grant 74-2592.

In this note we will pursue this relationship between analytic functions and solutions of the heat equation and outline some recent results we have obtained in developing an analogue of Walsh's theorem for solutions of the heat equation. In particular we want to approximate classical solutions $u(\underset{\sim}{x},t)$ of the heat equation by a linear combination of heat polynomials such that $u(\underset{\sim}{x},t)$ is uniformly approximated in the maximum norm over the closure of its domain of definition. In addition to being of interest in the general area of function theoretic methods in partial differential equations, such a result also has application to the numerical solution of initial-boundary value problems for the heat equation (c.f. [1]).

II. <u>One Space Variable</u>.

Let $u(x,t)$ be a classical solution of (1.1) in the domain D bounded by the analytic arcs $x = s_1(t)$ and $x = s_2(t)$ (where $s_1(t) < s_2(t)$) and the characteristics $t = 0$ and $t = T > 0$. Let D^* be the mirror image of D reflected across the arc $x = s_1(t)$, i.e. $(x,t) \in D^*$ if and only if $(2s_1(t) - x,t) \in D$. In the present case of one space dimension Walsh's theorem can be obtained as a consequence of the following reflection principle for solutions to the heat equation:

<u>Reflection Principle</u> ([1]): Let $u(x,t) \in C^2(D) \cap C^o(\bar{D})$ be a solution of (1.1) in D such that $u(x,t) = f(t)$ on $\sigma : x = s_1(t)$ where $f(t)$ is analytic on $[0,T]$. Then $u(x,t)$ can be uniquely continued into D^* as a solution of (1.1) such that $u(x,t) \in C^2(D \cup D^* \cup \sigma) \cap C^o(\bar{D} \cup \bar{D}^*)$.

<u>Walsh's Theorem (n = 1)</u>: Let $u(x,t) \in C^2(D) \cap C^o(\bar{D})$ be a solution of (1.1) in D. Then given $\varepsilon > 0$ there exist constants $a_o, a_1, \ldots a_N$ such that

$$\max_{(x,t)\in\bar{D}} \left| u(x,t) - \sum_{n=0}^{N} a_n h_n(x,t) \right| < \varepsilon$$

where the $h_n(x,t)$ are defined by (1.2).

Proof: Without loss of generality we can assume that $u(x,t)$ assumes analytic boundary data. Using the reflection principle to repeatedly reflect $u(x,t)$ across the arcs $x = s_1(t)$ and $x = s_2(t)$ we can continue $u(x,t)$ into a closed rectangle $\bar{R} \supseteq \bar{D}$. On $\bar{R}$ $u(x,t)$ can be approximated by a linear combination of heat polynomials and a partial eigenfunction expansion, i.e. by a solution of (1.1) which is an entire function of its independent complex variables. This now implies that $u(x,t)$ can be approximated on $\bar{R}$, and hence on $\bar{D}$, by a linear combination of heat polynomials. For full details see [1].

III. Several Space Variables.

Let D be a bounded simply connected domain in $\mathbb{R}^n$ with ∂D in class C^{2i+2} where $i = 1 + [\frac{n}{4} + \frac{1}{2}]$. We first prove a generalized version of Walsh's theorem for solutions of (1.4) defined in the cylindrical domain $D \times (0,T)$.

Generalized Walsh Theorem: Let $u(\underset{\sim}{x},t) \in C^2(D \times (0,T)) \cap C^0(\bar{D} \times [0,T])$ be a solution of (1.4) in $D \times (0,T)$. Then given $\varepsilon > 0$ there exists a solution $u_1(\underset{\sim}{x},t)$ of (1.4) which is an entire function of its independent complex variables such that

$$\max_{\bar{D}\times[0,T]} |u(\underset{\sim}{x},t) - u_1(\underset{\sim}{x},t)| < \varepsilon \quad .$$

Proof: Without loss of generality we can assume that $u(\underset{\sim}{x},0) = 0$. Then there exists a bounded simply connected domain $D_1 \supset \bar{D}$ with D_1 in class C^{2i+2} and a solution $u_0(\underset{\sim}{x},t) \in C^2(D_1 \times (-1,T)) \cap C^0(\bar{D} \times [-1,T])$ of (1.4) in $D_1 \times (-1,T)$ such that

$$\max_{\bar{D}\times[0,T]} |u(\underset{\sim}{x},t) - u_0(\underset{\sim}{x},t)| < \varepsilon \quad . \tag{3.1}$$

The solution $u_0(\underset{\sim}{x},t)$ can also be assumed to have analytic boundary data on ∂D_1. On compact subsets of $D_1 \times [-1,T]$ we can now approximate

$u_o(\underset{\sim}{x},t)$ by the sum of two solutions, one of which is a polynomial in t with coefficients satisfying the polyharmonic equation $\Delta_n^k u = 0$ and the other being a partial eigenfunction expansion. Representing the polyharmonic functions in terms of harmonic functions and applying the Runge approximation property for elliptic equations to these harmonic functions and also to each of the terms in the partial eigenfunction expansion now yields the desired entire function $u_1(\underset{\sim}{x},t)$. Details of this proof will appear in [2].

We are now in a position to prove Walsh's theorem for solutions to (1.4) defined in cylindrical domains.

<u>Walsh's Theorem (n > 1):</u> Let $u(\underset{\sim}{x},t) \in C^2(D \times (0,T)) \cap C^o(\bar{D} \times [0,T])$ be a solution of (1.4) in $D \times (0,T)$. Then given $\varepsilon > 0$ there exists an integer M and constants a_m, $|m| \leq M$, such that

$$\max_{\bar{D}\times[0,T]} \left| u(\underset{\sim}{x},t) - \sum_{|m|\leq M} a_m h_m(\underset{\sim}{x},t) \right| < \varepsilon$$

where the $h_m(\underset{\sim}{x},t)$ are defined by (1.3).

<u>Proof</u>: By the generalized Walsh theorem it suffices to prove the theorem for $u(\underset{\sim}{x},t)$ an entire function of its independent complex variables. From the results of [4] we can represent $u(\underset{\sim}{x},t)$ in the form

$$u(\underset{\sim}{x},t) = (\underset{\sim}{I} + \underset{\sim}{G})h \tag{3.2}$$

$$= h(\underset{\sim}{x},t) + \frac{1}{2\pi i} \oint_{|t-\tau|=\delta} \int_0^1 \sigma^{n-1} G(r^2, 1-\sigma^2, \tau-t) h(\underset{\sim}{x}\sigma^2, \tau) d\sigma d\tau$$

where $\delta > 0$, $h(\underset{\sim}{x},t)$ is an entire function of its independent complex variables such that $\Delta_n h = 0$ for each fixed t, and

$$G(r,\xi,t) = \frac{r^2}{2t^2} \exp\left(\frac{\xi r^2}{4t}\right). \tag{3.3}$$

Using the Runge approximation property for elliptic equations and the representation (3.2) we can now approximate $u(\underset{\sim}{x},t)$ on $\bar{D} \times [0,T]$ by a linear combination of the polynomial solutions of (1.4) defined by

$$u_{jk}(\underset{\sim}{x},t) = (\underset{\sim}{I} + \underset{\sim}{G})h_j(\underset{\sim}{x})t^k \tag{3.4}$$

where the $h_j(\underset{\sim}{x})$ are harmonic polynomials. Since the $u_{jk}(\underset{\sim}{x},t)$ are polynomials there exists an integer M_o and constants $b_m = b_m(j,k), |m| \leq M_o$, such that

$$u_{jk}(\underset{\sim}{x},0) = \sum_{|m|\leq M_o} b_m h_m(\underset{\sim}{x},0) , \tag{3.5}$$

and hence from the uniqueness theorem for Cauchy's problem for the heat equation each $u_{jk}(\underset{\sim}{x},t)$ can be expressed as a linear combination of the $h_m(\underset{\sim}{x},t)$. The theorem now follows. For full details see [2].

IV. Other Complete Sets

In the application of Walsh's theorem to the approximation of solutions to initial-boundary value problems for the heat equation it is often desirable to have available a complete set of solutions such that each member of the set is bounded for $t \geq 0$ instead of having the polynomial growth rate of the heat polynomials. In order to construct such a set it follows from Walsh's theorem that it suffices to show that on compact subsets $h_n(x,t)$ or $u_{jk}(\underset{\sim}{x},t)$ can be approximated by a finite linear combination of solutions that are bounded for $t \geq 0$. We first consider the case n = 1. From [1] we have the representation

$$\begin{aligned} h_{2n}(x,t) &= \underset{\sim}{P}_1\{t^n,0\} \\ t_{2n+1}(x,t) &= \underset{\sim}{P}_1\{0,t^n\} \end{aligned} \tag{4.1}$$

where

$$\underset{\sim}{P}_1\{f,g\} = -\frac{1}{2\pi i}\oint_{|t-\tau|=\delta} E^{(1)}(x,t-\tau)f(\tau)d\tau - \frac{1}{2\pi i}\oint_{|t-\tau|=\delta} E^{(2)}(x,t-\tau)g(t)d\tau \tag{4.2}$$

with

$$E^{(1)}(x,t) = \frac{1}{t} + \sum_{j=1}^{\infty} \frac{x^{2j}(-1)^j j!}{(2j)!\,t^{j+1}}$$

$$E^{(2)}(x,t) = \frac{x}{t} + \sum_{j=1}^{\infty} \frac{x^{2j+1}(-1)^j j!}{(2j+1)!\,t^{j+1}} \quad . \tag{4.3}$$

Since $(\mathrm{Log}\,\frac{1}{z})^n$ is analytic in a neighbourhood of $[1,e^T]$, by Runge's theorem we can approximate $(\mathrm{Log}\,\frac{1}{z})^n$ by a polynomial on compact subsets of this neighbourhood, and hence setting $z = e^{-t}$ we can approximate t^n in a neighbourhood of $[0,T]$ by a finite linear combination of functions taken from the set $\{e^{-kt}\}_{k=0}^{\infty}$. From (4.1) we can now approximate $h_{2n}(x,t)$ on compact subsets of $\{(x,t): -\infty < x < \infty,\ t \geq 0\}$ by a linear combination of functions taken from the set $\{\cos\sqrt{k}\,x\,e^{-kt}\}_{k=0}^{\infty}$ and $h_{2n+1}(x,t)$ by a linear combination of functions taken from the set $\{\sin\sqrt{k}\,x\,e^{-kt}\}_{k=1}^{\infty}$. Hence $\{\cos\sqrt{k}\,x\,e^{-kt},\ \sin\sqrt{k}\,x\,e^{-kt}\}_{k=0}^{\infty}$ is a complete set of solutions for the heat equation in one space dimension defined in a domain D as described in section II.

For $n > 1$ we can apply the same reasoning as above to the complete set $\{u_{jk}\}_{j,k=0}^{\infty}$ where we now use the representation (3.4). This leads to the complete set of solutions $\{v_{jk}\}_{j,k=0}^{\infty}$ for solutions of the n dimensional heat equation defined in a cylindrical domain where $v_{jk}(\underset{\sim}{x},t)$ is defined by

$$v_{jk}(\underset{\sim}{x},t) = \Big[\, h_j(\underset{\sim}{x}) - \sqrt{k}\,r\int_0^1 \sigma^{n-1} J_1(r\sqrt{k(1-\sigma^2)})\,\frac{h_j(\underset{\sim}{x}\sigma^2)}{\sqrt{1-\sigma^2}}\,d\sigma\,\Big]e^{-kt} \tag{4.4}$$

with $h_j(\underset{\sim}{x})$ a harmonic polynomial and $J_1(z)$ a Bessel function. From [3] we recognize that the quantity in brackets is equal to

$r^{-\frac{1}{2}(n-2)} J_{\frac{1}{2}(n-2)+j}(\sqrt{k}r) S_j(\theta;\phi)$ where $S_j(\theta;\phi)$ denotes a spherical harmonic $(\theta = (\theta_1, \theta_2, \ldots \theta_{n-2}))$, and hence

$$v_{jk}(\underset{\sim}{x}, t) = r^{-\frac{1}{2}(n-2)} J_{\frac{1}{2}(n-2)+j}(\sqrt{k}r) S_j(\theta;\phi) e^{-kt} . \qquad (4.5)$$

References

1. D.Colton, The Solution of Boundary Value Problems by the Method of Integral Operators, Pitman Press Lecture Note Series, Pitman Press, London, to appear.
2. D.Colton and W.Watzlawek, Complete families of solutions to the heat equation and generalized heat equation in $\mathbb{R}^n$, to appear.
3. R.P. Gilbert, Constructive Methods for Elliptic Equations, Springer-Verlag Lecture Note Series, Vol.365, Springer Verlag, Berlin, 1974.
4. W.Rundell and M.Stecher, A method of ascent for parabolic and pseudoparabolic partial differential equations, SIAM J. Math.Anal., to appear.
5. J.L. Walsh, Interpolation and Approximation by Rational Functions in the Complex Domain, American Mathematical Society, Providence, 1965.
6. D.V. Widder, Some analogies from classical analysis in the theory of heat conduction, Arch.Rat.Mech.Anal. 21(1966),108-119.

A New Statement of Wazewski's Theorem and an Example

C. Conley

The new statement of Wazewski's theorem given here is somewhat more general than Wazewski's original version; however, it is not generality that is the point, but simplicity of statement and proof. The example is simple and has been treated in the literature already but not by these methods. (See [1]).

§1. Wazewski's Theorem

Let R denote the real numbers and let R^+ denote the non-negative reals.

1.1 **Definition.** Let Γ be a topological space and let a continuous function on $\Gamma \times R \to \Gamma$ be given by $(\gamma, t) \to \gamma \cdot t$. This function is called a flow if for all $s, t \in R$ and $\gamma \in \Gamma$, $\gamma \cdot (s+t) = (\gamma \cdot s) \cdot t$. For $R' \subset R$ and $\Gamma' \subset \Gamma$, $\Gamma' \cdot R' \equiv \{\gamma \cdot t \mid \gamma \in \Gamma' \text{ and } t \in R'\}$.

1.2 **Definition.** Given $W \subset \Gamma$, let $W^* \equiv \{\gamma \in W \mid \text{ for some } t \in R^+, \gamma \cdot t \notin W\}$ and let $W^- \equiv \{\gamma \in W \mid \text{ if } \varepsilon > 0 \text{ then } \gamma \cdot [0, \varepsilon] \not\subset W\}$.

A set W is called a Wazewski set if:

a. $\gamma \in W$ and $\gamma \cdot [0, t] \subset \mathrm{cl}\, W$ (= closure W) imply $\gamma \cdot [0, t] \subset W$.

b. W^- is closed relative to W^*.

1.3 Theorem (Wazewski, [3]). If W is a Wazewski set then W^- is a strong deformation retraction of W^* and $W \setminus W^*$ is closed in W.

Proof: Define $\tau : W^* \to R^+$ by $\tau(\gamma) = \sup\{t \geq 0 \mid \gamma \cdot [0, t] \subset W\}$. Observe first that $\gamma \cdot \tau(\gamma) \in W^-$. Namely, $\tau(\gamma) < \infty$ by definition of W^*, and by definition of $\tau(\gamma)$, $\gamma \cdot [0, \tau(\gamma)] \subset \mathrm{cl}(W)$. By condition a. of 1.2, $\gamma \cdot \tau(\gamma)$ is then in W and by definition of $\tau(\gamma)$, $\gamma \cdot \tau(\gamma)$ is then in W^-.

Now $\tau(\gamma)$ is upper semi-continuous on W^*. Namely, given $\gamma \in W^*$ and $\varepsilon > 0$, $\gamma \cdot [\tau(\gamma), \tau(\gamma) + \varepsilon] \not\subset W$, so (by a.) there exists $t' \in [\tau(\gamma), \tau(\gamma) + \varepsilon]$ such that $\gamma \cdot t' \notin \mathrm{cl}(W)$. Let U be a neighborhood of $\gamma \cdot t'$ disjoint from W and let V be a neighborhood of γ such that $V \cdot t' \subset U$. Then for $\gamma' \in U \cap W$, $\gamma' \cdot t' \notin W$. Thus $V \cap W \subset W^*$ and $\tau \mid V \cap W < t'$. This proves $W \setminus W^*$ is closed in W and τ is upper semi-continuous.

To prove that τ is lower semi-continuous, for a fixed $\gamma \in W^*$, let $\bar{t} \equiv \inf\{t \mid$ if U is a neighborhood of γ, there exists $\gamma' \in U \cap W^*$ with $\tau(\gamma') \leq t\}$. Because $\tau(\gamma)$ is in the set, $\bar{t} \leq \tau(\gamma)$. It will be shown that $\bar{t} = \tau(\gamma)$. Let V be a neighborhood of $\gamma \cdot \bar{t}$ and choose $\varepsilon > 0$ and a neighborhood U of γ such that $U \cdot (\bar{t} - \varepsilon, \bar{t} + \varepsilon) \subset V$. Then there is a $\gamma' \in U$ such that $\tau(\gamma') \in (\bar{t} - \varepsilon, \bar{t} + \varepsilon)$ and so $\gamma' \cdot \tau(\gamma') \in V$. Thus $W^- \cap V \neq \phi$. Since V was an arbitrary neighborhood of $\gamma \cdot \bar{t}$ and W^- is closed relative to W^*, $\gamma \cdot \bar{t} \in W^-$. By definition of $\tau(\gamma)$, $\bar{t} = \tau(\gamma)$. This proves τ is lower semi-continuous, so continuous on W^*.

Now define $r : W^* \times [0, 1] \to W^*$ by $r(\gamma, 0) = \gamma \cdot (\sigma \times \tau(\gamma))$. Then r is continuous, and for all $\gamma \in W^*$, $r(\gamma, 0) = \gamma \cdot 0 = \gamma$ and $r(\gamma, 1) = \gamma \cdot \tau(\gamma) \in W^-$. Also, if $\gamma \in W^-$, $\tau(\gamma) = 0$ so for all $\sigma \in [0, 1]$, $r(\gamma, \sigma) = \gamma \cdot 0 = \gamma$. This proves r is a strong deformation retraction of W^* to W^- and so the theorem.

This theorem of Wazewski is, in practice, an existence theorem. In applications one chooses W so that the orbits of $W \setminus W^*$ are the ones whose existence one wants to prove. One then finds W^- and shows that W^- is not a strong deformation retraction of W. By Wazewski's theorem, $W \setminus W^*$ must therefore be non-empty.

In general there are many ways to show W^- is not a strong deformation retraction of W. In the following example, W^- is not connected while W is; thus there is not even a retraction of W to W^- (were $W^* = W$, $\bar{r}(\gamma) \equiv r(\gamma, 1)$ would be such a retraction).

In the version of the theorem given here, the set W does not generally have any interior (as it does in the original version) and the role of the exit points is played by W^-. Even if condition b. of 1.2 is not satisfied, a theorem can be made if W^- is replaced by a quotient space of the smallest relatively closed subset of W^* which contains W^- and is positively invariant with respect to W. However the power of Wazewski's method relies on being able to determine W^- easily and the more general statement does not seem to be real value (at least not in the examples with which this author is familiar).

§2. A problem concerning the range of a non-linear operator

The following problem is treated by M. G. Crandall and L. C. Evans in [1], and there the relation to the non-linear equation $u_t - (\phi(u))_{xx} = 0$ is mentioned (see also [2]). Aside from details of treatment, the problem differs only in that it is not natural in the present case to assume β is monotone increasing (while in [1], β is considered to be a monotone operator because of the connection to the partial differential equation) and it is assumed here that solutions of the equations, (1), depend continuously on the initial data.

Let $\beta(u)$ be a positive function of the real variable u which is integrable near $-\infty$ and non-integrable near $+\infty$. (The author is indebted to H. Kurland for pointing out an error in an earlier "proof" in which the non-integrability condition was omitted).

Let f be in $L'(R)$ with $\int_{-\infty}^{\infty} f > 0$. Then (assuming solutions exist and depend continuously on initial data) the equation $\beta(u) - u'' = f(x)$ admits a solution $u(x)$ such that $u'(x) \to 0$ as $x \to \pm\infty$.

A feeling for the problem comes from examination of solutions of the equation $\beta(u) - u'' = 0$, the phase portrait of which appears in Figure 1. With $u' = v$, the solution curves depicted are the level curves of the (Hamiltonian) function $v^2/2 - B(u) = v^2/2 - \int_{-\infty}^{u} \beta(s)\, ds$. Observe there are two (distinct) distinguished solutions, namely those determined by $v = \pm\sqrt{B(u)}$. Each of these satisfies one of the boundary conditions, but no solution satisfies both boundary conditions. One might say that the presence of f forces these two solutions to hook together to make the desired solution.

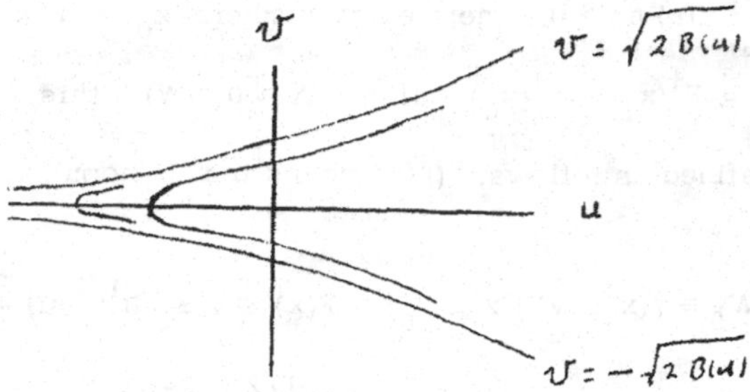

Figure 1

The proof of the result is given in terms of the equivalent system in $\mathbb{R}^3$; namely,

$$\dot{x} = 1 \qquad\qquad \cdot = \frac{d}{dt}$$
$$\dot{u} = v$$
$$\dot{v} = \beta(u) - f(x)$$

The problem becomes that of finding a solution $u(t)$, $v(t)$, $x(t)$ such that $v \to 0$ as $t \to \pm\infty$. The method is to find two Wazewski sets, W_0 and W_1 such that solutions in $W_0 \setminus W_0^*$ and $W_1 \setminus W_1^*$ (respectively) satisfy the boundary conditions at $-\infty$ and $+\infty$ (respectively). Then it is shown that $W_0 \setminus W_0^* \cap W_1 \setminus W_1^* \neq \phi$ and the result is proved.

These Wazewski sets are defined in terms of the following functions.

$B(u) = \int_{-\infty}^{u} \beta(s)\, ds.$

f^+ and f^- are (respectively) the positive and negative parts of f.

$F^{\pm}(x) = \int_{-\infty}^{x} f^{\pm}(s)\, ds$ and $G^{\pm}(x) = \int_{x}^{\infty} f^{\pm}(s)\, ds$. This defines six functions, those with no superscript as well as those with superscript $+$ or $-$.

Because $\int_{-\infty}^{\infty} f(s)\,ds > 0$, there exist numbers x_0 and x_1 such that $\int_{x_0}^{x_1} f(s)\,ds - \|f^-\| = F^+(x_1) - F^+(x_0) - \|f^-\| = \Delta > 0$. With this value of x_0, W_0 and W_1 are defined as follows. ($\|f\|$ means the L' norm).

$$W_0 \equiv \{(x, u, v) \mid x \le x_0;\ -F(x) \le v \le 2B^{1/2}(u) + F^-\}$$

$$W_1 \equiv \{(x, u, v) \mid x_0 \le x;\ -2B^{1/2} - G^- \le v \le G\} \quad .$$

Then W_1 is a Wazewski set for increasing t while W_0 is a Wazewski set for decreasing t. Also both W_0^- and W_1^- have two components given by:

$$W_0^- = (W_0 \cap \{-F(x) = v\}) \cup (W_0 \cap \{v = 2B^{1/2}(u) + F^-\})$$

$$W_1^- = (W_1 \cap \{-2B^{1/2} - G^- = v\}) \cup (W_1 \cap \{v = G\}) \quad .$$

Observe that, since $-F(x) < 2B^{1/2}(u) + F^-(x)$ and $-2B^{1/2} - G^- < G$, the sets W_0^- and W_1^- described above do have two components. Also note that both of W_0 and W_1 are closed in $\mathbb{R}^3$ so condition a. of 1.2 is satisfied. Furthermore the second condition is satisfied if W_0^- and W_1^- are correctly described since these sets are even closed in W. Now points interior to W_1 cannot be in W_1^- because they stay in W_1 for small positive time. Similarly points in the interior of W_0 stay in W_0 for small negative time so cannot be in W_0^-. Also in either case, if v lies strictly between its bounds, the points in W_1 with $x = x_0$ enter W_1 as time increases ($\dot{x} = 1$) so cannot be in W_1^-. A similar statement holds for W_0 with time decreasing.

From the equations one computes that $(v - G)^{\cdot} = \beta - f + f = \beta > 0$; thus $v - G$ increases with t. This implies the points of $W_1 \cap \{v = G\}$ are in W_1^-. Similarly, since $(v + F)^{\cdot} = \beta - f + f > 0$, one concludes that as t decreases, points of $W_0 \cap \{-F = v\}$ leave W_0.

Let $H_1 \equiv (v + G^-)^2/2 - 2B$. Then $\dot{H}_1 = (v + G^-)(\beta - f - f^-) - 2\beta v = -\beta v + G^-\beta - (v + G^-)(f + f^-)$. If $v + G^- = -2B^{1/2}$, both v and $v + G^-$ are negative and so $\dot{H}_1 > 0$. This implies points of $W_1 \cap \{-2B^{1/2} - G^- = v\}$ are in W_1^-. With $H_0 \equiv (v - F^-)^2/2 - 2B$, $\dot{H}_0 = (v - F^-)(\beta - f - f^-) - 2\beta v = -\beta v - F^-\beta - (v - F^-)(f + f^-)$. Again if $v - F^- = 2B^{1/2}$, $\dot{H}_0 < 0$ and it follows that points of $W_0 \cap \{v = 2B^{1/2} + F^-\}$ are in W_0^-. This proves W_0 and W_1 are Wazewski sets for decreasing and increasing time respectively. Wazewski's theorem now implies that neither of $W_0 \setminus W_0^*$ or $W_1 \setminus W_1^*$ can be empty. In fact, since W_0 is connected, $W_0 \setminus W_0^*$ must separate the two components of W_0^- and a similar statement holds for $W_1 \setminus W_1^*$.

Now points in $W_1 \setminus W_1^*$ satisfy the boundary condition $v(t) \to 0$ as $t \to +\infty$. Thus since $(v - G)^{\cdot} = \beta > 0$, $v - G$ has a limit at $+\infty$ so v does. The limit of v is non-positive since $v \le G$ in W_1 and G goes to zero as $t \to +\infty$. Were the limit negative, u would go to $-\infty$ ($\dot{u} = v$) so $B(u) \to 0$. Then, since $v \ge -G^- - 2B^{1/2}$ and $G^- \to 0$, v would go to zero. Thus $v \to 0$.

Also points of $W_0 \setminus W_0^*$ satisfy the boundary condition at $-\infty$. Namely, $(v + F)^{\cdot} = \beta - f + f \ge 0$ so $v + F$ and v have limits as $t \to -\infty$. The limit of v is non-negative because $v \ge -F(x)$ and $F \to 0$. Were the limit positive, u would decrease to $-\infty$ with t and again $B(u) \to 0$, and since $2B^{1/2}(u) + F^- \ge v$, v must go to zero.

Observe also that on half orbits in W_0, respectively W_1, $u \to -\infty$ as $t \to -\infty$, respectively $t \to +\infty$. Thus, in either case, if $|u|$ had a limit other than ∞, $\beta(u)$ would have a positive limit and the equation $\dot{v} = \beta(u) - f$ then implies $|v|$ blows up. Now, in the case of W_1, $\dot{u} = v$ is bounded above. Coupled with the fact that β is not integrable near $+\infty$ this implies that if $u \to \infty$, $v \to \infty$ (since $\dot{v} = \beta(u) - f$ and f is integrable). A similar argument works in the case of W_0.

Now let C_0 be the set of points in the plane $\{x = x_0\}$ which lie in $W_0 \setminus W_0^*$ and let C_1 be those in $W_1 \setminus W_1^*$. These sets (being closed) can be considered as graphs of upper semi-continuous set valued functions from u to v. The aim is to show that $C_0 \cap C_1 \neq \phi$.

To this end, it is shown that if u_1 is large enough $C_1 \cap \{u = u_1\} \cap W_0 \cap W_1 = \phi$ and if u_0 is small enough, $C_1 \cap \{u = u_0\} \cap W_0 \cap W_1 = \phi$. This means C_1 is above C_0 at u_0 and below C_0 at u_1 so that the graphs must intersect.

Thus, choose u_1 so that $B(u_1) > \|f^+\|^2/2$. Let $H(x, u, v) \equiv (v - \|f^+\| + F(x))^2/2 - B$. From the equations, $\dot{H} = (v - \|f^+\| + F)\beta - \beta v = (-\|f^+\| + F)\beta \leq 0$. Thus H decreases with increasing t. On orbits in C_1, both v and $B(u)$ go to zero as $t \to +\infty$, so $H \to (-\|f_+\| + F(\infty))^2/2 = \|f^-\|^2/2$.

On the other hand, in $S \equiv \{u = u_1\} \cap W_0 \cap W_1$, $x = x_0$ so $-F(x_0) \leq v \leq G(x_0)$. Then $-\|f^+\| \leq v - \|f^+\| + F(x_0) \leq G(x_0) - \|f^+\| + F(x_0) = -\|f^-\|$. Since $\|f^-\| < \|f^+\|$, $H \mid S \leq \|f^+\|^2/2 - B(u_1) < 0 \leq \|f^-\|^2/2$. It follows that no points $S = \{u = u_1\} \cap W_0 \cap W_1$ are in C_1.

Recalling now that x_0 was determined so that for some $x = x_1$, $F(x_1) - F(x_0) - \|f^-\| = \Delta > 0$, choose u_0 (close to $-\infty$) so that

$$\sup\{\beta(s) \mid s \le u_0 + \|f^+\| (x_1 - x_0) < \Delta/3(x_1 - x_0)\}$$

and so that

$$2B^{1/2}(u_0 + (x_1 - x_0)\|f^+\|) < \Delta/3 \quad .$$

It will be shown that orbits initiating in $S = \{u = u_0\} \cap W_0 \cap W_1$ leave W_1 by the time $x = x_1$. Thus, in W_1, v is bounded by $\|f^+\|$ so if a solution initiating in S stays in W_1 for $x \in [x_0, x_1]$, then for x in that interval, $u(x) \le u_0 + \|f^+\| (x_1 - x_0)$ and $\beta(u(x)) \le \Delta/3(x_1 - x_0)$.

It then follows that $v(x_1) \le v(x_0) + (\Delta/3(x_1 - x_0))(x_1 - x_0) - F(x_1) + F(x_0)$. In S, $v \le 2B^{1/2}(u_0) + F^-(x_0) < \Delta/3 + F^-(x_0)$ (since $S \subset W_0$); thus $v(x_1) < 2\Delta/3 + F^-(x_0) - F(x_1) + F(x_0) = 2\Delta/3 - F(x_1) + F^+(x_0)$. Now $v(x_1) + 2B^{1/2}(u(x_1)) + G^-(x_1) < \Delta - F(x_1) + F^+(x_0) + G^-(x_1) = \Delta + \|f^-\| + F^+(x_0) - F^+(x_1) = 0$. But this last statement shows that at x_1, v violates one of the W_1 conditions so solutions initiating in S cannot stay W_1.

Now Wazewski's theorem implies that C_0 separates the curves $v = -F(x_0)$ and $v = 2B^{1/2}(u) + F^-(x_0)$ in the plane $\{x = x_0\}$. Likewise C_1 separates the curves $v = -2B^{1/2}(u) - G^-(x_0)$ and $v = G(x_0)$ in the same plane.

Since C_1 does not meet the set $S = \{u = u_0, -2B^{1/2}(u_0) - G^-(x_0) \le v \le 2B^{1/2}(u_0) + F^-(u_0)\}$, C_1 must lie above C_0 when $u = u_0$.

(Observe that $G(x_0) > 2B^{1/2}(u_0) + F^-(x_0)$ since

$$G(x_0) - 2B^{1/2}(u_0) - F^-(x_0) = G^+(x_0) - G^-(x_0) - F^-(x_0) - 2B^{1/2}(u_0)$$
$$= \|f^+\| - F^+(x_0) - \|f^-\| + F^-(x_0) - F^-(x_0) - 2B^{1/2}(u_0)$$
$$= \|f^+\| - F^+(x_0) - \|f^-\| - 2B^{1/2}(u_0) > F^+(x_1) - F^+(x_0) - \|f^-\| - \Delta = 0.$$

This had to be true but it is comforting to check it).

Similarly, since C_1 does not meet the set $\{u = u_1, -F(x_0) \le v \le G(x_0)\}$, C_1 must lie below C_0 when $u = u_1$. (Again one might want to check that this is indeed possible). Now since C_0 and C_1 are closed sets they make up graphs of upper semi-continuous set valued functions from u to v and so must intersect. This concludes the existence proof.

References

[1] M. G. Crandall and L. G. Evans, A singular semi-linear equation in $L^1(\mathbb{R})$. Technical Summary Report #1566. University of Wisconsin - Madison Mathematics Research Center.

[2] M. G. Crandall, An introduction to evolution governed by accretive operators, Proceedings of the International Symposium on Dynamical Systems, Brown University, 1974.

[3] T. Wazewski, Sur une methode topologique de l'examine de l'allure asymptotique des integrales des equation differentielles, Proc. of the International Congress of Mathematicians, Vol. III, Amsterdam, 1954.

ON THE ABSENCE OF SQUARE-INTEGRABLE SOLUTIONS OF THE STURM-LIOUVILLE EQUATION

M. S. P. EASTHAM

1. We consider the Sturm-Liouville equation

$$y''(x) + \{\lambda - q(x)\}y(x) = 0 \quad (0 \leqslant x \leqslant \infty) \tag{1.1}$$

in which

$$q(x) = r(x) + s(x),$$

where

$$r(x) \to 0 \text{ and } s(x) \to 0 \tag{1.2}$$

as $x \to \infty$. We assume further that $s(x)$ is differentiable and that there are constants K and L, with $0 \leqslant K < \infty$ and $0 \leqslant L < \infty$, such that

$$\limsup x|r(x)| = K \tag{1.3}$$

and

$$\limsup xs'(x) = L \tag{1.4}$$

as $x \to \infty$. We denote by Λ the least non-negative number depending only on K and L such that (1.1) has no non-trivial square-integrable solution for $\lambda > \Lambda$.

A number of estimates for Λ are known, dating back it seems to 1948 although the fact that the possibility $\Lambda > 0$ can actually occur goes back to 1929 (14). These estimates, together with some generalizations, are as follows.

(i) Wallach (18) proved that $\Lambda \leqslant K^2$ in the case $s(x) = 0$. One may conjecture that $\Lambda = K^2$ in this case; Wallach gave an example of (1.1) with a square-integrable solution for a value of $\lambda > 0$ which shows that $\Lambda \geqslant \frac{1}{4}K^2$. This example was based on a construction of Wintner (20).

(ii) Borg (3) considered a more general condition than (1.3), namely

$$\limsup (\log x)^{-1} \int_0^x |r(t)| \, dt = K, \tag{1.5}$$

and proved that $\Lambda \leqslant K^2$ again in the case $s(x) = 0$ (See also Note (b) in §3 below.) Borg gave an example which shows that $\Lambda = K^2$ under the condition (1.5). However, the $r(x)$ in this example does not satisfy (1.3) and so it does not settle the conjecture in (i).

(iii) Kato (9) extended the results in (i) to the Schrödinger equation in two or more dimensions.

(iv) Weidmann (19) proved that $\Lambda \leqslant L$ if

$$\limsup (\log x)^{-1} \int_0^x |ds| = L \tag{1.6}$$

and, this time, $r(x) = 0$. Note that (1.6) is, in one sense, more general than (1.4) in that it does not require the existence of

$s'(x)$. On the other hand, of course, (1.6) is less general than (1.4) in that (1.4) is only an upper bound for $xs'(x)$. One may conjecture that $\Lambda = \frac{1}{2}L$ in the situation considered by Weidmann and he gave an example which shows that $\Lambda \geqslant \frac{1}{4}L$. This example can be improved to show that $\Lambda \geqslant \frac{1}{2}L$ - see Note (c) in §3 below.

(v) Simon (17) proved that $\Lambda \leqslant L$ under the conditions (1.3) and (1.4) in the case $K = 0$. There is also the earlier paper (15) by Odeh for the case $K = L = 0$.

(vi) Rohde (16) and Agmon (1) improved Simon's result to $\Lambda \leqslant \frac{1}{2}L$ again under the conditions (1.3) and (1.4) with $K = 0$. In fact, Rohde took $r(x) = 0$ and Agmon $r(x) = O(x^{-1-\epsilon})$ ($\epsilon > 0$) as $x \to \infty$.

The papers referred to in (v) and (vi) cover not only (1.1) but also the Schrödinger equation in two or more dimensions. The papers by Kalf (8) and Müller-Pfeiffer (12, 13) should also be mentioned in this brief survey.

The above results, when supplemented by Note (c) in §3 below can be summarized as follows.

I. Let $s(x) = 0$. Then

$$\tfrac{1}{4}K^2 \leqslant \Lambda \leqslant K^2 \text{ if (1.3) holds}$$

$$\Lambda = K^2 \text{ if (1.5) holds.}$$

II. Let $r(x) = 0$. Then

$$\tfrac{1}{4}L \leqslant \Lambda \leqslant \tfrac{1}{2}L \text{ if (1.4) holds}$$

$$\tfrac{1}{2}L \leqslant \Lambda \leqslant L \text{ if (1.6) holds.}$$

2. All the results quoted in §1 had either $L = 0$ or $K = 0$. The following theorem is a result in which both K and L appear.

THEOREM. Let (1.2), (1.3), and (1.4) hold and denote by Λ the least non-negative number such that (1.1) has no non-trivial square integrable solution for $\lambda > \Lambda$. Then

$$\Lambda \leqslant \tfrac{1}{2}\{L + K^2 + K\sqrt{(2L + K^2)}\}. \tag{2.1}$$

In the proof we shall denote by K_1, K_2, K_3 and L_1, L_2, L_3 constants which lie within an arbitrary fixed ϵ of K and L respectively and we shall assume that $x \geqslant X(\epsilon)$.

We define

$$F(x) = y'^2(x) + \{\lambda - s(x)\}y^2(x),$$

where $y(x)$ is a non-trivial real-valued solution of (1.1) and $\lambda > 0$. Then

$$\begin{aligned} F' &= 2y'\{y'' + (\lambda - s)y\} - s'y^2 \\ &= 2ryy' - s'y^2 \\ &\geqslant -x^{-1}(2K_1|yy'| + L_1y^2) \end{aligned}$$

by (1.3) and (1.4). Also, for any constant c,

$$c(x^{-1}yy')' = cx^{-1}\{y'^2 - (\lambda - r - s)y^2\} - cx^{-2}yy'. \tag{2.2}$$

Hence

$$(F - cx^{-1}yy')' \geqslant -x^{-1}\{cy'^2 + 2K_2|yy'| + (L_2 - c\lambda)y^2\}$$
$$\geqslant -x^{-1}\{(c + \alpha K_2)y'^2 + (L_2 - c\lambda + \alpha^{-1}K_2)y^2\},$$

where α is an arbitrary positive constant. Hence we can write

$$(F - cx^{-1}yy')' \geqslant -Ax^{-1}(F - cx^{-1}yy'), \tag{2.3}$$

where A is a positive constant, if

$$A > c + \alpha K_3 \quad \text{and} \quad \lambda A > L_3 - c\lambda + \alpha^{-1}K_3. \tag{2.4}$$

We wish to be in a position to choose

$$A < 1. \tag{2.5}$$

By (2.4) and bearing in mind the proximity of K_3 and K and of L_3 and L, (2.5) holds if

$$1 - c > \alpha K \quad \text{and} \quad \lambda - L + c\lambda > \alpha^{-1}K.$$

These inequalities hold for some choice of α provided that

$$c < 1 \tag{2.6}$$

and

$$(1 - c)(\lambda - L + c\lambda) > K^2. \tag{2.7}$$

Solving (2.7) for λ, we obtain

$$\lambda > K^2(1 - c^2)^{-1} + L(1 + c)^{-1} \tag{2.8}$$

as long as $c > -1$. It is easily verified that the right-hand side of (2.8), considered as a function of c $(-1 < c < 1)$, takes its minimum value of

$$\tfrac{1}{2}\{L + K^2 + K\sqrt{(2L + K^2)}\} \tag{2.9}$$

when

$$c = L^{-1}\{L + K^2 - K\sqrt{(2L + K^2)}\}.$$

This, then, is our choice of c and we have shown that, if (2.8) holds, then (2.3) holds with $A < 1$ as required in (2.5). Of course, there is a slight and obvious modification of this analysis if either $K = 0$ or $L = 0$.

Integration of (2.3) gives

$$F(x) - cx^{-1}y(x)y'(x) \geqslant Cx^{-A},$$

where C is a positive constant. By (1.2), this gives

$$y'^2(x) + \lambda_1 y^2(x) \geqslant C_1 x^{-A} \tag{2.10}$$

for large x, say $x \geqslant X$, where λ_1 and C_1 are positive constants.

The proof is completed by a standard argument (10, 15). If y(x) were $L^2(0,\infty)$, we would have $(yy')(x_n) < 0$ for some sequence $x_n \to \infty$. Then integration of

$$y\{y'' + (\lambda - r - s)y\} = 0$$

gives

$$-\int_X^{x_n} y'^2 + \int_X^{x_n}(\lambda - r - s)y^2 = -(yy')(x_n) + (yy')(X).$$

Hence

$$-\int_X^{x_n} y'^2 + \lambda_1 \int_X^{x_n} y^2 \geqslant (yy')(X). \tag{2.11}$$

Integration of (2.10) gives

$$\int_X^{x_n} y'^2 + \lambda_1 \int_X^{x_n} y^2 \geqslant C_2 x_n^{1-A}$$

and addition of this to (2.11) gives

$$2\lambda_1 \int_X^{x_n} y^2 \geqslant C_3 x_n^{1-A}.$$

Here C_2 and C_3 are positive constants. Since $A < 1$ we have a contradiction to $y(x)$ being $L^2(0,\infty)$. Hence (1.1) has no non-trivial $L^2(0,\infty)$ solution when λ exceeds (2.9). This proves the theorem.

3. We make some notes on the results in §§1 and 2.

(a) Special cases of (2.1) are $L = 0$, $\Lambda \leqslant K^2$ and $K = 0$, $\Lambda \leqslant \frac{1}{2}L$. These are in line with the conjectures in (i) and (iv) in §1. If (2.2) had not been introduced, that is, if $c = 0$, (2.8) gives only $\Lambda \leqslant K^2 + L$. The above proof represents a simplification of those of (1) and (16) in the case of one dimension and it may extend to cope with the Schrödinger equation. It is not clear whether (2.1) holds under the integral conditions (1.5) and (1.6).

(b) I take this opportunity to point out that there is an error in Borg's proof (3) that $\Lambda \leqslant K^2$ under the condition (1.5). However, an alternative proof is indicated on pp.1560-1 of (5). I am grateful to W. N. Everitt for a discussion on this point. The error is in the statement at the foot of p.123 of (3) that, on the basis of (1.5), the inequality

$$\int_{x_n-\delta}^{x_n+\delta} |r(t)| \, dt > \epsilon$$

will hold only for certain indices $n = m_j$ ($j = 1, 2, 3, \ldots$) such that $m_j < n$ implies $j = O(\log n)$. Here ϵ and δ are given positive constants and $\{x_n\}$ is a sequence of positive numbers such that $x_n = O(n)$ as $n \to \infty$. For a counterexample to this statement, write $X_m = \exp(m^2)$ ($m = 1, 2, \ldots$), let I_m be the interval $[X_m - 1, X_m]$, and define $r(x) = m$ in I_m, $= 0$ elsewhere. Then

$$\int_0^x |r(t)| \, dt \leqslant \sum_{m=1}^{O\{(\log x)^{\frac{1}{2}}\}} m = O(\log x).$$

Thus (1.5) holds. Now let the points x_n ($X_m \leqslant n < X_{m+1}$) be equally spaced in I_m. Then $x_n \leqslant n$ and, for any given positive δ (<1) and ϵ,

$$\int_{x_n-\delta}^{x_n+\delta} |r(t)| \ dt \geqslant \tfrac{1}{2}(\log n)^{\frac{1}{2}}\delta > \epsilon$$

for <u>all but a finite number</u> of the x_n.

(c) To improve Weidmann's example on p.292 of (19), we replace his factor $\{(n+2)/(n+1)\}^2$ by $\{(n+2)/(n+1)\}^k$, where $k > 1$. Then, as in (19), (1.1) with $\lambda = 1$ has an $L^2(0,\infty)$ solution. Also $\int_0^x |ds|$ is now asymptotic to $2k\log x$. Since k can be near to 1, this shows that $\Lambda \geqslant \frac{1}{2}L$ in the situation considered in (iv) of §1. For a discussion of a general class of examples of this kind, see Eastham and Thompson (6). The kind of example considered by Wallach (18) gives $\Lambda \geqslant \frac{1}{4}L$, but this has the additional feature that $s'(x)$ exists.

(d) Examples of (1.1) with a square-integrable solution for a positive value of λ can be obtained from the inverse spectral theory of Gelfand and Levitan - see Moses and Tuan (11), Chaudhuri and Everitt (4), and Everitt (7). These examples have the same basic nature as those of Wallach (18), von Neumann and Wigner (14), and Kato (9). Further examples arise from the delicate asymptotic analysis of Atkinson (2) (see also (7)).

References

1. S.Agmon, J. d'Analyse Math. 23 (1970) 1-25.
2. F.V.Atkinson, Ann. Mat. Pur. Appl. (4) 37 (1954) 347-78.
3. G.Borg, Amer. J. Math. 73 (1951) 122-6.
4. J.Chaudhuri and W.N.Everitt, Proc. Roy. Soc. Edinburgh (A) 68 (1968) 95-119.
5. N.Dunford and J.T.Schwartz, Linear operators, Part 2 (Interscience, 1963).
6. M.S.P.Eastham and M.L.Thompson, Quart. J. Math. (Oxford) (2) 24 (1973) 531-5.
7. W.N.Everitt, Applicable Analysis 2 (1972) 143-60.
8. H.Kalf, Israel J. Math. 20 (1975) 57-69.
9. T.Kato, Commun. Pure Appl. Math. 12 (1959) 403-25.
10. K.Kreith, Proc. Amer. Math. Soc. 14 (1963) 809-11.
11. H.E.Moses and S.F.Tuan, Nuovo Cimento 13 (1959) 197-206.
12. E.Müller-Pfeiffer, Math. Nachr. 60 (1974) 43-52, 62 (1974) 163-78, 67 (1975) 255-63.
13. E.Müller-Pfeiffer, Czechoslovak Math. J., to appear.
14. J. von Neumann and E.Wigner, Z. Physik 50 (1929) 465-7.
15. F.Odeh, Proc. Amer. Math. Soc. 16 (1965) 363-6.

16. H.-W.Rohde, Math. Z. 112 (1969) 375-88.
17. B.Simon, Commun. Pure Appl. Math. 22 (1969) 531-8.
18. S.Wallach, Amer. J. Math. 70 (1948) 833-41.
19. J.Weidmann, Math. Z. 98 (1967) 268-302.
20. A.Wintner, Amer. J. Math. 68 (1946) 385-97.

On limit-point and Dirichlet-type results for second-order differential expressions

W. D. Evans

1. Introduction

Let τ denote the formally self-adjoint second-order differential expression given by

$$\tau u = -(pu')' + qu \tag{1.1}$$

on $[0,\infty)$, where the coefficients p,q are real and satisfy the conditions:-

i) $p(x) > 0$ on $[0,\infty)$ and $p \in AC_{loc}[0,\infty)$, the set of functions which are locally absolutely continuous on $[0,\infty)$,

ii) $q \in L^1_{loc}[0,\infty)$,

so that τ is regular at x = 0.

At the singular end point ∞, τ is said to be limit-point (LP) if there exists at least one solution of $\tau u=0$ which is not in $L^2(0,\infty)$. If Δ denotes the set of functions u which satisfy the conditions

i) $u \in L^2(0,\infty)$, ii) $u' \in AC_{loc}[0,\infty)$, iii) $\tau u \in L^2(0,\infty)$,

then it follows from the Green's formula for τ that for all $u,v \in \Delta$

$$\lim_{x\to\infty} p(x)\{u(x)v'(x) - u'(x)v(x)\} \tag{1.2}$$

exists and it is well-known that τ is LP at ∞ if and only if this limit is zero. The linear space Δ is the domain of the maximal operator associated with τ in $L^2(0,\infty)$, this being the adjoint of the minimal operator associated with τ and τ is LP at ∞ if and only if this minimal operator has deficiency indices (1.1).

The vanishing of the limit (1.2) when τ is LP at ∞ prompted the definition that τ is strong limit-point (SLP) at ∞ if

$$u,v \in \Delta \Rightarrow \lim_{x\to\infty} p(x)\, u(x)\, v'(x) = 0 \tag{1.3}$$

and weak limit point (WLP) if τ is LP at ∞ but (1.3) does not hold. Both the SLP and WLP cases exist as was shown in [8] and [17].

Closely associated with the SLP property at ∞ are the so-called Dirichlet (D) and conditional Dirichlet (CD) properties. τ is said to be D at ∞ if

$$u \in \Delta \quad \Rightarrow \quad p^{1/2}u' \text{ and } |q|^{1/2}u \quad \in L^2(0,\infty), \tag{1.4}$$

and to be CD at ∞ if

$$u, v \in \Delta \quad \Rightarrow \quad p^{1/2}u' \in L^2(0,\infty) \text{ and } \lim_{X\to\infty} \int_0^X quv \quad \text{exists and is finite.} \tag{1.5}$$

Again both these cases are possible (see [3] and [7]). Also it was shown by Kalf in [13] that if τ is D at ∞ it is SLP at ∞. However, the converse is false (see [7] and [17]). It is not known if CD implies SLP at ∞ but again the converse is false (see §2 in [7]).

The objective in this article is to determine conditions on the coefficients p and q for τ to have one or more of the above properties. This problem has attracted a great deal of attention during recent years and many strong results are known. In many of these investigations, the method of approach is based on the same basic idea. For instance, in establishing LP conditions the method usually involves contradicting the non-vanishing of the limit in (1.2). Our aim here is to try to fully utilise the scope of this method and we do in fact obtain two main theorems which include many well-known results, the first theorem for LP criteria and the second for D, CD and SLP criteria. The technique we employ is based on the use of a suitably chosen sequence of functions with compact supports, the purpose of these being to eliminate some of the tedious complications that arise in the analysis due to integration by parts and which often tend to obscure the underlying method. A consequence of this simplification and clarification is that the same technique can be used effectively and comparatively easily for the analogous problems for diffeıential expressions τ of any order (see e.g. [4]). Also, in [6], a similar technique was used to establish the essential self-adjointness of Schrödinger-type operators in $\mathbb{R}^n$, giving both new and extensions of known results.

The results in this paper are confined to a τ which is regular at the finite end point of the interval on which it is defined. In [13] Kalf obtains a result which covers the case when τ has a finite singular point.

For a fuller discussion of the concepts introduced above, we refer to [7] and [18].

2. Limit-point criteria

Throughout the paper K will denote various absolute positive constants not necessarily the same on each occurrance. Any dependence on some parameter will be indicated by $K(\varepsilon)$, K_m etc. Note that it is sufficient to verify the conditions (1.2) - (1.5) for real-functions $u,v \in \Delta$ only. Thus all the functions considered hereafter will be real. Also, (1.5) holds if and only if it holds for $u = v$.

Theorem 1 Let (a_m,b_m), $m=1,2,\ldots$, be a sequence of disjoint intervals in $(0,\infty)$. Suppose that in each $I_m = [a_m,b_m]$ there exist a real locally integrable function Q_m, a non-negative function k_m, a positive absolutely continuous function W_m and positive constants δ, K, G_m such that with $q=q_1-q_2$, $(q_1,q_2 \in L^1_{loc}[0,\infty))$, we have in I_m:-

A) $q_1(x) \geqslant (1+\delta) H_m^2(x)/p(x) - k_m(x)$, $H_m = \int Q_m$,

B) $\int_\alpha^\beta (q_2 - Q_m) W_m \leqslant K\, G_m\, P_m^{1/2}$ for all $a_m \leqslant \alpha < \beta \leqslant b_m$, where

$$P_m = \inf \{ p(t) \mid t \in I_m \}.$$

Suppose also that there exists a real, piecewise continuously differentiable function φ_m with support in I_m such that

C) $$\sum_{m=1}^{\infty} \Phi_m^{-1} \int_{I_m} W_m^2 (q_1 + k_m)^{1/2} p^{-1/2} \varphi_m^2 = \infty$$

where

$$\Phi_m = \sup_{I_m} \left\{ \left(G_m^2 + P_m^{1/2} G_m |W_m'| + k_m W_m^2 \right) \varphi_m^2 + p (W_m \varphi_m)'^2 \right\}.$$

Then τ is LP at ∞.

Proof We first prove three subsidiary results. Let $u \in \Delta$ be real and put $\psi_m = w_m \varphi_m$, $u_m = \psi_m u$. Then, on integration by parts,

$$\int_{I_m} \psi_m^2 u \tau u = \int_{I_m} p u' (\psi_m^2 u' + 2\psi_m \psi_m' u) + \int_{I_m} q u_m^2$$
$$= \int_{I_m} p (u_m'^2 - \psi_m'^2 u^2) + \int_{I_m} q u_m^2$$

so that

$$\int_{I_m} p u_m'^2 = \int_{I_m} \psi_m^2 u \tau u + \int_{I_m} (p \psi_m'^2 u^2 - q u_m^2). \qquad (2.1)$$

Secondly, on integrating by parts, we have for any $\varepsilon_1 > 0$,

$$\left| \int_{I_m} Q_m u_m^2 \right| = 2 \left| \int_{I_m} H_m u_m u_m' \right|$$
$$\leq \varepsilon_1 \int_{I_m} p u_m'^2 + (1/\varepsilon_1) \int_{I_m} (H_m^2/p) u_m^2 \qquad (2.2)$$

For the third result needed we appeal to a result of Atkinson (Lemma 5 in [1]). This is that condition B) implies that there exists in I_m an absolutely continuous function g_m such that

$$(q_2 - Q_m) W_m \leq g_m', \quad |g_m| \leq K G_m P_m^{1/2} \qquad (2.3)$$

When $p = 1$, this result plays a similar role to Lemma 3 in Brinck's paper [3]. We therefore get, for any $\varepsilon_2 > 0$,

$$\int_{I_m} (q_2 - Q_m) u_m^2 \leq \int_{I_m} W_m^{-1} g_m' u_m^2$$
$$= - \int_{I_m} g_m (2 W_m^{-1} u_m u_m' - W_m' W_m^{-2} u_m^2)$$
$$\leq \varepsilon_2 \int_{I_m} P_m u_m'^2 + K(\varepsilon_2) \int_{I_m} (G_m^2 / W_m^2 + P_m^{1/2} G_m |W_m'| W_m^{-2}) u_m^2$$
$$\leq \varepsilon_2 \int_{I_m} p u_m'^2 + K(\varepsilon_2) \int_{I_m} (G_m^2 + P_m^{1/2} G_m |W_m'|) \varphi_m^2 u^2 \qquad (2.4)$$

We now substitute (2.2) and (2.4) in (2.1) to get

$$\int_{I_m} p u_m'^2 \leq \int_{I_m} \psi_m^2 |u \tau u| + \int_{I_m} p \psi_m'^2 u^2 + \int_{I_m} (-q_1 + q_2 - Q_m + Q_m) u_m^2$$
$$\leq \int_{I_m} \psi_m^2 |u \tau u| - \int_{I_m} (q_1 - 1/\varepsilon_1 \, H_m^2/p) u_m^2 + (\varepsilon_1 + \varepsilon_2) \int_{I_m} p u_m'^2$$
$$+ K(\varepsilon_2) \int_{I_m} \{ (G_m^2 + P_m^{1/2} G_m |W_m'|) \varphi_m^2 + p \psi_m'^2 \} u^2 .$$

Now we choose $1/\varepsilon_1 = (1-\eta)(1+\delta)$ where $0 < \eta < 1$ and $\varepsilon = \varepsilon_1 + \varepsilon_2 < 1$. Then, from A)

$$(1-\varepsilon)\int_{I_m} p u_m'^2 \leq \int_{I_m} \psi_m^2 |u\tau u| - \eta\int_{I_m}(q_1+k_m)u_m^2 + K(\varepsilon)\int_{I_m}\Phi_m u^2$$

and so

$$\int_{I_m}\{p u_m'^2 + (q_1+k_m)u_m^2\} \leq K\int_{I_m}(\psi_m^2|u\tau u| + \Phi_m u^2). \qquad (2.5)$$

Suppose now that there are two real linearly independent $L^2(0,\infty)$ solutions u,v of $\tau y = 0$ and let $p(uv'-u'v) = 1$. Then, with $v_m = \psi_m v$,

$$\psi_m^2(q_1+k_m)^{1/2}p^{-1/2} = p^{1/2}(q_1+k_m)^{1/2}\{u_m v_m' - v_m u_m'\}$$
$$\leq \{p(u_m'^2+v_m'^2)\ (q_1+k_m)(u_m^2+v_m^2)\}^{1/2}$$

Hence, from (2.5)

$$\int_{I_m}\psi_m^2(q_1+k_m)^{1/2}p^{-1/2} \leq K\int_{I_m}\Phi_m(u^2+v^2).$$

Thus,

$$\sum_{m=1}^{\infty}\Phi_m^{-1}\int_{I_m}\psi_m^2(q_1+k_m)^{1/2}p^{-1/2} \leq K\int_0^{\infty}(u^2+v^2) < \infty$$

and as this contradicts C) the result is established.

By choosing the functions q_1, q_2, Q_m, k_m, w_m and Φ_m appropriately we get from Theorem 1 the following corollaries which include many well-known results for τ to be LP at ∞.

<u>Corollary 1</u> Let (a_m,b_m), m=1,2,..., be disjoint intervals in $(0,\infty)$ and put $d_m=b_m-a_m$. Suppose that with $p_m = \inf\{p(t)\mid t\in[a_m,b_m]\}$

a) $d_m\int_\alpha^\beta q \geq -Kp_m$ for all $a_m \leq \alpha < \beta \leq b_m$,

b) $\sum_{m=1}^{\infty}(p_m^2/d_m)\left(\int_{a_m}^{b_m}1/p\right)^3 = \infty.$

Then τ is LP at ∞.

<u>Proof</u> Let

$$\varphi_m(x) = \int_{a_m}^{x}1/p\int_x^{b_m}1/p,\quad a_m\leq x\leq b_m,\quad 0 \text{ otherwise.}$$

Then, writing $J(x) = \int_{a_m}^{x}1/p$, $J_m = J(b_m)$, we have

$$0 \leq \varphi_m \leq J_m^2,\qquad p^{1/2}|\varphi_m'| \leq p^{-1/2}J_m \leq p_m^{-1/2}J_m.$$

We now substitute $q_1=0,\ q_2=-q,\ Q_m=0,\ G_m=p_m^{1/2}/d_m,\ k_m=G_m^2,\ W_m=1$ in Theorem 1. Then A) and B) in Theorem 1 are clearly satisfied and also

$$\Phi_m \leq K(p_m \delta_m^2/d_m^2 + 1/p_m)\delta_m^2 \leq K\delta_m^2/p_m$$

since $\delta_m \leq d_m/p_m$. Hence,

$$\Phi_m^{-1}\int_{a_m}^{b_m}(q_1+k_m)^{1/2}p^{-1/2}g_m^2 \geq K\, p_m^2/d_m\delta_m^2 \int_{a_m}^{b_m} p^{-1}g_m^2$$

$$= K\, p_m^2\,\delta_m^3/d_m$$

and the corollary follows from Theorem 1.

If in Corollary 1 we also suppose that $p \leq K p_m$ on $[a_m, b_m]$ then b) is equivalent to

$$\sum_{m=1}^{\infty} d_m^2/p_m = \infty \tag{2.6}$$

This is Atkinson's Theorem 11 in [1]. When p=1 this result is similar to that obtained by Eastham in [5] but Eastham's result does not appear to be a consequence of Theorem 1.

The following examples are special cases of Corollary 1 when p=1:-

Example 1 (Ismagilov [11]). τ is LP at ∞ if in $[a_m,b_m]$, m=1,2,...

$$q \geq -K d_m^{-2} \quad , \quad \sum_{m=1}^{\infty} d_m^2 = \infty \tag{2.7}$$

Example 2 (Brinck [3]). τ is LP at ∞ if $b_m = a_m + 1 \leq a_{m+1}$, m=1,2,... and

$$\int_J q \geq -K \quad \text{for all} \quad J \subseteq [a_m, a_m+1] \tag{2.8}$$

This includes Theorem 3 in [3] when w=1, since Brinck requires the condition on q in (2.8) to be satisfied for all intervals J of length ≤ 1 (see also Corollary 2 in §3 below).

A particular case of example 2 is the result of Hartman in [10] that τ is LP at ∞ if p=1 and q is bounded below on a sequence of intervals whose lengths are bounded away from zero. Also, observe that (2.8) is satisfied if $q_- = \min(0,q) \in L^r(0,\infty)$, $1 \leq r \leq \infty$.

Corollary 2 (Atkinson and Evans [2]). Let (a_m,b_m), m=1,2,... be disjoint and $[c_m,e_m] \subset (a_m,b_m)$. Suppose that there exist a positive, piecewise continuously differentiable function σ and positive constants K_1, K_2, δ, such that in the intervals $[a_m,b_m]$,

a) $p\sigma^{-1/2} \le K_1$,

b) $q\sigma^{-2} \ge -K_2$,

c) $\int_{a_m}^{c_m} p^{-1/2}\sigma^{-1} \ge \delta > 0$, $\int_{e_m}^{b_m} p^{-1/2}\sigma^{-1} \ge \delta > 0$,

d) $\sum_{m=1}^{\infty} \int_{c_m}^{e_m} \sigma p^{-1/2} = \infty$.

Then τ is LP at ∞

<u>Proof</u> In Theorem 1 we choose $q_1 = q$, $q_2 = Q_m = 0$, $G_m = 0$, $k_m = \frac{K_3}{\sigma^2}$ $(K_3 > K_2)$ and $w_m = 1$. Also, we define

$$\varphi_m(x) = \begin{cases} \sigma(x) & , \quad x \in [c_m, e_m] , \\ 0 & , \quad x \notin [a_m, b_m] , \end{cases}$$

and for $x \in [a_m, c_m)$

$$\varphi_m(x) = \sigma(x)\left\{1 - \alpha_m \int_x^{c_m} \sigma^{-1}p^{-1/2}\right\}$$

where α_m is chosen so that $\varphi_m(a_m)=0$, with a similar definition in $(e_m, b_m]$.

We then have

$$\Phi_m \le K\left(\varphi_m^2/\sigma^{2} + p\varphi_m'^2\right) \le K$$

and

$$\Phi_m^{-1} \int_{a_m}^{b_m} (q+k_m)^{1/2} p^{-1/2}\varphi_m^2 \ge K \int_{c_m}^{e_m} \sigma p^{-1/2}$$

which gives the result.

Corollary 2 implies that the well-known criterion of Levinson [16] for τ to be LP at ∞ needs only to hold on a suitable sequence of intervals. This then, in common with all the above interval-type results, allows for oscillatory coefficients.

<u>Corollary 3</u> (Ismagilov [12], Knowles [14]). Let (a_m, b_m) $m=1,2\ldots$ be disjoint intervals and suppose that in $[a_m, b_m]$ $q(x) \ge q_m > 0$ and $p(x) \ge p_m > 0$. Then τ is LP at ∞ if

$$\sum_{m=1}^{\infty} p_m^{3/2} q_m^{1/2} \left(\int_{a_m}^{b_m} 1/p\right)^3 = \infty .$$

<u>Proof</u> We put $q_1 = q$, $q_2 = Q_m = 0$, $w_m = 1$, $G_m = k_m = 0$ and

$$\varphi_m(x) = \int_{a_m}^{x} 1/p \int_x^{b_m} 1/p \quad , \quad a_m \le x \le b_m , \quad 0 \text{ otherwise.}$$

Then, as in the proof of Corollary 1, $\Phi_m \le K J_m^2/p_m$ and

$$\Phi_m^{-1}\int_{a_m}^{b_m} q^{1/2}p^{-1/2}\varphi_m^2 \geq K\,(P_m^{3/2}q_m^{1/2}/J_m^2)\int_{a_m}^{b_m} p^{-1}\varphi_m^2$$
$$= K\,P_m^{3/2}q_m^{1/2}J_m^3.$$

The Corollary therefore follows.

3. Dirichlet and conditional Dirichlet results.

Theorem 2 Suppose that there exists a function $Q \in L^1_{loc}[0,\infty)$, a positive function $w \in AC_{loc}[0,\infty)$ and positive constants δ, K_1, K_2, K_3, such that

A) $q_1(x) \geq (1+\delta)H^2(x)/p(x) - K_1/w^2(x)$, $H = \int Q$,

B) $\int_J (q_2 - Q)w \leq K_2 \inf_I p^{1/2}$ whenever $\int_I p^{-1/2}w^{-1} \leq 1$ and $J \subseteq I$,

C) $p^{1/2}|w'| \leq K_3$,

D) $\int_0^\infty p^{-1/2}w = \infty$.

Then we have:-

i) τ is LP at ∞.

ii) If w is bounded $u \in \Delta \Rightarrow wp^{1/2}u'$ and $w|q_1|^{1/2}u \in L^2(0,\infty)$.

iii) If w is bounded and there exist positive constants K_4 and K_5 such that

E) either $|H(x)| < K_4 x$ and $x < K_5\,p^{1/2}(x)\,|q_1(x)|^{1/2}w^2(x)$, or $Q=0$,

then for $u \in \Delta$

$$\lim_{X\to\infty}\int_0^X Qw^2u^2 \quad , \quad \lim_{X\to\infty}\int_0^X (q_2 - Q)w^2u^2 \; ,$$

exist and are finite. If w=1 τ is CD and SLP at ∞.

iv) If Q is of one sign, w=1 and instead of B)

B′) $\int_I |q_2 - Q| \leq K_2 \inf_I p^{1/2}$ whenever $\int_I p^{-1/2} \leq 1$

then τ is D at ∞.

Proof We first observe (c.f. Lemma 2 in [15]) that C) and D) imply that $\int_0^\infty p^{-1/2}w^{-1} = \infty$. For, otherwise

$$\log\,(w(x)/w(0)) = \int_0^x w'/w \leq K_3\int_0^x p^{-1/2}w^{-1} < K$$

and so w(x)<K. This leads to the contradiction

$$\int_0^\infty p^{-1/2}w^{-1} \geq K\int_0^\infty p^{-1/2}w = \infty .$$

Hence, there exists a sequence of positive numbers c_m,m=1,2,... tending to infinity, which are such that $\int_{c_m}^{c_{m+1}} p^{-1/2}w^{-1} = 1/3$.

Now define

$$\varphi_m(x) = \begin{cases} \int_{c_m}^{x} p^{-1/2} w^{-1} , & c_m \le x < c_{m+1} , \\ 1/3 , & c_{m+1} \le x \le c_{m+2} , \\ \int_{x}^{c_{m+3}} p^{-1/2} w^{-1} , & c_{m+2} < x \le c_{m+3} . \end{cases}$$

Then $0 \le \varphi_m \le 1/3$ and $w p^{1/2} |\varphi_m'| \le 1$. Thus, from (2.5) with $w_m = w$, $\psi_m = w\varphi_m$, $u_m = \psi_m u$, $I_m = [c_m, c_{m+3}]$, $Q_m = Q$, $G_m = 1$, $k_m = K_1/w^2$ and since

$$\Phi_m = \sup_{I_m} \{ (1 + p^{1/2}|w'| + K_1) \varphi_m^2 + p (\varphi_m w)'^2 \} \le K ,$$

$$\int_{c_m}^{c_{m+3}} (p u_m'^2 + [q_1 + K_1/w^2] u_m^2) \le K \int_{c_m}^{c_{m+3}} (w^2 |u \tau u| + u^2) . \tag{3.1}$$

Hence, since $|q_1| \le q_1 + 2K_1/w^2$ and $\varphi_m = 1/3$ in $[c_{m+1}, c_{m+2}]$,

$$\int_{c_{m+1}}^{c_{m+2}} w^2 (p u'^2 + |q_1| u^2) \le K \int_{c_m}^{c_{m+3}} (w^2 |u \tau u| + u^2) . \tag{3.2}$$

i) To prove i) we suppose that u and v are real $L^2(o,\infty)$ solutions of $\tau u = 0$ which satisfy $p(uv' - u'v) = 1$. Then

$$\begin{aligned} \int_{c_{m+1}}^{c_{m+2}} p^{-1/2} w &\le \int_{c_{m+1}}^{c_{m+2}} \{ w^2 p (u'^2 + v'^2)(u^2 + v^2) \}^{1/2} \\ &\le 1/2 \int_{c_{m+1}}^{c_{m+2}} \{ w^2 p (u'^2 + v'^2) + (u^2 + v^2) \} \\ &\le K \int_{c_m}^{c_{m+3}} (u^2 + v^2) \end{aligned}$$

from (3.2). Hence, summing over m we get a contradiction to D) and so i) is proved.

ii) If w is bounded, we get from (3.2)

$$\int_{c_{m+1}}^{c_{m+2}} w^2 (p u'^2 + |q_1| u^2) \le K \int_{c_m}^{c_{m+3}} (|\tau u|^2 + |u|^2)$$

and hence (ii) follows on summing over m.

iii) For any X

$$\int_0^X Q w^2 u^2 = H w^2 u^2 \Big|_0^X - 2 \int_0^X H (wu)(wu)' . \tag{3.3}$$

From E

$$| (H w^2 u^2)(X) | \le K X u^2(X)$$

and

$$X u^2(X) = \int_0^X u^2 + 2 \int_0^X (x p^{-1/2} u)(p^{1/2} u')$$

The last integral is convergent as $X \to \infty$ from E) and ii) and so $\lim_{X \to \infty} X u^2(X)$

exists. This limit must be zero for $u \in L^2(0,\infty)$ and so for $u \in \Delta$ $\lim_{X\to\infty} (Hw^2u^2)(X) = 0$. The last integral in (3.3) converges since, from A) and C),

$$|H(wu)(wu)'| \leq K\, p^{1/2}(q_1 + K_1/w^2)^{1/2}(wu)(wu)'$$
$$\leq K\{w^2(pu'^2 + |q_1|u^2) + u^2\}.$$

Hence, from (3.3) $\lim_{X\to\infty} \int_0^X Qw^2u^2$ exists and is finite.

For any X in $[c_{m+1}, c_{m+2}]$ we have from (2.3), B) and C),

$$\int_{c_{m+1}}^{X} (q_2 - Q)w^2u^2 \leq \int_{c_{m+1}}^{X} g_m' \, wu^2$$
$$\leq g_m wu^2 \Big|_{c_{m+1}}^{X} + K\int_{c_{m+1}}^{c_{m+2}} (w^2pu'^2 + u^2).$$

Also, with $p_m = \inf\{p(t) \mid t \in [c_m, c_{m+3}]\}$,

$$|(g_m wu^2)(X)| \leq K p_m^{1/2}(wu^2)(X) \leq K\int_{c_m}^{X} p^{1/2}|(w^{-1}u_m^2)'|$$
$$\leq K\int_{c_m}^{c_{m+2}} (pu_m'^2 + \varphi_m^2 u^2). \qquad (3.4)$$

Hence, from (3.1) and (3.2), for $X \in [c_{m+1}, c_{m+2}]$,

$$\int_{c_{m+1}}^{X} (q_2 - Q)w^2u^2 \leq K\int_{c_m}^{c_{m+3}} (|\tau u|^2 + u^2) \qquad (3.5)$$

and similarly

$$\int_{X}^{c_{m+2}} (q_2 - Q)w^2u^2 \leq K\int_{c_m}^{c_{m+3}} (|\tau u|^2 + u^2) \qquad (3.6)$$

We now write

$$\frac{1}{9}\int_{c_{m+1}}^{c_{m+2}} (q_2 - Q)w^2u^2 = \int_{c_m}^{c_{m+3}} (q_2 - Q)u_m^2 - \left(\int_{c_m}^{c_{m+1}} + \int_{c_{m+2}}^{c_{m+3}}\right)(q_2 - Q)u_m^2 \qquad (3.7)$$

Again from (2.3)

$$9\left(\int_{c_m}^{c_{m+1}} + \int_{c_{m+2}}^{c_{m+3}}\right)(q_2 - Q)u_m^2 \leq g_m w^2u^2 \Big|_{c_{m+2}}^{c_{m+1}} + K\int_{c_m}^{c_{m+3}} p^{1/2}|u_m u_m'|$$
$$\leq K\int_{c_m}^{c_{m+3}} (|\tau u|^2 + u^2) \qquad (3.8)$$

from (3.4) and (3.1). Also, from (2.2) with $I_m = [c_m, c_{m+3}]$, $Q_m = Q$, $1/\varepsilon_1 < 1+\delta$,

$$\left|\int_{c_m}^{c_{m+3}} Qu_m^2\right| \leq K\int_{c_m}^{c_{m+3}} (|\tau u|^2 + u^2) \qquad (3.9)$$

Hence, from (2.1), (3.1) and (3.9)

$$\left|\int_{c_m}^{c_{m+3}} (q_2 - Q)u_m^2\right| \leq K\int_{c_m}^{c_{m+3}} (|\tau u|^2 + u^2)$$

and so, from (3.5), (3.7) and (3.9)

$$|\int_{c_{m+1}}^{c_{m+2}} (q_2-Q)w^2u^2| \leq K\int_{c_m}^{c_{m+3}} (|\tau u|^2+u^2). \tag{3.10}$$

Lastly we have from (3.6) and (3.10), for $X \in [c_{m+1}, c_{m+2}]$

$$\int_{c_{m+1}}^{X} (q_2-Q)w^2u^2 = \int_{c_{m+1}}^{c_{m+2}} (q_2-Q)w^2u^2 - \int_{X}^{c_{m+2}} (q_2-Q)w^2u^2$$

$$\geq -K\int_{c_m}^{c_{m+3}} (|\tau u|^2+u^2)$$

and this, together with (3.5) gives

$$|\int_{c_{m+1}}^{X} (q_2-Q)w^2u^2| \leq K\int_{c_m}^{c_{m+3}} (|\tau u|^2+u^2). \tag{3.11}$$

Now let $c_{M+1} < X \leq c_{M+2}$. Then

$$\int_{c_2}^{X} (q_2-Q)w^2u^2 = \sum_{m=1}^{M-1}\int_{c_{m+1}}^{c_{m+2}} (q_2-Q)w^2u^2 + \int_{c_{M+1}}^{X} (q_2-Q)w^2u^2.$$

For $u \in \Delta$ the first term on the right converges as $M \to \infty$ by (3.10) and hence in view of (3.11) it follows that $\lim_{X\to\infty} \int_0^X (q_2-Q)w^2u^2$ exists and is finite.

When w=1 τ is thus CD at ∞ and, as a consequence, for $u,v \in \Delta$, $\lim_{X\to\infty}(puv')(x)$ exists and has a value L say. If $L \neq 0$ then for m sufficiently large

$$\tfrac{1}{2}|L|\int_{c_{m+1}}^{c_{m+2}} p^{-1/2} \leq \int_{c_{m+1}}^{c_{m+2}} p^{1/2}|uv'| \leq K\int_{c_m}^{c_{m+3}} (|\tau u|^2+u^2)$$

from (3.2). This leads to a contradiction of D) and hence L=0. Thus τ is SLP at ∞ and iii) is proved.

iv) If Q is of one sign and w=1 then $|Q|^{1/2}u \in L^2(0,\infty)$ for $u \in \Delta$ from (3.9). Also, B') gives

$$\int_{c_{m+1}}^{c_{m+2}} |q_2-Q|u^2 \leq 9\int_{c_m}^{c_{m+3}} |q_2-Q|u_m^2$$

$$\leq K\int_{c_m}^{c_{m+3}} p^{1/2}u_m u_m' \leq K\int_{c_m}^{c_{m+3}} (|\tau u|^2+u^2)$$

and hence iv).

<u>Corollary 1</u> (Everitt [7]). Suppose that

a) $q_1 \geq (1+\delta)H^2/p - K_1$, $H = \int q_2$,

b) $\int_0^\infty p^{-1/2} = \infty$

Then:-

i) τ is LP at ∞.

ii) $u \in \Delta \Rightarrow p^{1/2}u'$ and $|q_1|^{1/2}u \in L^2(0,\infty)$.

iii) If

c) $|H(x)| < K_2 x$ and $x < K_3\, p^{1/2}(x)\, |q_1(x)|^{1/2}$

then τ is CD and SLP at ∞.

iv) If q_2 is of one sign (eventually) then τ is D at ∞.

This is the case $w=1$ and $Q=q_2$ of Theorem 2. It generalises Theorems 1 and 2 of [7] as Everitt puts additional conditions on the coefficients.

If in Corollary 1 we put $q_2=0$ and $q_1=q$ we get that τ is D and SLP at ∞ if $q \geq -K_1$ and $\int_0^\infty p^{-1/2} = \infty$.

<u>Corollary 2</u> (Brinck [3], Knowles [15]). Suppose $p=1$ and

a) $\int_J wq \geq -K$ whenever $\int_J w^{-1} \leq 1$,

b) w' is bounded,

c) $\int_0^\infty w = \infty$.

Then

i) τ is LP at ∞.

ii) If w is bounded $wu' \in L^2(0,\infty)$ for $u \in \Delta$.

iii) If $w=1$, τ is CD and SLP at ∞.

iv) If $w=1$ and instead of a) we have, with $q_- = \min(0,q)$,

a') $\int_J q_- \geq -K$ whenever J is of length ≤ 1

then τ is D at ∞.

<u>Proof</u> Parts i) - iii) follow from Theorem 2 with $p=1$, $q_1=Q=0$ and $q_2=-q$. For iv) we put $q_1=q_+=\max(q,0)$, $Q=0$ and $q_2=-q_-$ in Theorem 2 (iv).

From Corollary 2 (iv) we get that τ is D and SLP at ∞ if

$$p = 1 \;, \quad q_- \in L^r(0,\infty) \;, \quad 1 \leq r \leq \infty \tag{3.12}$$

(see [9] and [13]).

<u>Corollary 3</u> (Levinson [16], Brown and Evans [4]). Suppose that there exists a positive continuously differentiable function M and positive

constants K_1, K_2 such that

a) $q(x) \geq -K_1 M(x)$,

b) $p M'^2 M^{-3} \leq K_2$,

c) $\int_0^\infty (pM)^{-1/2} = \infty$.

Then,

i) τ is LP at ∞

ii) If M is bounded away from zero then

$$u \in \Delta \Rightarrow M^{-1/2} p^{1/2} u' \ , \ M^{-1/2} |q|^{1/2} u \in L^2(0,\infty).$$

Proof Let $W = M^{-\frac{1}{2}}$, $Q = 0$, $q_2 = 0$ in Theorem 2. The assumption in ii) that M is bounded away from zero can be made without loss of generality as was proved by Read in [19].

References

1. F.V.Atkinson,: Limit-n criteria of integral type. Proc.Roy.Soc. Edinburgh (A), 73, 11, 1975, 167-198.
2. F.V.Atkinson and W.D.Evans,: Solutions of a differential equation which are not of integrable square. Math.Z. 127 (1972), 323-332.
3. L.Brinck,: Self-adjointness and spectra of Sturm-Liouville operators, Math.Scand. 7 (1959), 219-239.
4. B.M.Brown and W.D.Evans,: On the limit-point and strong limit-point classification of 2nth order differential expressions with wildly oscillating coefficients. Math.Z. 134 (1973), 351-368.
5. M.S.P.Eastham,: On a limit-point method of Hartman. Bull. London Math. Soc. 4 (1972), 340-344.
6. M.S.P.Eastham, W.D.Evans and J.B.McLeod,: The essential self-adjointness of Schrödinger-type operators, (to appear in Arch.Rat.Mech. and Analysis).
7. W.N.Everitt,: On the strong limit-point condition of second-order differential expressions. Proceedings of International Conference on Differential Equations (Los Angeles 1974). (Academic Press, New York,

1975) 287-307.

8. W.N.Everitt, M.Giertz and J.B.McLeod,: On the strong and weak limit-point classification of second-order differential expressions. Proc. London Math.Soc. (3) 29 (1974) 142-158.

9. W.N.Everitt, M.Giertz and J.Weidmann,: Some remarks on a separation and limit-point criterion of second-order ordinary differential expressions. Math.Ann. 200, (1973), 335-346.

10. P.Hartman,: The number of L^2 solutions of $x'' + q(t)x = 0$. Amer.J.Math. 73, (1951) 635-645.

11. R.S.Ismagilov,: Conditions for self-adjointness of differential equations of higher order. Soviet Math. 3, (1962) 279-283.

12. R.S.Ismagilov,: On the self-adjointness of the Sturm-Liouville operator. Uspehi Mat. Nauk. 18, No.5 (113), (1963), 161-166.

13. H.Kalf,: Remarks on some Dirichlet-type results for semi-bounded Sturm-Liouville operators. Math.Ann. 210, (1974), 197-205.

14. I.Knowles,: Note on a limit-point criterion. Proc.Amer.Math.Soc. 41 (1973), 117-119.

15. I.Knowles,: A limit-point criterion for a second-order linear differential operator. J.London Math.Soc. (2), 8 (1974), 719-727.

16. N.Levinson,: Criteria for the limit-point case for second-order linear differential operators. Časopis pro pěsto ványi matematiky a fysiky. 74, (1949), 17-20.

17. J.B.McLeod,: The limit-point classification of differential expressions. Spectral theory and asymptotics of differential equations (Mathematics Studies 13, North-Holland, Amsterdam, 1974), 57-67.

18. M.A.Naimark,: Linear differential operators. Part II (Ungar, New-York, 1968).

19. T.T.Read,: A limit point criterion for expressions with oscillating coefficients. (To appear).

"See note overleaf"

Note Dr T. T. Read has pointed out that the result of Atkinson mentioned after Corollary 1 in §2 does in fact include the result of Eastham [5]. Also note that it has been proved recently by Everitt in the article below that if τ is CD at ∞ it is SLP at ∞ so that we have the implications

$D \Rightarrow CD \Rightarrow SLP \Rightarrow LP$.

Additional Reference

W. N. Everitt, : A note on the Dirichlet condition for second-order differential expressions. Canadian J. Math. 28 (1976), 312-320.

Spectral theory of the Wirtinger inequality

W. N. Everitt

1. The inequality ascribed to Wirtinger may be described as follows: let f be a complex-valued function defined and absolutely continuous on the closed interval $[0,2\pi]$, such that the derivative f' is of integrable-square on $[0,2\pi]$ and the following boundary conditions are satisfied

$$f(0) = f(2\pi) \quad \underline{\text{and}} \quad \int_0^{2\pi} f(x)dx = 0; \tag{1.1}$$

then

$$\int_0^{2\pi} |f'(x)|^2 dx \geq \int_0^{2\pi} |f(x)|^2 dx \tag{1.2}$$

with equality if and only if for some complex numbers A and B

$$f(x) = A \cos x + B \sin x \qquad (x \in [0,2\pi]). \tag{1.3}$$

This inequality has a long and interesting history and here reference should be made to the now classic text by Hardy, Littlewood and Pólya [6, Section 7.7], and the detailed account to be found in the recent book by Mitrinović (and Vasić) [9, Section 2.23.1] which lists no fewer than 55 references in the section concerned with the Wirtinger inequality. For some recent historical comments on this inequality see the papers by Janet [7], and by Mitrinović and Vasić [10].

For a discussion on the method of proof of the inequality (1.2) reference should be made to two sources. Firstly to [6] where in Section 7.6 may be found a discussion of a proof using the calculus of variations (although all the details are not given); and then in Section 7.7 an elegant and elementary (in the technical sense) proof which, however, by its very nature fails to bring out the structural reasons for the validity of the inequality. Secondly to the book by Beckenbach and Bellman [2] where in Chapter 5, Sections 10 to 13 an interesting account is given, however with many details excluded, of three proofs of the Wirtinger inequality; the first by means of Fourier series, the second by Sturm-Liouville theory, and the third following the ideas in [6, Section 7.7]; of these the most interesting is the Fourier series proof; indeed the second proof really discusses a different inequality and not the Wirtinger inequality (1.2).

Of all these methods of proof the simplest and most effective method for the Wirtinger inequality, and one of its extensions given below as Theorem 2, is undoubtedly the proof based on the use of Fourier series and the Parseval identity; see again [2, Chapter 5, Section 11]. This proof is also mentioned in [6, Section 7.7]. On the other hand the Fourier series method does not readily lend itself to more general inequalities.

The 'elementary' proof of the Wirtinger inequality, given in [6, Section 7.7], has much to commend itself but it does require an insight into the existence of certain integral identities which are far from obvious; this form of proof is used in a variety of other cases considered in [6, Chapter 7]. In [3] Beesack has shown that this method may be extended to more general inequalities; see also [2, Chapter 5, Section 13].

The method of the calculus of variations, see [6, Section 7.6], does require a detailed background knowledge of variation theory; moreover this method does have in general certain other disadvantages, as may be seen in the interesting account given in [6, Sections 7.1 and 2]. On the other side it is a method which extends very successfully to the consideration of inequalities in integrable-p spaces, see [6, Sections 7.5 and 6], an extension not possible or, at the very least, difficult by other methods.

Finally there is the spectral theory method considered in this paper. The name 'spectral theory' is to be preferred to the so-called "Sturm-Liouville' method of [2, Chapter 5, Section 12] since it is applicable to both regular and singular symmetric differential expressions. In any case we pointed out above that the proof given in [2, Chapter 5, Section 12] needs amendment if it is to give a proof of the Wirtinger inequality (1.2). The importance of spectral theory in certain integrable-square inequalities is discussed by Everitt in [5]. The method used in this paper depends on an approximation technique developed by Bradley and Everitt in [4] which has the advantage of working for very general symmetric differential expressions with minimal conditions on the coefficients; we discuss this point briefly at the end of this paper. The method does have the interest, but possibly also the disadvantage, of requiring a special argument for the cases of equality, see [4, Section 5] where however the argument is unnecessarily complicated; this point may also be made in the calculus of variations method, see [6, Pages 183 and 4]. The spectral theory method does show that the best possible constants in these inequalities are determined by the spectrum of certain well-defined differential operators; moreover it provides a framework in which the various types of boundary conditions, and (1.1) above is a good example, find their most natural setting.

In Section 2 of this paper we give some standard notations and then state the two theorems to be discussed. The subsequent sections contain the spectral theory proof of Theorem 1, a brief discussion on Theorem 2 and some remarks concerning generalizations.

2. Let [a,b] denote a compact interval of the real line R; let C denote the complex field. Let $L^2(a,b)$ denote the integrable-square Lebesgue integration space of complex-valued functions defined on [a,b]. In the usual way we also let $L^2(a,b)$ denote the Hilbert function space of equivalence classes, with norm $\|\cdot\|$ and inner-product $(\cdot,\cdot)$. Also AC[a,b] denotes the class of complex-valued functions which are absolutely continuous on [a,b]. The expression '$(f \in D)$' is to be read as 'for all f in the set D'. The symbol $C_0^\infty(0,2\pi)$ denotes the class of infinitely differentiable complex-valued functions with support contained in the open interval $(0,2\pi)$.

Let D be defined as the linear manifold of $L^2(a,b)$ given by

$$D \overset{\text{def}}{=\!=} \{f \in AC[a,b] : f' \in L^2(a,b)\}. \tag{2.1}$$

Wirtinger type inequalities are concerned with the 'comparability', see [6, Section 1.6] of the norms $\|f\|$ and $\|f'\|$ when $f \in D$. In particular with inequalities of the form

$$\|f'\| \geq k\|f\| \qquad (f \in D) \tag{2.2}$$

where k is a non-negative real number. Clearly (2.2) is always satisfied when $k = 0$ and this case is excluded. On the other hand (2.2) is clearly false when $k > 0$ and D is given by (2.1); for we need only take f to be a constant function on [a,b].

An inequality of the form (2.2) with $k > 0$ results if the functions in D are restricted to satisfy a set of boundary conditions. Examples are as follows (in each case the value of k is best possible and there is equality if and only if f takes the form shown (where $A \in C$)):

(i) see [6, Section 7.7]

$a = 0 \quad b = \frac{1}{2}\pi \quad f(0) = 0$

then $k = 1 \quad f(x) = A \sin x \quad (x \in [0,\frac{1}{2}\pi])$

(ii) see [6, Section 7.7]

$a = 0 \quad b = \pi \quad f(0) = f(\pi) = 0$

then $k = 1 \quad f(x) = A \sin x \quad (x \in [0,\pi])$

(iii) see [2, Chapter 5, Section 12]

$a = 0 \quad b = 2\pi \quad f(0) = f(2\pi) = 0$

then $k = \frac{1}{2} \quad f(x) = A \sin x \quad (x \in [0,2\pi])$.

A fourth example is the Wirtinger inequality given in (1.2) which we state as

Theorem 1. *Let D_1 be the linear manifold of $L^2(0,2\pi)$ determined by*

$$D_1 \overset{\text{def}}{=\!=} \{f \in AC[0,2\pi] : f' \in L^2(0,2\pi) \text{ and } \text{(i) } f(0) = f(2\pi) \quad \text{(ii) } \int_0^{2\pi} f = 0\}; \tag{2.3}$$

then the following inequality holds

$$\|f'\| \geq \|f\| \quad \text{i.e.} \quad \int_0^{2\pi} |f'|^2 \geq \int_0^{2\pi} |f|^2 \quad (f \in D_1) \tag{2.4}$$

with equality if and only if for some A and $B \in C$

$$f(x) = A \cos x + B \sin x \quad (x \in [0,2\pi]). \tag{2.5}$$

Proof See [6, Section 7.6] or [2, Chapter 5, Section 11] or the spectral theory method given below.

An extension of this result may be seen in

Theorem 2. Let p *be a given positive integer and let* D_p *be the linear manifold of* $L^2(0,2\pi)$ *determined by*

$$D_p = \{f \in AC[0,2\pi] : f' \in L^2(0,2\pi) \text{ and}$$
$$\text{(i)}\quad f(0) = f(2\pi)$$
$$\text{(ii)}\quad \int_0^{2\pi} f(x)\cos nx\,dx = \int_0^{2\pi} f(x)\sin nx\,dx = 0$$
$$\text{for } n = 0,1,2,\ldots,p-1.\}; \tag{2.6}$$

then the following inequality is valid

$$\|f'\| \geq p\|f\| \quad \text{i.e.} \quad \int_0^{2\pi} |f'|^2 \geq p^2 \int_0^{2\pi} |f|^2 \qquad (f \in D_p) \tag{2.7}$$

with equality if and only if for some A *and* $B \in C$

$$f(x) = A\cos px + B\sin px \qquad (x \in [0,2\pi]). \tag{2.8}$$

Proof. This follows from a straight forward extension of the Fourier series method in [2, Chapter 5, Section 11] but see also the remarks on the corresponding spectral theory method, given below.

It is clear that Theorem 2 reduces to Theorem 1 when $p = 1$.

3. In this and the two subsequent sections we give the spectral theory proof of Theorem 1.

Let the operator $T : D(T) \to L^2(0,2\pi)$ be determined as follows:

$$D(T) \stackrel{\text{def}}{=} \{f \in L^2(0,2\pi) : f' \in AC[0,2\pi] \quad f'' \in L^2(0,2\pi)$$
$$f(0) = f(2\pi) \quad f'(0) = f'(2\pi)\}$$

and

$$T(f) \stackrel{\text{def}}{=} -f'' \qquad (f \in D(T)). \tag{3.1}$$

Then it is known that T is a self-adjoint, unbounded operator in $L^2(0,2\pi)$ with a discrete spectrum

$$\{\lambda_n = n^2 : n = 0,1,2,\ldots\}; \tag{3.2}$$

the first eigenvalue $\lambda_0 = 0$ is simple with eigenfunction $\psi_0(x) = 1 \quad (x \in [0,2\pi])$; for $n \geq 1$ the eigenvalues $\lambda_n = n^2$ are all double with corresponding eigenfunctions

$$\psi_{n,1}(x) = \cos nx \text{ and } \psi_{n,2}(x) = \sin nx \qquad (x \in [0,2\pi]). \tag{3.3}$$

For some details of these results see the book of Titchmarsh [11, Chapter 1 and Section 1.14].

We note that the boundary condition (i) of (2.3), the definition of D_1, is included in the definition of the domain $D(T)$. To introduce the boundary condition (ii) of (2.3) we employ the technique of reduction of the operator T to a 'smaller' operator T_1 (for the essential ideas of reduction of operators, and in particular self-adjoint operators, see the book by Akhiezer and Glazman [1, Section 40, Theorem 1; and Section 44]). Let $L_1^2(0,2\pi)$ be the Hilbert subspace of $L^2(0,2\pi)$ determined by considering in $L^2(0,2\pi)$ the orthogonal complement of $\{\psi_0\}$, the eigenspace generated by the single eigenfunction ψ_0 of T, <u>i.e.</u>

$$L_1^2(0,2\pi) = L^2(0,2\pi) \ominus \{\psi_0\}.$$

(For the notation $\ominus$ see [1, Section 7]). We see that

$$L_1^2(0,2\pi) = \{f \in L^2(0,2\pi) : (f,\psi_0) = 0, \text{ i.e. } \int_0^{2\pi} f\bar{\psi}_0 = \int_0^{2\pi} f = 0\};$$

it is clear that this restriction introduces the boundary condition (ii) of (2.3) as required.

Let T_1 be the reduction of T to $L_1^2(0,2\pi)$. Then

$$D(T_1) = \{f \in D(T) : (f,\psi_0) = 0 \text{ i.e. } \int_0^{2\pi} f = 0\}$$

and

$$T_1 f = -f'' \qquad (f \in D(T_1)).$$

It is known that, see [1, Section 44, Page 93], T_1 is self-adjoint in $L_1^2(0,2\pi)$ with spectrum $\{\lambda_n = n^2 : n = 1,2,3,\ldots\}$ and with corresponding eigenfunctions $\{\psi_{n,1}$ and $\psi_{n,2} : n = 1,2,3,\ldots\}$ as given by (3.3).

Since the spectrum of T_1 is bounded below by the first eigenvalue $\lambda_1 = 1$, the operator T_1 is bounded below by $\lambda_1 I$, where I is the identity operator in $L_1^2(0,2\pi)$. From a known result for such self-adjoint operators, see the book by Kato [8, Section 10, page 278], we obtain the inequality

$$(T_1 f,f) \geq \lambda_1 (f,f) \qquad (f \in D(T_1)) \tag{3.4}$$

with equality if and only if f is in the eigenspace of T_1 at λ_1, i e. for some A and $B \varepsilon C$ it is the case that $f = A\psi_{1,1} + B\psi_{1,2}$.

Now on integration by parts and using the boundary conditions satisfied by $f \varepsilon D(T_1)$ at the end-points 0 and 2π, we find

$$(T_1 f,f) = -\int_0^{2\pi} f''\bar{f} = -[f'\bar{f}]_0^{2\pi} + \int_0^{2\pi} |f'|^2$$

$$= \int_0^{2\pi} |f'|^2 \qquad (f \varepsilon D(T_1)).$$

With this result, the fact that $\lambda_1 = 1$, and the explicit form of $\psi_{1,1}$ and $\psi_{1,2}$ as given by (3.3), we see that (3.4) is equivalent to

$$\int_0^{2\pi} |f'|^2 \geq \int_0^{2\pi} |f|^2 \qquad (f \varepsilon D(T_1)) \tag{3.5}$$

with equality if and only if for some A and $B \varepsilon C$

$$f(x) = A \cos x + B \sin x \qquad (x \varepsilon [0,2\pi]). \tag{3.6}$$

It follows from the definitions of D_1, $D(T)$ and $D(T_1)$ given above that $D(T_1) \subset D_1$ in view of the second derivative existence for elements of $D(T_1)$. We see then that (3.5) and (3.6) give the required inequality of Theorem 1 except for this restriction to $D(T_1)$.

4. To extend the inequality (3.5) from $D(T_1)$ to D_1 as required for the proof of Theorem 1 we now follow the approximation technique introduced in [4, Section 3].

Let f be any element of D_1. For n = 1,2,3,... let $\phi_n \varepsilon C_0^\infty(0,2\pi)$ be chosen so that

$$\int_0^{2\pi} |f' - \phi_n|^2 < \frac{1}{n^2} \qquad (n = 1,2,3,\ldots); \tag{4.1}$$

this is possible since $C_0^\infty(0,2\pi)$ is dense in $L^2(0,2\pi)$. Now define Φ_n on $[0,2\pi]$ by

$$\Phi_n(x) \stackrel{\text{def}}{=} \alpha_n + \beta_n x + \int_0^x \phi_n(t)dt \qquad (x \varepsilon [0,2\pi])$$

with

$$\alpha_n \stackrel{\text{def}}{=} -\frac{1}{2\pi}\left\{\int_0^{2\pi}(2\pi - t)\phi_n(t)dt + \frac{1}{2}\beta_n(2\pi)^2\right\}$$

and

$$\beta_n \overset{\text{def}}{=\!=} -\frac{1}{2\pi}\int_0^{2\pi} \phi_n(t)dt.$$

A direct calculation shows that

$$\Phi_n(0) = \Phi_n(2\pi) = \alpha_n \qquad \Phi_n'(0) = \Phi_n'(2\pi) = \beta_n \tag{4.2}$$

and

$$\Phi_n' \in AC[0,2\pi] \qquad \Phi_n'' = \phi_n' \in L^2(0,2\pi). \tag{4.3}$$

If Ψ_n is defined on $[0,2\pi]$ by

$$\Psi_n(x) = \int_0^x (x-t)\phi_n(t)dt + \tfrac{1}{2}\beta_n x^2 + \alpha_n x \qquad (x \in [0,2\pi])$$

then we may verify that $\Psi_n(0) = \psi_n(2\pi) = 0$ and $\Psi_n' = \Phi_n$. Thus

$$\int_0^{2\pi} \Phi_n(x)dx = \int_0^{2\pi} \Psi_n'(x)dx = [\Psi_n(x)]_0^{2\pi} = 0. \tag{4.4}$$

Hence from (4.2,3 and 4) we see that $\Phi_n \in D(T_1)$ for $n = 1,2,3,\ldots$.

To see that Φ_n approximates to the given $f \in D_1$ we proceed as follows.

$$\int_0^{2\pi} (f' - \phi_n) = f(2\pi) - f(0) - \int_0^{2\pi} \phi_n = -\int_0^{2\pi} \phi_n = 2\pi\,\beta_n$$

i.e. $$|\beta_n| \le \frac{1}{2\pi}\left|\int_0^{2\pi}(f' - \phi_n)\right| \le \frac{1}{2\pi}\left\{2\pi\int_0^{2\pi}|f' - \phi_n|^2\right\}^{1/2} < \frac{1}{n\sqrt{(2\pi)}} \tag{4.5}$$

on using (4.1). Hence

$$\begin{aligned}\left\{\int_0^{2\pi}|f' - \Phi_n'|^2\right\}^{1/2} &= \left\{\int_0^{2\pi}|f' - \phi_n - \beta_n|^2\right\}^{1/2} \\ &\le \left\{\int_0^{2\pi}|f' - \phi_n|^2\right\}^{1/2} + \{2\pi|\beta_n|^2\}^{1/2} \\ &< 1/n + 1/n = 2/n \end{aligned} \tag{4.6}$$

on using (4.1) and (4.5). Also

$$f(x) - \Phi_n(x) = \int_0^x (f' - \Phi_n') + f(0) - \Phi_n(0) \qquad (x \in [0,2\pi]) \tag{4.7}$$

and so

$$0 = \int_0^{2\pi} f - \int_0^{2\pi} \Phi_n = \int_0^{2\pi}\left\{\int_0^x (f' - \Phi_n')\right\}dx + 2\pi\big(f(0) - \Phi_n(0)\big),$$

i.e. $$|f(0) - \Phi_n(0)| \leq \frac{1}{2\pi} \int_0^{2\pi} \left\{ \int_0^x |f' - \Phi_n'| \right\} dx$$

$$\leq \int_0^{2\pi} |f' - \Phi_n'| \leq \left\{ 2\pi \int_0^{2\pi} |f' - \Phi_n'|^2 \right\}^{1/2}$$

$$\leq 2\surd(2\pi)/n \qquad (4.8)$$

Thus from (4.7 and 8) we have for all $x \in [0,2\pi]$

$$|f(x) - \Phi_n(x)| \leq 2\surd(2\pi)/n + 2\surd(2\pi)/n = 4\surd(2\pi)/n$$

and so

$$\left\{ \int_0^{2\pi} |f - \Phi_n|^2 \right\}^{1/2} \leq \{2\pi \,.\, 16(2\pi)/n^2\}^{1/2} = 8\pi/n \qquad (4.9)$$

for $n = 1,2,3,\ldots$.

Returning now to the inequality (3.5) we see that, since $\Phi_n \in D(T_1)$,

$$\|\Phi_n'\| \geq \|\Phi_n\| \qquad (4.10)$$

i.e. $$\|f'\| = \|f' - \Phi_n' + \Phi_n'\|$$

$$\geq \|\Phi_n'\| - \|f' - \Phi_n'\|$$

$$\geq \|\Phi_n\| - \|f' - \Phi_n'\| \qquad \text{(on using (4.10))}$$

$$\geq \|f\| - \|f' - \Phi_n'\| - \|f - \Phi_n\|$$

$$\geq \|f\| - \{2/n + 8\pi/n\} \quad \text{(on using (4.6) and (4.9))}.$$

Since this last result is valid for $n = 1,2,3,\ldots$ and $\|f'\|$ and $\|f\|$ do not depend on n we may let $n \to \infty$ to obtain, for f is any element of D_1,

$$\|f'\| \geq \|f\| \qquad (f \in D_1) \qquad (2.4)$$

which is the required inequality (2.4) of Theorem 1. Note however that although we know the conditions of equality in (3.5) on $D(T_1)$, *i.e.* (3.6), the use of the limit process above does not allow the immediate deducation of all the cases of equality in (2.4). The passage to the limit could introduce additional elements in D_1 but not in $D(T_1)$ which give cases of equality in (2.4). We shall in fact see in the next section that this is not the case. This point in the proof also occurs in the calculus of variations proof of such inequalities; see [6, Section 7.6].

5. To settle the cases of equality in (2.4) of Theorem 1 we proceed as follows.

If f is of the form (2.5) then a calculation shows that equality holds in (2.4).

Conversely let $F \in D_1$ and suppose that $\|F'\| = \|F\|$. For any $f, g \in D_1$ define the functional $J(f,g)$ by, compare with [4, Section 4],

$$J(f,g) = \int_0^{2\pi} \{f'\overline{g}' - f\overline{g}\}.$$

Then for any $\alpha \in C$, any $f \in D_1$ and F given above, on using (2.4) since $F + \alpha f \in D_1$,

$$\begin{aligned} 0 \leq \|F' + \alpha f'\|^2 - \|F + \alpha f\|^2 &= J(F + \alpha f, F + \alpha f) \\ &= J(F,F) + \alpha J(f,F) \\ &\quad + \overline{\alpha} J(F,f) + |\alpha|^2 J(f,f) \end{aligned}$$

which gives, since $\|F'\| = \|F\|$, i.e. $J(F,F) = 0$,

$$0 \leq \alpha J(f,F) + \overline{\alpha} J(F,f) + |\alpha|^2 J(f,f)$$

valid for all $\alpha \in C$. By making suitable choices of α with $|\alpha|$ small enough this last result yields a contradiction unless we assume

$$J(f,F) = 0 \qquad (f \in D_1),$$

i.e. since $D(T_1) \subset D_1$,

$$J(\Phi,F) = 0 \qquad (\Phi \in D(T_1)).$$

From the definition of J this yields

$$\int_0^{2\pi} (\Phi'\overline{F}' - \Phi\overline{F}) = 0 \qquad (\Phi \in D(T_1))$$

and on integration by parts, as in the proof of (3.5), this gives

$$\int_0^{2\pi} (T_1\Phi - \Phi)\overline{F} = 0 \qquad (\Phi \in D(T_1))$$

or

$$(T_1\Phi - \Phi, F) = 0 \qquad (\Phi \in D(T_1)).$$

Thus F is in the orthogonal complement of the range set $\{T_1\Phi - \Phi : \Phi \in D(T_1)\}$ in $L_1^2(0,2\pi)$. Since T_1 is self-adjoint in $L_1^2(0,2\pi)$ and since $\lambda_1 = 1$ is an eigenvalue of T_1 it follows from a property of self-adjoint operators, see [1, Section 43,

Theorem 2*], that F must be in the eigenspace of T_1 for the eigenvalue λ_1. From the explicit form of this eigenspace, see (3.3), we conclude that for some A and B ε C we have

$$F(x) = A \cos x + B \sin x \qquad (x \in [0,2\pi]).$$

This completes the spectral theory proof of Theorem 1.

6. Some remarks are now in order:

(i) compared with the Fourier series proof of [2] or the integral identity proof of [6], both the calculus of variations proof of [6] and the spectral theory proof given above are long and complicated; however the two shorter proofs both hide the reason for the form of the inequality; the calculus of variations proof does not bring out the reason for the form of the inequality but does work effectively in general L^p spaces; the spectral theory proof works only in L^2 spaces but in this case does identify the form of the inequality with certain spectral parameters of an underlying differential operator

(ii) we now see that the Wirtinger inequality (2.4), as compared with the general type of this inequality (2.2), the reason that the number k = 1 follows from the facts that (a) integral boundary condition $\int_0^{2\pi} f = 0$ requires the reduction of the operator T to the operator T_1, (b) that the first eigenvalue of T_1 is $\lambda_1 = 1$ which determines the value of k, and (c) that all the cases of equality arise from the eigenspace of T_1 at λ_1 and <u>vice versa</u>

(iii) all the other Wirtinger type inequalities mentioned in Section 2 above are also associated with the spectral properties of certain differential operators; in this sense the spectral theory approach does unify these inequalities in the case of the L^2 spaces.

7. We omit the spectral theory proof of Theorem 2 but point out that it depends on, given the positive integer p, the reduction of the operator T of Section 3 above to an operator T_p by removing the eigenspaces generated by the eigenvectors ψ_0 and $\psi_{r,s}$ (r = 1,2,..., p - 1; s = 1,2). It follows that T_p is then bounded below by

$p^2(I)$, i.e. $(T_p f,f) \geq p^2(f,f)$ $(f \in D(T_p))$, and the eigenspace at $\lambda_p = p^2$ determines all the cases of equality. The approximation technique of Section 4 above extends to this case but care has to be taken to ensure that all the integral boundary conditions are used in the definition of the approximating function Φ.

The Fourier series proof works well for Theorem 2 but there are complications in applying the proof using integral identities and the proof based on the calculus of variations.

8. The extension of these inequalities to more general differential expressions takes the form

$$\int_a^b \{p|f'|^2 + q|f|^2\} \geq k \int_a^b w|f|^2 \qquad (f \in D) \tag{8.1}$$

where p, q and w are real-valued coefficients on the compact interval [a,b], and D is a linear manifold of the w-weighted space $L_w^2(a,b)$. Here p and w are required to satisfy some form of positivity condition, and the elements of D may have to satisfy certain boundary conditions.

The number k in (8.1), which may be negative if q takes negative values, can then be identified with spectral properties of certain differential operators generated in $L_w^2(a,b)$ by the differential expression $w^{-1}\left(-(pf')' + qf\right)$. Some results in this direction have already been obtained in [4].

References

1. N. I. Akhiezer and I. M. Glazman. Theory of linear operators in Hilbert space: Volume I. (New York: Ungar, 1961).

2. E. F. Beckenbach and R Bellman. Inequalities. (Berlin: Springer-Verlag, 1961).

3. P. R. Beesack. Integral inequalities of the Wirtinger type. Duke Math. J. 25 (1958), 477-498.

4. J. S. Bradley and W N Everitt. Inequalities associated with regular and singular problems in the calculus of variations. Trans. Amer. Math. Soc. 182 (1973), 303-321.

5. W. N. Everitt. Integral inequalities and spectral theory. Lecture Notes in Mathematics, 448 (Berlin: Springer-Verlag, 1975).

6. G. H. Hardy, J. E. Littlewood and G. Pólya. Inequalities. (Cambridge: University Press, 1934).

7. M. Janet. Sur l'inégalité classique du problème isopérimétrique. Univ. Beograd. Publ. Elektrotehn. Fak. Ser. Mat. Fiz. No. 274-301 (1969), 9-10.

8. T. Kato. Perturbation theory for linear operators. (Berlin: Springer-Verlag, 1966).

9. D. S. Mitrinović. Analytic inequalities. (Berlin: Springer-Verlag, 1970).

10. D. S. Mitrinović and P M Vasić. An inequality ascribed to Wirtinger, and its variations and generalizations. Univ. Beograd. Publ. Elektrotehn.Fak. Ser. Mat. Fiz. No. 247-273 (1969), 157-170.

11. E. C. Titchmarsh. Eigenfunction expansions associated with second-order differential equations: Part 1. (Oxford: University Press, 1962).

Nonlinear evolution operators and delay equations

W. E. Fitzgibbon

1. Introduction

Our objective is to investigate existence and stability of solutions to nonlinear Banach space Volterra integral equations with delay. This study is undertaken to provide techniques which can guarantee the existence of mild solutions and stability properties for a class partial functional differential equations. The equation which we consider is:

$$(1.1) \qquad x(\varphi)(t) = W(t,\tau)\varphi(0) - \int_\tau^t W(t,s)F(s,x_s(\varphi))ds \quad t \geq \tau, x_t(\varphi) = \varphi \in C.$$

Our notation follows that of J. Hale [9] and C. Travis and G. Webb [12]. X denotes a Banach space and W(t,s) is a linear evolution operator mapping X to X. $C = C([-r,0],X)$, $r > 0$, is the Banach space of continuous X-valued functions with supremum norm, $||\cdot||_C$. $F(\cdot,\cdot)$ is a continuous function from $R^+ \times C$ to X. If u is a continuous function from an interval $[a-r,b]$ to X then u_t is that element of C having pointwise definition $u_t(\theta) = u(t+\theta)$ for $\theta \in [-r,0]$. If the linear evolution operator W(t,s) is generated by a densely defined linear operator $-A(t)$ then equation (1.1) may be seen to be the integrated form of solutions to the abstract functional differential equation:

$$(1.2) \qquad \dot{x}(\varphi)(t) = -A(t)x(\varphi)(t) - F(t,x_t(\varphi)), \; x_\tau(\varphi) = \varphi \in C$$

In the autononomous case (ie A and F independent of t) our results may be obtained by the nonlinear semigroup approach of Travis and Webb [12]; in [14] Webb treats an autononomous Hilbert space version of (1.2) with A nonlinear.

Recently, J. Dyson and R.V. Bressan [4] allow $A(t)$ to be nonlinear and treat a Hilbert space version of (1.2) by employing the product integral techniques of M. Crandall and T. Liggett [2] and M. Crandall and A. Pazy [3].

2. The nonlinear evolution operator $U(t,s)$

This section is concerned with the existence of solutions to (1.1) Our existence result is used to define a nonlinear evolution operator $U(t,s)$ which describes the action of solutions to (1.1) on initial functions $\varphi \in C$ and yields segments of length r of the solution $x(\varphi)(t)$. It is now convenient to make precise our notion of a linear evolution operator $W(t,s)$.

Definition 2.1 A family of linear operators $\{W(t,\tau) | 0 \le \tau \le t \le \infty\}$ defined on a Banach space X is said to be a linear evolution system of type E on X provided the following are true:

i) $W(t,\tau)$ is jointly continuous in τ and t,
$W(\tau,\tau) = I$, and $W(t,s)W(s,\tau) = W(t,\tau)$
for $0 \le \tau \le s \le t$.

ii) There exists a continuous real valued function
$\alpha(\ )$ defined on R^+ so that,

$$||W(t,\tau)|| \le \exp(\textstyle\int_\tau^t \alpha(s)ds).$$

We have the following result for the existence and uniqueness of solutions to (1.1). Its proof is a straightforward adaptation of Picard-type arguments.

Proposition 1. Let $\{W(t,s) | 0 \le s \le t\}$ be a linear evolution system of type E on X and suppose that $\beta(\cdot):R^+ \to R^+$ and $F(\cdot,\cdot):R^+ \times C \to X$ are continuous and satisfy,

$$||F(t,\varphi) - F(t,\Psi)|| \le \beta(t)||\varphi - \Psi||_C \quad \text{for } \varphi, \Psi \in C.$$

If $\varphi \in C$ and $\tau \ge 0$ then there exists a unique $x(\varphi):[\tau-r,\infty) \to X$ which satisfies (1.1).

If $\alpha(t) < 0$ for $t \in [\tau,T]$ and $\sup_{t\in[\tau,T]} (-\beta(t)/\alpha(t)) < \gamma$ and $x(\varphi)$ and $y(\Psi)$ satisfy (1.1) with initial histories φ and $\Psi \in C$ then straightforward computation yields the inequality:

$$\text{(2.2)} \qquad \exp(\textstyle\int_\tau^t - \alpha(s)ds)||x(\varphi)(t) - y(\Psi)(t)|| \le ||\varphi - \Psi||_C$$

$$+ \int_\tau^t - \gamma\, \alpha(s)\exp(\textstyle\int_\tau^s - \alpha(u)du)||x_s(\varphi) - y_s(\Psi)||_C ds \quad \text{for } t \in [\tau,T].$$

At this point we associate a nonlinear evolution operator with solutions to (1.1). Let $\varphi \in C$ and let $x(\varphi)(t)$ denote the solution having prescribed initial history $x_\tau(\varphi) = \varphi$ we define:

$$U(t,\tau)\varphi = x_t(\varphi) \qquad \text{for } t \ge \tau.$$

Thus for $0 \le \tau \le t$ we see that $U(t,\tau):C \to C$; examination of the integral equation yields the observation that $U(t,s)U(s,\tau)\varphi = U(t,\tau)\varphi$ and that $U(\tau,\tau) = I$ on C. Continuity properties of $U(t,\tau)$ may be deduced directly from the integral equation.

3. Stability properties of $U(t,\tau)$

This section contains our main stability results for $U(t,\tau)$. The following simple application of Gronwall's lemma provides a maximum principle which is crucial to our subsequent development. The details appear in [6].

Lemma 3.1. Suppose that $f(\cdot)$ is a continuous nonnegative function on an interval $[a,b]$. If t_0 denotes the maximum point of $f(\cdot)$ on $[a,b]$ and there exists a continuous nonnegative function $g(\cdot)$ on $[a,b]$ such that

$$(\exp\int_a^{t_o} g(s)ds)f(t_0) \le f(a) + \int_a^{t_o} g(s)(\exp\int_a^s g(u)du)f(s)ds$$

then $f(a) = \sup_{t\in[a,b]} f(t)$.

We have the following result which provides for the uniform stability of $U(t,\tau)$.

Theorem 1. Let $\{W(t,s)|0 \le s \le t < \infty\}$ and $F(\cdot,\cdot)$ satisfy the conditions of Proposition 1 and suppose that for all $t > 0$, $\alpha(t) < 0$ and $\sup_{t>0}(-\beta(t)/\alpha(t)) \le 1$. If $U(t,\tau)$ denotes the nonlinear evolution operator associated with solutions to (1.1) then for $t \ge \tau$ and $\varphi, \Psi \in C$ we have $||U(t,\tau)\varphi - U(t,\tau)\Psi||_C \le ||\varphi - \Psi||_C$.

Indication of proof. One establishes the above result for $|t-\tau| \le r$ and uses the iterative property of the nonlinear evolution operator to extend the inequality for all $t \ge \tau$. If $||U(t,\tau)\varphi - U(t,\tau)\Psi||_C \le ||\varphi-\Psi||_C$ it may be argued that $\sup_{t\in[\tau,\tau+r]} ||x_t(\varphi) - y_t(\Psi)||_C = \sup_{t\in[\tau,\tau+r]} ||x(\varphi)(t) - y(\Psi)(t)||$. This fact may be used in conjunction with (2.2) and Lemma 3.1 to reach a contradiction.

Placing additional requirements on $\alpha(\cdot)$ and $\beta(\cdot)$ leads to the asymptotic stability of $U(t,\tau)$.

Theorem 2. Let $\{W(t,s)|0 \le s \le t\}$ and $F(\cdot,\cdot)$ satisfy the conditions of Theorem 1. Further suppose that:

i) There is a $\lambda > 0$ so that

$$\sum_{k=0}^{\infty} \exp\left(\int_{t_0+k(r+\lambda)}^{t_0+k(r+\lambda)+\lambda} \alpha(s)ds\right) \text{ is convergent}$$

for each $t_0 \in [0,r+\lambda]$.

ii) $\sup(-\beta(t)/\alpha(t)) \le \gamma < 1$.

Then there exists a $K > 0$ so that for $t-\tau > r+\lambda$ we have,

$$||U(t,\tau)\varphi - U(t,\tau)\Psi||_C \le K||\varphi - \Psi||_C \exp((\log \gamma)|t-\tau|/(r+\lambda)). \tag{3.2}$$

We remark that a satisfactory constant K may be computed:

$$K = \frac{1}{\gamma} \lim_{n\to\infty} \prod_{k=0}^{n} \left(1 + \frac{1-\gamma}{\gamma} \exp\left(\int_{\tau+k(r+\lambda)}^{\tau+k(r+\lambda)+\lambda} \alpha(s)ds\right)\right)$$

We remark that condition (i) requires that $\int^t \alpha(s)ds$ approaches $-\infty$ rapidly enough to insure that the infinite product converges, c.f. [4]. This condition is satisfied if $\sup_{0\le t\le\infty} \int_t^{t+\lambda} \alpha(s)ds < 0$ for some $\lambda > 0$.

A point φ_0 is said to be a rest point of a nonlinear evolution system $\{U(t,\tau)|0 \le \tau \le t\}$ if $U(t,\tau)\varphi_0 = \varphi_0$ for all $t \ge \tau \ge 0$. If the conditions of Theorem 2 are satisfied and $\varphi_0 \in C$ is a rest point for the nonlinear evolution operator then (3.2) insures that $\lim_{t\to\infty} U(t,\tau)\varphi = \varphi_0$ for all $\varphi \in C$.

The nonlinear evolution operator will have a rest point if $F(t,0) = 0$ for all $t \geq 0$.

4. Infinite delay.

The foregoing techniques may also be applied to equations with infinite delay. If we let $C_u = C((-\infty,0];X)$ the space of uniformly continuous functions from $(-\infty,0]$ to X equipped with the supremum norm; require that $F(\ ,\ ):R^+ \times C_u \to X$ satisfy the inequality $||F(t,\varphi) - F(t,\Psi)|| \leq \beta(t)||\varphi - \Psi||_{C_u}$, we can solve the integral equation

$$(4.1) \qquad x(\varphi)(t) = W(t,\tau)\varphi(0) - \int_\tau^t W(t,s)F(s,x_s(\varphi))ds, \quad \varphi \in C_u, \quad t \geq \tau.$$

The solutions to (4.1) give rise to a nonlinear evolution operator $U(t,\tau):C_u \to C_u$. If we require that $\sup(-\beta(t)/\alpha(t)) \leq 1$ then arguments of the type employed in Theorem 1 will show that $||U(t,\tau)\varphi - U(t,\tau)\Psi||_{C_u} \leq ||\varphi - \Psi||_{C_u}$; however, since φ and Ψ are initial segments of $U(t,\tau)\varphi$ and $U(t,\tau)\Psi$ we conclude that $||U(t,\tau)\varphi - U(t,\tau)\Psi||_{C_u} = ||\varphi - \Psi||_{C_u}$. Thus it is impossible to obtain an asymptotic stability result similar to Theorem 2 for $U(t,\tau)$ in C_u.

5. Example.

We conclude by briefly outlining an example which will hopefully illustrate the applicability of our material to partial function differential equations. We consider a simple delay equation of the form:

$$(5.1) \qquad \partial u(x,t)/\partial t = a(t)u_{xx}(x,t) + b(t)u(x,t) + f(u(x,t-r));$$
$$u(x,t) = \varphi(x,t), - r \leq t \leq 0.$$

Here $a(\cdot)$, $-b(\cdot)$ are positive and continuous from R^+ to R^+ and $f(\cdot)$ is a Lipschitz continuous scalar function. We set (5.1) in the Banach space of continuous real valued functions which vanish at ∞ and formally define the operator,

$$(L(t))u(x) = a(t)u_{xx}(x) + b(t)u(x).$$

We define a family of operators $\{A(t)|t > 0\}$ so that $A(t)g = L(t)g$ for $g \in C_0^2$. In [8] it is shown that $A(t)$ generates a linear evolution system of type E which solves initial value problems $dW(t,\tau)g/dt = A(t)W(t,\tau)g$ for $g \in C_0^2$, $t \geq 0$.

We define $F(\varphi)(x) = f(\varphi(-r)(x))$ for $\varphi \in C$. Thus we see that equation (1.1) provides an integrated form of (5.1) or in the terminology of F. Browder [1] guarantees mild solutions to (5.1). Stability properties may be obtained by further specification of the relationship between $b(t)$ and the Lipschitz constant of f.

Bibliography

1. F. Browder, "Nonlinear equation of evolution", Ann. Math. 80(1964),485-523.
2. M. Crandall and T. Liggett, "Generation of semigroups of nonlinear transformations on general Banach spaces", Amer. J. Math. 93(1971),265-298.
3. __________ and A. Pazy, "Nonlinear evolution equations in Banach spaces, Israel J. Math. 11(1972) 57-94.
4. J. Dyson and R.V. Bressan, "Functional differential equations and nonlinear evolution operators" Edinburgh J. Math. (to appear).
5. H. Flaschka and M. Leitman, "On semigroups of nonlinear operators and the solution of the functional differential equation $x(t) = F(x_t)$" J. Math. Anal. Appl. 49(1975), 649-658.
6. W.E. Fitzgibbon, "Stability for abstract nonlinear Volterra equations involving finite delay" (to appear).
7. ______________, "Nonlinear Volterra equations with infinite delay." (to appear).
8. J. Goldstein, "Abstract evolution equations", Trans. Amer. Math. Soc. 141 (1969), 159-185.
9. J. Hale, Functional Differential Equations, Appl. Math. Series, Vol. 3, Springer-Verlag, New York 1971.
10. V. Laksmikantham and S. Leola, Differential and Integral Inequalities, Vol. II, Academic Press, New York, 1972.
11. T. Saaty, Modern Nonlinear Equations, McGraw Hill, New York, 1967.
12. C. Travis and G. Webb, "Existence and stability for partial functional differential equations", Trans. Amer. Math. Soc. (to appear).
13. G. Webb, "Autononomous nonlinear functional differential equations", J. Math. Anal. Appl. 46(1974), 1-12.
14. ______, Asymptotic stability for abstract nonlinear functional differential equations", Proc. Amer. Math. Soc.
15. ______, Functional differential equations and nonlinear semigroups in L^P spaces, J. Diff. Equations (to appear).

A Singular Functional Differential Equation Arising in an Immunological Model

by

Juan A. Gatica

and

Paul Waltman[+]

1. Introduction.

The task of developing a mathematical model of the antigen stimulated antibody response of the immune system has recently attracted considerable attention. See, for example, [1], [2], [3], [4], [6], [8]. In [4] the following system of equations was proposed as a model:

$$x'(t) = -rx(t)y(t) - sx(t)z(t), \quad x(0) = x_0$$

$$y'(t) = -rx(t)y(t) + \alpha rx(\tau_1(t))y(\tau_1(t))H(t-t_1), \quad y(0) = y_0$$

$$z'(t) = -sx(t)z(t) - \gamma z(t) + \beta rx(\tau_2(t))y(\tau_2(t))H(t-t_2), \quad z(0) = 0$$

$$w'(t) = rx(t)y(t), \quad w(0) = 0 \tag{1.1}$$

$$\int_{\tau_1(t)}^{t} f_1(x(s),y(s),w(s))ds = m_1, \quad t \geq t_1$$

$$\tau_1(t) = 0, \quad t \leq t_1$$

$$\int_{\tau_2(t)}^{t} f_2(y(s),w(s))ds = m_2, \quad t \geq t_2$$

$$\tau_2(t) = 0, \quad t \leq t_2,$$

[+]Research supported by Public Health Service grant IR01CA18639-01 from the National Cancer Institute and National Science Foundation grant BMS74 18648.

where t_1 and t_2 are given by

$$\int_0^{t_1} f_1(x(s),y(s),w(s))ds = m_1$$

$$\int_0^{t_2} f_2(y(s),w(s))ds = m_2$$

or by $t_1 = +\infty$, $t_2 = +\infty$, if no such t_i are defined by the above integrals.

The relevant variables are:

$x(t)$, concentration of free antigen molecules at time t,

$y(t)$, concentration of free receptor molecules at time t,

$z(t)$, concentration of free antibody molecules at time t,

$w(t)$, concentration of antigen bound to receptors.

$r, s, \alpha, \beta, \gamma, m_1, m_2$ are constants of the system and H is the usual Heavyside function. τ_1 and τ_2 provide delays derived from threshold considerations. The biological background and a detailed derivation can be found in [4]. Note that the integral equations involving τ_1 and τ_2 can be differentiated to provide a system of functional differential equations. The choice of the functions f_1 and f_2 correspond to various biological hypotheses concerning the triggering of cells towards differentiation and triggering antibody secretion. It seems reasonable to take f_2 to be a function of the total receptor population, i.e., to take $f_2(y,w) = f_2(y+w)$. With this assumption the following theorem was proved in [4].

<u>Theorem</u> 1.1. <u>Let</u> $r, s, \alpha, \beta, \gamma, m_1, m_2$ <u>be nonnegative constants</u>. <u>Let</u> $R^+ = [0,\infty)$, $f_1 : R^+ \times R^+ \times R^+ \to R^+$ <u>be continuous, locally Lipschitzian and</u> $f_1(x,y,w) > 0$ <u>if</u> $x > 0$, $y > 0$. <u>Let</u> $f_2 : R^+ \to R^+$ <u>be continuous</u>,

locally Lipschitzian, and $f_2(\xi) > 0$ if $\xi > 0$. Then there exists a solution of (1.1) which depends continuously on the initial conditions and parameters.

2. Statement of Results.

The sign condition on f_1 in the theorem of the preceding section rules out the possibility that $f_1(x,y,w) = w$. This in turn corresponds to the very attractive biological hypothesis that triggering towards differentiation by a cell is initiated when a requisite number of surface receptors become bound to antigen. A technical difficulty arose in the proof of the theorem in that, since $w(0) = 0$, a system of singular functional differential equations was encountered—singular in the sense that a function multiplying the derivative was zero at the initial point. The sample computation of [4] used $f_1 = xy + w$ which avoids this difficulty since x and y are always positive. The purpose of this paper is to treat the system (1.1) with $f_1 = w$ and show that there is a unique solution to this problem. Differentiating the integral involving τ_1 and using $f_1 = w$ produces the singular problem mentioned in the title. The principal result is the following:

Theorem 2.1. Let $f_1(x,y,w) = w$. Let $r, s, \alpha, \beta, m_1, m_2, \gamma$ be nonnegative constants. Let $f_2 : R^+ \to R^+$ be continuous and locally Lipschitzian, with $f_2(\xi) > 0$ if $\xi > 0$. Then there exists a unique solution of (1.1) which depends continuously on the initial conditions and parameters.

3. Proof of the Theorem.

First there exists a solution of the initial value problem of the system of ordinary differential equations

(3.1)
$$\begin{aligned} x' &= -rxy \\ y' &= -rxy \\ w' &= rxy \end{aligned}$$

$$x(0) = x_0, \quad y(0) = y_0, \quad w(0) = 0,$$

valid on $[0,\infty)$. Clearly, $w(t)$ is strictly increasing. If

$$\int_0^\infty w(s)ds \le m_1,$$

then the solutions of (3.1) and $\tau_1(t) = \tau_2(t) = z(t) \equiv 0$ form a solution of the system (1.1).

Suppose there exists a point t_0 such that

$$\int_0^{t_0} w(s)ds = m_1.$$

We seek then to find a solution of the system of functional differential equations

(3.2)
$$\begin{aligned} x'(t) &= -rx(t)y(t) \\ y'(t) &= -rx(t)y(t) + \alpha rx(\tau(t))y(\tau(t)) \\ w'(t) &= rx(t)y(t) \end{aligned}$$

$$w(\tau)\tau'(t) = w(t)$$

with initial conditions

$$x(t), y(t), w(t) \text{ solutions of (3.1)}, \quad t \in [0, t_0],$$

$$\tau(t) \equiv 0, \quad t \in [0, t_0],$$

valid on an interval $[0, t_0+h]$, $h > 0$. This system is singular because $\tau(t_0) = 0$ and $w(0) = 0$. The techniques of Driver [5] cannot be applied as was done in [4] and a special argument will be given. Note that if a solution can be found on $[0, t_0+h]$ for any $h > 0$ then since $w(t)$ is strictly monotone, the arguments given in [4] will apply to continue the solution until the second threshold is reached. Thus we need to be concerned only with the technical problem of taking the "first step."

The basic idea is to define a set of mappings such that a fixed point of the composite will yield a solution to our problem, as was done for example in [7]. Symbolically, the map may be represented

$$w \longrightarrow \tau \longrightarrow \binom{x}{y} \longrightarrow w.$$

Define

$$\mathcal{M} = \{\phi \mid \phi \in C[0, t_0+h], \varphi(t) = w(t),\ t \in [0, t_0],\ 0 \le \varphi(t) \le x_0,$$
$$\varphi(t) \text{ nondecreasing}\}$$

where $C[a,b]$ denotes the continuous functions on $[a,b]$ with uniform norm (abbreviated hereafter by C when the domain is clear) and a priori take $h < m_1/x_0$. (Further restrictions on the size of h will be encountered as the proof proceeds.)

Define $U : \mathcal{M} \longrightarrow C$ by

$$U(\varphi)(t) = \begin{cases} 0, & t \in [0,t_0], \\ \text{the unique number } \eta \text{ such that} \\ \int_\eta^t \varphi(s)ds = m_1, & t > t_0. \end{cases}$$

(Note that η exists since $\varphi(t) = w(t)$ on $0 \le t \le t_0$.) Let $\varphi_1, \varphi_2 \in \mathfrak{M}$. Then for each $t \in [t_0, t_0+h]$,

$$\int_{U\varphi_1(t)}^t \varphi_1(s)ds = \int_{U\varphi_2(t)}^t \varphi_2(s)ds$$

or

$$(3.3) \qquad \int_{t_0}^t (\varphi_1(s)-\varphi_2(s))ds = \int_{U\varphi_2(t)}^{U\varphi_1(t)} w(s)ds$$

since if $h < \frac{m_1}{x_0}$, $(U\varphi)(t) < t_0$ for every $\varphi \in \mathfrak{M}$ and every $t \in [t_0, t_0+h]$. Note that

$$w'(t) > 0$$

and

$$w''(t) = r(x'(t)y(t) + x(t)y'(t)) < 0, \quad t \in (0,t_0),$$

so $w'(t)$ is monotone decreasing. Further, from Taylor's theorem,

$$w(t) = w(0) + w'(ct)t \ge w'(t_0)t, \qquad 0 < c < 1.$$

Using these facts in (3.3) yields

$$\left|\int_{t_0}^t (\varphi_1(s) - \varphi_2(s))ds\right| \ge \frac{w'(t_0)}{2}\left|(U\varphi_1(t))^2 - (U\varphi_2(t))^2\right|$$

or

$$(3.4) \qquad (t-t_0)\|\varphi_1 - \varphi_2\| \geq \frac{w'(t_0)}{2}|U\varphi_1(t) - U\varphi_2(t)|\,|U\varphi_1(t) + U\varphi_2(t)|.$$

If $t > t_0$, $U\varphi$ is differentiable at t, and

$$(U\varphi)'(t) = \frac{\varphi(t)}{w(U\varphi(t))} \geq 1$$

since $U\varphi(t) < t$ and w is increasing and nonnegative. Hence

$$U\varphi(t) \geq U\varphi(t_0) + t - t_0$$

for $\varphi \in \mathcal{M}$ or since $U\varphi(t_0) = 0$, for every φ in $\mathcal{M}$,

$$U\varphi_1(t) + U\varphi_2(t) \geq 2(t-t_0).$$

Thus, from 3.4, it follows that

$$(3.5) \qquad \|U\varphi_1 - U\varphi_2\| \leq L\|\varphi_1 - \varphi_2\|,$$

where $L = 1/w'(t_0)$.

Now on the range of U, denoted $\mathcal{R}U$, define a mapping $V : \mathcal{R}U \to C \times C$ by $V(\varphi) = (\psi_1, \psi_2)$ where in $C \times C$ we use $\max(\|\psi_1\|, \|\psi_2\|)$ as norm and where (ψ_1, ψ_2) is given by $(\psi_1(t), \psi_2(t)) = (x(t), y(t))$, $t \in [0, t_0]$, and on $[t_0, t_0+h]$ by the unique solution of

$$(3.6) \qquad \begin{aligned} u' &= -ruv \\ v' &= -ruv + \alpha ru(\varphi(t))v(\varphi(t)) \\ u(t_0) &= x(t_0), \\ v(t_0) &= y(t_0). \end{aligned}$$

Since φ is given, (3.6) is an ordinary differential equation. We have immediately the apriori bounds

$$0 \le u(t) \le x_0$$

$$0 \le v(t) \le \alpha r x_0 y_0 (t_0 + h) + y_0 .$$

Suppose now that φ_1, φ_2 are in the domain of V and that the images are $(u_1, v_1), (u_2, v_2)$, respectively. We seek to show that V also is Lipschitzian. First of all,

$$u_1(t) - u_2(t) = -r\int_{t_0}^{t} (u_1(s)v_1(s) - u_2(s)v_2(s))ds$$

$$= -r\int_{t_0}^{t} \left[u_1(s)(v_1(s) - v_2(s)) + v_2(s)(u_1(s) - u_2(s))\right]ds.$$

Taking absolute value of both sides and using the above bounds yields

$$|u_1(t) - u_2(t)| \le r x_0 h \|v_2 - v_1\| + (\alpha r x_0 y_0 (t_0+h) + y_0) r h \|u_1 - u_2\|.$$

Let $\delta(h) = \alpha r x_0 y_0 (t_0+h) + y_0$ and note that $h\delta(h)$ is strictly increasing and zero for $h = 0$. Choose h so small that $rh\delta(h) < 1$. Then,

$$\|u_1 - u_2\| \le \frac{r x_0 h}{1-\delta(h)hr} \|v_1 - v_2\|. \tag{3.7}$$

Similarly,

$$|v_1(t) - v_2(t)| \le r\|u_1 - u_2\|h + x_0 r\|v_1 - v_2\|h$$

$$+ \alpha r\int_{t_0}^{t} |x(\varphi_1(s)y(\varphi_1(s)) - x(\varphi_2(s)y(\varphi_2(s))|ds.$$

Using the above bound on $\|u_1 - u_2\|$, the mean value theorem, and consolidating terms this may be written

$$\|v_1 - v_2\| \le h\rho\|v_1 - v_2\| + \alpha r^2 x_0 y_0 (x_0 + y_0)\|\varphi_1 - \varphi_2\|$$

where

$$\rho(h) = \left(\frac{r\delta x_0 h}{1-\delta h} + x_0\right) r.$$

If h is further chosen so small that $h\rho(h) < 1$ (this is possible since $\rho \to x_0 r$ as $h \to 0$), then

$$\|v_1 - v_2\| \le \frac{\alpha r^2 x_0 y_0 (x_0 + y_0) h}{1 - h\rho} \|\varphi_1 - \varphi_2\|. \tag{3.8}$$

Using the maximum of each component norm (3.7) and (3.8) shows that there exists an M such that

$$\|V\varphi_1 - V\varphi_2\| \le Mh\|\varphi_1 - \varphi_2\|. \tag{3.9}$$

On the range of V define $W : \mathcal{R}V \to C$ by

$$W(\psi_1, \psi_2)(t) = x_0 - \psi_1(t).$$

Clearly W is Lipschitzian with Lipschitz constant one.

Define $T : \mathcal{M} \to C$ by

$$T\varphi = W(V(U\varphi)).$$

(3.5) and (3.9) imply that

$$\|T\varphi_1 - T\varphi_2\| \le \|V(U\varphi_1) - V(U\varphi_2)\|$$
$$\le Mh\|U\varphi_1 - U\varphi_2\|$$
$$\le LMh\|\varphi_1 - \varphi_2\|.$$

Thus for h so small that $LMh = \alpha < 1$, T is a contraction. It remains to show that $T : \mathfrak{M} \longrightarrow \mathfrak{M}$.

First of all, it is clear that T preserves continuous functions. Moreover, the mapping V always has, as the first component of a point in its range, a monotone decreasing function. It is monotone decreasing on $0 \le t \le t_0$ because it agrees with $x(t)$ there and on $t_0 \le t \le t_0 + h$ because it solves the first differential equation of (3.6). This insures that a point in the image of W is monotone increasing and will provide the necessary bounds to guarantee that $T : \mathfrak{M} \longrightarrow \mathfrak{M}$. Clearly from what has been said the first component of a point in the range of V satisfies $0 < (V\varphi) \le x_0$. Hence for an arbitrary point in the domain of W one has $0 \le (W\varphi)(t) < x_0$ and thus that $T : \mathfrak{M} \longrightarrow \mathfrak{M}$.

By the contraction mapping theorem, T has a unique fixed point in M. We denote this fixed point by w. Let $\tau = Uw$ and $(x,y) = T(\tau)$. The quadruple (x,y,w,τ) satisfies the system (3.2). Further, it is unique since the fixed point is unique and the mappings U and V single valued.

The problem on $[t_0 + h, \infty)$ is no longer singular and the proof of Theorem 2.1 may proceed as in the proof given in [4].

Finally, we note that the choice $f_1(x,y,w) = w$ is more restrictive than the proof requires. If $f_1(x,y,w) = g(w)$ where $g(0) = 0$, $g'(w) > 0$, $g''(w) \le 0$, all of the proofs will go through.

REFERENCES

[1] Bell, G. I. "Mathematical Model of Clonal Selection and Antibody Production," J. Theor. Biol. 29(1970), 191-232.

[2] ________. "Mathematical Model of Clonal Selection and Antibody Production. II," J. Theor. Biol. 33(1971), 339-378.

[3] ________. "Mathematical Model of Clonal Selection and Antibody Production. III," J. Theor. Biol. 33(1971), 379-

[4] Waltman, P. and Butz, E. "A Threshold Model of Antigen-Antibody Dynamics,"

[5] Driver, R. "Existence Theory for a Delay-Differential System," Contributions to Differential Equations 1(1963), 317-336.

[6] Hoffman, G. W. "A Theory of Regulation and Self-Nonself Discrimination in an Immune Network," European J. of Immunology (1975), In Press.

[7] Hoppensteadt, F. and Waltman, P. "A Problem in the Theory of Epidemics II," Math. Biosci. 12(1971), 133-145.

[8] Richter, P. H. "A Network Theory of the Immune System," European J. of Immunology 5(1975), 350-354.

On linear partial integro differential equations with a small parameter

Hans Grabmüller

Let Ω be the open rectangle $(0,1)\times(0,+\infty)$. Consider for each $\varepsilon > 0$ the initial boundary value problem

$$
(P_\varepsilon)\begin{cases} M_\varepsilon[u] \equiv (k+\frac{\partial}{\partial t})u + \varepsilon\{c\frac{\partial^2}{\partial x^2}u + \int_0^\infty [k_0(t-s)\frac{\partial^2}{\partial x^2} + k_1(t-s)(k+\frac{\partial}{\partial s})]u(x,s)ds\} = \varepsilon r(x,t) \quad \text{in } \Omega, \\ u(j,t) = f_j(t), \quad t \geqq 0, \quad j = 0,1, \\ u(x,0) = h(x), \quad 0 \leqq x \leqq 1, \end{cases}
$$

where $k > 0$ and $c \neq 0$ are given constants and where the kernels k_j $(j=0,1)$ are assumed to have certain properties explained in the sequel.

The purpose of this paper is to give an expansion for the solution of (P_ε) which is valid as $\varepsilon \to 0+$. Problems of this kind have been the subject of studies by many authors. We only mention the pioneering paper of M.I.Vishik and L.A.Lyusternik [8] concerning linear ODE, and the systematic treatment [1] of W.Eckhaus. The more related problems for linear integro-ODE and for parabolic PDE have been studied by N.P. Vekua [7] resp. by F.Hoppensteadt [4].

However, the study of singular perturbation problems for partial integro differential equations on a half line, involving kernels of convolution type, seems to be a new approach. It should be pointed out that our method of proving the uniform validity of the resulting expansion is based on the method of successive approximation. Here the usual way of using a maximum principle does not apply. Our approach may give some insight to the treatment of more general problems.

We confine ourselves to give only a zeroth order approximation for the solution of (P_ε). Our results contain the proof of existence and uniqueness for such a solution. We are also able to treat the more general case of an arbitrary order approximation. These results shall be published in a forthcoming paper.

The main result of this paper is

THEOREM 1: *Let the conditions (H1) - (H3) of Section 3 be satisfied. Then for each small* $\varepsilon > 0$ *the problem* (P_ε) *has a unique solution* u *which can be represented by*

(1) $u(x,t;\varepsilon) = U(x,t) + V^o(x,t;\varepsilon) + V^1(x,t;\varepsilon) + Z(x,t;\varepsilon).$

Here U *is the solution of the regular problem*

(P_o) $(k+\frac{\partial}{\partial t})U = 0, \qquad U(x,0) = h(x), \quad 0 \leq x \leq 1.$

The functions V^j (j=0,1) *are boundary layer terms having the properties that, for any fixed* $\eta \in (o,1)$,

$$\max_{x\in[\eta,1]} \|V^o(x,\cdot;\varepsilon)\|_2 \to 0 \quad \text{and} \quad \max_{x\in[o,1-\eta]} \|V^1(x,\cdot;\varepsilon)\|_2 \to 0 \quad \text{as } \varepsilon \to 0+,$$

with an exponential rate of decay. Moreover, Z *is small of order* ε *in the sense that*

(2) $\max_{x\in[o,1]} \|Z(x,\cdot;\varepsilon)\|_2 = O(\varepsilon)$ *as* $\varepsilon \to 0+$.

The proof of this theorem will be given in Section 4. Section 1 contains some preliminary results concerning integral equations of convolution type on a half line. In Section 2 the formal expansion (1) of the solution u is derived, and Section 3 is devoted to the determination of U, V^o and V^1.

Problems of the type (P_ε) are of considerable interest in the theory of generalized heat conduction in material with fading memory (cf. e.g. [6], [3]). In that context the parameter ε is related to the coefficient of heat capacity, and the limiting case $\varepsilon \to 0+$ represents a conductor with an infinite heat capacity.

This work was motivated by a lecture of Professor G.C.Hsiao on singular perturbation methods, and I am indebted to him for many helpful discussions.

1. *Preliminaries.* Let $\mathbb{R}_+$ $(\overline{\mathbb{R}_+})$ be the semi infinite interval $(0,+\infty)$ (resp. $[0,+\infty)$), and let us denote by E the space $L^2(\mathbb{R}_+)$ equipped with the usual $\|\cdot\|_2$-norm. We shall use the notations $D_t = \frac{\partial}{\partial t}$ and $D_x = \frac{\partial}{\partial x}$. Let the Sobolev spaces E^m, $m \geq 1$, consist of those functions $v \in E$ with absolutely continuous derivatives $D_t^j v \in E$ of order $j \leq m-1$, for which $D_t^m v \in E$. Then $E_o^m = E^m \cap \{v \in E \mid D_t^j v(0)=0, \ j=0,1,\dots,m-1\}$ is a linear subspace of E^m.

Let K_o, K_1 and G be the linear integral operators

$$(K_j v)(t) = {}_0\int^\infty k_j(t-s)v(s)ds \quad \text{and} \quad (Gv)(t) = {}_0\int^t e^{-k(t-s)}v(s)ds, \quad k > 0,$$

with kernels $k_j \in L^1(\mathbb{R})$. Let us denote by $B(E)$ the Banach algebra of bounded linear operators from E to E. It is well known that K_j, $G \in B(E)$ (cf. e.g. [5]). Moreover, the operator G maps bijectively from E onto E_o^1, and the inverse map $G^{-1} = kI+D_t$ is defined as a closed operator on the domain of definition, E_o^1 (see [2]).

Consider for a fixed complex number c the integral operator $K_c = cI+K_o$. According to [5], this operator maps bijectively from E onto E, if and only if the winding number, $w(K_c) = -\frac{1}{2\pi}\int_{\mathbb{R}} d_\xi \arg(c+\hat{k}_o(\xi))$, equals zero. Here $\hat{k}_o(\xi)$ denotes the Fourier transform of k_o. The inverse operator K_c^{-1} can be represented in the form $c^{-1}I+\Gamma \in B(E)$, where Γ is generated by a kernel $\gamma \in L^1(\mathbb{R})$.

Suppose $K_c^{-1} \in B(E)$ exists with $\gamma \in L^1(\mathbb{R}) \cap E$. This is the case, if $|c| > \|K_o\|$ and $k_o \in L^1(\mathbb{R}) \cap E$. Let A be the closed operator $K_c^{-1}G^{-1}: E_o^1 \to E$, and denote by $R(\lambda;A)$ the resolvent $(\lambda I-A)^{-1}$, λ being an element of the resolvent set $\rho(A)$. From [5] it follows that $\lambda \in \rho(A)$, if and only if

$$(3) \qquad w(\lambda I-A) = -\frac{1}{2\pi}\int_{\mathbb{R}} d_\xi \arg\left[\lambda \frac{c+\hat{k}_o(\xi)}{i\xi-k} + 1\right] = 0.$$

This condition implies the validity of

$$(4) \qquad \lambda \frac{c+\hat{k}_o(\xi)}{i\xi-k} + 1 \neq 0 \quad \text{for any } \xi \in \mathbb{R}.$$

Furthermore, suppose the resolvent set $\rho(A)$ contains a sector $S_\theta = \{\lambda \mid \lambda = re^{i\phi},\ 0 \leq r,\ -\theta \leq \phi \leq +\theta\}$ of width 2θ, $0 < \theta < \pi$. Consider for instance the case where $k_o = 0$, and where $c < 0$. Then $\rho(A)$ contains the open set $\operatorname{Re}\lambda > -k/|c|$. From a simple computation we can derive

$$(5) \qquad A^{-1}(\lambda A^{-1}-I)^{-1} = R(\lambda;A) = (\lambda A^{-1}-I)^{-1}A^{-1}, \quad \lambda \in \rho(A).$$

Since $A^{-1} = GK_c$ is bounded, we deduce from the Neumann series that the open disk $D = \{\lambda \mid |\lambda| < \|A^{-1}\|^{-1}\}$ is contained in $\rho(A)$ too.

Next, we define an oriented path L in the complex plane by

$$(6) \qquad L = \Gamma_\rho \cup \Gamma_+ \cup \Gamma_-, \quad \Gamma_\rho = \{\lambda \mid \lambda = \rho e^{i\phi},\ \theta \leq \phi \leq 2\pi-\theta\}, \quad \Gamma_\pm = \{\lambda \mid \lambda = re^{\pm i\theta},\ \rho \leq r < \infty\},$$

where ρ is any fixed number in the open set $0 < \rho < \|A^{-1}\|^{-1}$. L runs in the positive direction from the lower half plane to the upper half plane. Clearly, $L \subset \rho(A)$ by the assumptions made on $\rho(A)$.

An important condition for our work is the existence of a constant $C_o > 0$, such that

(7) $$||R'(\lambda;A)|| \leqq C_o \quad \text{uniformly for } \lambda \in S_\theta .$$

Here $R'(\lambda;A)$ denotes the reduced resolvent $(\lambda A^{-1}-I)^{-1}$. The condition $||A^{-1}||^{-1} < 1$, for example, is sufficient to ensure such a boundedness property, as can be seen by applying the Uniform-Boundedness-Principle.

2. The formal derivation of the asymptotic expansion. Assume for a moment that the problem (P_ε) has a unique solution, $u = u(x,t;\varepsilon)$. An asymptotic expression for u is obtained formally by using the Taylor expansion about $\varepsilon = 0$. If $U = u(x,t;0)$ denotes the zeroth order term of this expansion, then U satisfies the reduced equation $(k+D_t)U = 0$ in Ω. On the other hand, if U is the solution of the initial value problem (P_o), then U represents the solution u of (P_ε) asymptotically as $\varepsilon \to 0+$ for x away from zero and from 1, i.e. for x outside some "boundary layer".

At this stage of approximation it is important to introduce boundary correction terms $v^j(\tilde{x}_j,t)$ $(j=0,1)$ in order to take care for the given boundary data. By following the "matching principle" of W.Eckhaus [1], we introduce new (stretching or rapidly varying) variables $\tilde{x}_o = x/\sqrt{\varepsilon}$, and $\tilde{x}_1 = (1-x)/\sqrt{\varepsilon}$. We wish to find an expression for the solution u in the form (1), with $v^j(x,t;\varepsilon) = v^j(\tilde{x}_j,t)$ $(j=0,1)$. Substituting this in (P_ε), we obtain

$$(Q_j)\quad \begin{cases} (k + D_t + K_c D^2_{\tilde{x}_j})v^j(\tilde{x}_j,t) = 0, & \tilde{x}_j > 0, \quad t > 0, \\ v^j(0,t) = f_j(t) - U(j,t), & t \geqq 0, \\ v^j(\tilde{x}_j,0) = 0, & \tilde{x}_j \geqq 0, \\ \lim\limits_{\tilde{x}_j \to +\infty} v^j(\tilde{x}_j,t) = 0, & \end{cases}$$

where the last condition is the so called matching condition. From this condition it follows that the influence of the boundary correction terms $v^j(\tilde{x}_j,t)$ on the opposite boundary is kept as small as possible.

Assume that the problems (Q_j) can be solved. Then we can form a zeroth order approximation $u_o = U(x,t) + v^o(\tilde{x}_o,t) + v^1(\tilde{x}_1,t)$ of the solution u. In Section 4 we shall prove that a unique solution $Z = Z(x,t;\varepsilon) = u - u_o$ of the remaining problem (Z) exists having the property (2). Here (Z) is defined to be the problem

$$(Z)\quad \begin{cases} M_\varepsilon[Z] = -M_\varepsilon[u_o] + \varepsilon r(x,t) & \text{in } \Omega, \\ Z(0,t;\varepsilon) = -V^1(1/\sqrt{\varepsilon},t), \quad Z(1,t;\varepsilon) = -V^o(1/\sqrt{\varepsilon},t), & t \geqq 0, \\ Z(x,0;\varepsilon) = 0, & 0 \leqq x \leqq 1. \end{cases}$$

3. Evaluation of the approximate solution u_o. We assume that M_ε, f_j and h have the following properties denoted by H1, H2, H3:

(H1) K_c, $R'(\lambda;A)$ and L have the properties listed in Section 1.

(H2) The compatibility conditions $f_j(0) = h(j)$ $(j=0,1)$ hold for the given data $f_j \in E^1$ and $h \in C[0,1]$.

(H3) $f_j \in E^2$, $h \in C^2[0,1]$, $r \in C([0,1];E^1)$, and $k_o, k_1 \in L^1(\mathbb{R}) \cap E$.

Conditions (H1) and (H2) ensure the unique solvability of (P_o) and (Q_j), while the condition (H3) involve certain regularity properties needed for the given data in Section 4.

Clearly, problem (P_o) is solved uniquely by

(8) $\quad U(x,t) = e^{-kt}h(x), \qquad U \in C([0,1];E^m),$ m being an arbitrary natural number.

Concerning the problems (Q_j), we have the following result on existence and uniqueness.

THEOREM 2: Under the hypotheses (H1) and (H2), the problems (Q_j) admit, for both $j = 0$ and $j = 1$, a unique solution $V^j: \overline{\mathbb{R}}_+ \to E^1_o$ with the regularity properties $V^j \in C^\infty(\mathbb{R}_+;E) \cap C_o(\overline{\mathbb{R}_+};E)$. Here $C_o(\overline{\mathbb{R}_+};E)$ denotes the class of continuous functions $w: \overline{\mathbb{R}}_+ \to E$ having the limiting property $\lim_{y\to+\infty} \|w(y)\|_2 = 0$. The solutions V^j can be represented by a Dunford-Taylor integral

$$(9)\qquad V^j(y,t) \equiv V_o(y)\left[f_j - U(j,\cdot)\right](t) = \frac{1}{2\pi i} \int_L e^{y\sqrt{-\lambda}} R'(\lambda;A)\left[f_j - U(j,\cdot)\right](t)\frac{d\lambda}{\lambda},$$

where the square root of the complex number $-\lambda = -re^{i\phi}$ is defined by $\sqrt{-\lambda} = i\sqrt{r}e^{\frac{i\phi}{2}}$, $0 \leqq r < \infty$, $0 \leqq \phi < 2\pi$.

The proof of Theorem 2 can be derived from various properties of the operator family $V_o(y)$, $y \geqq 0$, defined by (9). We shall summarize these properties first together with some properties needed later on for the following operator families

$$(10) \qquad V_1(y) = \frac{1}{2\pi i} \int_L e^{y\sqrt{-\lambda}} R'(\lambda;A)\frac{d\lambda}{\lambda\sqrt{-\lambda}} \quad \text{and} \quad V_2(y) = \frac{1}{2\pi i} \int_L \frac{e^{y\sqrt{-\lambda}}}{e^{2\delta\sqrt{-\lambda}}-1} R'(\lambda;A)\frac{d\lambda}{\lambda\sqrt{-\lambda}},$$

where $\delta > 0$ is an arbitrary constant.

LEMMA: Let the conditions (H1) be satisfied. Then, if $\delta_o > 0$ is an arbitrarily fixed number, V_o, V_1, and V_2 have the following properties

(i) $V_p \in C^\infty(\mathbb{R}_+;B(E)) \cap L^1(\mathbb{R}_+;B(E))$, $p = 0,1,2$. There are positive constants C_1 and $C_2(\delta_o)$ such that $\|V_1(y)\| \leqq C_1$ and $\|V_2(y)\| \leqq C_2$ uniformly for $y \geqq 0$ and for $\delta \geqq \delta_o > 0$.

(ii) $V_p \in C_o(\overline{\mathbb{R}}_+;B(E))$, $p = 1,2$, with $C_o(\overline{\mathbb{R}}_+;B(E))$ being defined as $C_o(\overline{\mathbb{R}}_+;E)$.

(iii) $\lim_{y\to o+} \|V_o(y)v - v\|_2 = 0 = \lim_{y\to+\infty} \|V_o(y)v\|_2$ for any $v \in E^1_o$.

(iv) $V_p(y) + GK_cD^2_yV_p(y) = 0$, $p = 0,1,2$, for any $y > 0$.

(v) $V_o(y) = D_yV_1(y)$ for any $y > 0$, $V_1(y) = V_2(2\delta - y) - V_2(y)$ for any $y \geqq 0$ and any $\delta > 0$.

Proof: The properties (i) and (ii) follow immediately from the definitions of V_p. Thanks to the uniform boundedness of $\|R'(\lambda;A)\|$, and to the fact that

$$|e^{y\sqrt{-\lambda}}| \leqq e^{-y\sqrt{r}\sin\frac{\theta}{2}} \quad \text{for } \lambda = re^{\pm i\theta} \in \Gamma_+ \cup \Gamma_-,$$

all the appearing improper integrals exist. Next, from (i) the properties (iv) and (v) follow by a straigthforward computation. The proof of (iii) is more involved. Recall that E^1_o is the domain of definition of A. Thus, $R'(\lambda;A)v = \lambda^{-1}Av + \lambda^{-1}R'(\lambda;A)Av$, for any $v \in E^1_o$ and any $0 \neq \lambda \in \rho(A)$. Hence, the integral $V_o(y)v$ can be decomposed into two parts

$$V_o(y)v = \frac{1}{2\pi i} \int_L e^{y\sqrt{-\lambda}} \frac{d\lambda}{\lambda^2} Av + \frac{1}{2\pi i} \int_L e^{y\sqrt{-\lambda}} R'(\lambda;A)Av \frac{d\lambda}{\lambda^2}.$$

The first part vanishes for any $y \geqq 0$. In the second part the function under the integral sign can be dominated by an integrable function uniformly with respect to $y \geqq 0$. With this observation, the properties (iii) are obtained by an application of the Lebesgue - Dominated - Convergence Theorem. This completes the proof of the Lemma.

REMARK: The properties (iii) can be extended to functions $v \in E^1$. Indeed, define $\tilde{v}$ by $\tilde{v}(t) = v(t) - v(0)g(t)$, $g(t) = e^{-kt}$. Then we have $\tilde{v} \in E_o^1$, and hence (iii) is true for $\tilde{v}$. From Parseval's relation and from (4) it follows that

$$\|R'(\lambda;A)g\|_2 = \|(i\xi-k+\lambda(c+\hat{k}_o(\xi)))^{-1}\|_2 \leq C_3|\lambda|^{-1/2}\ \||\xi+ik|^{-2/3}\|_2 = O(|\lambda|^{-1/2})$$

for any $\lambda \in L$ and with a constant C_3 defined by

$$C_3^{-2} = \inf\{\left|\frac{i\xi-k+\lambda(c+\hat{k}_o(\xi))}{|\lambda|^{1/2}|\xi+ik|^{2/3}}\right| \mid \xi \in \mathbb{R},\ \lambda \in L\} > 0.$$

This leads to $\lim_{y\to o+} \|V_o(y)g\|_2 = 0 = \lim_{y\to+\infty} \|V_o(y)g\|_2$.

Now we are able to sketch the proof of Theorem 2. From (iv) it follows that $V_o(y)(E) \subset E_o^1$, or equivalently that $G^{-1}V_o(y)(E) \subset E$. Thus, for any $v \in E$, $V_o(y)v$ satisfies the equation $(k+D_t+K_cD_y^2)V_o(y)v = 0$. Define v to be $f_j - U(j,\cdot) \in E_o^1$, then it follows from the properties (iii) that $V_o(y)v$ is a solution of (Q_j). Now, any solution $W: \overline{\mathbb{R}_+} \to E_o^1$ of the inhomogeneous equation $W(y) + GK_cD_y^2W(y) = GK_c\tilde{r}(y)$, such that $W \in C^\infty(\mathbb{R}_+;E) \cap C_o(\overline{\mathbb{R}_+};E)$ and $W(0) = 0$, can be represented by

$$W(y) = \frac{1}{2}\Big[{}_y\!\int^\infty V_1(z-y)\tilde{r}(z)dz + {}_o\!\int^y V_1(y-z)\tilde{r}(z)dz - {}_o\!\int^\infty V_1(y+z)\tilde{r}(z)dz\Big],$$

provided that $\tilde{r}: \overline{\mathbb{R}_+} \to E_o^1$ is sufficietly smooth and vanishes at infinity. Choose $\tilde{r}$ to be zero, then it follows that $W = 0$. Hence, the problems (Q_j) are uniquely solvable. This completes the proof of Theorem 2.

4. Proof of the main result. In this section we shall show that the initial boundary value problem (Z) admits a unique solution $Z(x,t;\varepsilon)$ having the property (2).

The basic method to be used in our approach is the method of successive approximation, applied to an equation of the form $(R_o^{-1} + \varepsilon R_1)T = \tilde{r}$. The linear operators R_o and R_1R_o are assumed to be bounded. Suppose $0 < \varepsilon \leq \varepsilon_o < \|R_1R_o\|^{-1}$. Then this equation admits a unique solution T which can be represented by an infinite series

$$(11) \qquad T = \sum_{j=o}^{\infty} (-\varepsilon)^j R_o(R_1R_o)^j\tilde{r} \qquad \text{with a bound} \quad \|T\| \leq (1-\varepsilon_o\|R_1R_o\|)^{-1}\|R_o\|\,\|\tilde{r}\|.$$

With this goal in mind, we make a change of variables in (Z) in order to obtain a more convenient form of the problem. Let $\delta_o > 0$ be a fixed number, and let $y = \delta x$ for any $\delta = \varepsilon^{-1/2} \geqq \delta_o > o$. Then, substituting

(12) $$Z = T + W = T(y,t;\varepsilon) + (\frac{y}{\delta} - 1)V^1(\delta,t) - \frac{y}{\delta} V^o(\delta,t)$$

in (Z) we obtain

$$(T)\quad \begin{cases} (k+D_t)T + K_c D_y^2 T + \varepsilon K_1(k+D_t)T = - M_\varepsilon\left[W+u_o\right] + \varepsilon r(\frac{y}{\delta},t) \equiv \tilde{r}(y,t;\varepsilon), \\ T(0,t;\varepsilon) = 0 = T(\delta,t;\varepsilon) \text{ for } t \geqq 0, \\ T(y,0;\varepsilon) = 0 \qquad\qquad \text{for } y \geqq 0 . \end{cases}$$

The problem of solving (T) now reduces to the problem of finding an element T in the set

$$\overset{o}{D} = \{T \in C([0,\delta];E_o^1) \cap C^2((0,\delta);E) \mid T(0) = 0 = T(\delta)\},$$

such that the following equation holds

$$(T_\varepsilon)\qquad G^{-1}T + K_c D_y^2 T + \varepsilon R_1 T = \tilde{r} \quad \text{with } R_1 = K_1 G^{-1}.$$

A straigthforward calculation shows that the reduced equation (T_o) (i.e. the equation (T_ε) without the term $\varepsilon R_1 T$) admits a unique solution $T_o \in \overset{o}{D}$ in terms of $\tilde{r}$ as

$$\begin{aligned} T_o(y,t;\varepsilon) &\equiv \left[R_o(y)\tilde{r}\right](t;\varepsilon) \\ &= \frac{1}{2}\int_o^\delta \left[V_2(y+z) - V_2(2\delta+y-z) + V_2(2\delta-y-z) - V_2(2\delta-y+z)\right]K_c^{-1}\tilde{r}(z,t;\varepsilon)dz \\ &+ \frac{1}{2}\left[\int_o^y V_1(y-z)K_c^{-1}\tilde{r}(z,t;\varepsilon)dz + \int_y^\delta V_1(z-y)K_c^{-1}\tilde{r}(z,t;\varepsilon)dz\right], \\ &0 \leqq y \leqq \delta , \end{aligned}$$

provided that $\tilde{r} \in C((0,\delta);E^1)$.

Next, we observe that there exist two constants $C_4 = C_4(\delta_o)$ and $C_5 = C_5(\delta_o)$ such that

$$\|R_o(y)\| \leqq C_4 \quad\text{and}\quad \|R_1 R_o(y)\| \leqq C_5 \quad \text{uniformly for } y \geqq 0 \text{ and for } \delta \geqq \delta_o > 0.$$

These constants may be computed directly from the definitions of V_1 and V_2. Consequently, if $\varepsilon_o = \min\{\delta_o^{-2}, qC_5^{-1}\}$ with an arbitrarily fixed $q \in (0,1)$, we can apply the method of successive approximation to (T_ε). In view of (11), we then obtain a uniquely determined solution $T \in \overset{o}{D}$ which is bounded by

$$\|T(y,\cdot;\varepsilon)\|_2 \leqq C_6\|\tilde{r}(y,\cdot;\varepsilon)\|_2, \qquad C_6 = C_6(\delta_o) = (1-\varepsilon_o C_5)^{-1} C_4 .$$

Using this information in (12), we obtain

$$(13) \qquad \max_{x\in[0,1]} \|Z(x,\cdot;\varepsilon)\|_2 \leqq \|V^o(\delta,\cdot)\|_2 + \|V^1(\delta,\cdot)\|_2 + \max_{x\in[0,1]} \|\tilde{r}(\delta x,\cdot;\varepsilon)\|_2 .$$

Here the function $\tilde{r}$ is given by

$$\tilde{r}(\delta x,t;\varepsilon) = (I+\varepsilon K_1)K_c\left[xD_y^2V^o(\delta,t) + (1-x)D_y^2V^1(\delta,t)\right]$$
$$+\varepsilon\left[r(x,t) - K_cD_x^2U(x,t) + \varepsilon K_1K_cD_x^2\left[V^o(\delta x,t) + V^1(\delta(1-x),t)\right]\right].$$

Because of the hypotheses (H3) it follows from the Remark of Section 3 that $\tilde{r}\in C((0,\delta);E^1)$.

Moreover, the right hand side of the inequality (13) can be shown to be bounded at least of order $O(\varepsilon)$ as $\varepsilon \to 0+$. Most of the estimates can be derived without making any effort. Only for the terms $\varepsilon D_x^2V^j$ is there more work needed. So we shall do it.

Let $v(t) = f_j(t) - U(j,t)$. From (H2) and (H3) it follows that $v\in E_o^1\cap E^2$, hence

$$\varepsilon D_x^2V^j = D_y^2V^j(y,t) = -\frac{1}{2\pi i}\int_L e^{y\sqrt{-\lambda}}R'(\lambda;A)v\,d\lambda = -\frac{1}{2\pi i}\int_L e^{y\sqrt{-\lambda}}R'(\lambda;A)Av\,\frac{d\lambda}{\lambda}.$$

Now, define $\tilde{v}$ by $\tilde{v}(t) = (Av)(t) - (Av)(0)e^{-kt}$ and proceed likewise as in the Remark of Section 3. Then it turns out that $\|D_y^2V^j(y,\cdot)\|_2 = O(1)$ uniformly for $y \geqq 0$. This completes the proof of Theorem 1.

References

[1] Eckhaus, W., Matched Asymptotic Expansions and Singular Perturbations. North-Holland Mathematical Studies 6. Amsterdam: North-Holland 1973.

[2] Gerlach, E., Zur Theorie einer Klasse von Integrodifferentialgleichungen. Dissertation. Berlin 1969.

[3] Grabmüller, H., On linear theory of heat conduction in materials with memory. Existence and uniqueness theorems for the final value problem. Submitted for publication in the Proceedings of the Royal Society Edinburgh.

[4] Hoppensteadt, F., On quasilinear parabolic equations with a small parameter. Comm. Pure and Appl. Math. 24 (1971), 17-38.

[5] Krein, M. G., Integral equations on a half line with kernel depending upon the difference of the arguments. Uspehi Mat. Nauk (N.S.) 13 (1958), 3-120.

[6] Nunziato, J. W., On heat conduction in materials with memory. Quart. Appl. Math. 29 (1971), 187-204.

[7] Vekua, N. P., Linear integro-differential equations with small parameters for higher derivatives. Problems of Continuum Mechanics. SIAM (1961), 592-601.

[8] Vishik, M. I., Lyusternik, L. A. Regular degeneration and boundary layer for linear differential equations with small parameter. Uspehi Mat. Nauk 12 (1957), 3-122.

Smoothness of the solution of a monotonic boundary value problem for a second order elliptic equation in a general convex domain

Pierre Grisvard

Abstract:

We prove the square integrability of the second derivatives of the solution of an elliptic second order equation in a general convex domain, bounded in the n-dimensional Euclidean space, under monotonic boundary conditions. Our boundary conditions are general enough to include strongly non-linear conditions as for instance Signorini's. There is no restriction concerning the singularities of the boundary of the convex domain in which the equation is considered; this domain is allowed for instance to be a two-dimensional polygon or a three-dimensional polyhedron.

CONTENTS:

1. Introduction.
2. A priori estimates
3. Varying the domain.
4. Proof of the main result.
5. Several remarks.

1. INTRODUCTION

In this paper we shall always use the following notations: Ω is an open subset of the euclidean n dimensional space $\mathbb{R}^n$, Γ is its boundary ; assuming that Γ is locally lipschitz (1), we denote by γ the normal exterior vector field on Γ (which is defined a.e. on Γ) ; $L^2(\Omega)$ is the space of all square integrable functions in Ω and $H^m(\Omega)$ the usual Sobolev space of order m in Ω (i.e. the space of all functions having their derivatives up to the order m in $L^2(\Omega)$), Δ is the Laplace operator, ∇ is the gradient operator and finally β and γ are two maximal monotone graphs in $\mathbb{R}$ such that

$$0 \in \beta(0) \cap \gamma(0).$$

We are going to solve the equation

$$(1.1) \qquad -\Delta u + u + \gamma(u) \ni f \quad \text{in } \Omega$$

under the boundary condition

$$(1.2) \qquad -\frac{\partial u}{\partial \gamma} \in \beta(u) \quad \text{on } \Gamma .$$

Let us recall that for a given $f \in L^2(\Omega)$ this problem has a unique weak solution u in $H^1(\Omega)$, which is obtained by minimizing the functionnal $v \longrightarrow \varphi(v)$ defined on $H^1(\Omega)$ by

$$\varphi(v) = \frac{1}{2}\int_{\Omega} [|\nabla v|^2 + |v|^2]\, dx - \int_{\Omega} fv \, dx + \int_{\Omega} k(v)\, dx + \int_{\Gamma} j(v)\, d\sigma$$

where j and k are two convex lower semi-continuous functions from $\mathbb{R}$ into $\mathbb{R} \cup \{+\infty\}$, such that β and γ are the subdifferentials of j and k respectively (c.f. Lions [8]).

Furthermore when Ω is bounded and of class C^2 (1) then it is proved by Brezis [2]* that the weak solution is in $H^2(\Omega)$ and is strong in the sense that (1.1) and (1.2) hold a.e. on Ω and on Γ respectively.

We prove here that the same result holds when Ω is a general bounded convex domain :

(1) c.f. for instance in Necas [10] the exact meaning of this hypothesis.

*An earlier proof due to Fichera is given in Handbuch der Physik Volume 6.A, second contribution (see also Fichera Proceedings of the International Congress of Mathematics, Nice, 1970).

Theorem : Let Ω be bounded and convex in $\mathbb{R}^n$ and f be given in $L^2(\Omega)$, then problem (1.1) and (1.2) has a unique solution u in $H^2(\Omega)$.

It is worthwhile to observe that Γ is uniformly lipschitz(2) since Ω is convex and bounded.

However Γ is not better in general since Ω is allowed to be a two-dimensional polygon or a three-dimensional polyhedron or even something worse. In particular Γ is not C^2 or even $C^{1,1}$, the minimal hypothesis required for Brezis proof.

Our boundary conditions (1.2) include as particular cases the following :

(i) Dirichlet's b.c. : take $\beta(x) = \emptyset$ for $x \neq 0$ and $\beta(0) = \mathbb{R}$.

(ii) Neuman's b.c. : take $\beta(x) = \{0\}$ for all x.

(iii) The third boundary problem : take $\beta(x) = \{\lambda x\}$ with $\lambda > 0$.

(iv) Signorini's b.c. : take $\beta(x) = \emptyset$ for $x < 0$, $\beta(0) =]-\infty,0]$ and $\beta(x) = \{0\}$ for $x > 0$.

Thus we see that our result is a far reaching generalization of Kadlec's result [6] which applies only to Dirichlet's problem . By the way, the particular case of Dirichlet's boundary condition shows that our result is optimal in the following sense. First, our result does not remain true for a non convex uniformly lipschitz domain; indeed the solution of Dirichlet's problem in a non convex two dimensional polygon Ω, is not in $H^2(\Omega)$ (c.f. Grisvard [4]). Second, even when f is very smooth, the solution of Dirichlet's problem is not smoother in general, than H^2 (3) indeed for every real number $S > 2$ there exists a 2-dimensional convex polygon such that u is not in $H^s(\Omega)$ for every f in $H^{s-2}(\Omega)$ or even in $C^\infty(\bar{\Omega})$ (c.f. Grisvard [5]).

---

(2) Indeed it is easy to check geometrically that Ω has the uniform cone property (following Agmon [1] for instance) and it is proved in Chenais [3] that this property is equivalent to saying that Γ is uniformly lipschitz.

(3) A function u is in $H^s(\Omega)$ iff it is the restriction to Ω of a function U defined in R^n and such that $(1 + |\xi|^2)^{s/2}\, \hat{U}(\xi) \in L^2(R^n)$, where $\hat{U}$ is the Fourier transform of U.

Let us now describe briefly the proof which will be worked out in the following paragraphs : the singularities of the problem are of two kinds namely those of Γ and those of β and γ. We shall do away with both by taking approximations. First, following Brezis [2] , we approximate β and γ by their Yosida approximations β_λ and γ_λ ($\lambda > 0$) which are two uniformly lipschitz non decreasing functions from $\mathbb{R}$ into $\mathbb{R}$: Then we approximate Ω by an increasing sequence Ω_m (m = 1,2,...) of convex open subsets of Ω , with regular boundaries i.e. of class C^∞, and thus we extend to some variational inequalities the technique of varying the domain, introduced first by Kadlec [6] . The following a priori estimate makes it possible to take limits

$$(1.3) \qquad ||u||^2_{H^2(\Omega)} \leq ||\Delta u||^2_{L^2(\Omega)} + ||u||^2_{H^1(\Omega)} \quad ;$$

this inequality holds for all $u \in H^2(\Omega)$, fulfilling condition (1.2), assuming that Ω is a convex bounded set in $\mathbb{R}^n$ with a C^2 boundary and that β is a lipschitz non decreasing function. This inequality will be proved in & 2 ; it is of the same kind as the so called "second fundamental inequality" in Ladizenskaia-Uralceva [7] , and is obtained by a straightforward calculation of $||\Delta u||^2_{L^2(\Omega)}$ based on integration by parts. & 3 is concerned with taking the limit when the domain varies (i.e. when $m \to \infty$) making use of the fact that the sequence Ω_m (m = 1,2,...) may be chosen so as to have the "uniform extension property " of Chenais [3] . For the sake of brevity we shall detail the proofs only in the two-dimensional case.

Finally let us mention that our result may be extended to a general strongly elliptic real second order operator with smooth coefficients L, the vector field γ being replaced by the conormal vector field γ_L. It is also possible to consider graphs β and γ depending smoothly on x in Γ and Ω respectively.
Further more the same method applies
to the case of some non convex domains. Some of these generalizations will only be sketched in & 5, while the details to-gether with the proof of our main result when the dimension is more that 2, will appear elsewhere.

2. A priori estimates.

Here Ω will denote a two-dimensional bounded open set with a C^∞ boundary. At each point of Γ we denote by $\vec{\gamma}$ the unit exterior normal vector and by $\vec{\tau}$ the unit tangent vector (following the direct orientation) ; finally let s be a curvilinear coordinate along Γ such that ds is the length measure and let Θ be the angle between $\vec{\gamma}$ and the abcissa axis. Our basic identity is given in

LEMMA 2.1. : For every $u \in H^3(\Omega)$ we have

$$\int_\Omega |\Delta u|^2 \, dx - \sum_{j,k=1}^{2} \int_\Omega \left| \frac{\partial^2 u}{\partial x_j \, \partial x_k} \right|^2 dx = \int_\Gamma \frac{\partial \Theta}{\partial s} |\nabla u|^2 \, ds + 2 \int_\Gamma \frac{\partial u}{\partial \gamma} \frac{\partial^2 u}{\partial s^2} \, ds \tag{2.1}$$

This is an identity between quadratic forms, which is easy to derive from the corresponding identity between bilinear forms : for every u and $v \in H^3(\Omega)$ we have

$$\int_\Omega \Delta u \;\; \Delta v \, dx - \sum_{j,k=1}^{2} \int_\Omega \frac{\partial^2 u}{\partial x_j \, \partial x_k} \; \frac{\partial^2 v}{\partial x_j \, \partial x_k} \, dx = \int_\Gamma \frac{\partial \Theta}{\partial s} \nabla u . \nabla v \, ds + \int_\Gamma \left[\frac{\partial u}{\partial \gamma} \frac{\partial^2 v}{\partial s^2} + \frac{\partial^2 u}{\partial s^2} \frac{\partial v}{\partial \gamma} \right] ds \tag{2.2}$$

PROOF : It is easy to localize such an identity and thus it is enough to prove it assuming that u and v have their supports in a neighbourhood V of a point of Γ . We choose V sufficiently small so as to be able to define two families $\mathcal{F}$ and $\mathcal{G}$ of C^∞ curves in V, such that $\Gamma \cap V$ is a curve of $\mathcal{F}$ and such that each point x of V is on one and only one curve of $\mathcal{F}$ and on one and only one curve of $\mathcal{G}$. Those two curves being orthogonal at x. It is then possible to extend $\vec{\tau}$ and $\vec{\upsilon}$ to a couple of C^∞ vector fields defined on V. Now we introduce curvilinear coordinates s on the curves of $\mathcal{F}$ and n on the curves of $\mathcal{G}$ such that ds a and dn are the length measures and in such a way taht s extends to V the function already defined on Γ , while n increase in the direction of $\vec{\upsilon}$

Obviously we have

$$\nabla u = \frac{\partial u}{\partial s}\vec{\tau} + \frac{\partial u}{\partial n}\vec{\gamma}$$

and consequently

$$\nabla.\vec{\Phi} = \vec{\tau}.\frac{\partial\vec{\Phi}}{\partial s} + \vec{\gamma}.\frac{\partial\vec{\Phi}}{\partial n}$$

where u and $\vec{\Phi}$ are respectively a function and a vector field defined in V.

Now from the definition of Θ we have $\vec{\gamma} = \{\cos\Theta;\ \sin\Theta\}$ and $\vec{\tau} = \{-\sin\Theta\ ;\ \cos\Theta\}$, and consequently for every variable r we have

$$\frac{\partial\vec{\gamma}}{\partial r} = \frac{\partial\Theta}{\partial r}\vec{\tau}, \quad \frac{\partial\vec{\tau}}{\partial r} = -\frac{\partial\Theta}{\partial r}\vec{\gamma}.$$

From these identities it follows easily that

$$\Delta u = \nabla.\nabla u = \frac{\partial^2 u}{\partial s^2} + \frac{\partial^2 u}{\partial n^2} + \left(\frac{\partial u}{\partial n}\frac{\partial\Theta}{\partial s} - \frac{\partial u}{\partial s}\frac{\partial\Theta}{\partial n}\right)$$

while a tedious calculation shows that

$$\frac{\partial^2 u}{\partial n\,\partial s} - \frac{\partial^2 u}{\partial s\,\partial n} = -\nabla\Theta.\ \nabla u = -\frac{\partial\Theta}{\partial s}\frac{\partial u}{\partial s} - \frac{\partial\Theta}{\partial n}\frac{\partial u}{\partial n}.$$

It is now possible to prove (2.2) ; indeed applying Green's formula we obtain

$$\int_\Omega \Delta u\ \Delta v\ dx =$$

$$\int_\Gamma \frac{\partial u}{\partial n}\Delta v\ ds - \int_\Omega \nabla u.\ \nabla\Delta v\ dx =$$

$$\int_\Gamma \frac{\partial u}{\partial n}\Delta v\ ds - \int_\Gamma \nabla u.\ \frac{\partial u}{\partial n}\nabla v\ ds + \int_\Omega \nabla^2 u.\ \nabla^2 v\ dx$$

and since

$$\nabla^2 u.\ \nabla^2 v = \sum_{j,k=1}^{2} \frac{\partial^2 u}{\partial x_j\ \partial x_k}\ \frac{\partial^2 v}{\partial x_j\ \partial x_k},$$

we see that the left hand side term in (2.2) is equal to

$$I = \int_\Gamma \frac{\partial u}{\partial n}\ \Delta v\ ds - \int_\Gamma \nabla u.\ \frac{\partial}{\partial n}\ \nabla v\ ds.$$

Now using the formulae we established earlier, we obtain that

$$I = \int_\Gamma \left(\frac{\partial u}{\partial n}\left[\frac{\partial^2 v}{\partial s^2} + \frac{\partial \theta}{\partial s}\frac{\partial v}{\partial n}\right] - \frac{\partial u}{\partial s}\left[\frac{\partial^2 v}{\partial n\,\partial s} + \frac{\partial \theta}{\partial n}\ \frac{\partial v}{\partial s}\right] \right) ds$$

$$= \int_\Gamma \left[\frac{\partial u}{\partial n}\ \frac{\partial^2 v}{\partial s^2} + \frac{\partial \theta}{\partial s}\ \frac{\partial u}{\partial n}\ \frac{\partial v}{\partial s} - \frac{\partial u}{\partial s}\ \frac{\partial^2 v}{\partial s\,\partial n} + \frac{\partial \theta}{\partial s}\ \frac{\partial u}{\partial s}\ \frac{\partial v}{\partial s}\right] ds$$

$$= \int_\Gamma \frac{\partial \theta}{\partial s}\ \nabla u.\ \nabla v\ ds + \int_\Gamma \frac{\partial u}{\partial n}\ \frac{\partial^2 v}{\partial s^2}\ ds - \int_\Gamma \frac{\partial u}{\partial s}\ \frac{\partial^2 v}{\partial s\,\partial n}\ ds$$

and finally, integrating by parts in the third integral we get

$$I = \int_\Gamma \frac{\partial \theta}{\partial s}\ \nabla u.\ \nabla v\ ds + \int_\Gamma \left[\frac{\partial u}{\partial n}\ \frac{\partial^2 v}{\partial s^2} + \frac{\partial^2 u}{\partial s^2}\ \frac{\partial v}{\partial n}\right] ds$$

Q.E.D.

From lemma 2.1 we deduce the following

<u>LEMMA 2.2</u> : <u>Let Ω be a convex bounded set in $\mathbb{R}^2$ with a C^∞ boundary and $u \in H^2(\Omega)$ be a function fulfilling the boundary condition (1.2) with β a non decreasing uniformly lipschitz function; then inequality (1.3) holds.</u>

<u>PROOF</u> : From the convexity of Ω it follows that $\frac{\partial \theta}{\partial s} \geq 0$. Then for $u \in H^3(\Omega)$, (2.1) implies the inequality

$$\int_\Omega |\Delta u|^2\ dx \geq \sum_{j,k=1}^{2} \int_\Omega \left|\frac{\partial^2 u}{\partial x_j\,\partial x_k}\right|^2 dx + 2\int_\Gamma \frac{\partial u}{\partial \gamma}\ \frac{\partial^2 u}{\partial s^2}\ ds.$$

Now identifying the function $\frac{\partial^2 u}{\partial s^2}$ on Γ with the distribution $\frac{\partial^2 u}{\partial s^2}\ ds$ (4), our inequality may be rewritten as

$$\int_\Omega |\Delta u|^2\ dx \geq \sum_{j,k=1}^{2} \int_\Omega \left|\frac{\partial^2 u}{\partial x_j\,\partial x_k}\right|^2 dx + 2 < \frac{\partial^2 u}{\partial s^2}\ ;\ \frac{\partial u}{\partial \gamma} > ;$$

if we consider the last bracket as the duality bracket

---

(4) This is why we assumed that Γ is of class C^∞, instead of only C^2, an hypothesis which is enough for the rest of the proof.

between $H^{1/2}(\Gamma)$ and $H^{-1/2}(\Gamma)$; we see that all the terms of the last inequality are quadratic continuous forms on $H^2(\Omega)$ (recall that $u \longrightarrow \frac{\partial u}{\partial \gamma}$ is continuous from $H^2(\Omega)$ into $H^{1/2}(\Gamma)$) and that

$u \quad \longrightarrow \quad \frac{\partial^2 u}{\partial s^2}$ is continuous from $H^2(\Omega)$ into $H^{-1/2}(\Gamma)$, c.f. Lions-Magenes [9])

This allows us to extend our inequality to every $u \in H^2(\Omega)$ since $H^3(\Omega)$ is dense in $H^2(\Omega)$. If furthermore our function u fulfills (1.2), the inequality becomes

$$\int_\Omega |\Delta u|^2 \, dx - \sum_{j,k=1}^{2} \int_\Omega \left| \frac{\partial^2 u}{\partial x_j \, \partial x_k} \right|^2 dx \geq$$

$$- 2 < \frac{\partial^2 u}{\partial s^2} ; \beta(u) > =$$

$$- 2 \int_\Gamma \frac{\partial^2 u}{\partial s^2} \beta(u) \, ds =$$

$$2 \int_\Gamma \left| \frac{\partial u}{\partial s} \right|^2 \beta'(u) \, ds \geq 0$$

and consequently

$$\sum_{j,k=1}^{2} \int_\Omega \left| \frac{\partial^2 u}{\partial x_j \, \partial x_k} \right|^2 dx \leq \int_\Omega |\Delta u|^2 \, dx \; ;$$

this implies obviously (1.3). Q.E.D.

We shall now derive a complete estimate for the solution u of our problem.

LEMMA 2.3. Let Ω be a convex bounded open set in R^2 with a C^∞ boundary and let β and γ be non decreasing and uniformly lipschitz functions with $\beta(0) = \gamma(0) = 0$; then for $u \in H^2(\Omega)$ solution of (1.1) (1.2) we have

$$||u||_{H^2(\Omega)} \leq \sqrt{5} \, ||f||_{L^2(\Omega)}$$

PROOF : First multiplying equation (1.1) by u and integrating by parts we see easily that

$$||u||_{H^1(\Omega)} \leq ||f||_{L^2(\Omega)}$$

since we have $\beta(t).t \geq 0$ and $\gamma(t).t \geq 0$ for every $t \in R$. Then from (1.3) we deduce inequality

$$||u||^2_{H^2(\Omega)} \leq ||\Delta u||^2_{L^2(\Omega)} + ||f||^2_{L^2(\Omega)}$$

and thus in order to conclude we have only to estimate Δu in terms of f.

To do this we calculate the L^2 norm of u-f : we get

$$||u - f||^2_{L^2(\Omega)} = ||\Delta u - \gamma(u)||^2_{L^2(\Omega)} =$$

$$||\Delta u||^2_{L^2(\Omega)} + ||\gamma(u)||^2_{L^2(\Omega)} - 2\int_\Omega \Delta u\ \gamma(u)\ dx\ ;$$

integrating by parts and using Green's formula we see that

$$\int_\Omega \Delta u\ \gamma(u)\ dx =$$

$$-\int_\Gamma \frac{\partial u}{\partial \gamma}\ \gamma(u)\ d\sigma + \int_\Omega \nabla u.\ \nabla\gamma(u)\ dx =$$

$$\int_\Gamma \beta(u)\ \gamma(u)\ d\sigma + \int_\Omega |\nabla u|^2\ \gamma'(u)\ dx$$

and the last sum is non negative since $\beta(t)\ \gamma(t) \geq 0$ for every $t \in R$. Thus we conclude that

$$||\Delta u||_{L^2(\Omega)} \leq ||u - f||_{L^2(\Omega)} \leq ||u||_{L^2(\Omega)} + ||f||_{L^2(\Omega)}$$

$$\leq 2\ ||f||_{L^2(\Omega)}$$

and consequently

$$||u||^2_{H^2(\Omega)} \leq 5\ ||f||^2_{L^2(\Omega)}$$ Q.E.D.

3. Varying the domain

Now Ω is a general bounded convex open set in $\mathbb{R}^2$. We choose an increasing sequence Ω_m $(m = 1,2,\ldots)$ of open convex subsets in Ω, with C^∞ boundaries, which converges to Ω; here convergence means that the distance between Γ and Γ_m goes to zero where Γ and Γ_m are boundaries of Ω and Ω_m respectively(5).

LEMMA 3.1 : The family Ω_m $(m = 1,2,\ldots)$ has the uniform extension property

This means that there exists an operator P_m linear and continuous from $H^k(\Omega_m)$ into $H^k(\mathbb{R}^2)$, such that

(i) $\quad P_m u|_{\Omega_m} = u \quad$, for every $u \in H^k(\Omega_m)$

(ii) $\quad \|P_m\|_{H^k(\Omega_m) \longrightarrow H^k(\mathbb{R}^2)} \leq M_k \quad$,

for all $k = 1,2,\ldots$ where M_k does not depend on m.

PROOF : Following Chenais [3] we know that it is enough to check that the family $\Omega_m (m = 1,2,\ldots)$ satisfies the uniform cone property, namely that there exists $\theta \in]0, \pi/2[$, $h > 0$, $r > 0$ such that for every $x \in \Gamma_m$ there exists a unit vector ξ (depending on x and m) with $y + C \subset \Omega_m$ for every $y \in B(x,r)$ where $B(x,r)$ is the ball with radius r and center x and C is the cone of angle θ, height h and axis ξ.

Now choose $B(x_o, \rho) \subset \Omega$; for m great enough (say $m \geq m_o$) we have $B(x_o ; \rho/2) \subset \Omega_m$. Then for every $x \in \Gamma_m$ we have by convexity

$$x + C \subset \Omega_m$$

where C is the cone of angle $\omega = \text{arc cos } (\rho/_{d(x;x_o)})$, height $d(x ;x_o) \cos \omega$ and axis $\frac{x_o - x}{\|x_o - x\|}$. The uniform cone property follows by chosing h, r and θ small enough. Q.E.D.

We still assume that β and γ are two non decreasing and uniformly Lipshitz functions. Then we know from Brezis [2] that for every m there exists a unique $u_m \in H^2(\Omega_m)$ solution of

$$(3.1) \qquad \begin{cases} - \Delta u_m + u_m + \gamma(u_m) = f & \text{a.e. in } \Omega_m \\ - \dfrac{\partial u_m}{\partial \nu_m} = \beta(u_m) & \text{a.e. on } \Gamma_m \end{cases}$$

(5) Consequently the measure of $\Omega \setminus \Omega_m$ goes to zero.

Our aim is to prove that in some sense, the sequence u_m converges to some $u \in H^2(\Omega)$ which is solution of (1.1) (1.2) . Of course the basic tool is the estimate of Lemma 2.3 namely

$$\|u_m\|_{H^2(\Omega_m)} \leq \sqrt{5}\ \|f\|_{L^2(\Omega)}$$

Using the uniform extension property of Lemma 3.1 , let U_m(m = 1,2,...) be a bounded sequence of functions in $H^2(\Omega)$ such that $U_m|_{\Omega_m} = u_m$. By considering if necessary a subsequence we may assume that U_m converges to some $u \in H^2(\Omega)$ in the weak topology of $H^2(\Omega)$ (and also consequently in the strong topology of $H^1(\Omega)$) . We shall show that u is solution of (1.1)(1.2) and therefore prove :

<u>LEMMA 3.2</u> : <u>Let Ω be a bounded convex open set in $\mathbb{R}^2$ and let β and γ be two non decreasing and uniformly lipshitz functions , then for every $f \in L^2(\Omega)$ there exists a unique $u \in H^2(\Omega)$ solution of (1.1) (1.2) .</u>

<u>PROOF</u> : The uniqueness of u is clear since we already know the uniqueness of the weak solution without any hypothesis on Ω .

In order to prove that u is a solution , it is convenient to consider u_m as a weak solution of (3.1) i.e. that

$$\frac{1}{2}\|u_m\|^2_{H^1(\Omega_m)} - \int_{\Omega_m} f u_m dx + \int_{\Omega_m} k(u_m)dx + \int_{\Gamma_m} j(u_m)d\sigma$$

$$\leq \frac{1}{2}\|v\|^2_{H^1(\Omega_m)} - \int_{\Omega_m} f\ v\ dx + \int_{\Omega_m} k(v)dx + \int_{\Gamma_m} j(v)d\sigma$$

for every $v \in H^1(\Omega_m)$ where k and j are primitives of γ and β respectively. We can choose them such that $k(0) = j(0) = 0$. Now we restrict ourselves to those $v \in H^1(\Omega)$ and we try to take the limit of each term in (3.2).
We have

$$\|u_m\|^2_{H^1(\Omega_m)} = \int_{\Omega_m} [|u_m|^2 + |\nabla u_m|^2]dx$$

$$= \int_{\Omega} [|\tilde{u}_m|^2 + |\widetilde{\nabla u}_m|^2]dx \longrightarrow \int_{\Omega} [|u|^2 + |\nabla u|^2]\ dx$$

$$= \|u\|^2_{H^1(\Omega)}$$

since $\tilde{u}_m \longrightarrow u$ and $\widetilde{\nabla u}_m \longrightarrow \nabla u$ in $L^2(\Omega)$. We also have

$$\int_{\Omega_m} f\ u_m\ dx = \int_{\Omega} f\ \tilde{u}_m\ dx \longrightarrow \int_{\Omega} f\ u\ dx$$

and

$$\int_{\Omega_m} k(u_m)\, dx = \int_{\Omega} k(\tilde{u}_m)\, dx$$

and consequently

$$\int_{\Omega} k(u)\, dx \le \underline{\lim} \int_{\Omega_m} k(u_m) dx$$

since the functional $v \longmapsto \int_{\Omega} k(v)dx$ is lower semi-continuous on $L^2(\Omega)$.
It is harder to find the limit of the last terms ; for this we write

$$\int_{\Gamma_m} j(u_m) d\sigma - \int_{\Gamma} j(u)\, d\sigma =$$

$$\int_{\Gamma_m} [j(u_m) - j(u)]\, d\sigma - \int_{\Gamma_m} j(u) d\sigma - \int_{\Gamma} j(u)\, d\sigma$$

At this point we recall that the injection of $H^2(\Omega)$ in $C^0(\bar{\Omega})$(6) , is completely continuous, therefore U_m converges uniformly to u and also $j(U_m)$ converges uniformly to $j(u)$ since j is continuous. Consequently we see that $\int_{\Gamma_m} [j(u_m) - j(u)]\, d\sigma$ converges to zero since the length of the Γ_m is bounded independently of m . Finally we observe that u being in $H^2(\Omega)$, u is Hölder continuous and $j(u)$ too since j is continuously differentiable ; let α be the Hölder exponent of $j(u)$ then it is clear that

$$\int_{\Gamma_m} j(u)\, d\sigma - \int_{\Gamma} j(u)\, d\sigma = \mathcal{O}(d(\Gamma ; \Gamma_m)^{\alpha})$$

which converges to zero .

At this point we know that

$$\varphi(u) = \frac{1}{2} \|u\|^2_{H^1(\Omega)} - \int_{\Omega} f\, u\, dx + \int_{\Omega} k(u)\, dx + \int_{\Gamma} j(u)\, d\sigma$$

$$\le \underline{\lim} \left(\frac{1}{2} \|v\|^2_{H^1(\Omega_m)} - \int_{\Omega_m} f\, v\, dx + \int_{\Omega_m} k(v)\, dx + \int_{\Gamma_m} j(v)\, d\sigma \right)$$

But it is easier by using the same techniques to see that the limit of the right hand side term is $\varphi(v)$. This already shows that u is the weak solution of our problem ; however with our assumptions on β and γ , φ is differentiable on $H^1(\Omega)$ and furthermore since we know that $u \in H^2(\Omega)$ we can use Green's formula and it is easy to check that $\varphi'(u) = 0$ means (1.1) and (1.2) .
Q.E.D.

---

(6) The space of functions continuous up to the boundary of Ω .

4. Proof of the main result

We now prove the theorem stated in the introduction. Let β_λ and γ_λ be the Yosida approximations of β and γ i.e :

$$\beta_\lambda = \frac{1}{\lambda}\{1 - (1 + \lambda\beta)^{-1}\} \subset \beta \circ (1 + \lambda\beta)^{-1}$$

$$\gamma_\lambda = \frac{1}{\lambda}\{1 - (1 + \lambda\gamma)^{-1}\} \subset \gamma \circ (1 + \lambda\gamma)^{-1}$$

with $\lambda > 0$. Then from Lemma 3.2 we know that there exists a unique solution $u_\lambda \in H^2(\Omega)$ of

$$\begin{cases} -[-\Delta u_\lambda + u_\lambda - f] = \gamma_\lambda(u_\lambda) & \text{a.e. in } \Omega \\ -\dfrac{\partial u_\lambda}{\partial\nu} = \beta_\lambda(u_\lambda) & \text{a.e. on } \Gamma \end{cases}$$

From Lemma 2.3 we know that u_λ remains in a bounded set in $H^2(\Omega)$ so that by choosing a decreasing sequence $\lambda_j (j=1,2...)$ converging to zero, we may assume that $u_{\lambda_j} \longrightarrow u$ weakly in $H^2(\Omega)$, strongly in $H^1(\Omega)$ and a.e in Ω and that $\dfrac{\partial u_{\lambda_j}}{\partial\nu} \longrightarrow \dfrac{\partial u}{\partial\nu}$ strongly in $L^2(\Gamma)$ and a.e. on Γ.
Consequently we know that

$$\begin{cases} u_{\lambda_j} \longrightarrow u & \text{a.e. on } \Gamma \\ \beta_{\lambda_j}(u_{\lambda_j}) \longrightarrow -\dfrac{\partial u}{\partial\nu} & \text{a.e. on } \Gamma \end{cases} ;$$

but $\beta_{\lambda_j}(u_{\lambda_j}) \in \beta((1 + \lambda_j\beta)^{-1}u_{\lambda_j})$ and $u_{\lambda_j} - (1+\lambda_j\beta)^{-1}u_{\lambda_j} = \lambda_j\beta_{\lambda_j}(u_{\lambda_j}) \longrightarrow 0$ a.e. on Γ and therefore $(1 + \lambda_j\beta)^{-1}u_{\lambda_j} \longrightarrow u$, a.e. on Γ which implies (1.2) i.e. the Boundary condition

$$-\frac{\partial u}{\partial\nu} \in \beta(u) \quad \text{a.e. on } \Gamma$$

since β is maximal monotone. Notice that this proof follows exactly Brezis [2] ; however for taking the limit in the equation we follow a slightly different procedure since we know only the weak convergence of $-\Delta u_\lambda + u_\lambda - f$ towards $-\Delta u + u - f$ in $L^2(\Omega)$. We use the fact that

$$v \longmapsto \psi(v) = \int_\Omega k(v)\, dx$$

is a lower semi-continuous convex function on $L^2(\Omega)$, therefore its subdifferential $\partial\psi$ is maximal monotone ; on the other hand $\partial\psi$ is known to be defined by

$$\partial\psi(v) = \{w \in L^2(\Omega) \; ; \; w \in \gamma(v) \quad \text{a.e. in} \quad \Omega\} \quad ,$$

thus our equation reduces to

$$- [- \Delta u_{\lambda_j} + u_{\lambda_j} - f] \in \partial\psi \; ((1+\lambda_j\partial\psi)^{-1} u_{\lambda_j}) \quad ,$$

while we have

$$u_{\lambda_j} - (1+\lambda_j\gamma)^{-1} u_{\lambda_j} = \lambda_j \; \gamma_{\lambda_j}(u_{\lambda_j}) \longrightarrow 0$$

strongly in $L^2(\Omega)$ and therefore

$$\begin{cases} (1 + \lambda_j\partial\psi)^{-1} u_{\lambda_j} \longrightarrow u \quad \text{strongly in} \quad L^2(\Omega) \\ - [- \Delta u_{\lambda_j} + u_{\lambda_j} - f] \longrightarrow - [- \Delta u + u - f] \quad \text{weakly in } L^2(\Omega) \end{cases} .$$

$\partial\psi$ being maximal monotone this implies

$$- [- \Delta u + u - f] \in \partial\psi(u)$$

which means

$$- [- \Delta u + u - f] \in \gamma(u) \quad \text{a.e. in} \quad \Omega \; .$$

We have thus proved that $u \in H^2(\Omega)$ and u is solution of (1.1) (1.2) while uniqueness is known even for a weak solution. The proof of our theorem is complete.

5. Several remarks.

a. As we mentionned in the introduction it is possible to generalize our result to some operators L which differ from Δ : indeed we may consider L defined by

$$Lu(x) = \sum_{i,j=1}^{n} \frac{\partial}{\partial x_i} \left\{ a_{ij}(x) \frac{\partial u}{\partial x_j}(x) \right\}, \quad x \in \bar{\Omega}$$

where the $a_{i,j}$ are C^2 real functions up to the boundary of Ω and such that the matrix $\mathcal{A} = \{a_{ij}\}_{i,j=1}^{n}$ is symmetric positive definite and

$$\mathcal{A}(x) \geq \alpha I,$$

for all $x \in \bar{\Omega}$, with $\alpha > 0$. Then the following boundary value problem has a unique solution $u \in H^2(\Omega)$ for a given $f \in L^2(\Omega)$,

$$\begin{cases} -Lu + u + \gamma(u) \ni f & \text{a.e. in } \Omega \\ -\dfrac{\partial u}{\partial \gamma_L} \in \beta(u) & \text{a.e. on } \Gamma \end{cases}$$

where $\dfrac{\partial}{\partial v_L}$ is the so called "conormal derivative" defined by

$$\frac{\partial u}{\partial \gamma_L}(x) = \sum_{i,j=1}^{n} \gamma_i(x)\, a_{ij}(x) \frac{\partial u}{\partial x_j}(x), \quad x \in \Gamma.$$

In proving this result the basic a priori estimate is a consequence of the following analogue of identity (2.1)

$$\left| \int_\Omega |Lu|^2 \, dx - \sum_{i,j,k,\ell=1}^{2} \int_\Omega a_{ij} \frac{\partial^2 u}{\partial x_k \, \partial x_j} a_{k,\ell} \frac{\partial^2 u}{\partial x_i \, \partial x_\ell} \, dx - \int_\Gamma \det \mathcal{A} \frac{\partial \theta}{\partial s} |\nabla u|^2 \, ds - 2 \int_\Gamma \det \mathcal{A} \frac{\partial u}{\partial \nu} \frac{\partial^2 u}{\partial s^2} \, ds \right|$$
$$\leq C \left\{ \|u\|_{H^1(\Omega)} \|u\|_{H^2(\Omega)} + \|\nabla u\|^2_{L^2(\Gamma)} \right\}$$

where the constant C depends only on the upper bound of the coefficients a_{ij} together with their derivatives up to order 2 in $\bar{\Omega}$.

b. One unpleasant fact about our theorem is that it does not include the result of Brezis [2] for a non convex C^2 bounded open set Ω . We used our convexity hypothesis only to keep non negative the integral

$$\int_\Gamma \frac{\partial\theta}{\partial s} |\nabla u|^2 \, ds$$

in (2.1) . Now if we assume only that $\frac{\partial\theta}{\partial s}$ is bounded from below by some negative number $-M$ we have obviously

$$\int_\Gamma \frac{\partial\theta}{\partial s} |\nabla u|^2 \, ds \geq - M \int_\Gamma |\nabla u|^2 \, ds \quad ;$$

consequently the a priori estimate (1.3) becomes

$$\|u\|^2_{H^2(\Omega)} \leq \{\|\Delta u\|^2_{L^2(\Omega)} + \|u\|^2_{H^1(\Omega)} + M \|\nabla u\|^2_{L^2(\Gamma)}\} \quad .$$

Therefore our method of approximation works if we are able to choose the sequence Ω_m $(m = 1,2,\ldots)$ in such a way that the corresponding fonctions $\frac{\partial\theta}{\partial s}$ (which might be denoted by $\frac{\partial\theta_m}{\partial s}$) are bounded from below by some negative number $-M$ independant of m and that the compactness inequality

$$\|\nabla u\|_{L^2(\Gamma_m)} \leq \varepsilon \|u\|_{H^2(\Omega_m)} + C(\varepsilon) \|u\|_{H^1(\Omega_m)} \quad \forall \varepsilon > 0$$

holds with a constant $C(\varepsilon)$ non depending on m .

This obviously may be achived if Γ is a bounded curvilinear polygon whose boundary is a finite number of $C^{1,1}$ curves meeting at convex angles ; this hypothesis includes the C^2 open bounded set considered by Brezis.

B I B L I O G R A P H Y

[1] AGMON Lectures on elliptic boundary value problems, Van Nostrand, New-York, 1965.

[2] BREZIS Monotonicity methods in Hilbert space and some applications, Contributions to non linear functional analysis, Acad. Press 1971.

[3] CHENAIS On the existence of a solution in a domain identification pro-problem, J. of Math. Anal. and Appl. Vol 52, n°2, 1975.

[4] GRISVARD Alternative de Fredholm relative au problème de Dirichlet, Bollettino della U.M.I., (4) 5, 1972.

[5] GRISVARD Behaviour of the solutions of an elliptic boundary value problem, SYNSPADE III, Acad. Press 1975.

[6] KADLEC La régularité de la solution du problème de Poisson, Czechoslovak Mat. J. 89, 1964.

[7] LADYZENSKAIA-URALCEVA Equations aux dérivées partielles de type elliptique, Dunod, Paris, 1968.

[8] LIONS Quelques méthodes de résolution des problèmes aux limites non linéaires, Dunod-Gauthier-Villars, Paris, 1969.

[9] LIONS MAGENES Problèmes aux limites non homogènes tome I, Dunod, Paris, 1968.

[10] NECAS Les méthodes directes en théorie des équations elliptiques, Masson, Paris, 1967.

Pierre GRISVARD
I.M.S.P. Parc Valrose
06034 NICE CEDEX FRANCE

On the Method of Strained Coordinates

P. Habets

1. Introduction

M.J. Lighthill [2] introduced a technique to obtain uniform asymptotic expansions of solutions of some nonlinear equations. Let us consider a problem such as

$$x \frac{du}{dx} + f(x,u) = 0, \quad u(1) = b, \tag{1.1}$$

with u, f $\in \mathbb{R}$, whose linear part has a singular point of the first kind for $x = 0$ and whose solution $u = u_0(x)$ exists for $x : 0 < x \leqslant 1$. If we add a small perturbation such as in

$$(x + \varepsilon u) \frac{du}{dx} + f(x,u) = 0, \quad u(1) = b, \tag{1.2}$$

the singularity moves and there is some hope of obtaining a solution $u(x)$ of (1.2) for $x : 0 \leqslant x \leqslant 1$.

For an Euler's equation, (1.1) can be written

$$x \frac{du}{dx} + qu = r, \quad u(1) = b,$$

with q and r constant and its solution

$$u = u_0(x) = \frac{r}{q} + (b - \frac{r}{q})\, x^{-q}$$

has a pole at $x = 0$ if $q > 0$. Hence the convergence of the solution $u = u(x,\varepsilon)$ of (1.2) towards $u_0(x)$ cannot be uniform on $0 \leqslant x \leqslant 1$ and some boundary layer must appear for $x = 0$. This makes this problem a singular perturbation problem.

Lighthill's technique, known as the method of strained coordinates or PLK method, describes a solution of (1.2) using a parametric representation

$$u = u_0(t) + \sum_{i=1}^{\infty} u_i(t)\varepsilon^i, \quad x = t + \sum_{i=1}^{\infty} x_i(t)\varepsilon^i. \qquad (1.3)$$

Introducing (1.3) into (1.2) and collecting terms of equal power in ε, one gets for $n = 0$, problem (1.1) with x replaced by t, and for $n \geqslant 1$

$$t\frac{du_n}{dt} + x_n\frac{du_0}{dt} + f(t,u_0(t))\frac{dx_n}{dt} + \frac{\partial f}{\partial x}(t,u_0(t))x_n + \\ + \frac{\partial f}{\partial u}(t,u_0)u_n = F_n,$$

$$u_n\Big|_{t=1} = x_n\Big|_{t=1} = 0.$$

This can be solved using equations

$$t\frac{du_n}{dt} + \frac{\partial f}{\partial u}(t,u_0)u_n = F_{n1}, \; u_n(1) = 0,$$

$$f(t,u_0(t))\frac{dx_n}{dt} + (\frac{du_0}{dt} + \frac{\partial f}{\partial x}(t,u_0))x_n = F_{n2}, \; x_n(1) = 0,$$

where F_{n_1} and F_{n_2} are only constrained to be such that

$$F_{n_1} + F_{n_2} = F_n .$$

A description of this method together with applications to ODE as well as PDE can be found in A.H. Nayfey [3].

A first justification of this method was given by W. Wasow [5] for the problem

$$(x + \varepsilon u)\frac{du}{dx} + q(x)u - r(x) = 0, \; u(1) = b.$$

In this paper, which unfortunately is obscured by unessential errors, W. Wasow uses the parametric representation

$$u = \sum_{i=0}^{m} u_i(t)\varepsilon^i, \quad x = \sum_{i=0}^{\infty} x_i(t)\varepsilon^i \qquad (1.4)$$

This amounts to choosing $F_{i_1} = 0$ for $i > m$. A correction to W. Wasow's paper was made by Y. Sibuya and K. Takahasi [4]. These authors consider the parametric representation

$$u = u_0(t),\ x = \sum_{i=0}^{\infty} x_i(t)\ (\varepsilon t^{-q_0})^i \tag{1.5}$$

with $q_0 > 0$. This supposes $F_{i1} = 0$, for $i \geq 1$, which might be a disadvantage for computational purposes. Further in (1.5), the convergence is slower than in (1.4). Indeed on the given time interval, one can only prove $\varepsilon t^{-q_0} \leq \varepsilon^{\alpha}$ with $0 < \alpha < 1$, so that (1.5) can be thought of as a power series in ε^{α} as compared to the power series in ε of (1.4). Just as in W. Wasow's paper the proof is based on a majorant series argument. Hence one has to consider equations with C^{∞} coefficients.

In this paper we present an alternative proof of Lighthill's technique using Banach's fixed point theorem. This unables us to weaken the continuity assumptions. Further we give an asymptotic expansion.

$$u = u_0(t) + \varepsilon t^{-q_0}[\sum_0^{\infty} u_i(t,\varepsilon)\varepsilon^i],\ x = t + \varepsilon t^{-q_0}[\sum_0^{\infty} y_i(t,\varepsilon)\varepsilon^i]$$

of the solution $u = u(x)$ which contains infinitely many terms for u and is somewhat intermediate between W. Wasow's approach and the one of Y. Sibuya and K. Takahasi.

2. Assumptions and Notations

Consider the Cauchy problem

$$(x + \varepsilon u)\frac{du}{dx} + q(x)u - r(x) = 0 \tag{2.1}$$

$$u(1) = b \tag{2.2}$$

where $x \in [0,1]$, $u \in \mathbb{R}$ and $\varepsilon > 0$ and assume

(i) $q, r \in C^2([0,1],\mathbb{R})$;

(ii) $q(0) = q_0 > 0$.

We want to construct a representation

$$u = u(t),\ x = x(t)$$

of the solution of the initial value problem (2.1) (2.2) for $0 \leq x \leq 1$ and ε small enough.

3. The zero order solution

Let us consider first the problem (2.1) (2.2) with $\varepsilon = 0$

$$t \frac{du}{dt} + q(t)u - r(t) = 0 \qquad (3.1)$$

$$u(1) = b \qquad (3.2)$$

Equation (3.1) has a unique solution

$$w(t) = \int_0^t e^{-\int_s^t \frac{q(\tau)}{\tau} d\tau} \frac{r(s)}{s} ds$$

which is bounded for all $t \in [0,1]$. Let K be a generic constant. Then

$$|w(t)| \leq K \int_0^t \frac{1}{s} e^{-\int_s^t \frac{q_\circ}{\tau} d\tau} ds$$

$$\leq K \int_0^t \frac{s^{q_\circ - 1}}{t^{q_\circ}} ds = K/q_\circ .$$

Hence the solution $u_\circ(t)$ of (3.1) (3.2) is

$$u_\circ(t) = w(t) + (b - w(1))e^{-\int_1^t \frac{q(\tau)}{\tau} d\tau} , \quad 0 < t \leq 1, \qquad (3.3)$$

and one computes easily the following estimates

$$|u_\circ(t)| \leq |w(t)| + |b - w(1)| \, Kt^{-q_\circ} \leq Kt^{-q_\circ},$$

$$|u_\circ'(t)| = \left|\frac{q(t)u_\circ(t) - r(t)}{t}\right| \leq Kt^{-q_\circ - 1}.$$

In the sequel we shall need the following assumption

(iii) $|q(t)u_\circ(t) - r(t)| \geq ct^{-q_\circ}$, $0 < t \leq 1$.

Let us notice that this condition can be deduced from

(iii') $b \neq w(1)$ and

(iii") $q(t)u_\circ(t) - r(t) \neq 0$, $0 < t \leq 1$.

To show this, one computes first

$$qu_\circ - r = (qw - r) + q(b - w(1))e^{-\int_1^t \frac{q(\tau)}{\tau} d\tau}$$

$$= \{(qw - r)t^{q_0} + q(b - w(1))e^{-\int_1^t \frac{q(\tau) - q_0}{\tau} d\tau}\} t^{-q_0}.$$

It follows that for $t \in]\, 0, t_1\,]$ and t_1 small enough $|qu_0 - r| \geqslant Kt^{-q_0}$. Indeed $qw - r$ is bounded, $q(0)\ (b - w(1) \neq 0$ and $\exp\left(- \int_1^t \frac{q(\tau) - q_0}{\tau} d\tau\right)$ is bounded away from zero. Further assumption (iii'') implies

$$|qu_0 - r| \geqslant K > 0,\ t_1 \leqslant t \leqslant 1.$$

Hence

$$|qu_0 - r| \geqslant Kt^{-q_0},\ 0 < t \leqslant 1.$$

4. A fixed point problem

Let us introduce the new variables $v = v(t) = u(x(t)) - u_0(t)$, $y = y(t) = x(t) - t$ where $u(x)$ is the solution of (2.1) (2.2) and $x(t)$ will be chosen later.

Equation (2.1) implies

$$[\, t+y+\varepsilon(u_0+v)]\,(u'_0+v') + [\,(q+q'y+Q)\ (u_0+v) - (r+r'y+R)]\ (1+y') = 0,$$

where ' denotes the derivative with respect to t, $q = q(t)$, $q' = q'(t)$, $r = r(t)$, $r' = r'(t)$, $Q = q(t+y) - q - q'y$ and $R = r(t+y) - r - r'y$. This equation can be written

$$(tv'+qv) + [\,(qu_0-r)y' + (q'u_0+u'_0-r')y] = \mathcal{R}$$

with

$$\mathcal{R} = -\{yv'+\varepsilon\ (u_0+v)\ (u'_0+v') + qvy' + q'yv + q'y(u_0+v)y' + Q(u_0+v)\ (1+y') - r'yy' - R(1+y')\}.$$

A parametric representation of the solution of (2.1) (2.2) is then

$$u = u_0(t) + v(t),\ x = t + y(t)$$

where v and y are solutions of the system

$$tv' + qv = \mathcal{R}_1,\ v(1) = 0 \qquad (4.1)$$

$$(qu_0-r)y' + (q'u_0+u'_0-r')y = \mathcal{R}_2,\ y(1) = 0 \qquad (4.2)$$

and $\mathcal{R}_1 + \mathcal{R}_2 = \mathcal{R}$. There are infinitely many ways to write these equations depending upon the arbitrary choice we make for $\mathcal{R}_1 = \mathcal{R}_1(t, y, y', v, v', \varepsilon)$. Let us suppose this choice to be such that :

(a) $|\mathcal{R}_1\ (t, 0, 0, 0, 0, \varepsilon)| \leqslant K\varepsilon t^{-2q_0}$;

(b) if $|y_i| \leq kt^{-q_\circ}$, $|y'_i| \leq kt^{-q_\circ-1}$, $|v_i| \leq kt^{-2q_\circ}$ and $|v'_i| \leq kt^{-2q_\circ-1}$, $i = 1, 2$, then

$$|\mathcal{R}_1 (t, y_1, y'_1, v_1, v'_1, \varepsilon) - \mathcal{R}_1 (t, y_2, y'_2, v_2, v'_2, \varepsilon) \leq$$
$$\leq K\,(k+k\varepsilon t^{-q_\circ}+k^2\varepsilon t^{-2q_\circ}+\varepsilon)\; t^{-2q_\circ}$$
$$\cdot \max\,(|y_1-y_2|,\; t\,|y'_1-y'_2|,\; t^{q_\circ}\,|v_1-v_2|,\; t^{q_\circ+1}\,|v'_1-v'_2|).$$

An appropriate choice would be $\mathcal{R}_1 = 0$, but other cases might be interesting for computational purposes. It turns out that $\mathcal{R}$ satisfies some conditions similar to (a) and (b) but somewhat weaker. Indeed

(α) $|\mathcal{R}\,(t, 0, 0, 0, 0, \varepsilon)| = \varepsilon\,|u_\circ|\,|u'_\circ| \leq K\varepsilon t^{-2q_\circ-1}$;

(β) if $|y_i| \leq kt^{-q_\circ}$, $|y'_i| \leq kt^{-q_\circ-1}$, $|v_i| \leq kt^{-2q_\circ}$ and $|v'_i| \leq kt^{-2q_\circ-1}$, $i = 1, 2$, then

$$|\mathcal{R}\,(t, y_1, y'_1, v_1, v'_1, \varepsilon) - \mathcal{R}\,(t, y_2, y'_2, v_2, v'_2, \varepsilon)| \leq$$
$$\leq K\,(k+k\varepsilon t^{-q_\circ} + k^2 t^{-q_\circ}+\varepsilon)\,t^{-2q_\circ-1}$$
$$\cdot \max\,(|y_1-y_2|,\; t\,|y'_1-y'_2|,\; t^{q_\circ}\,|v_1-v_2|,\; t^{q_\circ+1}\,|v'_1-v'_2|)$$

This implies together with (a) and (b) that $\mathcal{R}_2 = \mathcal{R} - \mathcal{R}_1$ satisfies (α) and (β).

The integral form of (4.1) is

$$v = \mathcal{V}\,(y, v) = \int_1^t \frac{1}{s}\, e^{-\int_s^t \frac{q(\tau)}{\tau}\,d\tau}\; \mathcal{R}_1\,(s, y(s), y'(s), v(s), v'(s), \varepsilon)\,ds \tag{4.3}$$

If we notice that

$$p(t) = \frac{u'_\circ+q'u_\circ-r'}{qu_\circ-r}\; t = -1 + \frac{q'u_\circ-r'}{qu_\circ-r}\; t = -1 + \mathcal{O}(t^{q_\circ+1}) \tag{4.4}$$

the equations (4.2) can be written

$$ty' + p(t)y = \frac{t}{qu_\circ-r}\,\mathcal{R}_2,\quad y(1) = 0$$

and are equivalent to

$$y = \mathcal{Y}\,(y, v) = \int_1^t e^{-\int_s^t \frac{p(\tau)}{\tau}\,d\tau}\; \frac{\mathcal{R}_2\,(s, y(s), y'(s), v(s), v'(s), \varepsilon)}{q(s)u_\circ(s)-r(s)}\,ds \tag{4.5}$$

Consider now the Banach space

$$X = \{(y, v) : y \in C^1([t_1, 1], \mathbb{R}),\ v \in C^1([t_1, 1], \mathbb{R})\}$$

with the norm

$$\|(y, v)\| = \sup_{t_1 \leq t \leq t} \max (t^{q_o}|y(t)|,\ t^{q_o+1}|y'(t)|,\ t^{2q_o}|v(t)|,\ t^{2q_o+1}|v'(t)|),$$

and where $t_1 = \varepsilon^{(1-\alpha)/q_o}$, $0 < \alpha < 1$. Equations (4.3) (4.5) define the operator

$$T : X \to X,\ (y, v) \to (\mathcal{Y}(y, v),\ \mathcal{V}(y, v))$$

whose fixed points are solutions of (4.1) (4.2).

5. Asymptotic expansion

THEOREM 1. *Suppose assumptions* (i) (ii) (iii) *are satisfied. Then for* ε *small enough,* $n > 0$, $0 < \alpha < 1$ *and any choice of* R_1 *satisfying conditions* (a) *and* (b), *there exists a representation*

$$x = t + \varepsilon t^{-q_o}[\sum_{k=0}^{n+1} y_k(t, \varepsilon)\varepsilon^k],$$

$$u = u_o(t) + \varepsilon t^{-2q_o}[\sum_{k=0}^{n+1} v_k(t,\varepsilon)\varepsilon^k] \qquad (5.1)$$

of the solution of (2.1) (2.2), *defined for* $t \in [\varepsilon^{(1-\alpha)/q_o}, 1]$ $0<\alpha<1$, *and such that*

$$|y_k| \leq K,\ |y'_k| \leq Kt^{-1},\ |v_k| \leq K,\ |v'_k| \leq Kt^{-1}.$$

The functions y_k, v_k, $k = 0, \ldots, n$ *can be computed from the recurrence relations*

$$y_o = \mathcal{Y}(0, 0)/\varepsilon t^{-q_o},\quad v_o = \mathcal{V}(0, 0)/\varepsilon t^{-2q_o},$$

$$y_k = [\frac{1}{\varepsilon t^{-q_o}} \mathcal{Y}(\varepsilon t^{-q_o} \sum_{i=0}^{k-1} y_i\varepsilon^i,\ \varepsilon t^{-2q_o} \sum_{i=0}^{k-1} v_i\varepsilon^i) - \sum_{i=0}^{k-1} y_i\varepsilon^i] / \varepsilon^k,$$

$$v_k = [\frac{1}{\varepsilon t^{-q_o}} \mathcal{V}(\varepsilon t^{-q_o} \sum_{i=0}^{k-1} y_i\varepsilon^i,\ \varepsilon t^{-2q_o} \sum_{i=0}^{k-1} v_i\varepsilon^i) - \sum_{i=0}^{k-1} y_i\varepsilon^i] / \varepsilon^k.$$

Proof. Since R_1 satisfies condition (a) and R_2 condition (α) it follows for $\varepsilon^{(1-\alpha)/q_o} \leq t \leq 1$,

$$|Y(0,0)(t)| \leq K\int_t^1 e^{-\int_s^t \frac{p(\tau)}{\tau}d\tau}\, \varepsilon s^{-q_\circ-1}ds$$

$$\leq K\varepsilon t\int_t^1 s^{-q_\circ-2}\,ds \leq K\varepsilon t^{-q_\circ},$$

$$|Y'(0,0)(t)| = |-\frac{p(t)}{t}\,Y(0,0)(t) + \frac{R_2}{qu_\circ - r}| \leq K\varepsilon t^{-q_\circ-1}$$

$$|V(0,0)(t)| \leq K\int_t^1 e^{-\int_s^t \frac{q(\tau)}{\tau}d\tau}\, \varepsilon s^{-2q_\circ-1}ds$$

$$\leq K\varepsilon t^{-q_\circ}\int_t^1 s^{-q_\circ-1}ds \leq K\varepsilon t^{-2q_\circ}$$

$$|V'(0,0)(t)| = |-\frac{q(t)}{t}\,V(0,0)(t) + \frac{R_1}{t}| \leq K\varepsilon t^{-2q_\circ-1}$$

i.e.

$$||T(0,0)|| \leq K\varepsilon \tag{5.2}$$

In a similar way, one gets from conditions (b) and (β) that if $||(y_i,v_i)|| \leq k\varepsilon$, i = 1, 2, and $\varepsilon^{(1-\alpha)/q_\circ} \leq t \leq 1$,

$$|Y(y_1,v_1)(t) - Y(y_2,v_2)(t)| \leq K(k)\varepsilon t^{-q_\circ}||(y_1,v_1) - (y_2,v_2)||,$$
$$|Y'(y_1,v_1)(t) - Y'(y_2,v_2)(t)| \leq K(k)\varepsilon t^{-q_\circ-1}||(y_1,v_1) - (y_2,v_2)||,$$
$$|V(y_1,v_1)(t) - V(y_2,v_2)(t)| \leq K(k)\varepsilon t^{-2q_\circ}||(y_1,v_1) - (y_2,v_2)||,$$
$$|V'(y_1,v_1)(t) - V'(y_2,v_2)(t)| \leq K(k)\varepsilon t^{-2q_\circ-1}||(y_1,v_1) - (y_2,v_2)||.$$

These four inequalities prove

$$||T(y_1,v_1) - T(y_2,v_2)|| \leq K(k)\varepsilon\, ||(y_1,v_1) - (y_2,v_2)|| \tag{5.3}$$

It follows from (5.2) (5.3) that we can use Banach's fixed point theorem for ε small enough (cf. J. Dieudonné [1]). Further it is well-known the fixed point z = (y,v) of T can be obtained as limit of the sequence

$$z^{(0)} = (y^{(0)},v^{(0)}) = T(0,0)$$
$$z^{(i)} = (y^{(i)},v^{(i)}) = T(z^{(i-1)}),\quad i = 1, 2, \ldots$$

The following estimates hold

$$||z^{(i)} - z^{(i-1)}|| \leq K(k)\varepsilon||z^{(i-1)} - z^{(i-2)}||$$

$$\leq K(k)\varepsilon^{i}||z^{(0)}|| = O(\varepsilon^{i+1}) \quad (5.4)$$

$$||z^{(n)} - z|| \leq \frac{K(k)\varepsilon}{1-K(k)\varepsilon}||z^{(n)} - z^{(n-1)}|| = O(\varepsilon^{n+2}) \quad (5.5)$$

Define y_i, v_i from the relations

$$z^{(0)} = (\varepsilon t^{-q_\circ}y_\circ, \varepsilon t^{-2q_\circ}v_\circ)$$

$$z^{(i)} - z^{(i-1)} = (\varepsilon t^{-q_\circ}y_i, \varepsilon t^{-2q_\circ}v_i)\varepsilon^{i}, \quad i = 1, \ldots, n$$

$$z - z^{(n)} = (\varepsilon t^{-q_\circ}y_{n+1}, \varepsilon t^{-2q_\circ}v_{n+1})\varepsilon^{n+1}$$

The theorem follows then from

$$z = (y,v) = z^{(0)} + (z^{(1)} - z^{(0)}) + \ldots + (z^{(n)} - z^{(n-1)}) + (z - z^{(n)})$$

which can be written

$$y = \varepsilon t^{-q_\circ}[y_\circ + \varepsilon y_1 + \ldots + \varepsilon^{n+1}y_{n+1}]$$

$$v = \varepsilon t^{-2q_\circ}[v_\circ + \varepsilon v_1 + \ldots + \varepsilon^{n+1}v_{n+1}]$$

and the estimates (5.4) (5.5) imply

$$|y_i| \leq K, \; |y_i'| \leq Kt^{-1}, \; |v_i| \leq K, \; |v_i'| \leq Kt^{-1}.$$

Q.E.D.

If we wish (5.1) to be a uniform representation of the solution $u = u(x)$, $0 \leq x \leq 1$, of (2.1) (2.2) we must prove there exists $t_\circ \in [\varepsilon^{(1-\alpha)/q_\circ}, 1]$ such that $x(t_\circ) = 0$.

THEOREM 2. *Suppose assumptions* (i) (ii) *and* (iii) *are satisfied and* $\alpha < 1/(q_\circ + 1)$. *Suppose further* R_1 *is chosen such that for* $||(y,v)|| \leq k\varepsilon$

$$R_2 = \varepsilon u_\circ u_\circ' + O(\varepsilon^2 t^{-3q_\circ - 1}) \quad (5.4)$$

and

$$b - w(1) > 0.$$

Then there exists $t_\circ \in [\varepsilon^{(1-\alpha)/q_\circ}, 1]$ *such that*

$$x(t_\circ) = y(t_\circ) + t_\circ = 0.$$

Proof. It follows from (4.4) that

$$K_1 \frac{t}{s} \leqslant e^{-\int_s^t \frac{p(\tau)}{\tau} d\tau} = e^{\int_s^t (\frac{1}{\tau} + O(1))d\tau} \leqslant K_2 \frac{t}{s}$$

and from (4.5)

$$y(t) = \int_1^t e^{-\int_s^t \frac{p(\tau)}{\tau} d\tau} [\frac{\varepsilon u_\circ u_\circ'}{qu_\circ - r} + O(\varepsilon^2 s^{-2q_\circ - 1})]ds$$

$$\leqslant \int_1^t e^{-\int_s^t \frac{p(\tau)}{\tau} d\tau} \frac{\varepsilon u_\circ}{s} ds + K\varepsilon^2 t \int_t^1 s^{-2q_\circ - 2} ds$$

Using (3.3) and

$$K_3 \, s^{-q_\circ} \leqslant e^{\int_1^s \frac{q(\tau)}{\tau} d\tau} \leqslant K_4 \, s^{-q_\circ}$$

one gets

$$y(t) \leqslant \varepsilon(b - w(1)) \int_1^t e^{-\int_s^t \frac{p(\tau)}{\tau} d\tau} \frac{e^{-\int_1^s \frac{q(\tau)}{\tau} d\tau}}{s} ds$$

$$+ K \int_1^t e^{-\int_s^t \frac{p(\tau)}{\tau} d\tau} \frac{\varepsilon w(s)}{s} ds + K\varepsilon^2 t^{-2q_\circ}$$

$$\leqslant - K\varepsilon(b - w(1)) \; t \int_t^1 s^{-q_\circ - 2} ds + K\varepsilon t \int_1^t s^{-2} ds + K\varepsilon^2 t^{-2q_\circ}$$

$$y(t) \leqslant - K\varepsilon(b - w(1)) t^{-q_\circ} + K\varepsilon + K\varepsilon^2 t^{-2q_\circ}$$

and for $t = t_1 = \varepsilon^{(1-\alpha)/q_\circ}$

$$y(t_1) \leqslant -K\varepsilon^{\alpha}[(b - w(1)) + \varepsilon^{1-\alpha} + \varepsilon^{\alpha}] \leqslant - t_1 = - \varepsilon^{(1-\alpha)/q_\circ}$$

provided ε is small enough and $\alpha < (1-\alpha)/q_\circ$ (i.e. $\alpha < 1/(q_\circ+1)$). Hence there exists $t_\circ \in [t_1, 1]$ such that $y(t_\circ) = - t_\circ$.

Q.E.D.

Let us remark that if $R_2 = R$, (5.4) follows from condition (β)

$$
\begin{aligned}
|R - \varepsilon u_\circ u_\circ'| &= |R(t,y,y',v,v',\varepsilon) - R(t,0,0,0,0,\varepsilon)| \\
&\leqslant K(k\varepsilon + k\varepsilon^2 t^{-q_\circ} + k^2\varepsilon^2 t^{-q_\circ} + \varepsilon)t^{-3q_\circ-1}||(y,v)|| \\
&\leqslant K\varepsilon^2 t^{-3q_\circ-1}.
\end{aligned}
$$

6. References

[1] Dieudonné, J., Foundations of modern analysis, Academic Press, New York - London, 1960.

[2] Lighthill, M.J., A technique for rendering approximate solutions to physical problems uniformly valid, Phil. Mag. (7), *40*, 1949, 1179-1202.

[3] Nayfey, A.H., Perturbation method, J. Wiley, New York - London - Sydney - Toronto, 1973.

[4] Sibuya, Y. and Takahasi, K., On the differential equation $(x + \varepsilon u)du/dx + q(x)u - r(x) = 0$, Funkcialaj Ekvacioj, *9*, 1966, 71-81.

[5] Wasow, W., On the convergence of an approximation method of M.J. Lighthill, J. Rat. Mech. Anal., *4*, 1955, 751-767.

Nonlinear Diffusion Equations in Biology

K. P. Hadeler

0. Introduction
1. Fisher's model
2. General diffusion equations
3. Application to Fisher's equation
4. The convergence problem
5. Propagation of disturbances
6. Nonhomogeneous spatial domains
7. Propagation in higher dimensions
8. Nerve axon equations
9. The Fitzhugh and the Nagumo model
10. Pulse solutions starting from the axon hillock
11. Other problems

Interaction of species and diffusion are two features common to many biological phenomena. Here the word species comprises notions as different as taxonomic species in ecological models, genotypes in models from population genetics, or chemical species in reaction kinetics. Diffusion may occur in the physical sense as random movement of small particles, as active migration of individuals or as random flow of genes, or even as voltage change in nerve conduction models.

Interaction and diffusion in a model lead to a mathematical description by a system of nonlinear diffusion equations, possibly of a peculiar or degenerate form. Apparently nonlinear parabolic equations are of similar importance in biology as hyperbolic equations in some domains of classical physics. From a mathematical point of view the biological models provide examples for the application of general results on evolution equations, but even more they lead to a variety of questions of existence, uniqueness, stability, and convergence of solutions to partial and ordinary differential equations.

The aim of the paper is to provide a-necessarily short and incomplete – review of some major classes of models, to indicate the biological relevance and some of the results so far known.

1. Fisher's model

R.A. Fisher [1] considered selection at a single autosomal locus with two alleles in a large population with homogeneous genetic background, random mating and overlapping generations. Due to the overlap of generations an exact description of the development of the population can be given only in terms of genotype frequencies. Under specific assumptions on fertilities, mortalities, and the mode of replacement various systems of ordinary differential equations for the three genotype frequencies have been derived (see [7] , [6] , [5]). On the contrary, Fisher tried to describe the temporal changes of the gene frequencies. Although his equation, derived under the hypothesis that the whole population is in Hardy-Weinberg equilibrium at every moment, is unjustified, it is a good approximation for most models, at least in the neighborhood of an equilibrium or for small differences between the selection coefficients.

Let A,a denote the two genes and let ϱ, σ, and τ be the fertilities (viabilities) of the three genotypes AA, Aa, aa, respectively. Then the evolution of the frequency p of the gene A is governed by the ordinary differential equation

$$\dot{p}=p(1-p)(\varrho-\tau-(2\varrho-\sigma-\tau)p). \tag{1.1}$$

For constant selection coefficients ϱ, σ, τ we can normalize $\varrho=1$. Fisher [2] had also the idea to investigate a population with the selection law (1.1) in a spatial domain in the presence of migration. Most characteristic features of such selection-migration models can be exhibited in the case of a one-dimensional domain, simulating migration along a river or in a homogeneous strip. Then we have a parabolic equation

$$p_t=p(1-p)(\varrho-\tau-(2\varrho-\sigma-\tau)p)+Dp_{xx}. \tag{1.2}$$

With a suitable normalization of the space variable the migration coefficient becomes D=1. Thus we arrive at

$$p_t=p(1-p)(1-\tau-(2-\sigma-\tau)p)+p_{xx}\ . \tag{1.3}$$

Because of the symmetry between the genes A and a one can assume $\sigma\geq\tau$. There are two important cases

Case 1: Heterozygotes intermediate. Here $\sigma\geq 1>\tau$. By a substitution in x and t the equation is transformed into

$$u_t=u_{xx}+F(u)\qquad, \tag{1.4}$$

where

$$F(u)=u(1-u)(1+\nu u), \tag{1.5}$$

and the only remaining parameter is

$$\nu=(\sigma-1)/(1-\tau)-1, \qquad -1 \leq \nu < \infty . \qquad (1.6)$$

The source function F is positive in the interval (0,1). Important special cases are $\nu=-1$ for complete dominance and $\nu=0$ for additive fitness.

Case 2: Heterozygotes inferior. Here $\sigma>1$, $\tau>1$, and equation (1.3) is similar to (1.4), but such

$$F(u)=u(1-u)(u-\mu) \qquad (1.7)$$

where

$$\mu=(\tau-1)/(\sigma+\tau-2), \qquad 0<\mu<1 \quad . \qquad (1.8)$$

In this case the source function is negative for $0<u<\mu$ and positive for $\mu<u<1$.

A further case, heterozygotes superior, $\sigma<1, \tau<1$, will not be considered here in detail, because most of the propagation phenomena pertaining to this case are covered by case 1.

Fisher asked whether the solution of equation (1.4), (1.5) (with $\nu=-1$) which starts from the initial distribution

$$u(x,0)=\begin{cases} 1 & \text{for } x<0 \\ 0 & \text{for } x\geq 0 \end{cases} \qquad (1.9)$$

evolves into a "travelling population front". We apply a heuristic argument: For t=0 there are only superior individuals AA on the left half-axis, and only inferior on the right half axis. For small $t>0$ some aa-individuals will move to the left and become subsequently extinguished, whereas some AA-individuals move to the right and multiply. Thus the population front - the boundary between regions with predominant A and with predominant a - will shift to the right. The question arises whether it will travel to the right with asymptotically constant velocity, thereby assuming a constant shape. This idea leads to the general question whether Fisher's equation (1.3) has solutions in the form of travelling fronts or waves, and which initial data evolve into such fronts.

Some of these problems related to Fisher's equation have been solved in a contemporary paper by Kolmogorov, Petrovskij, and Piskunov [4] (in following quoted as KPP). However, these questions can be dealt with in a much wider frame-work.

2. General diffusion equations

We assume a general semilinear diffusion equation

$$u_t = u_{xx} + F(u) \tag{2.1}$$

where F is continuously differentiable in $[0,1]$ and satisfies

$$F(0) = F(1) = 0 \quad . \tag{2.2}$$

It can be easily shown ([4] ,[5]) that the solution to any piecewise continuous initial data $u(x,0)$ with $0 \leq u(x,0) \leq 1$ exists for all $t > 0$ and satisfies $0 \leq u(x,t) \leq 1$. As for Fisher's equation (1.4) we pose the problem of travelling fronts. A solution of equation (2.1) which is a front travelling with constant shape and constant velocity c is a solution of the form $u(x,t) = u(x-ct)$. Then the function of one variable $u = u(\xi)$, $\xi = x - ct$, is a solution of the ordinary differential equation (' denotes $d/d\xi$)

$$-cu' = u'' + F(u) \quad . \tag{2.3}$$

A travelling front arising from initial data (1.9) should satisfy the boundary conditions

$$u(-\infty)=1, \qquad u(+\infty)=0 \tag{2.4}$$

and the natural restriction

$$0 \leq u(\xi) \leq 1. \tag{2.5}$$

In a more general sense we shall call any non-trivial solution of (2.3) satisfying condition (2.5) a front travelling with speed c.

The second order equation (2.3) is equivalent with the system

$$u' = v \quad , \qquad v'=-cv-F(u) \quad . \tag{2.6}$$

The points (0,0) and (1,0) are stationary points in view of (2.2). Thus the boundary conditions (2.4) require one to find a trajectory of the system (2.6) connecting the two stationary points (1,0) and (0,0) and remaining in the domain (2.5).

For a general differential system a trajectory connecting two stationary points is rather improbable. Thus we expect that the boundary value problem (2.3), (2.4), (2.5) has a solution only for special values of the speed parameter c. Paralleling the discussion of Fisher's model we discriminate between several cases.

Case 1:

$$F(u) > 0 \quad \text{for} \qquad u \in (0,1) \quad . \tag{2.7}$$

If not otherwise stated we shall require also

$$F'(0) > 0 \quad . \tag{2.8}$$

Sometimes we use the additional hypothesis

$$F'(0) > F'(u) \qquad \text{for } u \in (0,1) \quad . \tag{2.9}$$

Case 2: Suppose $\mu \in (0,1)$.

$$F(u) < 0 \text{ for } u \in (0,\mu), \quad F(u) > 0 \text{ for } u \in (\mu,0). \tag{2.10}$$

Again, if not stated otherwise, we require

$$F'(0) < 0, \qquad F'(\mu) > 0.$$

Besides Fisher's model, some simplified models for the propagation of the nervous impulse also lead to case 2 (cf. section 9).

Case 3: Let $\mu \in (0,1)$,

$$F(u) \equiv 0 \text{ for } u \in (0,\mu) , \quad F(u) > 0 \text{ for } u \in (\mu,1) . \tag{2.11}$$

This case has been investigated in connection with flame propagation [11] , and also allows a biological interpretation: Assume the fitness parameters in equation (1.2) are frequency-dependent: If there is no selection ($\sigma = \varrho = \tau$) at low frequencies of A and intermediate heterozygotes ($\sigma \geqq \varrho > 1$) for high frequencies of A then (1.2) assumes the form (2.1), (2.11).

In these three cases the behavior of the differential system (2.6) is significantly different. In case 1 the point (1,0) is a saddle point for all values of c, and (0,0) is a stable focus for $c \geqq c^* := 2\sqrt{F'(0)}$, and a vortex, center or unstable focus for $c < c^*$. A travelling front is a trajectory starting at (1,0) and ending up at (0,0), thereby remaining in $0 \leqq u \leqq 1$. Thus there are no fronts for $c < c^*$. There is a minimal $c_0 \geqq c^*$ for which the unstable manifold of the saddle point enters (0,0). For $c > c_0$ it enters in the main direction, for $c = c_0$ in the side direction (or in the unique distinguished direction, if $c_0 = c^*$). The typical situation for $c = c_0 > c^*$ is depicted in fig. 1. For details of proof see [7] , [5] .

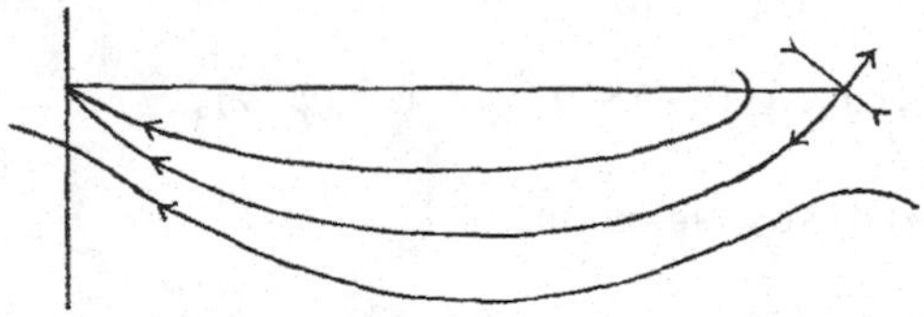

Fig 1

A heuristic argument can be extended to an exact proof for a variational principle characterizing the minimal speed. Suppose there is a differentiable arc $v=-q(u)$, $q: [0,1] \longrightarrow \mathbb{R}$, where

$$q(u) > 0 \text{ for } u \in (0,1) , \quad q(0)=0, \quad q'(0) > 0, \tag{2.12}$$

such that the vector field of equations (2.6) is pointing "upward" along this arc. Then the unstable manifold of the saddle point cannot leave the domain bounded by this arc, the u-axis and a piece of the line v=1, and thus must enter (0,0), i.e. it forms a front (fig 2).

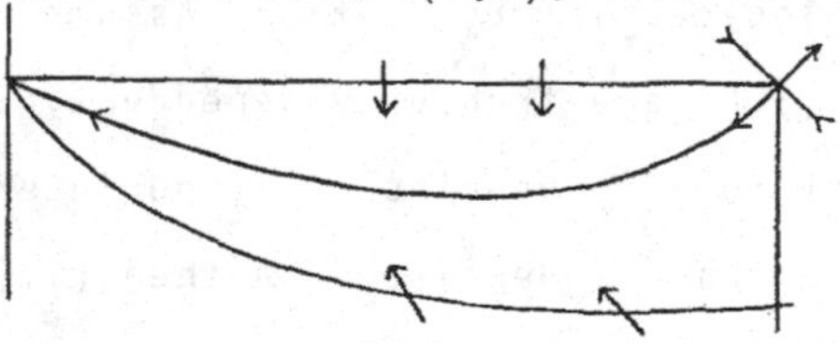

Fig 2

Therefore c_0 is the minimum of all numbers c, for which such an arc exists, or, equivalently

$$c_0 = \min_q \sup_{0<u<1} \left\{ q'(u) + \frac{F(u)}{q(u)} \right\} , \tag{2.13}$$

where the minimum is taken over all continuously differentiable functions q with properties (2.12).

In particular one can choose q linear and obtain an upper bound for c_0,

$$2\sqrt{F'(0)} \le c_0 \le 2\sqrt{L}, \text{ where } L = \sup_{0<u<1} F(u)/u. \tag{2.14}$$

Under condition (2.9) we have $F'(0)=L=c_0^2/4$.

For the convergence problem it is important to know that the trajectories arrive at (0,0) with the direction

$$\frac{dv}{du} = \begin{cases} \frac{1}{2}\sqrt{c^2-4F'(0)} - \frac{c}{2} & \text{for } c > c_0, \\ -\frac{1}{2}\sqrt{c^2-4f'(0)} - \frac{c}{2} & \text{for } c = c_0. \end{cases} \tag{2.15}$$

Thus we can say that, as functions of $\xi=x-ct$, fast fronts $u(\xi)$ decrease slowly to 0 for $\xi \longrightarrow +\infty$, whereas slow fronts fall steeply off to 0.

In <u>case 2</u> the stationary points (0,0), (1,0) are saddle points, whereas $(\mu,0)$ is a stable focus for $c \geq c^* = 2\sqrt{F'(\mu)}$, and a stable vortex, center, unstable vortex or unstable focus for $c < c^*$. Now there are various bounded solutions connecting two of the three stationary points. Because (2.1), (2.10) is carried into itself by the substutions $c \longrightarrow -c$, $v \longrightarrow -v$, $\xi \longrightarrow -\xi$ we need only consider speeds $c \geq 0$. Solutions remaining in one cf the domains $[0,\mu] \times \mathbb{R}$, $[\mu,1] \times \mathbb{R}$ are covered by case 1. For any c the phase pattern is mainly determined by the global behavior of the unstable manifolds of the saddle points and the character of the point $(\mu,0)$.

For large c the unstable manifold I of (1,0) enters $(\mu,0)$ in the main direction, and for $c < c_1$ it leaves the domain $u \geq \mu$. Since for $c << 0$ this trajectory I arrives at the negative v-axis there must be a number $c_0 < c_1$ for which I enters the stable manifold of (0,0). This number c_0 is unique (for proofs see [7]), c_0 is positive if $\int_0^1 f(u)du > 0$. In fig 3 the case $c_0 > 0$ is shown.

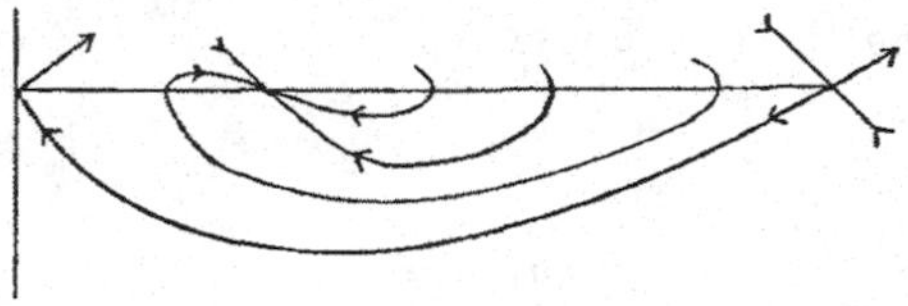

Fig 3

For $c > c_2$ the unstable manifold $\bar{I}$ of (0,0) enters $(\mu,0)$ in the main direction, for $c = c_2$ in the side direction, thereby always remaining in $u \leq \mu$. For $c < c_2$ it leaves $u \leq \mu$. Again, for a certain $\bar{c}_0 < c_2$ the manifold $\bar{I}$ enters (1,0). Because of the uniqueness of c_0, $\bar{c}_0$ and the afore-mentioned symmetry we have $\bar{c}_0 = - c_0$.

Now we can give a complete review of travelling fronts.

F1) For $c \geq c_1$ there is a monotonely decreasing front with $u(-\infty)=1$, $u(+\infty)=\mu$.

F2) For $c \geq c_2$ there is a monotonely increasing front with $u(-\infty)=0$, $u(+\infty)=\mu$.

F3) For $c = c_0$ there is a monotone front with $u(-\infty)=1$, $u(+\infty)=0$, for $c=-c_0$ there is a monotone front with $u(-\infty)=0$, $u(+\infty)=1$.

F4) For $\max(c^*,c_0)<c<c_1$ there is a front with $u(-\infty)=1$, $u(+\infty)=\mu$ which decreases to a certain value $u_0<\mu$ and then increases to μ.

F5) For $\max(c^*,-c_0)<c<c_2$ there is a front with $u(-\infty)=0$, $u(+\infty)=\mu$ which increases to a certain value $\bar{u}_0>\mu$ and then decreases to μ.

F6) For $\max(c_0,0)<c<c^*$ there are oscillating fronts with $u(-\infty)=1$, $u(+\infty)=\mu$.

F7) For max $(-c_0, 0) < c < c^*$ there are oscillating fronts with $u(-\infty)=1$, $u(+\infty)=\mu$.

F8) For $c=0$ there are periodic solutions with $|u| \leq 1$.

F9) For $c=0$ there is a front with $u(-\infty)=u(+\infty)=0$ (if $c_0 > 0$) and or a front with $u(-\infty)=u(+\infty)=1$ (if $c_0 < 0$). If $c_0=0$ then only periodic fronts and the two fronts F3 exist.

Case 3: Again $(1,0)$ is a saddle point. The segment $[0,\mu]$ of the u-axis consists of stationary points. In the strip $[0,\mu]\times\mathbb{R}$ the trajectories of (2.6) are the straight lines $v=-cu$. For $c \leq 0$ and small $c>0$ the unstable manifold of $(1,0)$ meets one of these lines on $u = \mu$ and then leaves $u \geq 0$ along this line. For large $c>0$ it ends up in one of the stationary points with $u\in(0,\mu]$. Exactly for one speed $c_0 > 0$ it meets the line through $(0,0)$. Thus we have a front with $u(-\infty)=1$, $u(+\infty)=0$ for $c=c_0$, and fronts with $u(-\infty)=1$, $u(+\infty)=u_c\in(0,\mu]$ for $c > c_0$ (see fig 4).

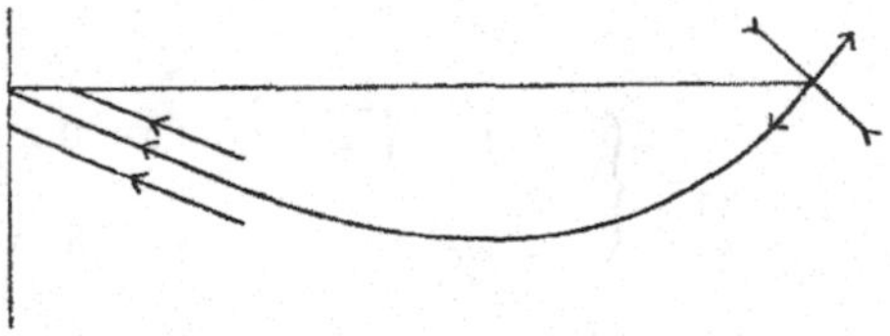

Fig 4

Also in case 2 and 3 the variational principle (2.13) characterizes the speed c_0, also the upper bound (2.14) is valid.

3. Application to Fisher's equation

The results of section 3 can be applied to Fisher's population genetic model. Here the function F is a cubic, and the distinguished speeds are explicitely known.

In <u>case 1</u> (intermediate heterozygotes) we have

$$c_0 = \begin{cases} 2 & \text{for} \quad -1 \le \nu \le 2 \\ \dfrac{\nu+2}{\sqrt{2\nu}} & \text{for} \quad \nu \ge 2 \end{cases} \quad . \tag{3.1}$$

This result has been obtained in [4] for $\nu=-1$, it is a simple consequence of [4] for $\nu \le 0$, for further details see [7]. For $\nu \ge 2$ the slowest travelling front is the Huxley pulse $u(t)=\left[1+\exp\left(\sqrt{\nu/2}\,t\right)\right]^{-1}$ for $\nu \in [-1,2)$ the front is not explicitely known. In [8] for the case of complete dominance ($\nu=0$) an asymptotic expansion for large c is given.

In case 2 we have $0<\mu \le 1/2$ in view of $\sigma \ge \tau$. The distinguished values of c are ([39], [7])

$$c^*=2\sqrt{\mu(1-\mu)} \quad ; \qquad c_1 = \begin{cases} (1+\mu)/\sqrt{2} & \text{for } 0<\mu \le 1/3 \\ c^* & \text{for } 1/3 \le \mu \le 1/2 \end{cases} \quad ;$$

$$c_2 = c^* \quad ; \qquad c_0 = 1/\sqrt{2}-\mu\sqrt{2}. \tag{3.2}$$

With $\bar{\mu}=(1-\sqrt{2/3})/2$ we have

$$c^* < c_0 \text{ for } 0<\mu<\bar{\mu} \quad , \qquad c^* > c_0 \quad \text{for } \bar{\mu}<\mu<1/2 \ .$$

Fronts F1, F2, F3 exist (with $c_0 \ge 0$). For $\mu > 1/3$ we have $\max(c^*,c_0)=c^*=c_1$, and fronts F4 do not exist. Fronts F5 never exist in view of $\max(c^*,-c_0)=c^*=c_2$. Fronts F6 can only exist for $\mu>\bar{\mu}$, fronts F7 exist for $0 < c < c^*$. The front with speed c_0 is explicitely known as

$$u(t) = \left[\exp(t/\sqrt{2})+1\right]^{-1} \quad , \qquad v=u(1-u)/\sqrt{2} \quad . \tag{3.3}$$

The representation of the minimal speed c_0 by the same variational principle suggests that c_0 depends in some sense continuously on the source function F. Indeed, if we substitute (3.1), (3.2) back in Fisher's equation (1.3), we obtain the speed c_0 as a continuous function of σ and τ , namely with $\Delta = \sigma + \tau - 2$

$$c_0 = \begin{cases} 2\sqrt{1-\tau} & \text{for } \sigma - 1 \leq 3(1-\tau), \\ (\sigma - \tau)/\sqrt{2\Delta} & \text{for } \sigma - 1 \geq 3(1-\tau), \tau \leq 1, \\ \sqrt{\Delta/2} - (\tau - 1)\sqrt{2/\Delta} & \text{for } \sigma \geq \tau \geq 1. \end{cases} \tag{3.4}$$

A surprising fact is that for constant τ the minimal propagation speed is independent of σ as long as $\sigma < 1+3(1-\tau)$.

4. The convergence problem

The question of stability of travelling fronts and of convergence towards travelling fronts are mathematically most interesting. They are important from a biological view-point since they single out those solutions which can appear in practical applications.

Consider again equation (2.1)

$$u_t = u_{xx} + F(u) \tag{4.1}$$

with $F(0)=F(1)=0$. Suppose ϕ is a solution of $\phi'' = -c\phi' - F(\phi)$, $\phi(-\infty)=1$, $\phi(+\infty)=0$, $0 \le \phi(\xi) \le 1$ for some value c. Then every function

$$u(x,t) = \phi(x-ct+k), \tag{4.2}$$

k an arbitrary constant, is a travelling front with speed c.

In the following we discuss the problem of which solutions to (4.1) converge to a front with a given speed c. For this purpose it is convenient to apply a substitution $\bar{u}(x,t) = u(x+ct,t)$ which leads to

$$\bar{u}_t = \bar{u}_{xx} + c\bar{u}_x + F(\bar{u}) \tag{4.3}$$

and carries fronts travelling with speed c into stationary solutions (standing fronts)

$$\bar{u}(x,t) = \phi(x+k) \quad . \tag{4.4}$$

The propagation effect by multiplication and diffusion is compensated by a drift $-c\bar{u}_x$.

Historically, the first convergence results were obtained by Kolmogorov et. al. [4] . They proved for a positive source term (case 1 with condition (2.9)) that the solution starting from a step function (1.9) converges to a front with minimal speed in the following sense: Let $\bar{u}$ be the solution of (4.3) , $c=c_0$, with initial condition (2.9). Fix $\varkappa \in (0,1)$. For each $t>0$ there is a unique $\zeta(t)$ such that $u(\zeta(t)t,t) = \varkappa$ (since $\partial\bar{u}/\partial x < 0$ for $t>0$). Define a function u^* by

$u^*(x,t)=\bar{u}(x+\zeta(t)t,t)$. Apparently the function u^* is obtained from u in such a way that the graphs of $u^*(\cdot,t)$ coincide for all t at x=0. The authors [4] prove (theorem 13)

$$u^*(x,t) \longrightarrow \Phi(x+k) \tag{4.5}$$

uniformly in x for $t \longrightarrow \infty$ (the constant k depending on $\varkappa$). One can even show that $u^*(x+\zeta(t_0)t_0, t+t_0)$ converges to the same $\Phi(x+k)$ as $t_0 \longrightarrow \infty$ uniformly in $\mathbb{R}\times[0,T]$, $T>0$. Moreover ([4], theorem 17) $\zeta'(t) \longrightarrow 0$, i.e. the speed of the close -to standing front u^* slows down for $t \longrightarrow \infty$.

In the terminology of section 1 the solution with initial values (2.9) approximates a travelling front (with speed c_0) <u>in shape,</u> and its speed, measured as the velocity of the point x with $\bar{u}(x,t)=\varkappa$ approximates c_0. Nevertheless, as we shall see later, the solution u(x,t) of (4.1) does not <u>converge</u> to a front $\Phi(x-c_0t+k)$, the solution lags behind the front.

Kanel' [13] proved a convergence theorem for arbitrary source terms and large speeds. He observed first the important fact that the asymptotic propagation speed (if the speed is not unique) is determined by the asymptotic behavior of the initial data for $x \longrightarrow +\infty$. An interesting result ([13], thm.4) is the following: Let F satisfy F(0)=F(1)=0 and be arbitrary otherwise. Suppose $c>0$ is such that

$$c^2 \geq 4 \sup_{0\leq u\leq 1} F'(u) \quad . \tag{4.6}$$

If $\bar{u},\bar{v}$ are any two solutions of equation (4.3) with

$$|\bar{u}(x,0)-\bar{v}(x,0)| = O(\exp(-\frac{c}{2}x-\varepsilon|x|^{\alpha})), \quad (4.7)$$

with $\alpha \in (1/2,1)$ and $\varepsilon > 0$, then $\bar{u}(x,t)-\bar{v}(x,t) \longrightarrow 0$ uniformly on every half-line $[x_0,\infty)$.

It is worthwhile to sketch the proof, which, being based on an analytical trick, cannot easily be generalized: The difference $\bar{w}=\bar{u}-\bar{v}$ satisfies a linear equation

$$\bar{w}_t=\bar{w}_{xx}+c\bar{w}_x+F'(\theta)\bar{w} \quad , \qquad 0\le \theta(x,t)\le 1.$$

Poisson's formula for the heat equation and the maximum principle show that $\bar{w}$ can be estimated as

$$|\bar{w}(x,t)| = O(\exp(-\frac{c}{2}x - \frac{\varepsilon}{2}|x|^{\alpha}))$$

in every bounded strip $\mathbb{R}\times[0,T]$, $0 < t < \infty$. The function $z(x,t)=\bar{w}^2(x,t)\exp(cx)$ satisfies

$$z_t=z_{xx}+ 2[F'(\theta)-\frac{c^2}{4}] z-2e^{cx}(w_x+cw/2)^2 \quad . \quad (4.8)$$

Since z decreases to zero for $|x|\longrightarrow\infty$ and the source term in (4.8) is nonpositive, z is majorized by the solution of the heat equation with the same initial data, and the latter uniformly tends to zero.

A related result is stated in [5] , thm. 4.1: Let F belong to case 1,2, or 3. For initial data $\bar{u}(x,0)$ with $u(x,0)\equiv 0$ for $x > x_0$ for some x_0 and $c > c_0$ holds $\bar{u}(x,t)\longrightarrow 0$ for all $x\in\mathbb{R}$.

We discuss some implications for the case of a positive source term (case 1). The asymptotic behavior of the front ϕ_c with speed c is

$$\phi_c(\xi) \sim \exp(-\lambda_c \xi)$$

where $-\lambda_c$ is given by the right-hand side of (2.15). According to Kanel's result a solution $\bar{u}(x,t)$ will converge to the front $\phi_c(x+k)$ if

$$|u(x,0)-\phi_c(x+k)| = O(\exp(-\tfrac{c}{2}x-\varepsilon|x|^{\alpha})). \qquad (4.9)$$

For $c > c_0$ (then $c > 2\lambda_c$) condition (4.9) requires that the difference $\bar{u}(\cdot,0)-\phi_c$ decreases significantly faster than ϕ_c itself. The same is true if $c=c_0$, $c_0=2\sqrt{F'(0)}$. Thus we have a negative result for the KPP problem (case 1, property (2.9), $c_0=2\sqrt{F'(0)}$): The step function and zero coincide for large x. Therefore the solution u arising from the step function goes to zero uniformly on every interval $[x_0,\infty)$. Consequently, it cannot converge to a front $\phi_{c_0}(x+k)$, although it converges to such a front in shape and speed.

For cases 2 and 3 Kanel' [13] proved that every solution $\bar{u}$ generated by a nonincreasing initial function $\bar{u}(x,0)$ with $\bar{u}(x,0)\equiv 1$ for $x \le x_1$, $\bar{u}(x,0)\equiv 0$ for $x \ge x_2 \ge x_1$ converges to a front $\phi(x+k)$ ($c=c_0$ is unique). Again, the proof uses direct analytic tools: Since ϕ and $\bar{u}(x,t)$ are monotone, $\bar{u}$ allows a representation $\bar{u}(x,t)=\phi(x+k(x,t))$, where k obeys the linear equation

$$k_t=k_{xx}+b(x,t)k_x=0, \qquad b(x,t)=c-q'(u)\left[1+u_x/q(u)\right],$$

where $v=-q(u)$ represents the trajectory of ϕ in the u,v-plane (see (2.6)). With estimates for $u_x/q(u)$ one proves convergence to a stationary solution $k(x,t)\longrightarrow k(x)$. Thus $\bar{u}(x,t)\longrightarrow \phi(x+k(x))$, and k(x) is constant. Although more general results have been obtained by now, this proof is of continuing interest. Closeness of initial data can be defined in various ways. In (4.7) we assumed that the difference decays sufficiently fast for $x \longrightarrow \infty$. A travelling front $\Phi_c(x)$ and the translated front $\Phi_c(x+k), k \neq 0$, are not necessarily close in this sense (they are close in cases 2,3, and in case 1 if $c = c_0 > 2\sqrt{F'(0)}$). This observation leads to the definition: If Φ_c is a front and k_0 is of bounded variation on the real axis then $u(x) = \Phi_c(x+k_0(x))$ is called a perturbation of Φ_c. Kanel' [13] proved that any solution starting from a non-increasing perturbed front converges to a front with the same speed (for case 1 with condition (2.9), and cases 2,3).

Kanel's result (4.9) says that a solution $u(x,t)$ converges to a front $\Phi_c(x)$ if $u(x,0)$ and $\Phi_c(x)$ have the same behavior for $x \longrightarrow +\infty$. A related result for the general case 1 has been shown by Rothe [14] : Let $c \geq c_0$, suppose the initial function $u_0 : \mathbb{R} \longrightarrow [0,1]$ has the following properties: u_0 is continuously differentiable with the possible exception of finitely many jumps, where u_0 is continuous from the right; $u_0(-\infty)=1$, $u_0(+\infty)=0$, and u_0 is strictly decreasing where $u_0(x) \in (0,1)$. If appropriately defined, the inverse function $\varphi_0 : [0,1] \longrightarrow \mathbb{R}$ is continuous and differentiable in neighborhoods of 0 and 1. Let φ_c be the inverse function of Φ_c. Suppose $u_0(x)$ does not go to 1 too slowly for $x \longrightarrow -\infty$ and approximates the front for $x \longrightarrow +\infty$,

$$\int_{1-\delta}^{1} |\varphi_0(v)|\, dv < \infty, \quad \int_{0}^{\delta} \left|\frac{d}{dv}(\varphi_0(v) - \varphi_c(v))\right| dv < \infty .$$

Then $u(x,t)$ converges to $\phi_c(x)$ uniformly for $t\longrightarrow\infty$. Further convergence results have been announced by Fife and McLeod [16].

5. Propagation of disturbances

A typical biological problem is the spread of a gene or a pest starting from a small area with a mutation or a few immigrants. This leads to the study of solutions of equation (4.1) starting from initial data with bounded support. Kanel' [17] showed for case 1 what is now called the "hair-trigger effect": For any initial function $u(x,0)$, $0\leq u(x,0)\leq 1$, which is positive on some interval (x_1,x_2), the solution converges to 1 uniformly on bounded intervals. A different proof has been given in [5], thm. 3.1. These authors cover also the case where $F(0)=F(1)=0$, $F'(0)>0$, $F(u)>0$ in $(0,\mu)$, $F(u)<0$ in $(\mu,1)$. Here one can show ([5], thm.3.1) that if $u(x,0)\not\equiv 0$, $u(x,0)\not\equiv 1$ then $\lim_{t\longrightarrow\infty} u(x,t)=\mu$. This case covers Fisher's model for superior heterozygotes: If the spatially distributed population is not completely homozygotic then it approaches the heterozygotic equilibrium everywhere.

Kanel' [17] demonstrated a threshold phenomenon for case 3 with the one parameter family of initial data

$$u(x,0) = \begin{cases} 1 & \text{for } |x| \leq \ell \\ 0 & \text{for } |x| > \ell \end{cases} ;$$

if

$$\ell < \mu\sqrt{\pi(2eL)} \tag{5.1}$$

(μ defined by (2.11), L by (2.14), and $e=2.72\ldots$) then u goes to zero uniformly in $\mathbb{R}$. On the other hand suppose $F(u)/u \geq k$ in $(\mu_1,\mu_2)\in(0,1)$.

If

$$\ell > \pi/(2\sqrt{k})+\mu_1/(\sqrt{k}(\mu_2-\mu_1)) \qquad (5.2)$$

then u tends to 1 uniformly on bounded sets.

A similar threshold property plays an important role in parabolic systems simulating the propagation of the nervous impulse. The threshold properties have been extended in [5] to case 2 and more general initial data. For case 2 a sufficient condition for decay $u(x,t) \longrightarrow 0$ is obviously $u(x,0) \leq \mu$. A weaker condition is the following: For some $\mu_1 \in [0,\mu)$ let $F(u) \leq L_1(u-\mu_1)$ for $u \in (\mu_1,1)$ and suppose

$$\int_{-\infty}^{\infty} \max\,[u(x,0)-\mu_1,0]\;dx < \left(\frac{2\pi}{L_1 e}\right)^{1/2} (\mu-\mu_1)\;. \qquad (5.3)$$

[5] ,thm.3.3 gives a sufficient condition for $u(x,t) \longrightarrow 1$. In [5] also the initial-boundary-value-problem in the quarter-plane $x \geq 0$, $t > 0$ with a boundary condition $u(0,t) = \psi(t)$ is discussed.

A result of Chafee [18] can be compared with (5.3). He allows more general functions F than in our previous case 2: Suppose F"' is continuous, $F(0)=0$, $F'(0) < 0$ and there is $\mu > 0$ such that $F(\mu) < 0$ and $\int_0^u F(v)dv < 0$ for $0 < u \leq \mu$. Then for any square integrable uniformly continuous initial function with $0 \leq u(x,0) \leq \mu$ for all x the solution converges to zero on bounded sets. A similar result has been proved for finite intervals with boundary conditions of the second kind [19] .

6. Nonhomogeneous spatial domains

Travelling fronts appear in spatially homogeneous domains under suitable initial conditions. On the other hand non-constant stationary distributions can arise when the source term of the diffusion equation depends explicitely on the space coordinate. E.g. in Fisher's model (1.2) the fitness parameter may depend on x,

$$p_t=p_{xx}+p(1-p)\left[\varrho(x)-\tau(x)-(2\varrho(x)-\sigma(x)-\tau(x)p)\right], \qquad (6.1)$$

such that in some areas the gene A is advantageous, in other areas the gene a. Models of this type have been discussed rather early ([23]), recently their study has been promoted by the development of discrete selection-migration models ([22]). Fleming[20] treats (cf. (1.2))

$$u_t=u_{xx}+g(x)u(1-u)(1+\nu u), \qquad -1<\nu<+\infty \qquad (6.2)$$

as a special case of

$$u_t=u_{xx}+\lambda g(x)F(u) \ , \qquad (6.3)$$

where F satisfies (2.7), (2.8), and $F'(1)<0$. Here the function g is piecewise continuous and takes both positive and negative values, λ is a positive parameter. If the habitat is the interval $[0,1]$ and genes do not cross the boundary, $u_x(0,t)=u_x(1,t)=0$, then a stationary state is a solution of the boundary value problem

$$u''(x)+\lambda g(x)F(u(x))=0, \qquad u'(0)=u'(1)=0 \quad . \qquad (6.4)$$

Fleming shows that if the trivial equilibria $u\equiv 0$, $u\equiv 1$ are isolated in $X=\{u\in H^1[0,1] : 0\leq u\leq 1\}$ then their stability is determined by the average relative fitness $G=\int_0^1 g(x)dx$ of the gene A.
One can assume $G<0$ (for $G>0$ interchange A and a). The equilibrium $u\equiv 1$ is unstable for any $\lambda>0$. There is a critical calue $\lambda_1>0$ such

that $u \equiv 0$ is stable for $\lambda \in (0, \lambda_1)$ and unstable for $\lambda > \lambda_1$. The biological interpretation is obvious: For small λ , corresponding to high diffusion rates, the average disadvantage of A causes the stability of 0. For large λ diffusion does not average the local effects.

Fleming established the existence of a third, nonconstant, equilibrium, with assistance of the appropriate Lyapunov functional

$$I[u] = \int_0^1 \left\{\frac{1}{2} u_x^2 - \lambda g(x) \mathcal{F}[u(x)]\right\} dx, \qquad \mathcal{F}[u] = \int_0^u F(v)dv \quad . \qquad (6.5)$$

Hoppensteadt [21] described the nontrivial equilibrium in the neighborhood of λ_1 by singular perturbation methods.

For a similar problem on the real line

$$u_t = u_{xx} + g(x)u(1-u) \qquad (6.6)$$

Conley [24] showed: If $xg(x) > 0$ for large $|x|$ then there is a stationary solution which is increasing and satisfies $u(-\infty)=0$, $u(+\infty)=1$. If g is negative and non-integrable at $\pm\infty$ and positive and sufficiently large on some finite interval then there is a nonvanishing stationary solution vanishing at $\pm\infty$.

7. Propagation in higher dimensions

From a biologist's view propagation of genes or pests can be studied in one, two or three dimensions, e.g. corresponding to propagation along a river, a surface domain or a water body. Although the phenomenon is intrinsically one-dimensional (since in spherical or plane fronts there is one distinguished coordinate) some interesting problems arise in higher dimensions.

Aronson and Weinberger [25] studied the equation

$$u_t = \Delta u + F(u). \qquad (7.1)$$

where F is subjected to the conditions of section 2, in $\mathbb{R}^n \times \mathbb{R}_+$, $n \geq 1$. From section 3 we see that the condition $F'(0) > 0$ causes the hair-trigger-effect. This condition is not necessary: we need only that in a neighborhood of 0 the diffusing matter multiplies faster than it is carried away by diffusion. This heuristic observation is concretized by the following result ([25]): If $F(u) > 0$ in $(0, \alpha)$ for some $\alpha \in (0,1)$ and

$$\liminf_{u \to 0} F(u) u^{-(1+2/n)} > 0 \qquad (7.2)$$

then the zero solution is unstable in the sense that any initial function $u(x,0) \not\equiv 0$ leads to a solution with $\liminf_{t \to \infty} u(x,t) \geq \alpha$ uniformly on bounded sets (in case 1 follows $u(x,t) \to 1$).

Condition (7.2) is close to being necessary: if $F(u) \leq ku^{\beta}$ in a neighborhood of 0 with $k > 0$, $\beta > 1 + 2/n$, then all solutions which are sufficiently close to zero at t=0 go to zero.
The propagation of disturbances with finite support can be discussed as in the one-dimensional case. Aronson and Weinberger [25] define that a disturbance with bounded support is propagated with speed less

than c if

$$\lim_{t\to\infty} \max_{|x|\geq ct} u(x,t) = 0, \tag{7.3}$$

thus avoiding the trivial "fast propagation" of small amounts of material in the linear diffusion equation. They show that in cases 1,2, 3 (7.3) is valid for all initial data with bounded support and for every $c > c_0$.

8. Nerve axon equations

In simulating the phenomena of nerve conduction we have to deal with three closely related problems, namely the excitation of a piece of axon, the propagation of an impulse along the axon, and the release of an impulse train at the begin of the axon (the axon hillock). The basic problem is the first: For a short piece of axon which can be assumed as spatially homogeneous the effects of voltage (or current) changes are measured. The state of the axon can be described by the voltage across the membrane and the permeabilities of the membrane for certain types of ions. In the Hodgkin-Huxley [26] model for the squid axon a good qualitative agreement is achieved with three such "assistant variables", Zeeman [27] uses two such variables, this model is general in the sense that every model based on the cusp singularity catastrophe is essentially equivalent (though perhaps in higher dimensions). The Fitzhugh [33] system contains one assistant variable, it produces a good qualitative modeling of all features of the nervous excitation.

These systems have the following common form. The voltage u_1 and the assistant variables $u_2,\ldots,u_m$, $m \geq 2$ are functions of the time t. The vector $u=(u_1,\ldots,u_m)$ satisfies a system of ordinary differential equations

$$\dot{u}=f(u)+DI \quad . \qquad (8.1)$$

Here I is the current density across the membrane and $D=(d_{jk})$ is an $m \times m$ matrix with $d_{11}=1$ und $d_{jk}=0$ otherwise. In the space clamp situation we have $I = 0$. The equation $f(u)=0$ has a unique solution u^* (by an appropriate translation we can achieve $u^*=0$), which is a stable stationary point of (8.1), the resting state of the axon. The threshold phenomena of nervous excitation can be explained by the peculiar phase picture (Fig. 5 depicts the situation for the Fitzhugh model or an appropriate two-dimensional projection of the Hodgkin-Huxley model). A rapid change of the voltage shifts the actual state parallel to the u_0-axis. If the voltage exceeds a certain threshold level then excitation takes place. The threshold is not a defined value, it is characterized by a zone of "densely packed trajectories".

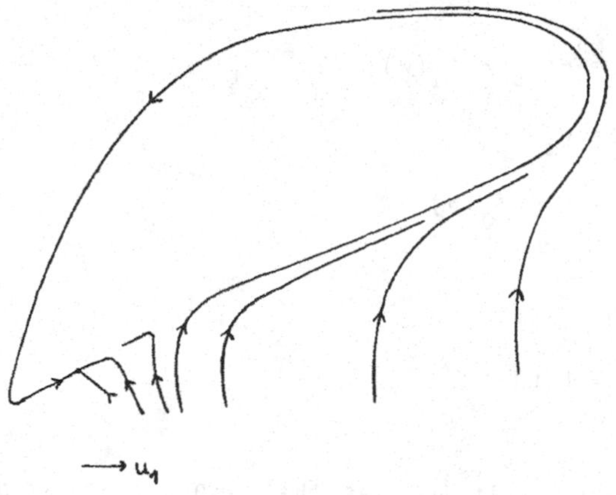

Fig 5

The model for nervous propagation is an immediate consequence of the excitation model: Along the axon (space coordinate x) the cross current density is proportional to the second spatial derivative of the voltage. Therefore, with an appropriate normalization of the space variable, u satisfies a degenerate parabolic system

$$u_t = Du_{xx} + f(u) \quad . \tag{8.2}$$

A pulse travelling with speed $c > 0$ is a solution $u(x,t) = \phi(x-ct)$, where the function of one variable $\phi = (\phi_1, \ldots, \phi_m)$, $\phi(\xi)$, $\xi = x-ct$, satisfies

$$\phi(-\infty) = \phi(+\infty) = 0 \quad . \tag{8.3}$$

The function ϕ is a solution of the system of ordinary differential equations

$$D\phi'' + c\phi' + f(\phi) = 0 \quad . \tag{8.4}$$

It is convenient to transform (8.4) into a first order system. With $w = (w_0, \ldots, w_m)$, $w_0 = \phi_1'$, $w_1 = \phi_1$, $w_j = \phi_j$ for $j = 2, \ldots, m$, $f = (f_1, \ldots, f_m)$ we have

$$\begin{aligned} w_0' &= -cw_0 - f_1(w) \quad , \\ w_1' &= w_0 \quad , \\ w_j' &= -\frac{1}{c} f_j(w) \quad , \end{aligned} \tag{8.5}$$

in shorter notation

$$w' = F(w) \tag{8.6}$$

From the hypothesis on f it follows that $w=0$ is the unique stationary state of (8.6). From the assumption that $\partial f/\partial u|_{u=0}$ has m eigenvalues in the left half-plane (stability of the resting state) follows that $\partial F/\partial w|_{w=0}$ has m eigenvalues with positive real part and one negative eigenvalue (Evans [31,II]). Thus w=0 is a saddle point with a one-dimensional stable manifold and an m-dimensional unstable manifold. The boundary value problem (8.3), (8.4) requires to find a trajectory of the system (8.6) which, when followed backward from $\xi=+\infty$ to $\xi=-\infty$ leaves the stable and enters the unstable manifold. In general such a "homoclinic" trajectory will exist only for special values of the speed parameter c. For the Nagumo system, derived from the Fitzhugh model (see section 9) it has been shown by numerical methods that normally two speeds c exist. The numerical values of these speeds, depending on the parameters of the system, are known [35]. The procedure is simple in principle: For given c the direction of the unstable manifold is known. The corresponding trajectory can be followed, e.g. by Runge-Kutta-method, until it eventually reaches w=0. For a certain class of models, Conley and Carpenter [28] have proved the existence of homoclinic orbits. Further results in this direction are due to Kopell and Howard [29]. There is numerical evidence that only one of the values of c corresponds to a stable pulse.

For the stability analysis (in the uniform norm) it is convenient to choose a fixed c and introduce a moving coordinate frame (cf. section 4). Then (8.2) becomes

$$\bar{u}_t = D\bar{u}_{xx} + c\bar{u}_x + f(\bar{u}) \tag{8.7}$$

and $\phi_{(\alpha)}$, $\phi_{(\alpha)}(x) = \phi(x+\alpha)$ is a one-parameter family of standing pulses (stationary solutions). A stationary state ϕ of (8.7) is not isolated and hence not asymptotically stable, in the uniform norm, say.

For this situation Evans [31,I-IV] introduced the concept of exponential stability: Let $\dot{V} = G(V)$ be an evolution equation in a normed space and let $\mathcal{M} = \{ V_\alpha : \alpha \in \mathbb{R} \}$ be a one-parameter family of stationary states, $G(V_\alpha)=0$ for all α. The system is called exponentially stable at V_0, if for every $\varepsilon > 0$ there is a $\delta > 0$ such that for all feasible initial data $V(0)$ with $|V(0)-V_0| < \delta$ the solution $V(t)$ converges to a stationary state $V_\alpha \in \mathcal{M}$ with $\| V_\alpha - V_0 \| < \varepsilon$ exponentially fast. The linearized system at V_0 is $\dot{V}=AV$, where $A = \partial G / \partial V |_{V=0}$. The linearized system has one a one-parameter family of stationary states, the linear space $\mathcal{M}'$ generated by $\partial V_\alpha / \partial \alpha |_{\alpha=0} = V'$, $AV'=0$. This general setting indicates which results can be expected: The nonlinear system is exponentially stable if the linear system is exponentially stable. The linear system is exponentially stable iff the spectrum of A, except $\lambda=0$, is contained in a half-plane $\mathrm{Re}\lambda \leq \varepsilon < 0$, and $\lambda = 0$ is a simple pole of the resolvent, i.e. there is no vector V with $AV=V'$. In the special case of equation (8.7) the system linearized at ϕ is

$$v_t = Dv_{xx} + cv_x + f'(\phi)v, \tag{8.8}$$

the eigenvalue problem of the operator A is

$$Dv'' + cv' + f'(\phi)v = \lambda v, \tag{8.9}$$

and $AV=V'$

$$Dv'' + cv' + f'(\phi)v = \phi' . \tag{8.10}$$

Similarly we can linearize at the resting state

$$v_t=Dv_{xx}+cv_x+f'(0)v \tag{8.11}$$

$$Dv''+cv'+f'(0)v=\lambda v \tag{8.12}$$

Indeed Evans showed that the system (8.7) is exponentially stable at ϕ iff (8.8) is stable at ϕ'. The system (8.8) is exponentially stable at ϕ' iff $\sup\{\text{Re}\,\lambda : \lambda\neq 0$, equation (8.9) has a bounded solution$\}<0$ and (8.10) has no bounded solution.
The eigenvalue equations (8.9), (8.12) can be written as first-order systems similar to (8.6),

$$w'=F(w,0,\lambda), \tag{8.13}$$

and

$$w'=F(w,\phi(x),\lambda) \quad . \tag{8.14}$$

Again, for (8.13) and $\text{Re}\,\lambda\geq 0$ one can prove that the unique stationary state 0 is a saddle point with a one-dimensional unstable manifold. Furthermore, at $\pm\infty$ we have $\phi(x)=0$ and (8.14) behaves like (8.13), in particular, there is a unique trajectory $\beta(x,\lambda)$ of (8.14) remaining bounded at $x = -\infty$. This function is the only candidate for an eigenfunction of (8.10). Evans shows that the function $D(\lambda)=\beta^*(x,\lambda)\beta(x,\lambda)$, where β^* is the corresponding solution of the adjoint problem is indeed independent of x and vanishes iff $\beta(x,\lambda)$ remains bounded for $x\longrightarrow\infty$. Thus the stability problem is reduced to the problem of locating the zeros of an analytic function: (Evans [31,IV]): The system (8.7) is exponentially stable at the pulse solution iff the function $D(\lambda)$ does not vanish for $\text{Re}\,\lambda\geq 0, \lambda\neq 0$ and

$\partial D/\partial\lambda|_{\lambda=0} = 0.$

9. The Fitzhugh and the Nagumo model

In this section we give a more detailed account of Fitzhugh's model. Fitzhugh's system is

$$\dot{u} = \varkappa w + \varkappa(u - u^3/3) - I, \qquad \dot{w} = (a - u - bw)/\varkappa \quad . \tag{9.1}$$

The parameters a,b, and $\varkappa$ are subject to the conditions

$$0 < b < 1, \qquad 1 - {}^{2b}/3 < a < 2 + {}^{2b}/3 \quad , \qquad b < \varkappa^2 \quad . \tag{9.2}$$

(For technical reasons large negative values of u correspond to nervous excitation). In Nagumo's paper [38] the equation has been transformed into

$$\dot{z} = z(1-z)(z-\gamma) - \delta v + \tilde{I} \tag{9.3}$$

$$\dot{v} = z - \nu v \quad .$$

In [48] it is shown that the Fitzhugh' model with the restrictions (9.4) corresponds to the Nagumo model with

$$0 < \gamma < 1, \; \nu > 0, \; \nu(1 - \gamma + \gamma^2) < 3\delta \, , \; 3\nu < 1 - \gamma + \gamma^2 \quad . \tag{9.4}$$

The models for the propagation of the nerve pulse corresponding to (9.1) and (9.4) are

$$u_t = \varkappa w + \varkappa(u - u^3/3) + u_{xx}, \tag{9.5}$$

$$w_t = (a - u - bw)/\varkappa \quad ,$$

and

$$z_t = z(1-z)(z-\gamma) - \delta v + z_{xx} \quad ,$$
$$v_t = z - \nu v. \tag{9.6}$$

For the special case $\nu=0$ (corresponding to the limiting case b=0 in (9.1)) the speed pattern seems to look approximately like that in fig.6, following the numerical work of Cohen [34] , Cooley and Dodge [35] , Knight [36] :

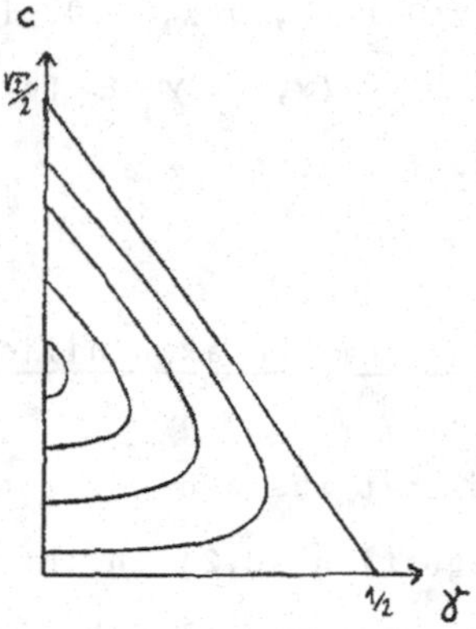

Fig 6

The axes correspond to γ and c, there are no pulse solutions for pairs γ ,c outside the triangle. The curves δ= const are indicated in the triangle (δ =0 is the boundary). Thus for given δ there are two possible speeds for small γ , exactly one speed for a maximal value of γ , and there is no speed above this value. There is no rigorous proof for this, but Rinzel and Keller [40] have solved explicitely the case where the cubic is replaced by the function

$$f(z) = \begin{cases} -z & \text{for } z < \gamma \\ 1-z & \text{for } z > \gamma \end{cases} \tag{9.7}$$

and found a similar speed picture. Conley, Sleeman (see [44]) showed

showed for (9.6) that there are no travelling waves for $\gamma \geq 1/2$. Sleeman and Green [43], [44] gave bounds for the speed in (9.6)

$$2\sqrt{\delta}/(1-\gamma) < c < 1-\gamma .$$

Levine and Sleeman [42] gave various convergence results on equations (9.6), $\nu = 0$, in the quarter plane $x \geq 0$, $t > 0$, predominantly negative: Let z,v be a solution such that for each t = 0 z is bounded and z, z_x are square-integrable. Let $z(0,t)=0$, $v(x,0)=0$. If $\sup|z(x,0)| < \gamma/(\gamma+1)$ then z decays exponentially. If $z(x,t) \leq \gamma - \varepsilon$ for some $\varepsilon > 0$ then z cannot be asymptotic to a pulse for large x and t.

10. Pulse solutions starting from the axon hillock

In section 8 it was assumed that the axon extends on both sides to infinity. Now we consider equation (8.2) on the half-line $[0,\infty)$ modeling the axon extended from the axon hillock. For a well-posed initial-boundary value problem we have to require a condition for u_1 at x=0. This condition is supplied by equation (8.1),

$$\dot{u}=f(u) + DI \quad , \tag{10.1}$$

where $u_1,\ldots,u_m$ are the membrane potential and the assistant variables at x=0, and I is the total synaptic current. Already Fitzhugh [33] and recently Knight [45] tried to explain the generation of nerve pulses in the following way: For small values of I the stationary state remains stable. If I exceeds a certain threshold then the stationary state becomes unstable and periodic oscillations occur. If u(t) is such a periodic solution of (10.1) for some value of I, then the first component $u_1(t)$ provides a boundary condition for u(x,t) at x = 0.

The existence and stability problems of these periodic solutions has been discussed by several authors. Troy [46] showed the existence of a Hopf bifurcation for a certain perturbation of the Hodgkin-Huxley equations. Hsü and Kazarinoff [47] showed a Hopf bifurcation for Fitzhugh's model (9.5) for certain values of the parameters.
If the stationary state (u_1, w_1) corresponding to the current I is shifted into zero by $u = u_I + x$, $w = w_I + y$, the system assumes the form

$$x = \varkappa(y + (1-u_I^2)x - u_I x^2 - \frac{1}{3}x^3 \quad , \tag{10.2}$$

$$y = -(x+by)/\varkappa \quad ,$$

where u_I ranges from $+\infty (I=-\infty)$ to $-\infty (I=+\infty)$. The stationary state is unstable for $|u_I| < \sqrt{1-b/\varkappa^2} = \bar{u}$. Hadeler, an der Heiden, Schumacher [48] proved the existence of closed orbits for all values of I, for which the stationary state is unstable.
For some values of the parameters there are closed orbits for $|u_I| > \bar{u}$ and also orbits with large amplitude which do not arise from a Hopf bifurcation ([48]).

The results are not yet satisfactory, in particular for the Hodgkin-Huxley system. Moreover there is no analytic proof that the solution of (8.1) with

$$u(x,0)=0 \text{ for } x>0, \quad u_1(0,t)=\tilde{u}_1(t)$$

is attenuated into a (spatially periodic) wave train for large x andt.

11. Reaction-diffusion systems and other problems

In the nerve axon models there are nonlinear diffusion equations with (spatially) non-homogeneous stationary states and closely related ordinary differential equations with periodically oscillating solutions. Similar phenomena occur in the theory of chemical reactions. Already Turing [51] proposed that spatially inhomogeneous structures could arise from interaction of species and diffusion, and Wiener and Rosenblueth [52] designed a system of excitable elements with diffusion as a model for the cardiac muscle.
Turing's ideas were justified by the discovery of the Belousov-Zhabotinskij reaction. Here temporal oscillations (Belousov 1958) and spatial patterns of concentrations (Busse 1969) can be found in experiments. Before the reaction scheme had been found (Field et.al. [53]), Prigogine and Glansdorff [53] showed by linear analysis of a hypothetical reaction scheme that the homogeneous stationary state may become unstable if the diffusion rates of the species differ significantly. The existence of periodic solutions has been shown by Hastings and Murray [56] . Winfree (see [55] also for further references) investigated rotating, spiral- and scroll-shaped moving fronts in two and three dimensions. His basic system is

$$u_t = -u - v + f(u) + \Delta u \quad , \qquad f(u) = \begin{cases} 1 & \text{for } u \quad 0{,}05 \\ 0 & \text{for } u \quad 0{,}05 \end{cases} \qquad (11.1)$$

$$v_t = ku + \Delta v ,$$

Fife [60] considered concentration fronts in chemical reaction systems (see also [57,61] for wave-type solutions in reaction-diffusion systems), pattern formation in slime molds are modeled by Segel [63]. Meinhard and Gierer [58], [59] explain pattern formation by the interaction of a short-range activator and a long-range inhibitor v

$$u_t = a+b\,\frac{u^2}{v} - \mu u + D_1 u_{xx},$$
$$v_t = cu^2 - \nu v + D_2 v_{xx}\quad, \tag{11.2}$$

or by the interaction of an activator u and a substance v which is consumed by the activation and carried away by rapid diffusion

$$u_t = a+bu^2v - \mu u + D_1 u_{xx}\quad,$$
$$v_t = c-du^2v - \nu v + D_2 v_{xx}\quad. \tag{11.3}$$

Various special cases have been investigated by numerical methods.

Oscillating or periodic stationary states can arise also with equal diffusion rates if Dirichlet boundary conditions are prescribed [65]. A simple example is the Lotka-Volterra model

$$u_t = au-buv-eu^2+D_1 u_{xx}\quad,$$
$$v_t = cuv-dv+D_2 v_{xx}\quad. \tag{11.4}$$

For equilibrium boundary conditions and von Neumann conditions convergence to equilibrium or to spatially constant periodic solutions has been discussed in [66], [67]; for non-equilibrium Dirichlet conditions stationary states have been established by numerical methods [65]. Existence can be proved by fixed point principles.

Models for the spread of epidemics lead to various types of diffusion equations. Kendall [69] introduced a diffusion term in the Kermack-McKendrick model (u,v are the densities of susceptible and infectuous individuals)

$$\dot{u} = -buv \quad , \qquad \dot{v} = cuv-dv \quad , \tag{11.5}$$

in the form

$$u_t = -bu(v_0+Dv_{xx}) \quad ,$$
$$v_t = cu(v_0+Dv_{xx})-dv \quad , \tag{11.6}$$

and Noble [70] proposed

$$u_t = -buv + D_1u_{xx} \quad ,$$
$$v_t = cuv-dv+D_2v_{xx} \quad . \tag{11.7}$$

For problem (11.6) the existence of travelling fronts with side conditions

$$u(-\infty)=u_0 \quad , \quad u(+\infty)=u_1 \quad , \quad v(-\infty)=v(+\infty)=0$$

can be shown by a reduction to the problem of section 3, case 1 ([69], [7],). Travelling bands of bacteria in certain media are described by diffusion models in [71].

Peculiar problems are provided by marine biology. Radach and Maier-Reimer [72], Wörz [73] investigate two interacting species of plankton with densities u,v (phytoplankton and zooplankton) which vary with time and water depth. Moreover the reproduction rate of plankton depends on the light intensity, the light intensity itself depends on the plankton density in the upper water layers. The resulting system (x in the distance from the water surface) has the form

$$u_t = D_1 u_{xx} + f(u,v,I) \quad ,$$

$$v_t = D_2 v_{xx} + g(u,v,I) \quad , \qquad (11.8)$$

$$I_x = -h(u,v)I \quad .$$

References

1. Fisher,R.A., The genetical theory of natural selection, Oxford University Press 1930

2. Fisher,R.A., The advance of advantageous genes,Ann. of eugenics 7, 355-369 (1937).

3. Hadeler, K.P., On the equilibrium states in certain selection models, J. of Math. Biol. 1, 51-56(1974).

4. Kolmogoroff,A., Petrovskij,I., and Piskunov,N., Etude de l'équation de la diffusion avec croissance de la quantité de matière et son application à une problème biologique. Bull. Univ. Moscou, Ser. Internat., Sec. A,1, 6,1-25 (1937).

5. Aronson,D.G., and Weinberger,H.F., Nonlinear diffusion in population genetics,combustion, and nerve propagation, Proceedings of the Tulane Program in partial differential equations, Lecture

Notes in Mathematics, Springer 1975.

6. Hoppensteadt,F., Mathematical theories of populations, demographics, genetics, and epidemics, Soc. Ind. Appl. Math. Regional Conference Series 20, 1975.

7. Hadeler, K.P., and Rothe,F., Travelling fronts in nonlinear diffusion equations, J.Math. Biol. 2, 251-263 (1975).

8. Canosa,J., On a nonlinear diffusion equation describing population growth, IBM-J. 17, 307-313 (1973).

9. Gazdag,J., and Canosa,J., Numerical solution of Fisher's equation, J. Appl. Prob. 11, 445-457 (1974).

10. Gelfand,I.M., Some problems in the theory of quasilinear equations, Uspeki Mat. Nauk (N.S.) 14, 87-158 (1959), Am. Math. Soc. Transl. (2) 29, 295-381 (1963).

11. Kanel',J.I., The behavior of solutions of the Cauchy problem when time tends to infinity, in the case of quasilinear equations arising in the theory of combustion, Dokl. Akad. Nauk SSSR 132, 268-271 (1961), Soviet Math. Dokl. 1, 533-536.

12. Kanel' J.I., Certain problems on equations in the theory of burning, Dokl. Akad. Nauk SSSR 136, 277-280 (1961), Soviet. Math. Dokl. 2, 48-51.

13. Kanel',J.I., Stabilization of solutions of the Cauchy problem for equations encountered in combustion theory, Mat. Sbornik (N.S.)59 (101), supplement, 245-288 (1962).

14. Rothe,F., Über das asymptotische Verhalten der Lösungen einer nichtlinearen parabolischen Differentialgleichung aus der Popu-

lationsgenetik, Dissertation Tübingen 1975

15. Rothe,F., Convergence to travelling fronts in semilinear parabolic equations, to appear.

16. Fife,P.C., and Mc Leod,J.B., The approach of solutions of nonlinear diffusion equations to travelling wave solutions, Bull. Am.Math.Soc. 81, 1076-1078 (1975).

17. Kanel',J.I'., On the stability of solutions of the equation of combustion theory for finite initial functions, Mat. Sbornik (N.S.) 65 (107), 398-413 (1964).

18. Chafee,N., A stability analysis for a semilinear parabolic partial differential equation, J.Diff.Equ. 15, 522-540 (1974)

19. Chafee,N., Asymptotic behavior for solutions of a one-dimensional parabolic equation with homogeneous boundary conditions, J. Diff. Equ. 18, 111-134 (1975).

20. Fleming,W.H., A selection migration model in population genetics, J.Math.Biol. 2, 219-234 (1975).

21. Hoppensteadt,F.C., Analysis of a stable polymorphism arising in selection migration model in population genetics, J.Math.Biol.2, 235-240 (1975).

22. Karlin,S., and Richter-Dyn,N., Some theoretical analysis of migration. Selection interaction in a cline: A generalized two range environment, in: Population genetics and ecology, ed. by S. Karlin and E. Nevo, Academic Press 1976.

23. Fisher,R.A., Gene frequencies in a cline determined by selection and diffusion, Biometrics 6, 359-361 (1950) .

24. Conley,C., An application of Wazewski's method to a nonlinear boundary value problem which arises in population genetics, J. Math. Biol. 2,241-249 (1975)

25. Aronson,D.G., and Weinberger,H.F., Multidimensional nonlinear diffusion arising in population genetics, to appear.

26. Hodgkin,A.L., and Huxley,A.F., A quantitative description of membrane current and its application to conduction and excitation in nerve, J. Physiol. 117, 500, (1952).

27. Zeeman,E.C., Differential equations for the heartbeat and nerve impulse, in: Towards a Theoretical Biology 4, Edinburgh University Press (1972).

28. Carpenter,G., Travelling wave solutions of nerve impulse equations, Thesis, Univ. of Wisconsin (1974).

29. Kopell,N., and Howard,L.N., Bifurcations and trajectories joining critical points, to appear.

30. Evans,J., and Shenk,N., Solutions to axon equations, Biophys. J. 10 ,1090-1101 (1970) .

31. Evans,J.W., Nerve axon equations I-IV, Indiana Univ. Math.J. 21, 877-885 (1972), 22, 75-90 (1972), 22, 577-593(1972), 24,1169-1190 (1975).

32. Rinzel,J., Neutrally stable wave solutions of nerve conduction equations, J.Math.Biol. 2,205-217 (1975).

33. Fitzhugh,R., Impulses and physiological states in theoretical models of nerve membrane, Biophys. J. 1, 445-466 (1961) .

34. Cohen,H., Nonlinear diffusion problems. In:Studies in Appl.Math. A.H.Taub, editor,Prentice Hall,Inc.,Englewood Cliffs, N.J.

35. Cooley,J.W., and Dodge,F.A., Digital computer solutions for excitation and propagation of the nerve impulse, Biophys. J. 6, 583-599 (1966).

36. Knight,B., Numerical results for non-linear diffusion systems, Courant Institute of Mathematical Sciences and IBM, T.J. Watson Research Center, Seminar on partial differential equations, summer, 1965.

37. Casten,R., Cohen,H., and Lagerstrom,P., Perturbation analysis of an approximation to Hodgkin-Huxley theory, Quart. Appl.Math. 32, 365-402(1975).

38. Nagumo,J., Arimoto,S. and Yoshizawa,S., An active pulse transmission line simulating nerve axon. Proceedings of the IRE 50, 2061-2071 (1962).

39. McKean,H.P., Nagumo's equation, Advances in Mathematics 4, 209-223 (1970).

40. Rinzel,J., and Keller,J.B., Travelling wave solutions of a nerve conduction equation, Biophysical J., 13, 1313-1336 (1973).

41. Hastings,S.P., On a third order differential equation form biology Quart. J. Math. Oxford (2) 23, 435-448 (1972).

42. Levine,H.A., and Sleeman,B.D., A note on the asymptotic behavior of solutions to Nagumo's equation. Batelle Advanced Studies Center, Geneva, Math. Report 69 (1972).

43. Green,M.W., and Sleeman,B.D., On Fitzhugh's nerve axon equations J.Math.Biol.1, 152-163 (1974).

44. Sleeman,B.D., Fitzhugh's nerve axon equations, J.Math.Biol.2, 341-349 (1975).

45. Knight,B.W., Some questions concerning the encoding dynamics of

neuron populations, IV Internat. Biophysics Congress, Academy of Sciences of the USSR, Pushchino 1973.

46. Troy,W.C., Oscillation phenomena in the Hodgkin-Huxley-equations, Univ. of Pittsburgh preprint 1975.

47. Hsü,I., and Kazarinoff,N.D., An applicable Hopf bifurcation formula and instability of small periodic solutions of the Field-Noyes model, J. Math. Anal. (to appear).

48. Hadeler,K.P., an der Heiden, U., and Schumacher,K., Generation of the nervous impulse and periodic oscillations, Biol.Cyb. to appear.

49. Poore,A.B., On the theory and applications of the Hopf-Friedrichs bifurcation theory, Arch. Rat. Mech. Anal., to appear.

50. Hastings,S.P., The existence of periodic solutions to Nagumo's equation, Quart. J. Math. Oxford (3),25,369-78 (1974).

51. Turing,A.M., The chemical basis of morphogenesis, Phil. Trans. R. Soc. London B 237, 37-72 (1952).

52. Wiener,N., and Rosenblueth,A., The mathematical formulation of conduction of impulses in a network of connected excitable elements, specifically in cariac muscle, Arch. Inst. Cardiologia de Mexico 16, 205-256 (1946).

53. Field,R.J., Körös,E., and Noyes,R.M., Oscillations in chemical systems II, Thorough analysis of temporal oscillation in the bromate-cerium-malonic acid system, J.Amer. Chem. Soc. 94, 8649-8664 (1972).

54. Glansdorff,P., and Prigogine,I., Thermodynamic theory of structure stability and fluctuations, Wiley London 1971.

55. Winfree,A.T., Rotating solutions to reaction-diffusion equation in simply-connected media, in: Mathematical aspects of chemical and biochemical problems and quantum chemistry, SIAM-AMS Proceedings vol. 8, Am. Math. Soc., Providence 1974.

56. Hastings,S.P., and Murray,J.D., The existence of oscillatory solutions in the Field-Noyes model for the Belousov-Zhabotinskij reaction, SIAM J. Applied Math. 28,678-688 (1975).

57. Tyson,J., Analytic representation of oscillations, excitability and travelling waves in a realistic model of the Belousov-Zhabotinskij reaction, J.Chem. Physics, to appear.

58. Gierer,A., and Meinhardt,A., A theory of biological pattern formation, Kybernetik 12, 30-39 (1972).

59. Meinhardt,H., and Gierer,A., Applications of a theory of biological pattern formation based on lateral inhibition J. Cell Science 15, 321-346 (1974).

60. Fife,P.C., Pattern formation in reacting and diffusing systems, J.Chem.Phys. 64, 554-564 (1976).

61. Maginu,K., Reaction-diffusion equation describing morphogenesis I, Wave form stability of stationary wave solutions in a one-dimensional model, Math. Biosciences 27, 17-98 (1975).

62. Kopell,N. and Howard,L.N., Plane wave solutions to reaction diffusion equations, Studies Appl. Math. 291-328 (1973).

63. Keller,E.F., and Segel, L.A., Initiation of slime mold aggregation viewed as an instability, J. theor. Biol. 26, 399-415 (1970).

64. Segel,L.A., and Jackson,J.L., Dissipative Structure; An explanation and an ecological example, J. Theor. Biol. 37, 545-559 (1972).

65. Hadeler,K.P., an der Heiden,U., and Rothe,F., Nonhomogeneous spatial distributions, J. Math. Biol. 2, 133-163 (1975).

66. Murray, J.D., Non-existence of wave solutions for the class of reaction-diffusion equations given by the Volterra interacting population equations with diffusion, preprint Oxford.

67. Rothe,F., Convergence to equilibrium in the Lotka-Volterra diffusion model, to appear.

68. Othmer,H.G., Nonlinear wave propagation in reacting systems, J. Math. Biol. 2, 133-163 (975).

69. Kendall,D.G., Mathematical models of the spread of infection, Mathematics and computer science in biology and medicine, Medical Research Council, 1965.

70. Noble,J.V., Geographical and temporal development of plagues, Nature 250, 726-728 (1974).

71. Keller,E.F., and Segel,L.A., Travelling bands of Chemotatic Bacteria: A theoretical analysis, J. Theor. Biology 30, 235-248(1971).

72. Radach,G. and Maier-Reimer,E., The vertical structure of phytoplankton growth dynamics, a mathematical model, Proceedings of the Sixth Colloquium on Ocean Hydrodynamics, Liège 1974, Mèmoirs de la Societé Royale des Sciences de Liége.

73. Wörz,A., On the solutions to a degenerate parabolic system from marine biology, to appear.

DISCRETE DISSIPATIVE PROCESSES

Jack K. Hale

1. Introduction. The theory of dissipative processes had its origin in a fundamental paper of Levinson [11] in 1944 dealing with periodic ordinary differential equations in the plane. By considering the period map T, he formulated very clearly the basic problems. First, one should characterize the set J which contains all the information about the limits of the trajectories and, secondly, discuss the properties of the transformation T restricted to J. Levinson [11] formulated the concept of point dissipative (i.e., there is a bounded set such that every orbit eventually lies in this set) and proved that point dissipative implies the existence of a maximal compact invariant set J. Using this fact, he was able to prove that some iterate of T has a fixed point; that is, the original differential equation has a periodic solution with the period equal to some multiple of the period in the equation. The behavior of the flow on J may be very complicated as the work of Cartwright and Littlewood [13] on van der Pol's equation and the work of Levinson [12] indicate.

Over the years, a tremendous literature on this subject accumulated and one may consult LaSalle [10], Pliss [14], Reissig, Sansone and Conti [15], and Yoshizawa [17, 18] for references. Continuing in the spirit of Levinson for finite dimensions, Pliss [14] showed that the maximal compact invariant set is globally asymptotically stable. For the special case of retarded functional differential equations for which the period ω is greater than the delay, Jones [8,9] and Yoshizawa [17] showed the existence of ω-periodic solutions by using Browder's asymptotic fixed point theorem. For a discrete point dissipative dynamical system T on an arbitrary Banach space X, the existence of fixed points of T were proved by Horn [7] and Gerstein and Krasnoselskii [4] when T is completely continuous. Billotti and LaSalle [1] proved the same result when T is completely continuous, and also characterized the maximal compact invariant set and proved it is globally asymptotically stable.

Gerstein [3] considered the case when T is point dissipative and is α-condensing on balls in X and showed the existence of a maximal compact invariant set, but concluded nothing about stability of this set or the existence of fixed points of T.

Hale, LaSalle and Slemrod [5] and Hale and Lopes [6] generalized all of the above results. The discussion in these latter two papers centered on trying to determine a "minimal" set of conditions which imply the existence of a maximal compact invariant set which is stable as well as "minimal" conditions which imply the existence of a fixed point of T. The present paper continues the spirit of these investigations. We show that the results in [5], [6] can be obtained by beginning with more elementary hypotheses and also point out some further implications of the theory.

Generalizations of the work of Pliss [14] on convergent systems is also given.

2. Definitions. Let X be a Banach space. If $T: X \to X$ is continuous, the family $\{T^k, k \geq 0\}$ of iterates of T is called a discrete dynamical system (dds) on X. Sometimes, we also say simply that T is a discrete dynamical system. If T is a dds on X the orbit $\gamma^+(x)$ through x is defined as $\gamma^+(x) = \{T^k x, k \geq 0\}$ and the orbit $\gamma^+(H)$ through a set $H \subset X$ is $\gamma^+(H) = \bigcup_{x \in H} \gamma^+(x)$. If T is a dds on X and H is a subset of X, then the ω-limit set of H is defined as

$$\omega(H) = \bigcap_{j \geq 0} \mathrm{Cl} \bigcup_{n \geq j} T^n H$$

where Cl designates closure. If H is a point $\{x\}$, this latter definition is equivalent to the following: $y \in \omega(x)$ if and only if there is a sequence $\{n_j\}$ of integers, $n_j \to \infty$ as $j \to \infty$, such that $T^{n_j} x \to y$ as $j \to \infty$. The concept of α-limit set is defined in an analogous manner using negative orbits. If T is a dds on X, a set $Q \subset X$ is invariant under T if $TQ = Q$. This latter definition is equivalent to the following: Q is invariant under T if and only if it is possible to extend the definition of T^k on Q to negative integers k and $T^k Q \subset Q$ for all $k \in (-\infty, \infty)$.

Lemma 2.1. If T is a dds on X, $H \subset X$ and $\gamma^+(H)$ is precompact then $\omega(H)$ is nonempty, compact and invariant and $T^k H \to \omega(H)$ as $k \to \infty$ in the Hausdorff metric. Also, H compact, $\gamma^+(H)$ precompact, $\omega(H) \subset H$ implies $\omega(H) = \bigcap_{n \geq 0} T^n H$.

Proof: The first part of the lemma is classical. To prove the second part, suppose H is compact, $\gamma^+(H)$ is precompact and $\omega(H) \subset H$. If $J(X) = \bigcap_{n \geq 0} T^n H$, then $J(H)$ is compact and $J(H) \subset \omega(H)$. If $y \in \omega(H)$, $T^{n_j} x_j \to y$ as $j \to \infty$ where $n_j \to \infty$ as $j \to \infty$ and each $x_j \in H$. Since $\gamma^+(H)$ is precompact, for any integer i, there is a subsequence, which we label as before, and a $y_i \subset \omega(H) \subset H$ such that $T^{n_j - i} x_j \to y_i$ as $j \to \infty$. Thus, $T^i y_i = y$ for all i and $y \in J(H)$. q.e.d.

Some more definitions are needed in later sections. Suppose $T: X \to X$ is continuous. We say a set $K \subset X$ attracts a set $H \subset X$ (with respect to the dds T) if, for any $\varepsilon > 0$, there is an integer $N(H,\varepsilon)$ such that $T^n H \subset \mathscr{B}(K,\varepsilon)$ for $n \geq N(H,\varepsilon)$, where $\mathscr{B}(K,\varepsilon)$ is the ε neighborhood of K. We say K attracts points of X or K is a global attractor if K attracts each point of X. We say K attracts compact sets of X if K attracts each compact set of X. We say K attracts neighborhoods of points of X (compact sets of X) if for each point x of X (each compact set H of X), there is a neighborhood O_x of x (H_0 of H) such that K attracts O_x (H_0). If there is a bounded K such that K attracts points (compact sets) (neighborhoods of points) (neighborhoods of compact sets) of X then we say T is <u>point</u> (<u>compact</u>) (<u>local</u>) (<u>local compact</u>) <u>dissipative</u>. The map T is said to be <u>asymptotically smooth</u> if, for any bounded set $B \subset X$, there is a compact set $B^* \subset X$ such that, for any $\varepsilon > 0$, there is an integer $n_0(B,\varepsilon)$ such that $T^n x \in B$ for $n \geq 0$ implies $T^n x \in \mathscr{B}(B^*,\varepsilon)$ for $n \geq n_0(B,\varepsilon)$.

A set $M \subset X$ is said to be <u>stable</u> (with respect to the dds T) if, for any $\varepsilon > 0$, there is a $\delta > 0$ such that $x \in \mathscr{B}(M,\delta)$ implies $T^n x \in \mathscr{B}(M,\varepsilon)$, $n \geq 0$. The set M is said to be <u>asymptotically stable</u> if it is stable and there is an $\varepsilon_0 > 0$ such that M attracts points of $\mathscr{B}(M,\varepsilon_0)$. The set M is said to be <u>uniformly asymptotically stable</u> if it is asymptotically stable and, for any $\eta > 0$, there is an integer $n_0(\eta,\varepsilon_0)$ such that $T^n \mathscr{B}(M,\varepsilon_0) \subset \mathscr{B}(M,\eta)$ for $n \geq n_0(\eta,\varepsilon_0)$.

3. <u>Maximal compact invariant sets</u>. In this section, we give sufficient conditions for the existence of a maximal compact invariant set for a dds T and discuss the stability properties of this set.

<u>Lemma 3.1</u>. If $T: X \to X$ is continuous and there is a compact set $K \subset X$ that attracts compact sets of X, then $\gamma^+(H)$ is precompact for any compact set $H \subset X$.

<u>Proof</u>: Let $\alpha(B)$ be the Kuratowskii measure of noncompactness of a bounded set $B \subset X$. For any compact set $H \subset X$, the set

$A = \bigcup_{n\geq 0} T^n H$ is bounded since K attracts compact sets of X. Since $T^j H$ is compact for any j, we have $\alpha(A) = \alpha(\bigcup_{n\geq j} T^n H)$ for any j. But, for any $\varepsilon > 0$, $\bigcup_{n\geq j} T^n H \subset \mathscr{B}(K,\varepsilon)$, $j \geq N(H,\varepsilon)$. Therefore, $\alpha(A) \leq 2\varepsilon$ for every $\varepsilon > 0$ and so $\alpha(A) = 0$. q.e.d.

Suppose T is a dds and there is a compact set K which attracts compact sets of X. In particular, this implies $\omega(K) \subset \mathscr{B}(K,\varepsilon)$ for every $\varepsilon > 0$ and, therefore, $\omega(K) \subset K$. From Lemma 2.1, it follows that

$$(3.1) \qquad J \overset{\text{def}}{=} \bigcap_{n\geq 0} T^n K = \omega(K)$$

is nonempty, compact and invariant. We claim J is independent of the set K which attracts compact sets of X. In fact, if we designate J by $J(K)$ and if K_1 is any other compact set which attracts compact sets of X, then there is an integer $n_0 = n_0(K_1,K,\varepsilon)$ such that $T^n J(K) \subset \mathscr{B}(K_1,\varepsilon)$, $T^n J(K_1) \subset \mathscr{B}(K,\varepsilon)$ for some $n \geq n_0$. Since $J(K), J(K_1)$ are invariant, this implies $J(K) \subset K_1$, $J(K_1) \subset K$ and $J(K) \subset T^n K_1$, $J(K_1) \subset T^n K$ for all $n \geq 0$. Therefore, $J(K) = J(K_1)$.

The main result of this section is

<u>Theorem 3.1.</u> Suppose $T: X \to X$ is continuous, there is a compact set $K \subset X$ which attracts compact sets of X and let J be defined by (3.1). Then the following conclusions hold:

(i) $J = \omega(K)$ is independent of K, is a nonempty, compact invariant set and is maximal with respect to this property;

(ii) J is a stable, global attractor;

(iii) For any compact set $H \subset X$, there is a neighborhood H_1 of H such that $\gamma^+(H_1)$ is bounded and J attracts H_1; in particular, J is uniformly asymptotically stable.

Proof: (i) It only remains to prove J is maximal. This is an easy consequence of the invariance of H and the fact that K attracts H.

(ii) The proof of this part can be modeled after the proof of Theorem 4.2 of [5].

(iii) Suppose $\varepsilon > 0$ is given and δ is the number associated with ε in the definition of stability. For any $x \in X$, there is an integer $N(x)$ such that $T^n x \in \mathcal{B}(J,\delta)$ for $n \geq N(x)$. Since T is continuous, there is a neighborhood O_x of x such that $T^{N(x)} O_x \subset \mathcal{B}(J,\delta)$ and $T^j O_x$ is bounded for $0 \leq j \leq N(x)$. Stability of J implies $\gamma^+(O_x)$ is bounded. If H is an arbitrary compact set, then a finite covering argument completes the proof. q.e.d.

Corollary 3.1. If $T: X \to X$ is continuous, the following are equivalent:

(i) there is a compact set which attracts compact sets of X

(ii) there is a compact set which attracts neighborhoods of compact sets of X.

In Theorem 3.1, the stability condition stated in (iii) is not as strong as the usual concept of ultimate boundedness in differential equations because J does not attract arbitrary bounded sets. It is, therefore, of interest to impose other conditions of T which will imply this latter property. We can prove

Theorem 3.2. If $T: X \to X$ is continuous, local dissipative and asymptotically smooth, then there is a set J satisfying all the properties stated in Theorem 3.1 and, in addition,

(iv) If $B \subset X$ is bounded and $\gamma^+(B)$ is bounded, then J attracts B.

Proof: It is shown in [5] that the hypotheses imply the existence of a compact set which attracts compact sets of X. Therefore, the first part of the theorem is proved. To prove (iv), suppose $B \subset X$, $\gamma^+(B) \subset S$ for some bounded set $S \subset X$. Let S^* be the corresponding compact set in the definition of asymptotically smooth. For any $\varepsilon > 0$, $T^n B \subset \mathscr{B}(C^*,\varepsilon)$ for $n \geq n_0(C,\varepsilon)$. Therefore, for any integer $m > 0$, $T^{n+m}B \subset T^m \mathscr{B}(C^*,\varepsilon)$ for $n \geq n_0(C,\varepsilon)$. From property (iii) of J, we may assume ε so small that J attracts $\mathscr{B}(C^*,\varepsilon)$. For any $\eta > 0$, there is an $m_0(C^*,\eta)$ such that $T^m \mathscr{B}(C^*,\varepsilon) \subset \mathscr{B}(J,\eta)$ for $m \geq m_0(C^*,\eta)$. If we let $N = n_0(C,\varepsilon) +$

$m_0(C^*,\eta)$, then it is easy to check that $T^n B \subset \mathscr{B}(J,\eta)$ for $n \geq N$. q.e.d.

We now give some sufficient conditions for the hypotheses of the above theorems to hold. To do this, some additional definitions are required. Let $\alpha(B)$ be the Kuratowskii measure of noncompactness of a bounded set B in X (see Sadovskii [16]). A continuous map $T: X \to X$ is said to be conditionally condensing if, for any bounded set $B \subset X$ for which $\alpha(B) > 0$ and $T(B)$ is bounded, it follows that $\alpha(TB) < \alpha(B)$. The map T is said to be a conditional α-contraction if there is a constant k, $0 \leq k < 1$, such that, for any bounded set $B \subset X$ for which TB is bounded, $\alpha(TB) \leq k\alpha(B)$. The map T is said to be conditionally completely continuous if, for any bounded set $B \subset X$ for which TB is bounded, the set TB is precompact. These definitions coincide with the usual definitions of condensing, α-contraction and completely continuous when T is a bounded map. They are the same definitions as in [6] with the term "weak" replaced by "conditional".

Theorem 3.3. If $T: X \to X$ is continuous, point dissipative and T^{n_0} is conditionally completely continuous for some n_0, then the following conclusions holds:

(i) There is a compact set $K \subset X$ such that for any compact set $H \subset X$, there is an open neighborhood H_0 of H and an integer $N(H)$ such that $\gamma^+(H_0)$ is bounded and

$T^n H_0 \subset K$ for $n \geq N(H)$.

(ii) If, in addition, T is a bounded map, then, for any bounded set $B \subset X$, the set $\gamma^+(B)$ is bounded and there is an integer $N(B)$ and a compact set K such that $T^j B \subset K$ for $j \geq N(B)$.

Proof: Conclusion (i) is contained in [1]. To prove (ii), suppose $B \subset X$ is bounded. Then $\text{Cl}\, T^{n_0} B \overset{\text{def}}{=} H$ is compact. Therefore, part (i) implies $T^n H \subset K$ for $n \geq N(H)$ where K is compact. Thus, $T^n B \subset K$ for $n \geq N(H) + n_0$. q.e.d.

If the conditions of Lemma 3.1 are satisfied, then (i) implies, in particular, that T is local dissipative and asymptotically smooths. Therefore, all of the conclusions of Theorem 3.2 are valid. The stability property is even stronger as stated in (ii) of Theorem 3.3.

Other sufficient conditions for the hypotheses of Theorems 3.1 and 3.2 to be satisfied are taken from [6] and stated as

Lemma 3.1 (i) If T is continuous, conditionally condensing and compact dissipative, then there is a compact invariant set which attracts compact sets.

(ii) If T is a conditional α-contraction, then T is asymptotically smooth.

From Lemma 3.1, Theorems 3.1 and 3.2, one obtains the following result.

Theorem 3.4. If $T: X \to X$ is an α-contraction and compact dissipative, then there is a set J satisfying the properties stated in Theorem 3.2.

It is not known whether Theorem 3.4 remains valid with compact dissipative replaced by point dissipative. If it turns out not to be generally true, it is of interest to determine that subset of the α-contractions for which it is true.

4. Fixed point theorems. The purpose of this section is to specify conditions in addition to the ones in Theorem 3.1 which will ensure that T or some iterate of T has a fixed point.

To motivate the discussion, let us examine some of the implications of Theorem 3.1. If we let $K = \overline{co}\, J$, where J is the maximal compact invariant set in Theorem 3.1, then K is compact and there is a convex neighborhood B of K such that $\gamma^+(B)$ is bounded and J (and, therefore, K) attracts B; that is, there exist convex subsets $K \subset B \subset S$ of X with K compact, S closed, bounded and B open in S such that $\gamma^+(B) \subset S$ and K attracts B. For nested sets of this type and even weaker properties of attraction, the following result was proved in [6].

Lemma 4.1. Suppose $K \subset B \subset S$ are convex sets of X with K compact, S closed, bounded and B open in S. If $T: S \to X$ is continuous, $\gamma^+(B) \subset S$ and K attracts points of B, then, for any integer $k \geq 1$, there is a closed bounded convex set $A_k \subset S$ such that

(4.1) $$A_k = \overline{co}\left[\bigcup_{j\geq k} T^j(B \cap A_k)\right].$$

The intriguing sets A_k have been further investigated by Chow and Hale [2] in the spirit of limit compact maps of Sadovskii [16].

The following result was also proved in [6].

<u>Theorem 4.1</u>. Suppose $K \subset B \subset S$ as in Lemma 4.1 and suppose K attracts compact sets of B. If the set A_k in (4.1) is compact, then T^k has a fixed point.

Theorem 4.1 asserts that sufficient conditions for the existence of a fixed point of T^k satisfying the conditions of Theorem 3.1 will be assured if the set A_k is compact. It is shown in [6] that the set A_k is compact if T^k is conditionally condensing. Therefore, if we combine this result with Lemmas 3.1 and 3.2, we obtain the following fixed point theorems [6].

<u>Theorem 4.2</u>. (i) If $T: X \to X$ is conditionally condensing, point dissipative, and T^{n_0} is conditionally completely continuous, then T has a fixed point.

(ii) If T is conditionally condensing and compact dissipative, then T has a fixed point.

<u>Corollary 4.1</u>. If $T: X \to X$ is continuous, $T = S + U$, where S is linear with spectrum contained in the open unit ball and

U is conditionally completely continuous, then

(i) T compact dissipative implies T has a fixed point.

(ii) T point dissipative implies T has a fixed point if S^{n_0} is completely continuous for some n_0.

It is not known if the above results are valid for point dissipative rather than compact dissipative.

Theorem 3.1 can also serve as the motivation for other types of fixed point theorems, the so called asymptotic fixed point theorems. To see this, we observe the following easy consequence of Theorem 3.1.

<u>Lemma 4.2</u>. If $T: X \to X$ is continuous and there is a compact set that attracts compact sets of X, then there is an integer m and convex bounded sets $S_0 \subset S_1 \subset S_2$ with S_0, S_2 closed and S_1 open, such that $\gamma^+(S_1) \subset S_2$, $\gamma^+(T^m S_1) \subset S_0$.

<u>Proof</u>: Let J be the maximal compact invariant set of Theorem 3.1 and let $K = \overline{co}\, J$. Theorem 3.1 (iii) implies the existence of a neighborhood K_1 of K such that $\gamma^+(K_1)$ is bounded and K attracts K_1. Let $\gamma^+(K_1) \subset S_2$ where S_2 is closed, bounded. Choose $\varepsilon > 0$ such that $Cl\ \mathscr{B}(K,\varepsilon) \subset K_1$ and let $S_0 = Cl\ \mathscr{B}(K,\varepsilon/2)$, $S_1 = \mathscr{B}(K,\varepsilon)$. It is now easy to verify the assertions in the lemma. q.e.d.

The famous asymptotic fixed point theorem of Browder asserts the existence of a fixed point of a map T satisfying the conditions of Lemma 4.2 provided that T is completely continuous.

Using Theorem 4.1 and arguments similar to the above, the following generalization of Browder's Theorem was obtained in [6].

<u>Theorem 4.3</u>. Suppose $S_0 \subset S_1 \subset S_2$ are convex bounded subsets of X, S_0, S_2 closed, S_1 open in S_2 and suppose $T: S_2 \to X$ is condensing in the following sense: if Ω, $T\Omega$ are contained in S_2 and $\alpha(\Omega) > 0$, then $\alpha(T\Omega) < \alpha(\Omega)$. If $\gamma^+(S_1) \subset S_2$ and for any compact set $H \subset S_1$, there is a number $N(H)$ such that $\gamma^+(T^{N(H)}H) \subset S_0$, then T has a fixed point.

5. <u>Convergent systems</u>. In this section, we investigate some of the implications of the hypothesis that $J = \{x_0\}$, a single point, in Theorem 3.1 and, also, determine some sufficient conditions to ensure that $J = \{x_0\}$.

<u>Lemma 5.1</u>. Suppose $T: X \to X$ is continuous and there is a compact set which attracts compact sets of X. If the set J in Theorem 3.1 consists of a single point $x_0 \in X$, then there is a bounded orbit $\{T^n x, -\infty < n < \infty\}$, and every orbit is stable and attracts neighborhoods of points of X.

<u>Proof</u>: Since J is invariant, $J = \{x_0\}$, $T^n x_0$ is defined for $-\infty < n < \infty$ and $T^n x_0 = x_0$. Therefore, $\gamma(x_0) = \{T^n x_0, -\infty < n < \infty\}$ is bounded. Suppose $x \in X$ is arbitrary.

Since γ is stable by Theorem 3.1, for any $\varepsilon > 0$, there is a $\delta = \delta(\varepsilon) > 0$ such that $|x-x_0| < \delta$ implies $|T^n x - T^n x_0| < \varepsilon$ for $n \geq 0$. Also, since γ satisfies property (iii) of Theorem 3.1, for any $x \in X$, there is an $n_0(x)$ and a neighborhood O_x of x such that $|T^n y - T^n x_0| < \delta/2$ for $n \geq n_0(x)$, $y \in O_x$. Therefore, $|T^n y - T^n x| < \delta$ for $n \geq n_0(x)$, $y \in O_x$ and the orbit $\gamma^+(x)$ is stable since T is continuous. The same type of argument also gives the fact that each orbit attracts neighborhoods of points of X. q.e.d.

Lemma 5.2. If $T: X \to X$ is continuous, if there is a bounded orbit of T on $(-\infty,\infty)$ and every trajectory is uniformly asymptotically stable, then T is local dissipative.

Proof: Following arguments similar to the ones in Pliss [14], one proves T is point dissipative. The fact that T is local dissipative follows easily from the hypothesis of uniform asymptotic stability.

Definition 5.1. A continuous map $T: X \to X$ is said to be convergent if

(i) there is a unique fixed point of T

(ii) this fixed point is stable and attracts neighborhoods of points of X.

We can now prove

Theorem 5.1. If T is conditionally condensing, then T is convergent if and only if there is a bounded orbit of T on $(-\infty,\infty)$ and every orbit is uniformly asymptotically stable.

Proof: If T is conditionally condensing and convergent, then T is condensing and local dissipative. Therefore, Lemma 3.1(i) implies there is a compact set which attracts compact sets of X. The set J in Theorem 3.1 consists of a single point. Lemma 5.1 implies the result.

Conversely, if there is a bounded orbit of T on $(-\infty,\infty)$ and every orbit is uniformly asymptotically stable, then Lemma 5.2 implies T is local dissipative. We use Lemma 3.1(i) again and conclude T is convergent directly from Theorem 3.1.

Using arguments similar to the ones used in the proofs of Theorems 3.1 and 3.3, one can prove the following

Lemma 5.3. If T is conditionally condensing and some iterate of T is completely continuous, then a bounded orbit on $(-\infty,\infty)$ is asymptotically stable if and only if it is uniformly asymptotically stable.

If the hypothesis of Lemma 5.3 is used in Theorem 5.1, one obtains the equivalence with every orbit being asymptotically stable.

REFERENCES

[1] Billotti, J.E. and J.P. LaSalle, Periodic dissipative processes. Bull. Am. Math. Soc. 6(1971), 1082-1089.

[2] Chow, S. and J.K. Hale, Strongly limit compact maps. Funk. Ekv. 17(1974), 31-38.

[3] Gerstein, V.M., On the theory of dissipative differential equations in a Banach space. Funk. Anal. i Prilozen. 4(1970), 99-100.

[4] Gerstein, V.M. and M.A. Kranoselskii, Structure of the set of solutions of dissipative equations. Dokl. Akad. Nauk SSSR 183(1968), 267-269.

[5] Hale, J.K., LaSalle, J.P. and M. Slemrod, Theory of a general class of dissipative processes. J. Math. Ana. Appl. 39(1972), 177-191.

[6] Hale, J.K. and O. Lopes, Fixed Point theorems and dissipative processes. J. Differential Eqns. 13(1973), 391-402.

[7] Horn, W.A., Some fixed point theorems for compact mappings and flows on a Banach space. Trans. Am. Math. Soc. 149(1970), 391-404.

[8] Jones, G.S., The existence of critical points in generalized dynamical systems, pp. 7-19. Seminar on Differential Equations and Dynamical Systems, Lecture Notes in Math. Vol. 60, 1968, Springer-Verlag.

[9] Jones, G. Stephen, Stability and asymptotic fixed-point theory. Proc. Nat. Acad. Sci. U.S.A. 53(1965), 1262-1264.

[10] LaSalle, J.P., A study of synchronous asymptotic stability. Annals of Mathematics, 65(1957), 571-581.

[11] Levinson, N., Transformation theory of non-linear differential equations of the second order. Annals of Math., 45(1944), 724-737.

[12] Levinson, N., A second order differential equation with singular solutions. Ann. Math. 50(1949), 126-153.

[13] Littlewood, J.E., On non-linear differential equations of the second order: IV. The general equation $\ddot{y} + kf(y)y + g(y) = bkp(\phi)$, $\phi = t + a$. Acta Mathematica vol. 98(1957).

[14] Pliss, V.A., Nonlocal Problems of the Theory of Nonlinear Oscillations. Academic Press, 1966 (Translation of 1964 Russian edition).

[15] Reissig, R., Sansone, G. and R. Conti, Nichtlineare Differential Gleichungen Höherer Ordnung. Cremonese, 1969.

[16] Sadovskii, B.N., Limit compact and condensing operators. Uspehi Mat. Nauk 271(1972), 81-146 (Russian). Russian Math. Surveys, 85-146.

[17] Yoshizawa, T., Stability theory by Liapunov's Second Method. Math. Soc. Japan, 1966.

[18] Yoshizawa, T., Stability Theory and the Existence of Periodic Solutions and Almost Periodic Solutions. Applied Math. Sciences, Vol. 14, 1975. Springer-Verlag.

INTEGRATING A DIFFERENTIAL EQUATION WITH A WEAK* CONTINUOUS VECTOR FIELD*

William S. Hall

When studying the wave equation

$$Z_{tt} - Z_{xx} = \varepsilon h(Z_t, Z_x) \tag{1}$$

$$Z(t,0) = Z(t,T/2) = 0$$

by one of the various averaging methods, one is lead to consider the initial value problem for the ordinary differential equation

$$\dot{u} = f(t,u) \tag{2}$$
$$u(0) = v$$

where v is T-periodic in x and

$$f(t,u(t))(x) = \frac{\varepsilon}{2} h(u(t,x) - u(t,2t-x), u(t,x) + u(t,2t-x)) \tag{3}$$

An indication of how (2) is derived from (1) is given at the end of this paper, and a more complete discussion can be found in [1].

To accommodate a possible loss of smoothness in the steady states of (1), solutions of (2) are desired in the space $C([0,T],X)$ of continuous functions of t in $[0,T]$ with values in $X = L_\infty(T)$, the T-periodic functions of x which are essentially bounded.

The usual approach of defining $\mathcal{J}$ by

$$\mathcal{J}u(t) = v + \int_0^t f(s,u(s))ds \tag{4}$$

---

*Support for this work was provided by a Type I grant from the University of Pittsburgh, and the Mathematics Institutes of the Université de Louvain and the Czech Academy of Sciences, the latter in cooperation with the International Research and Exchanges Board.

and applying a fixed point theorem is quite straightforward once it is understood what is meant by the integral on the right side of (4) and how to relate a fixed point of $\mathcal{J}$ with a solution to (2). The reasons these questions must be examined is that in L_∞, translation of the space variable x by t is not continuous in t. Consequently translation cannot even be strongly measurable [2]. Thus it is highly unlikely that the vector field in (3) can be Riemann or Bochner integrable since it is not even true when h is linear.

However translation is weak* continuous on X since if p is L_1 and is T-periodic,

$$(v(.+t),p) = \int_0^T v(x+t)p(x)dx = \int_0^T v(x)p(x-t)dx$$

and translation is strongly continuous on L_1. By a series of straightforward but extremely tedious calculations it can be shown that when h is smooth and locally bounded (such as a polynomial, for example) then the vector field (3) is also weak* continuous whenever u(t) is in C([0,T],X). Hence it is possible to construct a rather elegant and useful integral which comes complete with a fundamental theorem of calculus, and the purpose of this short note is to show how this can be done.

We model our approach on the classical Pettis integral [2] but with the important difference that the weak* topology replaces the weak topology. So we call the result the weak* Pettis integral. The problem can be done abstractly, so let f(t) be weak* continuous from I = [0,T] with values in X where X is the dual of another Banach space X_o.

The first step is to show $\|f(t)\|$ is measurable and bounded. Let $(\cdot,.)$ pair X and X_o and suppose B is the unit ball in X_o. Then

$$\|f(t)\| = \sup \{(f(t),p); p \in B\}$$

and so it is quite easy to see that

$$\{t \in I; \|f(t)\| > c\} = \bigcup_{p \in B} \{t \in I; (f(t),p) > c\} \qquad (5)$$

Because f is weak* continuous, each of the sets on the right side of (5) is open. Hence $\|f(t)\|$ is upper semicontinuous and measurable.

Next, let $T_t p = (f(t),p)$. For fixed $t \in I$, $T_t : X_o \to R$ is linear and bounded.

For fixed p in X_o, $T_t p$ is continuous on the compact set I. Hence there is a t_o in I such that $\|T_t p\| \leq \|T_{t_o} p\|$. By the principle of uniform boundedness, $\sup \|T_t\| \leq M$. But by definition,

$$\|T_t\| = \sup\{T_t p;\ p \in B\} = \|f(t)\|$$

Hence $\|f(t)\|$ is bounded. As a result, $\|f(t)\|$ is Lebesgue integrable on $[0,T]$.

Now let $p \in X_o$ and consider

$$(Jf)(p) = \int_0^T (f(t),p)dt \tag{6}$$

$(Jf)(p)$ certainly exists since the integrand is continuous. Also J is linear in p and

$$|(Jf)(p)|/\|p\| \leq \int_0^T \|f(t)\| dt \tag{7}$$

Hence Jf lies in $X_o^* = X$. We define

$$Jf = \int_0^T f(t)dt \tag{8}$$

as the integral of f. Obviously J is linear in f. In addition from (6) and (8),

$$\int_0^T (f(t),p)dt = (\int_0^T f(t)dt,\ p) \tag{9}$$

Thus we can interchange "$\int$ " with "(.,.)". Also, from (7),

$$\|\int_0^T f(t)dt\| \leq \int_0^T \|f(t)\| dt$$

so the norm of the integral is less than the integral of the norm.

For a fundamental theorem of calculus, let $0 < t \leq T$ and consider

$$g(t) = \int_0^t f(s)ds \tag{10}$$

We note that since $\|f(t)\|$ is bounded, g is strongly Lipschitz continuous in t. For p in X_o, we have by (9) that

$$(g(t),p) = (\int_0^t f(s)ds,p) = \int_0^t (f(s),p)ds \tag{11}$$

The integrand is continuous so

$$\frac{d}{dt}(g(t),p) = (f(t),p) \tag{12}$$

But then because $\|f(t)\|$ is bounded, the left side of (12) defines an element of $X_o^* = X$ which we denote by $Dg(t)$. We call this element the weak* derivative of g since

$$\frac{d}{dt}(g(t),p) = \lim_{\Delta t \to 0} (\frac{g(t+\Delta t) - g(t)}{\Delta t},p)$$

We note that

$$(Dg(t),p) = \frac{d}{dt}(g(t),p) \tag{13}$$

so we can interchange "D" with "(.,.)".

Hence we have shown that

$$D\int_0^t f(s)ds = f(t) . \tag{14}$$

Conversely, let g be weak* continuous with continuous weak* derivative. Integrating both sides of (13) gives

$$(g(t),p) = (g(0),p) + \int_0^t (Dg(s),p)ds$$

Now interchange "$\int$" with "(.,.)". Then

$$(g(t),p) = (g(0),p) + (\int_0^t Dg(s)ds,p)$$

Hence

$$g(t) = g(0) + \int_0^t Dg(s)ds \tag{15}$$

Let us interpret these results for the differential equation (2). Suppose f takes u(t) in C to the set of weak* continuous functions of t with values in X. Then $\mathcal{J}u(t)$ as given by (4) is strongly continuous in t. If J has a fixed point,

$$u(t) = v + \int_0^t f(s,u(s))ds \qquad (16)$$

and so by (14),

$$\dot{u}(t) = f(t,u(t)) \qquad (17)$$
$$u(0) = v$$

if by $\dot{u}$ we mean Du, and the integral is the weak* Pettis integral. Conversely, if (17) holds then (15) implies (16). Hence (15) and (16) are equivalent problems in so far as existence of solutions is concerned.

To see how (2) can be derived from (1) consider the transformation

$$y(t,x) = \frac{1}{2}(Z_x(t,x) + Z_t(t,x)) \qquad (18)$$

Then, proceeding formally, using (1),

$$\begin{aligned} y_t &= \frac{1}{2}(Z_{xt} + Z_{tt}) \\ &= \frac{1}{2}(Z_{tx} + Z_{xx} + \varepsilon h(Z_t,Z_x)) \\ &= y_x + \frac{\varepsilon}{2} h(Z_t,Z_x) \qquad (19) \end{aligned}$$

The boundary conditions are satisfied if Z and Z_t are 2T-periodic and odd in x. Hence

$$\begin{aligned} y(t, -x) &= \frac{1}{2}(Z_x(t, -x) + Z_t(t, -x)) \\ &= \frac{1}{2}(Z_x(t,x) - Z_t(t,x)) \qquad (20) \end{aligned}$$

Adding, then subtracing (18) and (20) gives,

$$\begin{aligned} Z_t(t,x) &= y(t,x) - y(t,-x) \\ & \qquad\qquad\qquad (21) \\ Z_x(t,x) &= y(t,x) + y(t,-x) \end{aligned}$$

Substituting into (19) we obtain the first order wave equation,

$$y_t(t,x) = y_x(t,x) + \frac{\varepsilon}{2} h(y(t,x) - y(t,-x), y(t,x) + y(t,-x)) \quad (22)$$

Letting y(t,x) = u(t,x+t) and then replacing x by x-t completes the derivation. Justifying this transformation here is out of the question, and the reader should see [1].

References

[1] W.S.Hall, The Rayleigh wave equation, in preparation.

[2] E. Hille and R. Phillips, Functional Analysis and Semi-groups, American Mathematical Society, Providence, R.I., 1957

ON ASYMPTOTIC INTEGRATION

W. A. Harris Jr.

1. Introduction.

In this note we shall describe and give applications of a method which has been utilized recently by Harris–Lutz [3, 4, 5] to give a unified treatment of asymptotic integration of the linear differential system

(1.1) $$x' = A(t)x$$

through the representation of a fundamental solution matrix X(t) in the form

(1.2) $$X(t) = P(t)\,(I + o(1)\exp \int^t \Lambda(s)ds,$$

with P(t) and $\Lambda(s)$ explicit and computable. Such a representation implies that the change of variables $x = P(t)y$ transforms equation (1.1) into

(1.3) $$y' = [\Lambda(t) + R(t)]y,$$

for which there exists a fundamental solution matrix of the form

(1.4) $$Y(t) = [I + o(1)]\exp \int^t \Lambda(s)ds.$$

If the linear differential system (1.3) is in L-diagonal form, i.e., $\Lambda(t) = \operatorname{diag}\{\lambda_1(t), \ldots, \lambda_n(t)\}$, $\|R(t)\| \in L^1(t_0, \infty)$, and if $\Lambda(t)$ satisfies the <u>dichotomy condition</u> that for each index pair $j \neq k$, not both

(1.5)
$$\text{(a)}\quad \limsup_{t\to\infty} \int_{t_0}^{t} \mathrm{Re}[\lambda_j(s) - \lambda_k(s)]ds = +\infty$$

and

$$\text{(b)}\quad \liminf_{t\to\infty} \int_{t_0}^{t} \mathrm{Re}[\lambda_j(s) - \lambda_k(s)]ds = -\infty \text{ hold,}$$

then there exists a fundamental solution matrix for (1.3) of the form (1.4). This is Levinson's Fundamental Theorem [7; 118-122], which is the basis for our theory. Hence, we will have achieved our goal if we can construct suitable P, Λ, and R.

Let

$$x = P(t)w \tag{1.6}$$

and consider the resultant linear differential system $w' = B(t)w$ under the assumption that $B(t) = \Lambda(t) + V(t)$, where Λ is a diagonal matrix and $V(t)$ is small in some suitable sense as $t \to \infty$, i.e., we consider

$$w' = [\Lambda(t) + V(t)]w. \tag{1.7}$$

If Levinson's Theorem does not apply to equation (1.7), we utilize the transformation

$$w = [I + Q(t)]u, \tag{1.8}$$

where $Q(t) = o(1)$ as $t \to \infty$, with the normalization $\operatorname{diag} Q(t) \equiv 0$, to obtain the linear differential system

$$u' = [\Lambda(t) + \hat{V}(t)]u. \tag{1.9}$$

If Levinson's Theorem applies, i.e., if $\hat{V}(t) - \operatorname{diag} \hat{V}(t) \in L^1$ and if $\Lambda(t) + \operatorname{diag} \hat{V}(t)$ satisfies the dichotomy condition, then we have effected the required asymptotic integration.

If $\hat{V}(t) - \operatorname{diag} \hat{V}(t) \notin L^1$, but it is better in some suitable sense, we may consider iterations of transformations (1.6) and (1.8).

The matrices $Q(t)$ and $\hat{V}(t)$ satisfy the equation

$$(I + Q)\hat{V} = \Lambda Q - Q\Lambda + V + VQ - Q'.$$

Our normalization $\operatorname{diag} Q \equiv 0$ implies that

$$\operatorname{diag} \hat{V} = [\operatorname{diag} V + \operatorname{diag}(VQ)]\,[I + o(1)].$$

If we can choose Q in an appropriate manner so that (in at most a finite number of repeated applications) $\hat{V}(s) - \operatorname{diag} \hat{V}(s) \in L^1$, we will have achieved our goal subject, of course, to $\Lambda + \operatorname{diag} \hat{V}$ satisfying the dichotomy conditions (1.5). Three specific choices of Q have proved useful.

I. $\Lambda Q - Q\Lambda + V + VQ - (I+Q)(\text{diag } V + \text{diag } (VQ)) = 0$

II. $V - \text{diag } V - Q' = 0$

III. $\Lambda Q - Q\Lambda + V - \text{diag } V - Q' = 0.$

2. Case I.

Consider the equation

$$(2.1) \qquad (I + Q)\hat{V} = \Lambda Q - Q\Lambda + V + VQ - Q'$$

in the case when Λ is a diagonal matrix for which $\Lambda \to \Lambda_o$ as $t \to \infty$, Λ_o is a constant diagonal matrix with distinct eigenvalues, and $V(t) \to 0$ as $t \to \infty$. For t sufficiently large, the matrix $\Lambda + V$ will have distinct eigenvalues, say $\lambda_i(t) + d_i(t)$, $i \le i \le n$, and there exists $Q_1(t)$, diag $Q_1(t) \equiv 0$, such that

$$(I + Q_1)^{-1}(\Lambda + V)(I + Q_1) = \Lambda + D,$$

where $D = \text{diag}\{d_1, \ldots, d_n\}$ and $\Lambda = \text{diag}\{\lambda_1, \ldots, \lambda_n\}$. Furthermore, Q_1 inherits the regularity properties of V, i.e., $Q_1 = 0(\|V\|)$ as $t \to \infty$, Q_1 is differentiable whenever V is differentiable and $Q_1' = 0(\|V'\|)$ as $t \to \infty$, etc.

The existence of a suitable diagonalizing matrix $I + Q_1$ is equivalent to the existence of a solution $Q_1 = o(1)$ as $t \to \infty$ of the equation

$$(2.2) \qquad (I + Q_1)D = \Lambda Q_1 - Q_1\Lambda + V + VQ_1 .$$

If we select $Q = Q_1$, diag $Q_1 \equiv 0$ satisfying equation (2.2), the resultant $\hat{V} = V_1$ of equation (2.1) has the form

$$V_1 = D - (I + Q_1)^{-1}Q_1', \qquad \text{where } Q_1' = 0(\|V'\|).$$

Thus we have achieved our goal for cases in which V' is more regular than V in the sense of absolute integrability.

Since $\operatorname{diag} Q_1 \equiv 0$, $D = \operatorname{diag}\{(I + Q_1)D\}$, and equation (2.2) yields

$$D = \operatorname{diag} V + \operatorname{diag} VQ_1$$

and

(2.3) $$(I + Q_1)(\operatorname{diag} V + \operatorname{diag} VQ_1) = \Lambda Q_1 - Q_1\Lambda + V + VQ_1.$$

We may obtain approximations for Q_1 by solving the simpler linear systems

$$\Lambda Q_2 - Q_2 \Lambda + V + VQ_2 = \operatorname{diag}\{V + VQ_2\} + Q_2 \operatorname{diag} V$$

$$Q_2 = Q_1 + o(\|Q_1\|^2)$$

or

$$\Lambda Q_3 - Q_3\Lambda + V - \operatorname{diag} V = 0$$

$$Q_3 = Q_1 + 0(\|Q_1\|^2).$$

These proceduces are the essence of the extensions of Levinson's Theorem by Devinatz [1] and Fedoryuk [2] and have been systematically exploited by Harris–Lutz [3].

3. Case II.

Consider the equation (2.1) when $\Lambda \equiv 0$ i.e.

(3.1) $$(I+Q)\hat{V} = V + VQ - Q'$$

and $V - \operatorname{diag} V$ is conditionally integrable for $t \geq t_0$. We utilize the solution

$$Q = Q_4(t) = \int_\infty^t [V(s) - \operatorname{diag} V(s)]ds$$

of the equation

(3.2) $$V - \text{diag } V - Q_4' = 0,$$

(which satisfies diag $Q_4 \equiv 0$ and $Q_4(t) = o(1)$ as $t \to \infty$)

to obtain

$$\hat{V} = [I + o(1)] [\text{diag } V + VQ_4]$$

which is an improvement to V. Clearly, this procedure is also applicable when $\Lambda(t) = o(1)$.

There are wide classes of problems of the form $y' = A(t)y$ which can be reduced to this case through a bounded linear transformation with a bounded inverse. For a systematic treatment of such problems including the physically important adiabatic oscillator, see Harris-Lutz [4].

4. Case III.

In equation (2.1) we now determine Q as a solution of the equation

(4.1) $$\Lambda Q - Q\Lambda + V - Q' = 0, \qquad \text{diag } Q \equiv 0.$$

If Λ a constant diagonal matrix $\Lambda = \text{diag}\{\lambda_1, \ldots \lambda_n\}$ where $\text{Re } \lambda_i \neq \text{Re } \lambda_j$ if $i \neq j$ and $V - \text{diag } V \in L^p$ for some p, $1 < p < \infty$, then setting $Q = (q_{ij})$, $1 \leq k, j \leq n$, $q_{ii} = 0$, $1 \leq i \leq n$, equation (4.1) becomes the (uncoupled) system of scalar differential equations

(4.2) $$q_{ij}' = (\lambda_i - \lambda_j) q_{ij} + v_{ij}, \qquad 1 \leq i \neq j \leq n.$$

If $\text{Re}(\lambda_i - \lambda_j) > 0$, we choose the solution

$$q_{ij}(t) = -\int_t^\infty e^{(\lambda_i - \lambda_j)(t-s)} v_{ij}(s)ds$$

and if $\text{Re}(\lambda_i - \lambda_j) < 0$, we choose the solution

$$q_{ij}(t) = \int_{t_0}^t e^{(\lambda_i - \lambda_j)(t-s)} v_{ij}(s)ds.$$

Note that by extending the functions in the integrands to the whole real line so that they are zero outside of their natuarl domains, both integrals can be expressed in the form

$$g(t) = \int_{-\infty}^{+\infty} h(t - s)v(s)ds,$$

where $h(u) \in L^1(-\infty, +\infty)$ and $v(s) \in L^p(-\infty, +\infty)$.

A standard result from real analysis, see e.g., Rudin [8; pp. 146-148], implies that $g(t) \in L^p(-\infty, +\infty)$. Hence there exists $Q(t) \in L^p$ satisfying (4.1), $\operatorname{diag} Q(t) \equiv 0$, and Holder's inequality shows that $Q(t) = o(1)$ as $t \to \infty$.

Clearly, similar results are valid when $\Lambda = \Lambda(t)$, which we formalize as <u>Lemma</u>. <u>Let</u> $\Lambda = \operatorname{diag}\{\lambda_1(t), \ldots, \lambda_n(t)\}$ <u>and for each index pair</u> $j \neq k$ <u>assume that</u> $|\operatorname{Re}(\lambda_j(t) - \lambda_k(t))| \geq \mu > 0$. <u>If also</u> $V(t) - \operatorname{diag} V(t) \in L^p(t \geq t_o)$ <u>for some</u> p, $1 < p < \infty$, <u>then there exists</u> $Q(t)$, <u>with</u> $\operatorname{diag} Q(t) \equiv 0$, <u>such that</u>

$$\text{(4.3)} \qquad \Lambda Q - Q\Lambda + V - \operatorname{diag} V - Q' = 0,$$

$Q(t) \in L^p(t \geq t_o)$ <u>and</u> $Q(t) \to 0$ <u>as</u> $t \to \infty$.

This lemma can be used to prove the following
<u>Theorem</u> (Hartman - Wintner [6; pp. 71 - 72]).

<u>Let</u> $\Lambda(t) = \operatorname{diag}\{\lambda_1(t), \ldots, \lambda_n(t)\}$ <u>be continuous for</u> $t \geq t_o$ <u>and assume that for each index pair</u> $j \neq k$, $|\operatorname{Re}(\lambda_j(t) - \lambda_k(t)| \geq \mu > 0$. <u>Furthermore, assume that</u> $V(t)$ <u>is continuous for</u> $t \geq t_o$ and $V(t) \in L^p(t_o, \infty)$, $1 < p \leq 2$. <u>Then the linear differential system</u> $x' = [\Lambda(t) + V(t)]x$ <u>has a fundamental solution matrix satisfying as</u> $t \to \infty$

$$\text{(4.4)} \quad X(t) = [I + o(1)]\exp\left(\int_{t_o}^{t} [\Lambda(s) + \operatorname{diag} V(s)]\, ds\right).$$

Proof: According to the Lemma, there exists $Q \in L^p(t_o, \infty)$ satisfying (4.3), $\operatorname{diag} Q \equiv 0$ and $Q(t) \to 0$ as $t \to \infty$. Utilizing the transformation $x = [I + Q(t)]y$, we obtain $y' = (\Lambda + \hat{V})y$, where $(I + Q)\hat{V} = \operatorname{diag} V + VQ$, hence for $t \geq t_1$ sufficiently large,

$$\hat{V}(t) = \operatorname{diag} V(t) + R(t),$$

where $R(t) = VQ - Q(I + Q)^{-1}(\operatorname{diag} V + VQ) \in L^1(t_1, \infty)$ since the product of two L^p functions is in $L^{p/2}$. Moreover, the dichotomy condition (1.5) is satisfied by $\Lambda(t)$ since $|\operatorname{Re}(\lambda_i(t) - \lambda_j(t))| \geq \mu > 0$ and therefore the integral in (1.5) tends at least exponentially either to $+\infty$ or to $-\infty$. Therefore (1.5) is satisfied by $\Lambda(t) + \operatorname{diag} V(t)$ since this corresponds to an additive change in the integrand by an L^p function. Therefore, we may apply Levinson's Basic Theorem to obtain the asymptotic integration of $y' = [\Lambda(t) + \operatorname{diag} V(t) + R(t)]y$ and the theorem of Hartman-Wintner is proven. Clearly, we may iterate this method, see Harris-Lutz [5] which includes many examples.

REFERENCES

[1] A. Devinatz, "An asymptotic theory for systems of linear differential equations", Trans. Amer. Math. Soc. 160 (1971), 353-363.

[2] M. Fedoryuk, "Asymptotic methods in the theory of one-dimensional singular differential operators", Trans. Moskow Math. Soc. (1966), 333-386.

[3] W. A. Harris, Jr. and D. A. Lutz, "On the asymptotic integration of linear differential systems". J. Math. Anal. Appl. 48 (1974), 1-16.

[4] ________________. "Asymptotic integration of adiabatic oscillators", J. Math. Appl. 51 (1975), 76-93.

[5] ________________. " A unified theory of asymptotic integration", J. Math. Anal. Appl. (to appear).

[6] P. Hartman and A. Wintner, "Asymptotic integration of linear differential equations", Amer. J. Math. 77 (1955), 45-86 and 932.

[7] N. Levinson, "The asymptotic nature of solutions of linear differential equations", Duke Math. J. 15 (1948), 111-126.

[8] W. Rudin, Real and Complex Analysis, McGraw-Hill, New York, 1966.

Supported in part by the United States Army under contract DAHCO4-74-6-0013.

EXISTENCE GLOBALE DES SOLUTIONS DE QUELQUES PROBLEMES AUX LIMITES

Gérard Hecquet

L'objet de cette Note et d'annoncer l'existence globale d'une solution des équations aux dérivées partielles :

$$(1) \qquad \frac{\partial^{r+s} u}{\partial x^r \partial y^s} = f_1\left(x,y,u, \frac{\partial u}{\partial x}, \frac{\partial u}{\partial y}, \ldots, \frac{\partial^{p+q} u}{\partial x^p \partial y^q}, \ldots\right) \qquad \begin{cases} 0 \leq p \leq r \\ 0 \leq q \leq s \\ p+q < r+s \end{cases}$$

$$(2) \qquad \frac{\partial^3 u}{\partial x^2 \partial y} - \frac{\partial^3 u}{\partial x \partial y^2} = f_2\left(x,y,u, \frac{\partial u}{\partial x}, \frac{\partial u}{\partial y}, \frac{\partial^2 u}{\partial x \partial y}\right)$$

soumises à différentes conditions initiales : G. Hecquet [1].

La méthode employée qui repose essentiellement sur le théorème de Tychonoff a déjà été utilisée par différents auteurs comme A.K. Aziz, J.P. Maloney [1], B. Palczewski [1], G. Teodoru [1] lors de la résolution de l'équation :

$$(3) \qquad u_{xy} = f(x,y,u,u_x,u_y), \quad u(x,0) = \sigma(x), \quad u(0,y) = \tau(y).$$

La première partie commence par l'examen du second problème de E. Picard

$$u_{xy} = f(x,y,u,u_x,u_y), \quad u_x(x,x) = \sigma(x), \quad u_y(x,x) = \tau(x), \quad u(0,0) = u_o$$

et se poursuit par l'étude des trois problèmes aux limites associés à l'équation (1) :

$$(\text{I.A}) \qquad \begin{cases} \dfrac{\partial^p u}{\partial x^p}(x,x) = \sigma_p(x) & 1 \leq p \leq r \qquad x \in \mathbb{R} \\ \qquad u(0,0) = u_o & \\ \dfrac{\partial^q u}{\partial y^q}(y,y) = \tau_q(y) & 1 \leq q \leq s \qquad y \in \mathbb{R} \end{cases}$$

$$(\text{I.B}) \qquad \begin{cases} \dfrac{\partial^p u}{\partial x^p}(0,y) = \mu_p(y) & 0 \leq p < r \qquad y \in \mathbb{R} \\ \dfrac{\partial^q u}{\partial x^q}(x,0) = \nu_q(x) & 0 \leq q < s \qquad x \in \mathbb{R} \end{cases}$$

$$(\text{I.C}) \quad \begin{cases} \dfrac{\partial^p u}{\partial x^p}(0,y) = \mu_p(y) & 0 \leq p < r \qquad y \in \mathbb{R} \\ \dfrac{\partial^q u}{\partial y^q}(x,g(x)) = \gamma_q(y) & 0 \leq q < s \qquad x \in \mathbb{R} \end{cases}$$

Nous supposerons que les données initiales sont compatibles entre elles et que la fonction f_1 de $C(\mathbb{R}^m,\mathbb{R})$ $m = 1 + (r+1)(s+1)$ vérifie sur $\mathbb{R}^m$ la relation :

$$|f_1(x,y,z_{oo},\ldots,z_{pq},\ldots)| \leq \phi(x,y, \sum_{p,q} |z_{pq}|) = \phi_{x,y}(\sum |z_{pq}|)$$

dans laquelle $\phi_{x,y}$ désigne une application continue sous-additive de $\mathbb{R}_+$. Le principal résultat obtenu dans cette première partie est que les trois problèmes (I.A), (I.B), (I.C) possèdent une solution définie sur tout $\mathbb{R}^2$ pourvu que la fonction f_1 soit lipschitzienne sur tout compact de $\mathbb{R}^m$ par rapport aux $r+s$ variables z_{ps} $(0 \leq p < r)$ et z_{rq} $(0 \leq q < s)$.

L'équation (2) fut examinée par M. Winants durant les années 1930-36 dans une série d'articles. Son étude ne concerne que l'existence locale et suppose que la fonction f_2 : $f_2(x,y,u,p,q,z)$ est lipschitzienne par rapport aux quatre variables u, p, q et z. Dans la seconde partie, nous examinons les quatre problèmes aux limites suivants :

$$(\text{II.A}) \quad \begin{cases} u(x,0) = \sigma(x), & u(0,y) = \tau(y), & \dfrac{\partial u}{\partial x}(0,y) = \tau_1(y), \\ \sigma(0) = \tau(0), & \sigma'(0) = \tau_1(0), & x,y \in \mathbb{R} \end{cases}$$

$$(\text{II.B}) \quad \begin{cases} \dfrac{\partial u}{\partial x}(x,x) = \mu(x), & \dfrac{\partial u}{\partial y}(x,x) = \nu(x), & \dfrac{\partial^2 u}{\partial x \partial y}(x,x) = \chi(x), \\ & u(0,0) = u_o & x,y \in \mathbb{R} \end{cases}$$

$$(\text{II.C}) \quad \begin{cases} u(x,0) = \sigma(x) & u(0,y) = \gamma(y), & \dfrac{\partial u}{\partial y}(x,0) = \sigma_1(x), \\ \sigma(0) = \gamma(0) & \sigma'(0) + \sigma_1(0) = \gamma'(0), & x \in \mathbb{R} \end{cases}$$

$$(\text{II.D})\quad \begin{cases} u(x,0) = \sigma(x), \qquad u(0,y) = \tau(y), \qquad u(x,x) = \gamma(x), \\ \sigma(0) = \tau(0) = \gamma(0), \quad \gamma'(0) = \sigma'(0) + \tau'(0) \qquad x,y \in \mathbb{R} \end{cases}$$

Si comme précédemment, la fonction f_2 vérifie une relation du type $|f_2(x,y,u,p,q,z)| \leq \psi(x,y,|u|+|p|+|q|+|z|)$ et est lipschitzienne sur tout compact par rapport aux variables p, q et z, les quatre problèmes précédents admettent une solution définie sur tout $\mathbb{R}^2$.

Dans ce qui suit nous n'exposerons que l'un de ces sept problèmes car la procédure utilisée est identique pour chacun d'eux et peut se résumer ainsi :

1°) définition de l'équation intégrale dont les solutions coïncident avec celles du problème examiné.

2°) définition d'une topologie à l'aide d'une famille de semi-normes et d'un sous-ensemble convexe compact.

3°) établissement du théorème d'existence comme application du théorème de Tychonoff.

Le problème que nous avons choisi, en l'occurence (II.B) est un de ceux dont la représentation intégrale et la topologie nécessaires s'obtiennent le plus facilement.

1°) Représentation intégrale du problème (II.B).

Recherchons comme M. Winants l'opérateur T permettant d'écrire sous forme intégrale les solutions du problème

$$(4)\quad \begin{cases} u_{x^2y} - u_{xy^2} = g(x,y) \qquad u(0,0) = u_0 \\ u_x(x,x) = \sigma(x), \quad u_y(x,x) = \nu(x), \quad u_{xy}(x,x) = \chi(x). \end{cases}$$

La solution générale de l'équation $u_{x^2y} - u_{xy^2} = g(x,y)$ étant

$$u(x,y) = p(x) + q(y) + \int_0^x \int_0^y \left[m(s,t) + P(s+t)\right] ds\, dt$$

avec $m(s,t) = \int_0^s g(\xi, s+t-\xi)d\xi$ déterminons les fonctions p, q et P grâce aux relations

$$(5) \quad \begin{cases} p(0) + q(0) = u_o \\ p'(x) + \int_0^x m(x,t)dt + \int_0^x P(x+t)dt = \mu(x) \\ q'(y) + \int_0^y m(s,y)ds + \int_0^y P(s+y)ds = \nu(y) \\ m(x,x) + P(2x) = \chi(x). \end{cases}$$

En posant $\mathfrak{S}(x,y) = u_o + \int_0^x \mu(s)ds + \int_0^y \nu(t)dt + \int_y^x \{\int_s^y \chi(\frac{s+t}{2})dt\}ds$

nous pouvons écrire la solution du problème (4) sous la forme

$$(6) \qquad u(x,y) = \mathfrak{S}(x,y) + \int_y^x \left[\int_s^y \left(\int_{\frac{s+t}{2}}^s g(r,s+t-r)dr\right)dt\right]ds .$$

Pour $u \in H(\mathbb{R}^2,\mathbb{R}) = \{u \in C(\mathbb{R}^2,\mathbb{R}) : u_x, u_y, u_{xy} \in C(\mathbb{R}^2,\mathbb{R})\}$ définissons $U = (u, u_x, u_y, u_{xy})$

$$(7) \quad F(x,y,U) = f_2(x,y,u(x,y),u_x(x,y),u_y(x,y),u_{xy}(x,y))$$

de sorte que la représentation intégrale du problème (II.B) cherchée est :

$$(8) \qquad (Tu)(x,y) = \mathfrak{S}(x,y) + \int_y^x \left[\int_s^y \left(\int_{\frac{s+t}{2}}^s F(r,s+t-r,U)dr\right)dt\right]ds.$$

Nous écrirons alors :

$$(Tu)_x(x,y) = \mu(x) + \int_x^y \chi(\frac{x+t}{2})dt + \int_x^y \left(\int_{\frac{x+t}{2}}^x F(r,x+t-r,U)dr\right)dt$$

$$(Tu)_y(x,y) = \nu(y) + \int_y^x \chi(\frac{s+y}{2})ds + \int_y^x \left\{\int_{\frac{y+s}{2}}^s F(r,s+y-r,U)dr\right\}ds$$

$$(Tu)_{xy}(x,y) = \chi(\frac{x+y}{2}) + \int_{\frac{x+y}{2}}^x F(r,x+y-r,U)dr.$$

2°) *Définition de la topologie de* $H(\mathbb{R}^2,\mathbb{R})$.

La fonction L_λ : $L_\lambda(x,y) = \sqrt{\operatorname{ch} \lambda(x-y)}$ définie sur $\mathbb{R}^2$ pour $\lambda \geq 0$ vérifie les propriétés suivantes :

$$\left|\int_{\frac{x+y}{2}}^{x} L_\lambda^2(s,x+y-s)ds\right| \leq \frac{1}{\lambda} L_\lambda^2(x,y) \qquad \left|\int_y^x \left(\int_{\frac{y+s}{2}}^{s} L_\lambda^2(r,s+y-r)dr\right)ds\right| \leq \frac{1}{\lambda^2} L_\lambda^2(x,y)$$

$$\left|\int_y^x \left(\int_{\frac{y+t}{2}}^{y} L_\lambda^2(r,x+t-r)dr\right)dt\right| \leq \frac{1}{\lambda^2} L_\lambda^2(x,y)$$

$$\left|\int_y^x \left\{\int_s^y \left(\int_{\frac{s+t}{2}}^{s} L_\lambda^2(r,s+t-r)dr\right)dt\right\}ds\right| \leq \frac{1}{\lambda^3} L_\lambda^2(x,y).$$

Pour $\Psi \in C(\mathbb{R}^2,\mathbb{R})$ définissons $\alpha_k^{\lambda_k}(\Psi) = \sup_{Q_k}\{|\Psi(x,y)|L_{\lambda_k}^{-1}(x,y)\}$

où $Q_k = \{(x,y) \in \mathbb{R}^2 : |x| \leq k, |y| \leq k\}$, $\lambda_k \geq 0$ et $k \in N$, puis

$$\beta_k^{\lambda_k}(\Psi) = \operatorname{Max}\{\alpha_k^{\lambda_k}(\Psi), \alpha_k^{\lambda_k}(\Psi_x), \alpha_k^{\lambda_k}(\Psi_y), \alpha_k^{\lambda_k}(\Psi_{xy})\}.$$

D'un autre côté s'il existe une fonction continue majorante de la fonction f_2 sous-additive de $\mathbb{R}_+$ dans $\mathbb{R}_+$ pour tout (x,y) : $|f_2(x,y,u,p,q,z)| \leq \vartheta_{x,y}(|u|+|p|+|q|+|z|)$ on peut en déduire l'existence d'une fonction $\bar{\vartheta}_{x,y}$ sous-additive encore mais croissante : $\bar{\vartheta}(x,y,t) = \sup_{0\leq\tau\leq t} \vartheta(x,y,\tau)$ majorant la fonction f_2. On pourra alors écrire que

$$|f_2(x,y,u,p,q,z)| \leq \bar{\vartheta}(x,y,1)\{1+|u|+|p|+|q|+|z|\},$$

et introduire les fonctions :

$$B_o(x,y) = |\mathcal{S}(x,y)| + \left|\int_y^x \left\{\int_y^s \left(\int_{\frac{s+t}{2}}^{s} \bar{\vartheta}(\xi,s+t-\xi,1)d\xi\right)dt\right\}ds\right|$$

$$B_1(x,y) = \operatorname{Max}\left\{B_o(x,y), \sqrt{\left|\int_y^x \left\{\int_y^s \left(\int_{\frac{s+t}{2}}^{s} \bar{\vartheta}^2(\xi,s+t-\xi,1)d\xi\right)dt\right\}ds\right|}\right\}$$

$$B_2(x,y) = \left|\frac{\partial}{\partial x}\,\mathfrak{S}(x,y)\right| + \left|\int_y^x\left(\int_{\frac{x+t}{2}}^x \bar{\vartheta}(\xi,x+t-\xi,1)d\xi\right)dt\right|$$

$$B_3(x,y) = \mathrm{Max}\{B_2(x,y), \sqrt{\left|\int_y^x\left(\int_{\frac{x+t}{2}}^x \bar{\vartheta}^2(\xi,x+t-\xi,1)d\xi\right)dt\right|}\}$$

$$B_4(x,y) = \left|\frac{\partial}{\partial y}\,\mathfrak{S}(x,y)\right| + \left|\int_y^x\left(\int_{\frac{y+s}{2}}^s \bar{\vartheta}(\xi,s+y-\xi,1)d\xi\right)ds\right|$$

$$B_5(x,y) = \mathrm{Max}\{B_4(x,y), \sqrt{\left|\int_y^x\left(\int_{\frac{y+s}{2}}^s \bar{\vartheta}^2(\xi,s+y-\xi,1)d\xi\right)ds\right|}\}$$

$$B_6(x,y) = \left|\frac{\partial^2}{\partial x\partial y}\,\mathfrak{S}(x,y)\right| + \left|\int_{\frac{x+y}{2}}^x \bar{\vartheta}(\xi,x+y-\xi,1)d\xi\right|$$

$$B_7(x,y) = \mathrm{Max}\{B_6(x,y), \sqrt{\left|\int_{\frac{x+y}{2}}^x \bar{\vartheta}^2(\xi,x+y-\xi,1)d\xi\right|}\}$$

$$\mathcal{B}(x,y) = \mathrm{Max}\{B_1(x,y),B_3(x,y),B_5(x,y),B_7(x,y)\}.$$

Ces fonctions permettent d'établir, comme A.K. Aziz et J.P. Maloney, que :

$$\alpha_k^{\lambda_k}(Tu) \leq N_k\{1 + \frac{4\beta_k^{\lambda_k}(u)}{\lambda_k\sqrt{\lambda_k}}\} \qquad \alpha_k^{\lambda_k}((Tu)_x) \leq N_k\{1 + \frac{4\beta_k^{\lambda_k}(u)}{\lambda_k}\}$$

$$\alpha_k^{\lambda_k}((Tu)_y) \leq N_k\{1 + \frac{4\beta_k^{\lambda_k}(u)}{\sqrt{\lambda_k}}\} \qquad \alpha_k^{\lambda_k}((Tu)_{xy}) \leq N_k\{1 + \frac{4\beta_k^{\lambda_k}(u)}{\sqrt{\lambda_k}}\}$$

pour $N_k = \alpha_k^o(\mathcal{B})$ et de conclure que le choix $\lambda_k = \mathrm{Max}\{\sqrt[3]{64}N_k^{2/3}, 8N_k, 64N_k^2\}$ implique la relation :

$$\beta_k^{\lambda_k}(u) \leq 2N_k \implies \beta_k^{\lambda_k}(Tu) \leq 2N_k.$$

D'un autre côté, l'hypothèse que les fonctions f_2, μ, ν et χ sont continues permet d'introduire des fonctions numériques croissantes, non négatives nulles à l'origine sous-additives telles que :

$$|f_2(x,y,u,p,q,z) - f_2(\bar{x},y,u,p,q,z)| \leq \omega_{1,k}(|x-\bar{x}|)$$
$$|f_2(x,y,u,p,q,z) - f_2(x,\bar{y},u,p,q,z)| \leq \omega_{2,k}(|y-\bar{y}|)$$
$$|f_2(x,y,u,p,q,z) - f_2(x,y,\bar{u},p,q,z)| \leq \omega_{3,k}(|u-\bar{u}|)$$
$$|f_2(x,y,u,p,q,z) - f_2(x,y,u,\bar{p},q,z)| \leq \omega_{4,k}(|p-\bar{p}|)$$
$$|f_2(x,y,u,p,q,z) - f_2(x,y,u,p,\bar{q},z)| \leq \omega_{5,k}(|q-\bar{q}|)$$
$$|f_2(x,y,u,p,q,z) - f_2(x,y,u,p,q,\bar{z})| \leq \omega_{6,k}(|z-\bar{z}|)$$
$$\operatorname{Max}\{|\mu(x) - \mu(\bar{x})|, |\nu(x) - \nu(\bar{x})|, |\chi(x) - \chi(\bar{x})|\} \leq \omega_{7,k}(|x-\bar{x}|)$$

pour $x,\bar{x},y,\bar{y} \in [-k,k], u,\bar{u},p,\bar{p},q,\bar{q},z,\bar{z} \in [-N_k,N_k]$.

Si à ces différentes fonctions, on adjoint une fonction majorante ρ_{2k} définie pour $x \in [-k,k]$ et $x+y, x+\bar{y} \in [-2k,2k]$

$$|u_{xy}(x,y) - u_{xy}(x,\bar{y})| \leq \rho_{2k}(x,|y-\bar{y}|)$$

on constate la possibilité d'obtenir une fonction Ω_{2k} définie sur $\mathbb{R}$ telle que :

$$|(Tu)_{xy}(x,y) - (Tu)_{xy}(x,\bar{y})| \leq \Omega_{2k}(|y-\bar{y}|) + \int_x^k \omega_{5,2k}\left(\int_s^k \rho_{2k}(r,|y-\bar{y}|)dr\right)ds + \\ + \int_x^k \omega_{6,2k}\{\rho_{2k}(r,|y-\bar{y}|)\}dr$$

si $x < y$ et $\bar{x} < y$

et $$|(Tu)_{xy}(x,y) - (Tu)_{xy}(x,\bar{y})| \leq \Omega_{2k}(|y-\bar{y}|) + \int_{-k}^x \omega_{5,2k}\left(\int_{-k}^s \rho_{2k}(r,|y-\bar{y}|)dr\right)ds + \\ + \int_{-k}^x \omega_{6,2k}\{\rho_{2k}(r,|y-\bar{y}|)dr\}$$

si $x > y$ et $\bar{x} > y$

et une fonction ρ_{1k} telle que :

$$|(Tu)_{xy}(x,y) - (Tu)_{xy}(\bar{x},y)| \leq \rho_{1k}(|x-\bar{x}|) \quad \text{pour} \quad \begin{cases} |x|,|\bar{x}| \leq k \\ |x+y|,|\bar{x}+y| \leq 2k \end{cases}$$

3°) *Théorème d'existence.*

Avec les hypothèses :

(h.1) $f \in C(\mathbb{R}^6,\mathbb{R}) \quad |f_2(x,y,u,p,q,z)| \leq \bar{\Phi}(x,y,1)\{1+|u|+|p|+|q|+|z|\}$

(h.2) $\mu,\nu,\chi \in C(\mathbb{R},\mathbb{R})$

(h.3) les équations intégrales :

$$\begin{cases} \rho_k(x,\delta) = \Omega_{2k}(\delta) + \int_x^k \omega_{5,2k}(\int_s^k \rho_k(r,\delta)dr)ds + \int_x^k \omega_{6,2k}(\rho_k(r,\delta))dr \\ \\ \rho_k(x,\delta) = \Omega_{2k}(\delta) + \int_{-k}^x \omega_{5,2k}(\int_{-k}^s \rho_k(r,\delta)dr)ds + \int_{-k}^x \omega_{6,2k}(\rho_k(r,\delta))dr \end{cases}$$

admettent des solutions positives tendant vers 0 avec δ uniformément sur le compact $[-k,k]$.

Le problème (II.B) admet une solution définie sur $\mathbb{R}^2$

La démonstration repose sur le fait que le sous-ensemble A :

$$A = \begin{cases} u \in H(\mathbb{R}^2,\mathbb{R}) : \beta_k^{\lambda_k}(u) \leq 2N_k \quad \text{pour} \quad k \in \mathbb{N} \\ |u_{xy}(x,y) - u_{xy}(x,\bar{y})| \leq \rho_{2k}(x,|y-\bar{y}|) \\ |u_{xy}(x,y) - u_{xy}(\bar{x},y)| \leq \rho_{1k}(|x-\bar{x}|) \\ \text{pour } |x|,|\bar{x}| \leq k, \ |x-\bar{x}| + |y-\bar{y}| \leq \delta, \ |x+y|,|\bar{x}+y|,|x+\bar{y}| \leq 2k \end{cases}$$

est convexe et compact pour la topologie de $H(\mathbb{R}^2,\mathbb{R})$ et sur le fait que T est un opérateur continu de A dans A.

Remarque 1.- Dès que la fonction f_2 est lipschitzienne sur tout compact par rapport aux variables q et z l'hypothèse (h.3) est satisfaite.

Remarque 2.- En remarquant que $(Tu)_{xy}$ peut s'écrire

$$(Tu)_{xy}(x,y) = \chi(\frac{x+y}{2}) + \int_y^{\frac{x+y}{2}} F(x+y-\xi,\xi,U)d\xi.$$

On pourra conclure en l'existence d'une solution définie sur tout $\mathbb{R}^2$ sous les hypothèses (h.1), (h.2) si de plus la fonction f_2 est lipschitzienne sur tout compact en p et z.

4°) *Conclusion.*

Le théorème qui vient d'être établi ne concerne nullement l'unicité comme le montre l'exemple suivant : $u_{x^2y} - u_{xy^2} = 15(y-x)|u|^{1/3}$ qui admet les deux solutions $u_1 \equiv 0$ et $u_2 = \frac{1}{64}(y-x)^6$. Mais si on suppose la fonction f_2 lipschitzienne en u, l'opérateur T devient une contraction et la solution est alors unique.

BIBLIOGRAPHIE

A.K. AZIZ, J.P. MALONEY - *An application of Tychonoff's fixed point theorem to hyperbolic partial differential equations,* Math. Annal. 162 (77-82), 1965-66.

G. HECQUET - *Existence globale des solutions de quelques problèmes aux limites de type hyperbolique,* (à paraître).

B. PALCZEWSKI - *On boundedness and stability of solutions of Darboux problem for abstract equations of hyperbolic type in an unbounded domain,* Zeszyty naukowe Politechniki gdanskiej, 1969, n° 150 (19-51).

G. TEODORU - *The Darboux problem for a hyperbolic partial differential equation of second order,* Buletinul Institutuliu politehnic din Iasi Tomul XIX (XXIII) fasc. 3-4 (1973).

M. WINANTS [1] - *Révolution du problème* (a_o, IV, 1°), Bull. de l'Acad. Roy. de Belgique, cl. des Sc. XXI, (376-384), 1934.

M. WINANTS [2] - *Résolution du problème* (a_o, IV, 2°), Bull. de l'Acad. Roy. de Belgique, cl. des Sc. XXI (495-503), 1934.

M. WINANTS [3] - *Chacun des deux problèmes* (a_o,III,3") *et* (a_o,III,2') *peut être résolu par le moyen d'une équation intégrale ayant un nombre infini de termes,* Bull. de l'Acad. Roy. de Belgique, cl. des Sc. XXII (8-25), 1935.

SINGULAR PERTURBATION PROBLEMS FOR A CLASS OF SINGULAR PARTIAL DIFFERENTIAL EQUATIONS*

George C. Hsiao and Richard J. Weinacht[x]

1. Introduction

In this paper we consider interior Dirichlet problems for the singular elliptic equation

(1.1) $$\mathcal{E}_k[u] \equiv \varepsilon^2 u_{xx} + u_{yy} + (k/y)u_y = 0$$

with k a real parameter and ε a small positive real parameter. Asymptotic expansions in ε of the solutions are developed and are proved to be uniformly valid for small ε.

When $\varepsilon = 1$ equation (1.1) is the equation of Weinstein's ([1-3] and bibliographies therein) Generalized Axially Symmetric Potential Theory. Of course, by a simple change of variables (1.1) can be transformed to a Schrödinger equation with potential proportional to y^{-2}. For $k \neq 0$ the line $y = 0$ is a singular line for (1.1) and the boundary value problems considered here have boundaries consisting, in part, of the singular line. For $\varepsilon \neq 0$ uniqueness of the solutions of the boundary value problems follows from Huber [4]. Existence theorems for regions for which the portion of the boundary lying in the open half plane $y > 0$ is smooth (which is not so in the characteristic case treated here in Sections 4 and 5) are stated in Huber [5] (for related results see Quinn and Weinacht [6]). For all problems treated here existence follows from the more general results of Moss [7]. If a portion of the boundary is in the direction of the characteristics of the reduced $(\varepsilon = 0)$ operator

(1.2) $$B_k[u] \equiv u_{yy} + (k/y)u_y$$

i.e. in the direction $x =$ constant, then the regular perturbation procedure breaks down and boundary layer corrections are needed. These corrections are typical for singular perturbation problems where the reduced operator is of lower order than the given operator (see e.g. Eckhaus and de Jager [8]). In the present case the reduced operator is of the same order as the original operator but in one less variable.

In the non-singular case $(k = 0)$ problems of this nature have been considered by Knowles [9] in a rectangle and by Lions [10] in cylindrical regions in higher dimensions. For $k = 0$ Jiji [11] derived matched asymptotic expansions in irregular regions as treated here but he gave no proof of uniform asymptotic validity.

*This research was supported by the Air Force Office of Scientific Research through AF-AFOSR Grant No. 74-2952, and in part by the Alexander von Humboldt-Stiftung.

Recently Ho and Hsiao [12] applied the technique developed here to investigate axi-symmetric $(k = 1)$ problems arising in catalytic reactions.

The formal perturbation procedure developed in the present paper is a variation of a technique due to Levinson [13] or, more precisely, a technique employed by Keller [14] for treating the initial-boundary value problem for the heat equation with a small parameter. The validity of the expansion is established by means of a Comparison Theorem (see Section 2) related to a maximum principle for (1.1).

Section 3 treats the non-characteristic case for $k \geq 1$ and the characteristic case for $k \geq 1$ is presented in Sections 4 and 5. The case $k < 1$ is reduced to the previous case by Weinstein's Correspondence Principle [1] and this reduction is indicated in [15].

2. Preliminaries

In the following Ω is a bounded simply connected region in the half plane $y > 0$, whose boundary $\partial\Omega$ consists of a closed segment $\overline{\Gamma}_o$ of the x axis and of an open arc Γ in the half-plane $y > 0$.

Lemma (Comparison Theorem): If w belongs to $C^2(\Omega) \cap C(\Omega \cup \Gamma)$ and $\mathcal{E}_k[w] \geq 0$ in Ω with $w \leq 0$ on Γ and

$$(2.1)^* \qquad \lim_{(x,y)\to(x^o,0)} \rho(y;k)w(x,y) = 0\ , \quad (x,y) \in \Omega \cup \Gamma\ , \quad (x^o,0) \in \overline{\Gamma}_o\ ,$$

then $w \leq 0$ in Ω. Here $\rho(y;k) = \begin{cases} y^{k-1} & , k > 1\ , \\ (\log y)^{-1}, & k = 1\ . \end{cases}$

Proof: The proof is based on Weinstein's Correspondence Principle [1] as follows. For $k > 1$ let $v = y^{k-1}w$. Then with (the extension of) v defined to be zero on $\overline{\Gamma}_o$ one has that v belongs to $C^2(\Omega) \cap C(\overline{\Omega})$, $\mathcal{E}_{2-k}[v] \geq 0$ in Ω and $v \leq 0$ on $\partial\Omega$. Then the usual (interior) maximum principle of Hopf [16] for the elliptic operator $\mathcal{E}_{2-k}$ in Ω guarantees $v \leq 0$ on Ω and hence also $w \leq 0$ on Ω, completing the proof for $k > 1$.

For $k = 1$, put $v = w \log\left(\frac{2Y}{y}\right)^{-1}$, where $y \leq Y$ for (x,y) in Ω. Now argue as in the case $k > 1$, noting that $\mathcal{E}_1[v] - 2[y \log(2Y/Y)]^{-1} v_y \geq 0$ in Ω.

Remarks: (1) The condition (2.1) is not superfluous as the example $\alpha\rho^{-1}(y;k)\{((x-1)^2+y^2)^\beta - 1\}$ shows when Γ is the semi-circle $(x-1)^2+y^2 = 1$, $y > 0$ with α and β chosen suitably.

(2) For $k \geq 1$ the Lemma is a refinement of the maximum principles given by Muckenhaupt and Stein [17] and Parter [18].

---

*Throughout the paper we shall refer (2.1) as the growth condition near $y = 0$ for the function under consideration, and it should be understood that w will be replaced by the corresponding function.

From the Lemma the following Corollary follows easily in a familiar way [19].

Corollary: Let w belong to $C^2(\Omega) \cap C(\Omega \cup \Gamma)$ and suppose $\mathcal{E}_k[w] = f$ in Ω and $w = \phi$ on Γ and further w satisfies (2.1). Then, if f and ϕ are bounded, we have the estimate

$$|w| \leq ||f||_\infty (e^Y - 1) + ||\phi||_\infty$$

where the supremum norms are used over Ω and Γ respectively and $y \leq Y$ for (x,y) in Ω.

Remark: Solutions of $\mathcal{E}_k[u] = 0$ satisfying the growth condition (2.1) can be continued analytically beyond Γ_o into the lower half plane [4,6] so that our solutions u of the boundary value problems considered below are well behaved on Γ_o.

3. Non-Characteristic Boundary

In this case the curve Γ has nowhere a vertical tangent and, for definiteness, it is assumed that Γ has the representation $\Gamma = \{(x,y): y = \gamma(x) , 0 < x < a\}$ where γ is a non-negative C^∞ function defined on $[0,a]$ and $\gamma(0) = \gamma(a) = 0$ with 0 and a the only zeros of γ. We consider here the boundary value problem (P_ε) consisting of (1.1) in Ω together with the boundary condition $u(x,y;\varepsilon) = \phi(x)$ on Γ and the growth condition (2.1) for u near $y = 0$. For simplicity, we assume that ϕ is a smooth function which can be differentiated as many times as needed.

A standard regular perturbation procedures based on the ansatz

$$u(x,y;\varepsilon) \sim \sum_{\ell=0}^{\infty} \varepsilon^{2\ell} U^\ell(x,y) \tag{3.1}$$

leads to the determination of the U^ℓ as solutions of the sequence of ODE problems in the variable y

$$B_k[U^\ell] = \begin{cases} 0 & , \ \ell = 0 \\ -U^{\ell-1}_{xx} & , \ \ell \geq 1 \end{cases} , \quad U^\ell|_\Gamma = \begin{cases} \phi(x) & \ell = 0 \\ 0 & , \ \ell \geq 1 \end{cases} \tag{3.2$_\ell$}$$

with U^ℓ satisfying the growth condition (2.1). Clearly,

$$U^o(x,y) \equiv \phi(x)$$

is the solution of the reduced problem (P_o) in $\bar{\Omega}$. Higher order terms U^ℓ can be obtained explicitly by making use of the Green's function $G(y;\eta;x)$ for the operator B_k. For convenience, introducing Green's operator $\mathcal{G}$,

$$\mathcal{G}[\psi](x,y) := \int_0^{\gamma(x)} \eta^k G(y;\eta;x)\psi(x,\eta)d\eta ,$$

we see that (3.1) takes the form:

$$(3.3)\qquad u(x,y;\varepsilon) \sim \phi(x) + \sum_{\ell=1}^{\infty} \varepsilon^{2\ell}(-1)^{\ell}\mathcal{G}[U_{xx}^{\ell-1}](x,y) \quad .$$

It is emphasized that in the present case there is no boundary layer correction term.

The uniform asymptotic validity of (3.3) is easily established as follows. A straightforward induction proof based on the explicit form of $G(y;\eta;x)$ yields for $\ell \geq 1$

$$U^{\ell}(x,y) = \sum_{i=2}^{2\ell} \sum_{j=1}^{\ell} a_{ij}(\gamma)y^{2j}\phi^{(i)}(x)$$

where a_{ij} is a polynomial in γ and its derivatives up through order $2(\ell-1)$. Hence, defining

$$U_N(x,y;\varepsilon) = \phi(x) + \sum_{\ell=1}^{N} \varepsilon^{2\ell}U^{\ell}(x,y)$$

and putting $Z_N \equiv u-U_N$, it follows that Z_N vanishes on Γ, satisfies (2.1) and in Ω

$$\mathcal{E}_k[Z_N] = \varepsilon^{2(N+1)}U_{xx}^N \quad .$$

By the Corollary of Section 2,

$$|Z_N(x,y)| \leq (e^Y-1)\varepsilon^{2(N+1)}||U_{xx}^N||_\infty$$

where, as before, $|\gamma(x)| \leq Y$ on $[0,a]$. Thus, we have proved the following theorem.

<u>Theorem 1. Suppose that</u> $\phi \in C^{N+2}(\overline{\Gamma})$ $(N = 0,1,2,\ldots)$. <u>Then for the solution</u> $u(x,y;\varepsilon)$ <u>of</u> (P_ε), <u>we have the asymptotic representation</u>

$$u(x,y;\varepsilon) = \sum_{\ell=0}^{N} \varepsilon^{2\ell}U^{\ell}(x,y)+O(\varepsilon^{2N+2}) \quad \text{as} \quad \varepsilon \to 0^+$$

<u>uniformly on</u> $\overline{\Omega}$, <u>where</u> U^{ℓ}'s <u>are the unique solutions of</u> $(3.2)_\ell$.

4. Characteristic Boundary-Formal Expansions

We now turn to the case where arc Γ contains segments which are characteristics of the reduced operator B_k in (1.2). To be more precise, let $\gamma = \gamma(x)$ be a positive C^∞ function defined on $[0,a]$ and consider the boundary value problem (P'_ε) defined by (1.1) in Ω, the growth condition (2.1) near $y = 0$ and the boundary conditions:

$$(4.1)\qquad \begin{aligned} u(x,\gamma(x);\varepsilon) &= \phi(x) , & 0 \leq x \leq a , \\ u(0,y;\varepsilon) &= \psi_1(y) , & 0 < y \leq \gamma(0) , \\ u(a,y;\varepsilon) &= \psi_2(y) , & 0 < y \leq \gamma(a) . \end{aligned}$$

Here ϕ, ψ_i are smooth functions which satisfy certain conditions to be specified. First we observe that in the present case the reduced problem (P'_0) is again defined by $(3.2)_o$ and (2.1), and hence has solution

$$U^o(x,y) = \phi(x)$$

for (x,y) in $\bar{\Omega}$. However, it is clear that in general U^o may not satisfy the boundary conditions (4.1) along the characteristics, i.e. $x = 0$ and $x = a$, and one has to consider the boundary layer correction terms along them. In the following we will develop a procedure for the construction of the formal asymptotic expansions for the solution of (P'_ε). A justification of these expansions will be discussed in the next section.

We assume the solution $u(x,y;\varepsilon)$ of (P'_ε) has the asymptotic form:

$$u(x,y;\varepsilon) = U(x,y;\varepsilon)+\tilde{V}(x,y;\varepsilon)+\tilde{W}(x,y;\varepsilon) \quad . \tag{4.2}$$

The first term $U(x,y;\varepsilon)$ corresponds to the outer solution, which takes the form

$$U(x,y;\varepsilon) \sim \sum_{\ell=0}^{\infty} \varepsilon^{2\ell} U^\ell(x,y) \quad , \tag{4.3}$$

where the U^ℓ's satisfy $(3.2)_\ell$ and hence are completely determined as in the previous section. The second and the third terms in (4.2) are the corresponding boundary layer terms near $x = 0$ and $x = a$ respectively. Their asymptotic developments are more involved. However, as we will see, both $\tilde{V}$ and $\tilde{W}$ admit a similar form:

$$\tilde{V}(x,y;\varepsilon) \sim \sum_{n=0}^{\infty} \varepsilon^n \tilde{V}^n(x,y;\varepsilon) \quad \text{and} \quad \tilde{W}(x,y;\varepsilon) \sim \sum_{n=0}^{\infty} \varepsilon^n \tilde{W}^n(x,y;\varepsilon) \quad . \tag{4.4}$$

We begin with the derivation of $\tilde{V}$. In view of the standard results now in the usual singular perturbation theory, it is natural to seek $\tilde{V}^n$ in the form:

$$\tilde{V}^n(x,y;\varepsilon) = e^{-s(x)/\varepsilon}\, V^n(x,y) \tag{4.5}$$

where s is a function to be determined such that $s(0) = 0$ and $s(x) > 0$ for $x > 0$. Formally substituting (4.4) (with 4.5) into (1.1) and taking into account (2.1) and the first equation in (4.1), we obtain

$$B_k[V^n]+[s'(x)]^2V^n = s''(x)V^{n-1}+2s'(x)V_x^{n-1}-V_{xx}^{n-2} \quad , \quad n \geq 0 \tag{4.6$_n$}$$

$$V^n(x,\gamma(x)) = 0 \quad , \quad 0 \leq x \leq a$$

with V^n also satisfying the condition (2.1). Here $V^{-1} = V^{-2} = 0$, and the boundary condition is homogeneous because of U^o.

For $n = 0$, the problem $(4.6)_o$ has a nontrivial solution iff $[s'(x)]^2$ is one of the eigenvalues defined by

$$s'(x) = \lambda_m(x) = \frac{j_m}{\gamma(x)} \quad , \quad (m = 1,2,\ldots) \tag{4.7}$$

where j_m denotes the m-th positive zero of the Bessel function J_p, $p = (k-1)/2$. Then for each m,

$$V_m^o(x,y) = a_m^o(x)\phi_m(x,y) \tag{4.8}$$

will be a solution of $(4.6)_o$ for arbitrary $a_m^o(x)$. Here the ϕ_m are the corresponding ortho-normalized eigenfunctions defined by

$$\phi_m(x,y) = \alpha_m(x)y^{-p}J_p\left(\frac{j_m}{\gamma(x)}\, y\right) \quad , \quad p = (k-1)/2 \tag{4.9}$$

with $\alpha_m(x)$ being so selected that $||\phi_m||_2^2 \equiv \int_0^{\gamma(x)} y^k\phi_m^2(x,y)dy = 1$. However, for $(4.5)_1$ to have a solution, the RHS of $(4.6)_1$ must be orthogonal to ϕ_m . This leads to the ODE for $a_m^o(x)$,

$$a_m^o(x) - \frac{\dot{\gamma}(x)}{2\gamma(x)}\, a_m^o(x) = 0 \quad , \tag{4.10}$$

and hence $a_m^o(x)$ is uniquely determined up to a multiplicative constant, say $\tilde{a}_m^o = a_m^o(0)$. To fix $\tilde{a}_m^o$, we now use the boundary condition (4.1) at $x = 0$, which has not been used so far. By following the usual Fourier analysis, it is clear that one should form the Fourier expansion of $\tilde{V}^o$ with respect to $\{\phi_m\}_{m=1}^{\infty}$. Based on (4.5) and (4.7), we define $\tilde{V}^o$ by the expansion:

$$\tilde{V}^o(x,y;\varepsilon) = \sum_{m=1}^{\infty} \exp\left\{-\frac{1}{\varepsilon}\int_0^x \lambda_m(\xi)d\xi\right\} V_m^o(x,y) \quad , \tag{4.11}$$

and require that $\tilde{V}^o$ satisfies the <u>matching condition</u> at $x = 0$,

$$u(0,y;\varepsilon) - U^o(0,y) = \tilde{V}^o(0,y;\varepsilon) \quad . \tag{4.12}$$

It follows that $\tilde{a}_m^o$ are directly related to the Fourier-Bessel coefficients of the function $y^p(\psi_1(\gamma(0))-\phi(0))$ and hence are known. This completes the derivation of $\tilde{V}^o$.

For higher order terms $\tilde{V}^n(x,y;\varepsilon)$, we can proceed in a similar manner. We have, in general,

$$\tilde{V}^n(x,y;\varepsilon) = \sum_{m=1}^{\infty} \exp\left\{-\frac{1}{\varepsilon}\int_0^x \lambda_m(\xi)d\xi\right\} V_m^n(x,y) \; ; \quad V_n^m(x,y) = \nu_m^n(x,y) + a_m^n(x)\phi_m(x,y). \tag{4.13}$$

Here $\nu_m^n(x,y)$ is a particular solution of (4.6) satisfying (2.1). The $a_m^n(x)$'s are determined by the solvability condition $(4.6)_{n+1}$ and the <u>matching condition</u> at $x = 0$:

$$u(0,y;\varepsilon) - \sum_{m=0}^{[N/2]} \varepsilon^{2n} U^n(0,y) = \sum_{n=0}^{N} \varepsilon^n \tilde{V}^n(0,y;\varepsilon) \quad . \tag{4.14}$$

The construction of $\tilde{W}$ is completely analogous to that of $\tilde{V}$. The general term $\tilde{W}^n$ in (4.4) has the representation

$$\tilde{W}^n(x,y;\varepsilon) = \sum_{m=1}^{\infty} \exp\left\{-\frac{1}{\varepsilon}\int_x^a \lambda_m(\xi)d\xi\right\} W_m^n(x,y) ; W_m^n(x,y) = \omega_m^n(x,y) + b_m^n(x)\phi_m(x,y) \quad , \tag{4.15}$$

where $\omega_m^n(x,y)$ and $b_m^n(x)$ are defined in an obvious way. The details can be found in [15].

5. Characteristic Boundary-Justification

The justification of the above formal expansion is long and technical, although, in principle, the idea involved is rather simple. From the construction of $\tilde{V}^n$ and $\tilde{W}^n$, clearly one has to verify that under some appropriate restrictions on the boundary data in (4.1), the expressions (4.13) and (4.15) are indeed in some sense the Fourier-Bessel expansions associated with the solution of the problem (P'_ε); hence the uniformity of the convergence of the series in (4.13) and (4.15) and the question of the permissibility of term-by-term differentiations of these series are main concerns for establishing the validity of the formal expansions.

Next, we notice that the remainder $Z_N(x,y;\varepsilon)$ defined by

$$(5.1)_N \qquad Z_N(x,y;\varepsilon) = u(x,y;\varepsilon) - \sum_{n=0}^{[N/2]} \varepsilon^{2n} U^n(x,y) - \sum_{n=0}^{N} \varepsilon^n \tilde{V}^n(x,y;\varepsilon) - \sum_{n=0}^{N} \varepsilon^n \tilde{W}^n(x,y;\varepsilon)$$

for each non-negative integer N, formally satisfies

$$(5.2)_N \qquad \begin{aligned} \mathcal{E}_k[Z_N] &= -\varepsilon^{N+1} h_N(x,y;\varepsilon) && \text{in } \Omega \\ Z_N(x,\gamma(x);\varepsilon) &= 0 , && 0 \le x \le a \\ Z_N(0,y;\varepsilon) &= -\sum_{n=0}^{N} \varepsilon^n \tilde{W}^n(0,y;\varepsilon) , && 0 < y \le \gamma(0) \\ Z_N(a,y;\varepsilon) &= -\sum_{n=0}^{N} \varepsilon^n \tilde{V}^n(a,y;\varepsilon) , && 0 < y \le \gamma(a) \end{aligned}$$

with the growth condition (2.1). Here the non-homogeneous term $h_N(x,y;\varepsilon)$ can be obtained explicitly. Now, after establishing the validity of $(5.2)_N$, it remains to show that $||h_N||_\infty$ is bounded in Ω. Then as a consequence of the Corollary of Section 2, Z_N will be of $O(\varepsilon^{N+1})$ uniformly in $\Omega \cup \Gamma$.

The general case is too lengthy to present here and so we shall only discuss the case $N = 0$, and focus our attention on

$$h_o(x,y;\varepsilon) = \varepsilon U^o_{xx} + \sum_{m=1}^{\infty} \exp\left\{-\frac{1}{\varepsilon}\int_0^x \lambda_m(\xi)d\xi\right\}(-2\lambda_m(x)a_m^o(x)\phi_{mx} + \varepsilon V^o_{mxx})$$

$$+ \sum_{m=1}^{\infty} \exp\left\{-\frac{1}{\varepsilon}\int_x^a \lambda_m(\xi)d\xi\right\}(2\lambda_m(x)b_m^o(x)\phi_{mx} + \varepsilon W^o_{mxx}) .$$

We present the general case in [15].

Our first step is to show that Z_o defined by (5.1) indeed satisfies $(5.2)_o$. Alternatively, we require that $\tilde{V}^o$ and $\tilde{W}^o$ should satisfy the conditions: (i) $\tilde{V}^o, \tilde{W}^o \in C(\Omega \cup \Gamma)$, (ii) $\tilde{V}^o, \tilde{W}^o \in C^2(\Omega)$ and twice termwise differentiations of the series (4.13) (and (4.15)) are valid, and (iii) $\tilde{V}^o$ and $\tilde{W}^o$ satisfy (2.1). From the properties of Bessel functions (see e.g. [20, Chap. XVIII]), in order to guarantee these conditions, it suffices to assume that

(5.4) $\psi_i \in C'(0,\gamma(x_i))$, $\int_0^{\gamma(x_i)} y^{\frac{1}{2}+p} |\psi_i(y)| dy$ exists and $\psi_i(\gamma(x_i)) - \phi(x_i) = 0$

with $x_1 = 0$ and $x_2 = a$. We note that the presence of the exponential term in $\tilde{V}^o$ and $\tilde{W}^o$ simplifies the justification of the differentiation of $\tilde{V}^o$ and $\tilde{W}^o$ under the summation sign, since we are examining differentiability in Ω and hence the uniform convergence of the corresponding series (obtained by termwise differentiating $\tilde{V}^o$ and $\tilde{W}^o$) in a neighborhood of each point is sufficient.

To establish the boundedness of h_o in Ω, we assume that ψ_i is smooth enough so that the Fourier-Bessel coefficients of $y^p f(y) = y^p(\psi_i(y) - \phi(x_i))$ drop off as $\lambda_m^{-\beta}$ for β large enough. This is possible, since under suitable restrictions on f, if we denote by f_m the Fourier-Bessel coefficients of $y^p f$, then we have the relation $f_m = \frac{-1^{\ell}}{\lambda_m^{2\ell}} (B_k[f])_m$.

We now summarize these results in the following theorem.

Theorem 2. Let ℓ be a positive integer, $\ell > 2+(p/2)$ and let $i = 1,2$. Under the assumptions: (a) $\phi \in C^2[0,a]$; (b) $\psi_i(\gamma(x_i)) = \phi(x_i)$; (c) $\psi_i^{(j)}(\gamma(\alpha_i)) = 0$ $j = 1,2,\ldots,2\ell-2$; (d) $\psi_i \in C^{2\ell}(0,\gamma(x_i)]$; (e) $y^{2p+1}\psi_i^{(2\ell-1)}(y) \to 0$ as $y \to 0^+$; and (f) $\int_0^{\gamma(x_i)} y^{p+V_2} |B_k[\psi_i](y)| dy < \infty$ the solution $u(x,y;\varepsilon)$ of (P'_ε) admits the asymptotic representation:

$$u(x,y;\varepsilon) = U(x,y;0) + \tilde{\mathcal{V}}(x,y;\varepsilon) + \tilde{\mathcal{W}}(x,y;\varepsilon) + O(\varepsilon) \quad \text{as} \quad \varepsilon \to 0^+$$

uniformly in $\Omega \cup \Gamma$.

In concluding this section we remark that in the special case where $\gamma(x) =$ constant, one can easily verify that both λ_m and ϕ_m are independent of x. Then ν_m^n and ω_m^n in (4.13) and (4.15) are identically equal to zero and hence one may expect $Z_N = O(\varepsilon^{N+2})$ instead of $O(\varepsilon^{N+1})$.

REFERENCES

[1] Weinstein, A., Discontinuous integrals and generalized potential theory, Trans. Amer. Math. Soc., 63 (1948), 342-354.

[2] _________, Generalized axially symmetric potential theory, Bull. Amer. Math. Soc., 59 (1953), 20-37.

[3] _________, Singular partial differential equations and their applications, in Fluid Dynamics and Applied Mathematics (Proc. Sympos., U. of Maryland, 1961), pp. 29-49, Gordon and Breach, New York, 1962.

[4] Huber, A., On the uniqueness of generalized axially symmetric potentials, Ann. of Math., (2) 60 (1954), 351-358.

[5] _________, Some results on generalized axially symmetric potentials, Proc. Conf. on Diff. Equn. (dedicated to A. Weinstein), U. of Maryland (1955), 147-155.

[6] Quinn, D. W., and Weinacht, R. J., Boundary value problems in generalized bi-axially symmetric potential theorem, Journal of Diff. Equs. (to appear).

[7] Moss, W. F., Boundary value problems and fundamental solutions for degenerate or singular second order linear elliptic partial differential equations, Ph.D. thesis, U. of Delaware (1974).

[8] Eckhaus, W., and de Jager, E. M., Asymptotic solutions of singular perturbation problems for linear differential equations of elliptic type, Arch. Rational Mech. and Anal., 23 (1966), 26-86.

[9] Knowles, J. K., The Dirichlet problem for a thin rectangle, Pro. Edinburgh Math. Soc., 15 (1967), 315-320.

[10] Lions, J. L., Perturbation Singulières dans les Problèmes aux Limites et en Contrôle Optimal, Lecture Notes in Math., No. 323, Springer-Verlag (1973), pp. 227-238.

[11] Jiji, L. M., Singular-perturbation solutions of conduction in irregular domains, Quart. J. Mech. Appl. Math., 27 (1974), 45-55.

[12] Ho, T. C., and Hsiao, G. C., A singular perturbation arising in catalytic reactions in finite cylindrical supports and some related problems, A. I. Ch. E. J. (to appear).

[13] Levinson, N., The first boundary value problem for $\varepsilon\Delta u+Au_x+Bu_y+Cu = D$ for small ε, Ann. of Math., 51 (1950), 227-238.

[14] Keller, J. B., Perturbation Theory, Lectures Notes, Michigan State Univ., East Lansing, Michigan (1968), pp. 55-59.

[15] Hsiao, G. C., and Weinacht, R. J., On a class of singular partial differential equations with a small parameter (in preparation).

[16] Hopf, E., A remark on linear elliptic differential equations of second order, Proc. Am. Math. Soc., 3 (1952), 791-793.

[17] Muckenhaupt, B., and Stein, E. M., Classical expansions and their relation to conjugate harmonic functions, Trans. Amer. Math. Soc., 118 (1965), 17-92.

[18] Parter, S. V., On the existence and uniqueness of symmetric axially symmetric potentials, Arch. Rational Mech. Anal., 20 (1965), 279-286.

[19] Courant, R., and Hilbert, D., Methods of Mathematical Physics, Vol. II, Interscience (1962), pp. 326-331.

[20] Watson, G. N., A Treatise On the Theory of Bessel Functions, Second Edition, Cambridge Univ. Press, 1944.

On the limit-n classification of ordinary differential operators with positive coefficients

Robert M Kauffman

0. Introduction. For some years the following question has been asked: "Let $L = \sum_0^N (-1)^i D^i p_i D^i$, with $D = d/dx$, each $p_i \geq 0$ and $p_0 \geq \varepsilon > 0$. Suppose that the p_i are, say, C^∞ and p_N is non-vanishing on $[a,\infty)$. Then is it true that there are always exactly N linearly independent square integrable solutions to $Lf = 0$?".

W. N. Everitt (7) posed this question in 1968, and since then some partial results have been obtained, especially in the fourth and sixth order cases. Devinatz (2) and (3), Eastham (5), Everitt (7) and (8), Kauffman (11) and others have investigated these cases, and the results have always previously been that, under the hypotheses of their theorems, the answer is "yes".

We will show here that the conjecture is false. A counter-example of type $L = -D^3 x^\alpha D^3 + \beta x^{\alpha-6}$ will be shown to exist. Nevertheless, a general 2Nth order theorem will be given (its proof will appear elsewhere) which states that, except for certain special situations, the conjecture is true when the p_i are finite sums of real multiples of real powers of x. Surprisingly, the maximal operators are even "separated". It is usually quite easy to check whether the hypotheses of the theorem are satisfied--one needs at worst only to examine the location of the roots of a certain polynomial equation, and usually this is not even necessary. Thus, the answer to the question asked by Everitt, for the case when the p_i are finite sums of real multiples of real powers of x, is "It is usually possible to prove that the answer is N. Nevertheless, certain special counterexamples exist."

It should be noted that, for a time, it was believed that counterexamples had been discovered for the case when $N = 3$ and $p_3 \equiv 1$. However, this is false; the conjecture is true when

$N = 3$ unless degree $p_3 > 6$ (and practically always even when degree $p_3 > 6$).

The author is indebted to Prof. W. N. Everitt and the University of Dundee for making possible his stay there during the academic year 1974-5. Much of the work in this paper was done at Dundee, and conversations with Prof. Everitt were very helpful to the author.

1. An Example. We introduce some notation and definitions.

Definition 1.1. An ordinary differential operator on $[1,\infty)$ will in this paper refer to an expression of type $M = \sum_0^m p_j D^j$, where $D = d/dx$, each p_j is a C^∞ complex valued function, and p_n is non-vanishing. All ordinary operators will be considered on $[1,\infty)$.

Definition 1.2. Let M be as in 1.1. m is called the order of M.

Definition 1.3. Let M be as in 1.1. Then the formal (Lagrange) adjoint of M, denoted by M^+, is the expression given by $M^+(f) = \sum_0^m (-1)^j D^j(\overline{p}_j f)$ for any m times differentiable f. Note that M^+ is also an ordinary differential operator.

Notation 1.4. C_0^∞ denotes the infinitely differentiable functions supported in a compact subset of $(1,\infty)$. Note that each such function vanishes in a neighborhood of 1.

Notation 1.5. L_2 denotes $L_2(1,\infty)$.

Definition 1.6. Let M be an ordinary differential operator. Then $T_0(M)$, the minimal operator associated with M, denotes the closure of the restriction of M to C_0^∞.

Definition 1.7. Let M be an ordinary differential operator of order m. The maximal operator $T_1(M)$ associated with M is the operator given by: domain $T_1(M)$ is the set of all f in L_2 with $f^{(m-1)}$ absolutely continuous and Mf in L_2. $T_1(M)f = Mf$ for all such f.

Remark: It is well known (see Dunford and Schwartz (4), p. 1294) that $(T_0(M))^* = T_1(M^+)$, where $(T_0(M))^*$ denotes the adjoint operator in L_2 of $T_0(M)$.

Definition 1.8. Let M be a differential operator of order m. M is said to be limit point if and only if $T_1(M)$ is an m dimensional extension of $T_0(M)$.

Remark: When $M = M^+$, this is equivalent to the usual definition (See Kauffman (10), p. 351, or Naimark (12).).

Lemma 1.9. Suppose $T_0(M)$ has closed range. Then M is limit point if and only if the nullity of $T_1(M)$ + the nullity of $T_1(M^+)$ equals the order of M.

Proof: This is theorem 7 of Kauffman (10).

Lemma 1.10. If f is in domain $T_0(M)$, then $f^{(i)}(1) = 0$ for all $i \leq m - 1$, where m is the order of M.

Proof: This follows immediately from variation of parameters. (See Goldberg (9), p. 139.)

Lemma 1.11. M is limit point if and only if, for any f in domain $T_1(M)$, there is a g in domain $T_0(M)$ such that $f - g$ is a C^∞ function supported in (say) $[1,2]$.

Proof: It is easy to see that there are at least m such functions, linearly independent modulo domain $T_0(M)$, since all functions in domain $T_0(M)$ have $f^{(i)}(1) = 0$ for all $i \leq m - 1$. Each C^∞ function supported in $[1,2]$ is in the domain of $T_1(M)$ so the conclusion follows.

Lemma 1.12. Let $M = \sum_0^{m/2} (-1)^j D^j p_j D^j$, with each $p_j \geq 0$. Suppose M is limit point. Then, for any f in domain $T_1(M)$, $(p_0)^{1/2} f$ is in L_2.

Proof: Integration by parts shows that, for any f in domain $T_0(M)$, $(p_0)^{1/2} f$ is in L_2. Lemma 1.12 then sufficies to complete the proof.

<u>Theorem 1.13</u>. There exist real numbers $\alpha > 6$ and $\beta > 0$ such that $(-D^3x^\alpha D^3 + \beta x^{\alpha-6})f = 0$ has at least 4 linearly independent L_2 solutions f; i.e. $-D^3x^\alpha D^3 + \beta x^{\alpha-6}$ is not limit point.

<u>Proof</u>: We show the existence of α and β by indicating how they may be constructed.

Let $M = -D^3x^\alpha D^3 + \beta x^{\alpha-6}$. Then $Mx^\lambda = 0$ if and only if $\lambda(\lambda-1)(\lambda-2)(\lambda+\alpha-3)(\lambda+\alpha-4)(\lambda+\alpha-5) = \beta$, where λ is a complex number.

Let $\lambda = a + ib$, where $a = -1$. We construct β,b and α such that $Mx^\lambda = 0$, but $(\alpha-6)/2 \geq 1$. Lemma 1.12 shows that, for such β and α, M is not limit point.

Let Θ_1 be the argument of the complex number λ, Θ_2 be the argument of $\lambda - 1$, Θ_3 of $\lambda - 2$, and let $\Theta_4(\alpha)$ be the argument of $\lambda + \alpha - 3$, $\Theta_5(\alpha)$ of $\lambda + \alpha - 4$, and $\Theta_6(\alpha)$ of $\lambda + \alpha - 5$. If $\sum_1^6 \Theta_i = 2\pi$, then $\lambda(\lambda-1)(\lambda-2)(\lambda+\alpha-3)(\lambda+\alpha-4)(\lambda+\alpha-5)$ is positive real, as we desire. We select b and α so that $\sum_1^6 \Theta_i = 2\pi$.

Note that, as b becomes large positively, Θ_1,Θ_2 and Θ_3 each approach $\pi/2$. When $\alpha = 8$, and b becomes large, Θ_4,Θ_5 and Θ_6 all approach $\pi/2$ as well, so that $\sum_1^6 \Theta_i$ approaches 3π. With any fixed a and b_1 however, $\Theta_4(\alpha)$, $\Theta_5(\alpha)$ and $\Theta_6(\alpha)$ approach 0 as α becomes large. We therefore see that, for any $\varepsilon > 0$ there are b and α such that $|\sum_1^3 \Theta_i - (3/2)\pi| < \varepsilon$, and $|\sum_4^6 \Theta_i(8) - (3/2)\pi| < \varepsilon$, and also $|\sum_4^6 \Theta_i(\alpha)| < \varepsilon$. It follows from continuity that there are real numbers $b > 0$ and $\alpha \geq 8$ such that $\sum_1^3 \Theta_i + \sum_4^6 \Theta_i(\alpha) = 2\pi$.

Setting $\lambda = -1 + ib$, we see that $D^3x^2D^3(x^\lambda) = \beta x^{\alpha-6}(x^\lambda)$, where $\beta = \lambda(\lambda-1)(\lambda-2)(\lambda+\alpha-3)(\lambda+\alpha-4)(\lambda+\alpha-5) > 0$. Since $\alpha \geq 8$, $x^{(\alpha-6)/2}x^\lambda$ is not in L_2. But x^λ is in domain $T_1(M)$. Thus by Lemma 1.12, M is not limit point.

2. The Theorem

<u>Definition 2.1</u>. Let $M = \sum_0^m p_iD^i$. $T_1(M)$ is said to be

<u>separated</u> if, for any f in domain $T_1(M)$, p_iD^if, $p_i^{(1)}D^{i-1}f$, $\ldots p_i^{(i)}f$ are all in L_2 for each $i \leq m$. $T_0(M)$ is said to be separated if the same condition holds for any f in domain $T_0(M)$.

<u>Definition 2.2</u>. An ordinary differential operator M is said to be disconjugate on an interval I if no solution has m or more zeros on I, where zeros are counted according to their multiplicity, and where m = order M.

<u>Theorem 2.3</u>. Let $M = \sum_0^{m/2}(-1)^iD^ip_iD^i$. Assume that, for each i, p_i is eventually non negative and p_i is a finite sum of real multiples of real powers of x. If p_i is not identically zero, let $p_i = c_ix^{n(i)}$ + lower order terms. Let S be the set of real numbers $k_i = n(i) - 2i$, and let c be the largest element of S. Let Q be the set of all i such that $k_i = c$. Let $R = \sum_{i\varepsilon Q}(-1)^ic_iD^ix^{n(i)}D^i$. Then, if $T_1(R)$ is separated, it follows that $T_1(M)$ is separated and M is limit point.

<u>Remark</u>: Note that R, the "essential part" of M, only involves those p_i for which degree $p_i - 2i$ is c, the maximum possible value, and the lower degree terms of p_i also do not show up in R. Very frequently, R only has one term, and, by the following corollary, M is then limit point and $T_1(M)$ is separated. Note also that the "essential part" of M usually has lower order than M.

<u>Corollary 2.4</u>. Let M and R be as in 2.3. Then $T_1(M)$ is separated and M is limit point when any of the following occurs:

a) $c \leq 0$

b) R has no solution x^λ, where $-c - 1/2 \leq \mathrm{Re}\lambda < -1/2$

c) R is disconjugate

d) Q has only one element

e) R is of order four or less

f) $n(i) \leq 2i$ for $i > 2$.

<u>Remark</u>: By f) of the corollary, in order for $-D^3p_3D^3 + D^2p_2D^2 - Dp_1D + p_0$ to fail to be limit point (where the p_i are

as above), p_3 must have degree greater than 6. In the fourth order case, all such operators are limit point.

Remark: The equation $Rx^\lambda = 0$ is equivalent to a polynomial equation in λ, so b) can be tested, at least in theory, by a computer. This is, of course, only necessary when none of the other criteria apply.

Theorem 2.5. Let $M = \sum_0^{m/2} (-1)^i D^i p_i D^i$, where, for each i, p_i is eventually non-negative and p_i is a finite sum of real multiples of real powers of x. Suppose that p_0 is not identically zero, and that degree $p_0 >$ degree $p_i - 2i$ for all i, and also that degree $p_0 \geq 0$. Then M is limit point, $T_1(M)$ is separated, and, for any f in L_2 such that $Mf = 0$, $x^r D^N f$ is in L_2 for all r and N. Further, if $Mf = 0$, and f is not in L_2, then $x^{-r} f$ is not in L_2 for any r.

Remark: The proofs of theorems 2.3 and 2.5 are too difficult to be indicated here. They are based upon the perturbation theory of Fredholm operators, not asymptotic methods. The methods of proof are new, and the author hopes they will have applications to other areas of differential operator theory.

Proof of Corollary 2.4:

a) When $c = 0$, R is an Euler differential operator, and the conclusion follows, since $T_1(R)$ is unitarily equivalent to $T_1(M)$, with M a constant coefficient operator on $[0,\infty)$. When $c < 0$, the proof will appear elsewhere.

b) will appear elsewhere.

c) When R is disconjugate, and $Rx^\lambda = 0$, λ must be real, since complex roots occur in conjugate pairs. If $Rx^\lambda = 0$, with $-c - 1/2 \leq \lambda < -1/2$, then $c_0 + \sum_{i=1}^N (-1)^i c_i(\lambda)(\lambda-1)\dots(\lambda-i+1)(\lambda+c+i)\dots(\lambda+c+1) = 0$. However, this is clearly impossible, because $\lambda, \lambda - 1, \dots \lambda - i + 1$ are all negative and $\lambda + c + i, \dots \lambda + c + 1$ are all positive.

d) follows from c).

e) is proved as follows: When R is of order four or less, we need only worry about the case when order $R = 4$. Let $R = c_2D^2x^{4+c}D^2 - c_1Dx^{2+c}D + c_0x^c$. We need only worry about the case $c > 0$. If $Rx^\lambda = 0$, then $c_2(\lambda)(\lambda-1)(\lambda+c+2)(\lambda+c+1) - c_1\lambda(\lambda+c+1) = -c_0$. If $c_0 = 0$, R is disconjugate, so we may assume $c_0 > 0$. Therefore $\lambda[c_2(\lambda-1)(\lambda+c+2)(\lambda+c+1) - c_1(\lambda+c+1)]$ is negative, so $\arg \lambda + \arg [c_2(\lambda-1)(\lambda+c+2)(\lambda+c+1) - c_1(\lambda+c+1)] = \pi$. If $-1/2 - c \leq \mathrm{Re}\lambda < -1/2$, we may suppose (since roots occur in conjugate pairs) that $\mathrm{Im}(\lambda) > 0$. (By the argument of c) there are no real roots λ). Thus $\pi > \arg \lambda > \pi/2$. But $\arg [c_2(\lambda-1)(\lambda+c+2)(\lambda+c+1)] = \arg(\lambda-1) + \arg(\lambda+c+2) + \arg(\lambda+c+1)$, and $\pi/2 < \arg(\lambda-1) < \pi$, and also $0 < \arg(\lambda+c+1), \arg(\lambda+c+2) < \pi/2$. Thus $\pi/2 < \arg [\lambda-1)(\lambda+c+2)(\lambda+c+1)] < 2\pi$. Also $\pi < \arg(-c_1(\lambda+c+1)) < 3\pi/2$. Therefore $\pi/2 < \Theta = \arg[c_2(\lambda-1)(\lambda+c+2)(\lambda+c+1) - c_1(\lambda+c+1)] < 2\pi$. Therefore $\pi < \arg \lambda + \arg \Theta < 3\pi$, a contradiction. e) is proved.

f) follows immediately from a) and e).

Bibliography

1. W.A. Coppel, Disconjugacy, Lecture Notes in Math., Vol. 220, Springer-Verlag, New York, 1971.

2. A. Devinatz, Positive definite fourth order differential operators, J. London Math. Soc. (2) 6 (1973), 412-416.

3. A. Devinatz, On limit-2 fourth order differential operators, J. London Math. Soc. (2) 7 (1973), 135-146.

4. N. Dunford and J. T. Schwartz, Linear Operators II: Spectral Theory. Selfadjoint operators in Hilbert space, Wiley-Interscience, New York, 1963.

5. M.S.P. Eastham, The limit-2 case of fourth order differential equations, Quart. J. Math. (Oxford), 22, 131-134.

6. W.N. Everitt, Integrable-square solutions of ordinary differential equations (III), Quart. J. Math. (Oxford)(2) 14 (1963) 170-180.

7. W.N. Everitt, Some positive definite differential operators, J. London Math. Soc. (1) 43 (1968), 465-473.

8. W.N. Everitt, On the limit-point classification of fourth-order differential equations, J. London Math. Soc. (1) 44 (1969) 273-281.

9. S. Goldberg, Unbounded linear operators: Theory and applications, McGraw-Hill, New York, 1966.

10. R.M. Kauffman, Polynomials and the limit point condition, Trans. Am. Math. Soc. 201, (1975), 347-366.

11. R.M. Kauffman, Disconjugacy and the limit point condition for fourth order operators (To appear)

12. M.A. Naimark, Linear Differential Operators, GITTL, Moscow, 1954, English transl., Part II, Ungar, New York, 1968.

13. A. Zettl, General theory of the factorization of ordinary linear differential operators, Trans. Amer. Math. Soc. 197 (1974) 341-353.

On the non-convergence of successive approximations in the theory of ordinary differential equations

Michał Kisielewicz

It is well known that continuity of f and the uniqueness of solution of the Cauchy problem $x' = f(t,x)$, $x(0) = x_0$ are not sufficient to guarantee the convergence of the successive approximations

$$y_m^f(t) = x_0 + \int_0^t f(s, y_{m-1}^f(s))\, ds .$$

For an example see Müller [2] and Coddington-Levinson [1],p.53. We show here that non-convergence of successive approximations is in any sense a rare case . Exactly , we shall prove that the set $\mathcal{A}$ of all f for which the sequence $\{y_m^f\}$ of successive approximations is not converging is of Baire's first category in any complete metric space .

1. **Notations** . We shall denote by

$B(x_0, \varepsilon_0)$ the ball of center x_0 and radius $\varepsilon_0 > 0$,

$D = [t_0 , T] \times B(x_0, \varepsilon_0)$,

$\mathfrak{F}(\bar{D})$ the set of all continuous functions $f : \bar{D} \to R^n$ such that $(T-t_0)\|f\|_{\mathfrak{F}} \le \varepsilon_0$, where $\|f\|_{\mathfrak{F}} = \sup\{\|f(t,x)\| : (t,x) \in \bar{D}\}$.

It is known $(\mathfrak{F}(\bar{D}), \rho)$, where $\rho(f_1, f_2) = \|f_1 - f_2\|_{\mathfrak{F}}$ is a complete metric space . The assumption $(T-t_0)\|f\|_{\mathfrak{F}} \le \varepsilon_0$ is necessary only in order that the successive approximations

$$y_n^f(t) = x_0 + \int_{t_0}^{t} f(s, y_{n-1}^f(s))\, ds \quad ; \quad y_0^f(t) = x_0$$

corresponding to $f \in \mathcal{F}(\bar{D})$ are well defined . Let us observe that for every $\delta > 0$ there is a vector-polynomal function $p^\delta : D \to R^n$ so that $\| f - p^\delta \|_{\mathcal{F}} \leq \delta$. It is clear that the sequence $\{ y_n^p \}$ corresponding to $p^\delta \in \mathcal{F}(\bar{D})$ is convergent . Let $\Delta(f,t) = \lim_{n\to\infty} \sup \{ \text{diam } E[y_n^f(t)] \}$, where $f \in \mathcal{F}(\bar{D})$, $t \in [t_0, T]$, $E[y_n^f(t)] = \{ y_n^f(t), y_{n+1}^f(t), \dots \}$ and diamA denotes the diameter of a set $A \subset R^n$. It is readily seen that for given $f \in \mathcal{F}(\bar{D})$ the sequence $\{ y_n^f \}$ is not converging in $[t_0, T]$ iff there is a $\tilde{t} \in [t_0, T]$ such that $\Delta(f, \tilde{t}) > 0$. Let $\{ t_\tau \}$ be a sequence dense in $[t_0, T]$ and such that $t_\tau \in [t_0, T]$ for every $\tau = 1, 2, \dots$. Then let $\Omega_{p\tau} = \{ f \in \mathcal{F}(\bar{D}) : \Delta(f, t_\tau) \geq 1/p \}$.

2. Basic lemmas .

Lemma 1. $\Omega_{p\tau}$ are closed subsets of $\mathcal{F}(\bar{D})$ for every $p, \tau = 1, 2, \dots$.

Proof. Suppose $\{ f_k \}$ is a sequence of $\Omega_{p\tau}$ such that $0 = \lim_{k\to\infty} \| f_k - f \|_{\mathcal{F}}$, where $f \in \mathcal{F}(D)$. It is not difficult to see that for every positive integer r and q there are numbers m_r , u_q and v_q so that

$$\| y^{f_k}_{n+m_r+u_q}(t_\tau) - y^{f_k}_{n+m_r+v_q}(t_\tau) \| > 1/p - 1/r - 1/q$$

for $n, k = 1, 2, \dots$. Furthermore it is readily to verify that the sequence $\{ y^{f_k}_{n+m_r+u_q} \}$ satisfies the hypotheses of Arzela's theorem . Then there is a subsequence of $\{ y^{f_k}_{n+m_r+u_q} \}$ which is uniformly converging in $[t_0, T]$ for every fixed n , m_r and u_q to any function , say $x_{n+m_r+u_q}$. It is not difficult to verify that $x_{n+m_r+u_q}$ satisfies

$$x_{n+m_r+u_q}(t) = x_0 + \int_{t_0}^{t} f(s, x_{n+m_r+u_q-1}(s))\, ds$$

and

$$\| x_{n+m_r+u_q}(t_\tau) - x_{n+m_r+v_q}(t_\tau) \| \geq 1/p - 1/r - 1/q$$

for every $n,r,q = 1,2,\dots$. Hence it follows that $x_{n+m_r+u_q} = y^f_{n+m_r+u_q}$ and $\Delta(f, t_\tau) \geq 1/p$. Therefore $f \in \Omega_{p\tau}$. This completes the proof .

Lemma 2. $\Omega_{p\tau}$ are non-dense in $\mathfrak{F}(D)$ for every $p,\tau = 1,2,\dots$.

Proof. Suppose $\Omega_{p\tau}$ is dense in a sphere $S_h(f_o) \subset \mathfrak{F}(D)$ with a center $f_o \in \mathfrak{F}(D)$ and a radius $h > 0$. In virtue of Lemma 1 we have $S_h(f_o) \subset \Omega_{p\tau}$. Acording to preceding remarks for every $\delta \in (0, h)$ and $f_o \in \mathfrak{F}(D)$ there is a vector-polynomal function p^δ such that $\| f_o - p^\delta \| \leq \delta$. Since $\Delta(p^\delta, t) = 0$ for every $t \in [t_o, T]$ then $p^\delta \notin \Omega_{p\tau}$. This completes the proof .

3. Main result . Let us denote by $\mathcal{A}$ the set of the form

$$\mathcal{A} = \{ f \in \mathfrak{F}(D) : \text{there is } t_f \in [t_o, T] ; \Delta(f, t_f) > 0 \} .$$

Theorem 3. $\mathcal{A}$ is of Baire's first category in the space $(\mathfrak{F}(D), \varrho)$.

Proof . In virtue of Lemma 2 it sufficies to show that

$$\mathcal{A} = \bigcup_{p=1}^{\infty} \bigcup_{\tau=1} \Omega_{p\tau} .$$

Since $\Omega_{p\tau} \subset \mathcal{A}$ for every $p,\tau = 1,2,\dots$ then we can only verify that $\mathcal{A} \subset \bigcup_{p=1}^{\infty} \bigcup_{\tau=1}^{\infty} \Omega_{p\tau}$. But this is not difficult to do . The proof is completed .

Corollary. The set $\mathcal{B}$ of all functions $f \in \mathfrak{F}(D)$ for which the sequence $\{ y^f_n \}$ is convergent is dense in $\mathfrak{F}(D)$ and of Baire's second category .

This result generalised the resut of the paper [3] .

References .

[1] E.A.Coddington,N.Levinson , Theory of ordinary differential equations , McGraw-Hill , New York , (1955) .

[2] M.Müller , Uber das Fundamentaltheorem in der Theorie der gewöhnlichen Differentialgleichungen , Math.Zeitschr. 26(1927) pp. 619-645 .

[3] G.Vidossich , Most of the successive approximations do converge , Jornal of Math.Anal.Appl. , 45(1)(1974),pp.127-131 .

COMMENTS ON NONLINEAR ELASTICITY AND STABILITY

R. J. KNOPS

INTRODUCTION

The theory of nonlinear elasticity was formulated in large part by mathematicians such as Euler, Cauchy, Navier, G. Green, Kelvin and Duhem. However, towards the end of the nineteenth century and for much of the first part of the twentieth century, the theory was neglected in favour of an intensive study of the linearised theory. This bias was probably due to the unavailability of appropriate theorems in nonlinear analysis and to contemporary technology being content with predictions and results based on the linear theory. Nevertheless, it had long been known that certain experimental effects could not be accounted for by the linear theory alone and the increasing demand for more precise results stimulated a revival of interest in the nonlinear theory. This return was led by people such as Rivlin whose contributions have done much to enrich and further extend the nonlinear theory both in its basic formulation and in its applications.

Stability has always been prominent in the study of elasticity and of the early contributors mention may be made of G. Green, Kirchhoff, Hadamard and Duhem, while more recent developments are due to Koiter [27,28], Hill [18,19] and Ericksen [9,10]. A survey of the subject is provided by Knops and Wilkes [26]. Of major interest has been justification in the nonlinear theory of the classical energy test, which is a generalisation of the Lagrange-Dirchlet criterion for discrete systems. This states that an equilibrium position is stable if the potential energy achieves its minimum at this position. The validity of this criterion for elasticity is still uncertain and a complete description of circumstances when it holds is apparently not yet available. Moreover, stability is intimately related to questions of existence and regularity of solutions, since without this knowledge, many of the computations must be regarded as purely formal.

We devote this article to a discussion of these and related topics. Since the subject is still under active development the description is necessarily incomplete and indeed more problems will be raised than actually solved. Nevertheless, the account is hoped to be of some interest.

We commence with an introductory description of basic nonlinear elasticity for both the equilibrium and dynamic problems. For simplicity, we treat only the mechanical theory, so that heat effects are omitted. Furthermore, we neglect body-force and include only Dirchlet boundary conditions and Cauchy initial data. We also confine ourselves to homogeneous hyperelastic bodies. In this way, it is hoped that sufficient essential features are retained to clearly convey an adequate idea of fundamental problems in the particular areas selected for study. After setting down the governing equations we briefly discuss existence of weak solutions to boundary and initial value problems. The final sections are devoted to an investigation of stability which is our primary concern.

As regards notation, we do not distinguish between scalars, vectors, matrices or tensors but instead rely upon the context to make clear the meaning. Likewise, the particular operation of multiplication is implied by the context in which it is being used. Other notation is explained as introduced.

Excellent surveys of nonlinear elasticity theory are provided in the review articles by Rivlin [36,37], and in the book by Wang and Truesdell [48]. The linear theory, whose fundamental development is almost complete is comprehensively described in the articles by Gurtin [13] and Fichera

[11]. Accounts of general continuum mechanics are available in several texts, but we again mention the articles by Rivlin [36,37] and also the work by Truesdell [44,45] and Truesdell and Noll [43].

2. BASIC THEORY

We consider a body composed of elastic material which is some configuration occupies a bounded region B of three-dimensional euclidean space. The region B, which may be multiply-connected, has a smooth surface ∂B and will be taken as the reference configuration. We suppose the body to be in equilibrium in a second configuration occupying the bounded region $B_0 \subset \mathbb{R}^3$. To describe motions of the body, we suppose there is a sequence of time-dependent configurations, one member of which occupies at fixed time t, say, the bounded region $B_1 \subset \mathbb{R}^3$.

The deformation of the body from B to B_0 and from B to B_1 may be described by specifying maps from B to B_0 and from B to B_1 respectively. Thus, let us take a fixed set of rectangular coordinates and identify each particle of the body by its position x in B with respect to these axes. Let z and y be the positions of the same particle in B_0 and B_1 respectively.

Then the deformation of the body in B_0 is completely determined by the map

$$\left.\begin{array}{l} z\colon B \to B_0 \\ x \to z(x), \end{array}\right\} \qquad (2.1)$$

and similarly the deformation of the body in the motion is completely determined by prescribing at fixed t the map

$$\left.\begin{array}{l} y\colon B \to B_1 \\ x \to y(x,t). \end{array}\right\} \qquad (2.2)$$

Appropriate smoothness properties of the maps z and y must be established from governing equations to be derived subsequently, so that at this stage computations are purely formal.

We define the deformation gradients Z and Y by the respective matrices

$$Z(x) = \nabla z(x) = \left(\frac{\partial z_i}{\partial x_\alpha}\right), \tag{2.3}$$

$$Y(x,t) = \nabla y(x,t) = \left(\frac{\partial y_i}{\partial x_\alpha}\right), \text{ at fixed } t. \tag{2.4}$$

In order to avoid local interpenetration of material, we ensure that the maps (2.1) and (2.2) are locally invertible and orientation preserving by requiring their Jacobians to satisfy

$$\det Z(x) > 0, \quad \det Y(x) > 0. \tag{2.5}$$

It is also convenient at this stage to introduce the displacement $u(x,t)$ defined by

$$u(x,t) = y(x,t) - z(x), \tag{2.6}$$

and the displacement gradient $U(x,t)$ defined by

$$U(x,t) = \nabla u(x,t) = \left(\frac{\partial u_i}{\partial x_\alpha}\right). \tag{2.7}$$

We denote the material time derivative by a superposed dot and define the velocity by

$$\dot{y}(x,t) = \frac{\partial y}{\partial t}(x,t) \text{ for fixed } x. \tag{2.8}$$

We shall consider only elastic materials which are homogeneous and hyper-elastic. These are characterised by the existence of a strain-energy function per unit volume given by $W: M^{3\times 3} \to \mathbb{R}$. The total strain-energy is then defined as

$$V(f) = \int_B W(\nabla f(x))dx, \tag{2.9}$$

Equations governing the deformations (2.1) and (2.2) may be obtained in either of two ways. The first stipulates balance laws for linear and angular momentum, while the second relies upon the first and second laws of thermodynamics. Both employ invariance postulates which in one form state that the theory must be invariant to superposed rigid body translations and rotations.

Important consequences of invariance are that the strain-energy W cannot depend upon the deformation z or y (although for non-homogeneous materials it will depend on position x) and that dependence on the deformation gradients Z and Y must be of the form

$$W = \hat{W}(ZZ^T),\ W = \hat{W}(YY^T),$$

where Z^T denotes the transpose of (the matrix) Z and $\hat{W}$: $M^{3\times 3} \to \mathbb{R}$ is different to the function W.

Furthermore, in line with most continuum mechanics theories, both approaches express the basic laws in integral, or global, form which therefore become statements about the behaviour of finite volumes of the materials. As such, there is an obvious relation with conservation laws studied for instance, by Lax [31], Glimm [12] and Dafermos [,7] and therefore the notion of a generalised solution arises quite naturally. Later, we shall introduce the notion of a weak solution. The precise connection between these two types of solution has still to be determined.

Provided respective quantities are sufficiently smooth, the basic integral laws yield pointwise equations which we now proceed to state along with appropriate initial and boundary conditions. For simplicity, we treat only Cauchy and Dirichlet data, although other forms are equally possible. Body-forces are similarly excluded.

Let us set

$$\frac{\partial W}{\partial F} = \frac{\partial W(\nabla f(x))}{\partial f_{i,\alpha}}\ ,\ (f_{i,\alpha}) = \frac{\partial f_i(x)}{\partial x_\alpha} \tag{2.10}$$

and

$$\left(\mathrm{Div}\ \frac{\partial W}{\partial F}\right)_i = \sum_{\alpha=1}^{3} \frac{\partial}{\partial x_\alpha}\left(\frac{\partial W(\nabla f(x))}{\partial f_{i,\alpha}}\right). \tag{2.11}$$

Then it may be shown that the equilibrium deformation z(x) satisfies the <u>displacement boundary value problem</u>.

$$\left.\begin{aligned} \mathrm{Div}\ \frac{\partial W}{\partial Z} &= 0 \quad , x \in B \\ z(x) &= g(x)\ , x \in \partial B \end{aligned}\right\} \tag{2.12}$$

where g is specified Dirichlet data.

Again, the motion $y(x,t)$ of the body may be shown to satisfy the <u>initial displacement boundary value problem</u>:

$$\left.\begin{aligned} \text{Div}\,\frac{\partial W}{\partial Y} &= \rho\,\ddot{y} \quad , \quad (x,t)\ \varepsilon\ B \times (O,T) \\ y(x,t) &= g(x) \quad , \quad (x,t)\ \varepsilon\ \partial B \times (O,T) \\ y(x,o) &= y_0(x)\, ,\ \dot{y}(x,o) = \bar{y}_0(x)\, ,\ x\ \varepsilon\ B. \end{aligned}\right\} \tag{2.13}$$

where $\rho = \rho(x) > 0$ is the material density of the body in configuration B, and y_0 and $\bar{y}_0$ is the specified Cauchy data. The maximal interval of existence of the solution to (2.13) is given by (O,T). Note that for the dynamic problem, the Dirichlet data $(2.13)_2$ could be time-dependent, although for later discussion this is not convenient: we prefer to maintain the surface ∂B fixed throughout the deformation of the body from B_0 to each B_1.

When W is twice differentiable, equation (2.13) reduces to the quasi-linear system

$$A_{\alpha\beta}\,\frac{\partial^2 y}{\partial x_\alpha \partial x_\beta} = \rho\,\ddot{y} \quad , \quad (x,t)\ \varepsilon\ B \times (O,T) \tag{2.14}$$

where $A_{\alpha\beta}$ are matrices defined by

$$\left\{\begin{aligned} A_{\alpha\beta}(y) &= \frac{\partial^2 W(\nabla y)}{\partial Y^2}\, , \\ A_{(\alpha\beta)ij} &\equiv A_{i\alpha j\beta}(\nabla y) = \frac{\partial^2 W \quad (\nabla y(x))}{\partial y_{i,\alpha}\partial y_{i,\beta}}\, . \end{aligned}\right. \tag{2.15}$$

A similar reduction holds for $(2.12)_1$. The hypothesis of hyperelasticity implies, formally at least, that $A_{i\alpha j\beta} = A_{j\beta i\alpha}$, or that the system (2.14) is formally self-adjoint. On the other hand, Cauchy elasticity, which does not postulate a strain-energy, yields equations of the same form as (2.14) but without $A_{i\alpha j\beta}$ having major symmetry.

In the usual way it may be easily shown from (2.13) that energy is conserved:

$$\mathcal{E}(t) \equiv \tfrac{1}{2}\int_B \rho\dot{y}\dot{y}\,dx + V(y) = \mathcal{E}(O). \tag{2.16}$$

On introducing the displacement (2.6) we may equivalently write (2.16) as

$$E(t) \equiv \tfrac{1}{2}\int_B \rho\dot{u}\dot{u}\,dx + I(u) = E(O) \tag{2.17}$$

where

$$I(u) = V(y) - V(z) \tag{2.18}$$

Finally, we alternatively express the initial value problem (2.13) in terms of the displacement:

$$\left.\begin{aligned} \operatorname{Div}\left(\frac{\partial W}{\partial Y} - \frac{\partial W}{\partial Z}\right) &= \rho\, \ddot{u} \quad , \ (x,t) \in B \times (0,T), \\ u(x,t) &= 0 \quad\ , \ (x,t) \in \partial B \times (0,T), \\ u(x,o) = u_0(x), \quad \dot{u}(x,o) &= u_0(x) \quad , \ x \in B \end{aligned}\right\} \tag{2.19}$$

3. THE EQUILIBRIUM PROBLEM

Because our main concern is not directly with qualitative properties of equilibrium solutions, we shall content ourselves with a brief description of some recent results in this area, and refer the interested reader to Wang and Truesdell [49] and Ball [3] for more detailed surveys and results.

Let us observe first that global uniqueness of the solution to boundary value problems in nonlinear elasticity is undesirable. Consider, for instance, the problem of buckling of a slender body. Until the critical load is reached, there is uniqueness, but at the critical load non-uniqueness must occur to allow bifurcation of a second equilibrium solution. Experience tells us that this lies outside a neighbourhood of the first. Other mathematical examples of nonuniqueness have been given by F. John [23]. These deal with the displacement value problem and apply to doubly-connected regions.

On the other hand, local existence and uniqueness are to be expected, and indeed lend justification to linearised theories. By means of the inverse function theorem Signorini [39], followed by Stopelli [40,41,42] and van Buren [46] proved that in the class of small displacement gradients satisfying specified Hölder continuity conditions, the traction, mixed and displacement boundary value problems process unique solutions in the neighbourhood of a known unique solution. Of course, in order that the Fréchet derivative of the corresponding nonlinear map is homeomorphic, it is necessary to impose conditions yielding a unique solution to the associated linearised problem. In the traction boundary value problem, local existence and uniqueness hold only under the additional requirement that the external loads do not have an axis of equilibrium. Such force-systems may have their equilibrium destroyed by small motions about the axis. An analysis of this situation, involving a sequence of dynamic solutions, has been given by Capriz and Guidugli [4].

Several proofs of existence in the nonlinear boundary value problem have been given based upon the monotonicity of the stress, or equivalently, the convexity of W, with respect to ∇z. However, these postulates are unsatisfactory since they not only lead to a contradiction of the non-uniqueness noted earlier, but also are not in accord with the invariance of W. (See Coleman and Noll [5] and Truesdell and Noll [43 p.163]). Indeed a fundamental problem in nonlinear elasticity is to discover constitutive restrictions on the strain-energy and stress that are acceptable mathematically and physically in the sense that solutions to boundary (and initial) value problems may be rigorously proved to possess qualitative behaviour compatible with experience. Obviously, convexity - strict or not - of W is unsuitable.

A standard technique for proving existence employs the direct method of the calculus of variations. It may be formally shown, by routine arguments, that the solution to the displacement boundary value problem (2.12) is a stationary value of the total energy V in the class of functions satisfying the boundary condition $(2.12)_2$. That is, we have

$$\left.\begin{aligned} \delta V(z)\ (v) &= 0 \\ v\big|_{\partial B} &= 0 \end{aligned}\right\} \tag{3.1}$$

where

$$\delta V(z)(v) = \int_B \frac{\partial W}{\partial z} \nabla v \, dx = -\int_B v \,\mathrm{Div}\, \frac{\partial W}{\partial z}\, dx. \tag{3.2}$$

If we suppose the solution is actually a global minimum, so that in particular

$$\delta^2 V(z)(v) = \int \nabla v \frac{\partial^2 W}{\partial z^2} \nabla v \, dx \geqslant 0,$$

the existence of an equilibrium solution becomes equivalent to establishing the existence of a minimiser to V. Morrey [33,34] has treated this problem, but unfortunately his proof requires growth assumptions on W that are not suitable for elasticity. Recently, however, Ball [4] has proved existence provided that W satisfies

$$W = g\,(\nabla z,\ (\nabla z)^*,\ \det \nabla z)$$

where g is convex in its arguments considered as elements of $\mathbb{R}^{19}$ and $(\nabla z)^*$ is the adjugate of ∇z. Ball's proof is valid for the displacement, mixed and traction boundary value problems, but so far only establishes existence in Sobolev spaces. Regularity of minimisers is still under investigation. Even so, his proof does not imply global uniqueness.

4. STABILITY

The notion of stability employed in this article is that introduced by Liapunov which is equivalent to the continuity of the map from the set of initial data to the solution space. It is therefore a concept based on the dynamics of the problem. At the same time, it is a local property, since continuity need hold only at the solution whose stability is under examination. We shall always assume this solution to be in equilibrium. Clearly, whether stability holds or not depends upon the underlying function spaces and measures used in the definition.

The customary device for proving stability employs Liapunov functions, (Hale [15], Hartman [17], Knops and Wilkes [26]) although when dissipation is present, an allied approach involving invariance principles has been successfuly used [Cp LaSalle [30], Hale [16], Dafermos and Slemrod [8], Dafermos [6], Infante and Slemrod [22]].

A classical criterion for stability is the so-called energy test which states that an equilibrium solution is stable if the potential energy achieves a minimum value at the solution. This test, whose validity for discrete conservative systems was proved by Dirichlet, does not hold in general for continuous systems. In elasticity theory much effort has been concentrated on discovering conditions under which the test is valid. This is due not only to the test's inherent simplicity but also because it is the criterion traditionally used by engineers and others in practical investigations.

Our interest in stability stems also from the help it might give in deciding constitutive relations. In this respect, conditions leading to instability are equally important since they identify materials unlikely to be experimentally observed and hence which may be excluded. Nevertheless, care must be taken to distinguish two broad categories of instability. There is inherent instability just described, and the instability that occurs in buckling or bursting balloons. The latter results from intricate cross effects of loads and material composition and cannot so easily be used in determining constitutive restrictions.

We shall later present some sufficient conditions and some necessary conditions for stability, and in so doing partially justify the energy test, at least for elasticity. Meanwhile, however, we comment briefly on the very important related question of existence to dynamic solutions without knowledge of which stability investigations must remain formal.

It is well-known that classical solutions to one-dimensional nonlinear hyperbolic conservation laws do not exist for all time, (Cp Dafermos [7]) and hence it can be expected that the same will hold for solutions of nonlinear elastodynamics. Shock waves and other singular surfaces will develop and hence stability analyses must be conducted in the class of generalised or weak solutions. In the case of systems of nonlinear hyperbolic conservation laws in one independent space variable, Glimm

[12] has proved global existence of the generalised solution for initial data of sufficiently small oscillation and total variation. However, his results cannot be immediately extended to elasticity since they require the stress to be either convex or concave and it is neither. (See Dafermos [7] for extended comments). Furthermore, generalised solutions may be nonunique and hence lead automatically to instability unless a selection procedure, such as the viscosity method is used to recover uniqueness.

We shall, in fact, not further consider generalised solutions, but instead confine ourselves to weak solutions, defined in the next section. Under specified conditions, these weak solutions cannot exist for all time. Such results are complementary to recent conclusions by Hughes, Kato and Marsden [21] who establish local existence of strong solutions in unbounded regions for sufficiently smooth initial data and under the hypothesis of strong-ellipticity of W. However, it is uncertain whether the latter condition of strong ellipticity is compatible with the conditions imposed for global non-existence of weak solutions and this combined with absence of a general existence theorem for weak solutions, emphasises once more the formal nature of our analysis.

5. ABSTRACT FORMULATION AND WEAK SOLUTION (Cf Levine and Page [32])

We abstract the equations of nonlinear elasticity and then define a weak solution.

Let H be a Hilbert space, with inner product (,) and suppose $D \subseteq H$. Let $D_1 \subseteq D_2 \subseteq D$ be dense linear subspaces of H and denote by L(H,H) the set of bounded linear operators from H to H. The operator $A: D \to H$ is linear, one-to-one and possibly unbounded while $B: R(A) \times R(A) \to \mathcal{R}$ is a positive-definite bilinear form which may be used to equip R(A) with an inner product and hence convert it into a Hilbert space. The adjoint of A is the operator $A^*: R(A)\big|_{D_1} \to H$ which satisfies $B(Ax,Ay) = -(A^*Ax,y)$,

$\forall x, y \in D_1$. The operator $P:D_1 \to H$ is linear, symmetric and positive-definite.

Finally, we let $V:R(A) \to \mathbb{R}$ be a nonlinear map with a Fréchet derivative continuous in the topology of $R(A)$ induced by $B(.,.)$. By the Riesz representation theorem the action of $V'(X)$ can be expressed by

$$V'(X)\,Y = B(Y,Z), \ \forall\, Y \in R(A)$$

where Z is uniquely determined by X. We set $Z = \mathcal{F}(X)$, where $\mathcal{F}:R(A) \to R(A)$.

Let $z \in D$ be the equilibrium solution defined by

$$B(A\phi, \mathcal{F}(Az)) = 0, \qquad \forall\, \phi \in D_2.$$

Let $u: [0,T) \to D_2$ be weakly continuous and define

$$y = u + z \in D. \tag{5.1}$$

Then u is a weak solution to the Cauchy problem

$$P\ddot{u} = A^*\mathcal{F}(Ay), \quad t\in[0,T) \tag{5.2}$$

$$u(0) = u_0,\ \dot{u}(0) = v_0,\ u_0, v_0 \in D_2$$

provided

(a) there exists a map $u_t: [0,T) \to D_2$, called the velocity, which is weakly continuous and satisfies

$$(u,\phi)\Big|_{t_1}^{t_2} = \int_{t_1}^{t_2} (u_s,\phi)\,ds, \ \forall\phi\in D_2,\ 0 \leqslant t_1 < t_2 < T; \tag{5.3}$$

(b) (Pu_t,u_t), (Pu,u) and $B(Au,(Au))$ are uniformly bounded on compact subintervals of $[0,T)$;

(c) for $\forall\phi: [0,T) \to D_2$ which are weakly continuous and possess a velocity in the sense of (a), there holds

$$(u_t,P\phi)\Big|_{t_1}^{t_2} = \int_{t_1}^{t_2} [\,(u_s,P\phi_s) - B(A\phi,\mathcal{F}(Ay))]\,ds; \tag{5.4}$$

(d) the total energy $E(t)$, defined by

$$E(t) \equiv \tfrac{1}{2}\,(u_t,Pu_t) + I(u), \tag{5.5}$$

where $I(u) = V(y) - V(z)$, satisfies the inequality

$$E(t) \leqslant E(0). \tag{5.6}$$

It would be preferable to omit the energy inequality (5.6) as a postulate and instead derive it as a consequence of the equations. This as yet has not been possible, especially as we wish not to impose any definiteness conditions. However, as Gurtin [14] has demonstrated, such an inequality arises when thermodynamics are introduced into the basic formulation and heat effects are admitted. In a purely mechanical elastic theory the status of (5.6) is less clear, since the (strict) inequality implies dissipation, absence in our nominally conservative system. It has been suggested that its presence in isothermal elasticity may be accounted for by the propagation of shock waves, but this has still to be demonstrated.

Certainly, the above definition of a weak solution admits acceleration waves and higher order singular surfaces but whether the conditions we are about to impose on V for stability are compatible with such singularities is another open question.

6. STABILITY ANALYSES

We here present some conditions which are sufficient for stability of the null solution $u = o$ to the abstract system (5.2). We shall also give conditions necessary for stability of $u = o$. Recall that the null solution, according to (5.1), corresponds to the equilibrium solution z. We emphasise again that our computations are formal, since suitable dynamic solutions are assumed to exist and conditions are unknown for which the assumption is justified.

To find conditions sufficient for stability, we employ the energy inequality (5.6), the well-known Liapunov theorem and the notion of a potential well which we now define. Let $\rho : D_2 \to \mathcal{R}$ be positive definite in the sense that

$$\begin{aligned} \rho(u) &> 0, \quad u \neq 0 \\ &= 0 \Leftrightarrow u = 0. \end{aligned}$$

Definition (Cp. Gurtin [14], Sattinger [38]). The set $\Omega \subset D_2$ is a potential well for $I(u)$ at the origin O if

(a) Ω is a neighbourhood of O, i.e. $u \in \Omega$ s.t. $\rho(u) < \varepsilon$, $\varepsilon > 0$;

(b) $I(u)$ has a minimum at O over Ω, i.e. for some $\varepsilon_0 > 0$

$$I(u) \geqslant 0, \quad \rho(u) < \varepsilon_0;$$

(c) $\inf_{\rho(u)=\varepsilon} I(u) = \psi(\varepsilon) > 0, \; 0 < \varepsilon < \varepsilon_0;$

This definition is similar to that given by Gurtin [14], who however replaces the role of the positive definite function by topological considerations.

It is easy to show by standard arguments (cp. Hale [15], Hartman [17]; Gurtin [14], Koiter [27]) that a potential well for I(u) at O is sufficient for the stability of the origin with respect to small disturbances. A slightly different approach is to make use of the following lemma together with the energy inequality (5.6):

Lemma The set $\Omega \subset D_2$ is a potential well for I(u) at O if and only if (i) I(u) is positive definite for $u \in D_2$, and (ii) the identity map from Ω_I to Ω_ρ is continuous.

Here, Ω_ρ denotes the set Ω equipped with positive-definite measure ρ.

In either case, stability is established with respect to initial data measured by E(O) and to perturbations measured by $\rho(\cdot)$. Upon taking I(u) as a measure of the potential energy, we see that the energy test, mentioned in Section 4 is partly justified. Gurtin [14] has proved stability with respect to other measures, but we wish to retain the (unspecified) function ρ in order to examine when a potential well exists. This clearly depends upon the function ρ. In one-dimension, several results are described by Morrey [34] which show the behaviour of I(u) may be predicted from its second variation. Expressed otherwise, these results state that nonlinear behaviour is determined from behaviour of the solution to the associated linearised problem. Similar results in three-dimensions would be extremely useful but unfortunately little so far appears to be known. As illustration, we recall a theorem due to van Hove [47], which in the notation of Section 2 states:

Theorem 6.1 Suppose W is twice differentiable at the origin and $u \in C_0^1(B)$. If

$$\delta^2 I(O)(u) \geqslant k \int_B \nabla u . \nabla u \, dx, \; k > 0 \tag{6.1}$$

then Ω is a potential well for $I(u)$ at O, where

$$\Omega = \{u: 0 \leqslant \|u\|_{W_0^{1,\infty}(B)} < \varepsilon_0\} \tag{6.2}$$

Other results are due to Naghdi and Trapp [35], while Koiter [29] has recently surveyed his own and other contributions to the problem.

In order to derive necessary conditions for stability, we invert the problem and determine sufficient conditions for the Liapounov instability of the origin. Some preliminary conclusions may be immediately established by defining a potential "peak" and adapting the standard Liapunov proof to show that this condition is sufficient for instability. However, for several reasons this is unsatisfactory and improvement is provided by the following theorem:

Theorem 6.2 (cp Knops and Straughan [25]). Set $J(t) = (Pu,u)$ and suppose (i) $\dot{J}(O) > O$ and (ii) $B(Au, \mathcal{J}(Ay)) \leqslant O$. Then on the interval of existence,

$$J(t) \geqslant J(O) + t\dot{J}(O) + \frac{t^2}{4}\left[\frac{\dot{J}(O)}{J(O)}\right]^2, \; t \in [O,T). \tag{6.3}$$

The proof of this theorem relies upon the differential inequality

$$\ddot{J}(t) \geqslant 2(Pu_t, u_t), \tag{6.4}$$

which may be derived from (5.3) and (5.4), together with the differential inequality

$$(J^{\frac{1}{2}}(t))'' \geqslant 0, \tag{6.5}$$

which follows from Schwarz's inequality. The energy inequality (5.6) is not required.

Clearly, this theorem may be used to establish the instability of the null solution $u = o$, corresponding to the equilibrium solution z. For if $u \in \Omega \subseteq D_2$, then either the weak solution u(t) ceases to exist after a finite time, or $u(t_1) \in \partial \Omega$, for some $t_1 > 0$. Furthermore, it may be shown that condition (ii) of the Theorem implies that the nonlinear function V achieves its maximum at z in the class $y \in D$. Thus we have partially confirmed the necessity of the classical energy test for stability.

Replacement of condition (ii) enables stronger statements to be made about the behaviour of J(t). For instance, when it is assumed that

$$2(1+2\alpha)\, I(u) - B(Au, \mathcal{F}(Ay)) \geqslant 0, \ \alpha > 0 \tag{6.6}$$

then it may be shown that J(t) possesses a lower bound becoming infinite in finite time. From this, it may be concluded that, subject to (6.6), a weak solution cannot exist for all time. Several other results of similar kind may be found in the articles by Hills and Knops [20], Knops, Levine and Payne [24], Levine and Payne [32] and Knops and Straughan [25]. However, none of these papers contains interpretations of (6.6) or analogous conditions. These are provided in the special case of some particular elastic theories by Andreussi [1,2]. More generally, they may be shown to say something about the behaviour of V at the equilibrium solution z and hence are restrictions on the strain energy. Full details will be presented elsewhere.

Acknowledgement

The author is grateful for helpful discussions with his colleagues Dr. J. M. Ball and Dr. N. S. Wilkes during the preparation of this article.

REFERENCES

[1] Andreussi, F., Su certe proprietá di evoluzione della soluzione in elastodinamica non lineare, Nota Interna B75-17, Istituto di Mathematica, Universita di Pisa, 1974.

[2] Andreussi, F., On certain evolution properties of the solution in nonlinear elastodynamics, Ist. Mat., University of Pisa, 1976.

[3] Ball, J. M., Convexity conditions and existence theorems in non-linear elasticity, Arch. Rat. Mechs. Anal. (submitted).

[4] Capriz, G. and Guidugli, P. Podio, on Signorini's perturbation method in finite elasticity, Arch. Rat. Mech. Anal. 57, 1-30, 1974.

[5] Coleman, B. D. and Noll, W., On the thermostatics of continuous media, Arch. Rat. Mech. Anal., 4, 97-128, 1959.

[6] Dafermos, C. M., Contraction semigroups and trend to equilibrium in continuum mechanics, Symposium on Applications of Methods of Functional Analysis to Problems of Mechanics, Marseille 1975.

[7] Dafermos, C. M., Quasilinear hyperbolic systems, Nonlinear waves, 83-100, Cornell Univ. Press, 1974.

[8] Dafermos, C. M. and Slemrod, M., Asymptotic behaviour of nonlinear contraction semigroups, J. Functional Analysis 13, 97-106, 1973.

[9] Ericksen, J. L., A thermo-kinetic view of elastic stability theory, Int. J. Sols. Structs., 2, 573-580, 1966.

[10] Ericksen, J. L., Thermoelastic stability, Proc. 5th U.S. Nat. Congr. Appl. Mech. ASME, 187-193, 1966.

[11] Fichera, G., Existence theorems in elasticity, Handbuch der Physik, Vol VIa/2, 347-389, Springer, 1974.

[12] Glimm, J., Solutions in the large for nonlinear hyperbolic systems of equations, Comm. Pure Appl. Math. 18, 697-715, 1965.

[13] Gurtin, M. E., The linear theory of elasticity, Handbuch der Physik, Vol VIa/2, 1-295, Springer 1974.

[14] Gurtin, M. E., Thermodynamics and stability, Arch. Rat. Mechs. Anal., 59, 63-96, 1975.

[15] Hale, J. K., Ordinary Differential Equations, Wiley-Interscience, New York 1969.

[16] Hale, J. K., Dynamical systems and stability, J. Math. Anal. Appl. 26, 39-59, 1969.

[17] Hartman, P., Ordinary Differential Equations, Corrected Reprint, Baltimore 1975.

[18] Hill, R., On uniqueness and stability in the theory of finite elastic strain, J. Mech. Phys. Sols., 5, 229-241, 1957.

[19] Hill, R., Eigenmodal deformations in elastic/plastic continua, J. Mech. Phys. Sols., 15, 371-386, 1967.

[20] Hills, R. N. and Knops, R. J., Qualitative results for a general class of material behaviour, S.I.A.M. J. Appl. Maths. (to appear).

[21] Hughes, T. J. R., Kato, T. and Marsden, J. E., Well-posed quasi-linear second order hyperbolic systems with applications to nonlinear elasticity and general relativity (to appear).

[22] Infante, E. F. and Slemrod, M., An invariance principle for dynamical systems on a Banach space: application to the general problem of thermoelastic stability, Symposium on Instability of Continuous Systems, 215-221, Springer 1970.

[23] John, F., Uniqueness of nonlinear elastic equilibrium for prescribed boundary displacements and sufficiently small strains, Comm. Pure Appl. Maths., 25, 617-634, 1972.

[24] Knops, R. J., Levine, H. A. and Payne, L. E., Nonexistence, instability and growth theorems for solutions of a class of abstract nonlinear equations with applications to nonlinear elastodynamics, Arch. Rat. Mech. Anal., 55, 52-72, 1974.

[25] Knops, R. J. and Straughan, B., Nonexistence of global solutions to nonlinear Cauchy problems arising in mechanics, Symposium on Trends in the Application of Pure Mathematics to Mechanics, Pitman (in press).

[26] Knops, R. J. and Wilkes, E. W., Theory of elastic stability, Handbuch der Physik VIa/3, 125-302, Springer, Berlin 1973.

[27] Koiter, W. T., The energy criterion of stability for continuous elastic bodies, Proc. Kon. Ned. Ak. Wet. B68, 178-202, 1965.

[28] Koiter, W. T., Thermodynamics of elastic stability, Proc. 3rd. Can. Congr. Appl. Mech. Calgary, 29-37, 1971.

[29] Koiter, W. T., A basic open problem in the theory of elastic stability, Symposium on Applications of Methods of Functional Analysis to Problems in Mechanics, Marseille, 1975.

[30] La Salle, J. P., An invariance principle in the theory of stability, Symposium on Differential Equations and Dynamical Systems, 277-286, Acad. Press. New York 1967.

[31] Lax P., Hyperbolic systems of conservation laws II, Comm. Pure Appl. Math. 10, 537-566, 1957.

[32] Levine, H. A. and Payne, L. E., On the nonexistence of global solutions to some abstract Cauchy problems of standard and non-standard type, J. Elast., 5, 273-286, 1975.

[33] Morrey, Jnr., C. B., Quasi-convexity and the lower semi-continuity of multiple integrals, Pacific J. Math. 2, 25-53, 1952.

[34] Morrey, Jnr., C. B., Multiple Integrals in the Calculus of Variations, Springer, Berlin, 1966.

[35] Naghdi, P. M. and Trapp, J. A., On the general theory of stability for elastic bodies, Arch. Rat. Mech. Anal., 51, 165-191, 1973.

[36] Rivlin, R. S., Some topics in finite elasticity, Structural Mechanics 169-198, Pergamon, 1960.

[37] Rivlin, R. S., The fundamental equations of nonlinear continuum mechanics, Dynamics of Fluids and Plasmas, 83-126, Academic Press, New York 1966.

[38] Sattinger, D. M., On global solutions of nonlinear hyperbolic equations, Arch. Rat. Mech. Anal., 30, 148-172, 1968.

[39] Signorini, A., Trasformazioni termoelastiche finite, Mem. 3ª Solidi incomprimibili, Ann. di Mat. Pur. Appl. 39, 147-201, 1955.

[40] Stopelli, F., Un teorema di esistenza e di unicità relative alle equazioni dell elastostatica isoterma per deformazioni finite, Richerche Mat. 3, 247-267, 1954.

[41] Stopelli, F., Sulla svrilluppibitità in serie di potenze di un parametro delle soluzioni delle equazioni dell'elastatica isoterma, Ricerche Mat. 4, 58-73, 1955.

[42] Stopelli, F., Sull'esistenza di soluzioni delle equazioni dell' elastostatica isoterma nel caso di sollectizioni dotate di assi di equilibrio, I, II, III, Richerche Mat. 6, 241-287, 1957; 7, 71-101, 138-152, 1958.

[43] Truesdell, C. and Noll, W., The non-linear field theories of mechanics, Handbuch der Physik, Vol III/3, Springer, Berlin 1965.

[44] Truesdell, C., The Elements of Continuum Mechanics, Springer, 1966.

[45] Truesdell, C., An introduction to Rational Mechanics, Academic Press, New York (1976).

[46] Van Buren, W., On the existence and uniqueness of solutions to boundary value problems in finite elasticity, Ph.D. thesis, Carnegie-Mellon Univ., 1968.

[47] Van Hove, L., Sur le signe de la variation seconde des intégrales multiples à plusieus functions inconnues, Acad. Roy. Belgique, 25, 3-68, 1949.

[48] Wang, C. C. and Truesdell, C., Introduction to Rational Elasticity, Noordhoff, Leyden, 1973.

A Mikusinski Calculus for the Bessel Operator B_μ

Eusebio L. Koh

Abstract. An operational calculus for the Bessel operator $B_\mu = t^{-\mu}Dt^{\mu+1}D(-1 < \mu < \infty)$ is developed. A convolution process is proposed which reduces to Ditkin's convolution when $\mu = 0$. Following Mikusinski, the construction is through the field extension of a commutative ring without zero divisors. The relationships between the calculus and those of Mikusinski and Ditkin are shown.

§ 1. Introduction. In [1] an operational calculus for the operator $B_n = t^{-n}\frac{d}{dt}t^{1+n}\frac{d}{dt}$ $(n = 0, 1, 2, \ldots)$ was constructed through the field extension of a commutative ring without zero divisors. By means of the Riemann-Liouville fractional derivative, the calculus is now extended to B_μ where μ is any real number greater than -1. When $\mu = 0$, our calculus reduces to Ditkin's calculus for $\frac{d}{dt}t\frac{d}{dt}$ (see [2]). When $\mu \in (-1, 1)$, we obtain results similar to Meller's (see [3]). The relation between the present calculus and those of Mikusinski [4] and of Ditkin [2] are established. A connection between the calculus and a modified Meijer transformation is indicated.

The Riemann-Liouville integral of order $\alpha > 0$ is defined in [5] :

$$I^\alpha f(t) = \frac{1}{\Gamma(\alpha)}\int_0^t (t-\xi)^{\alpha-1} f(\xi)d\xi . \qquad (1)$$

When α is a positive integer m, then $I^m f(x)$ is simply the m-fold integral of $f(x)$.
The following relations are easily verified:

$$\frac{d}{dt} I^{\alpha+1} f(t) = I^\alpha f(t) . \qquad (2)$$

$$I^\alpha I^\beta f(t) = I^{\alpha+\beta} f(t) , \quad \alpha, \beta > 0 . \qquad (3)$$

Equation (3) is called the index law and it follows from an interchange of integration and the definition of Beta function. Equation (2) suggests a definition of fractional differentiation. Let $\nu \in (0, \infty)$ and let n = the least integer greater than ν. For $f \in C^n[0, \infty)$, the νth order derivative of $f(t)$ is defined by

$$D^{\nu} f(t) = D^n I^{n-\nu} f(t) , \qquad D = \frac{d}{dt} . \tag{4}$$

As a nontrivial example, we have

$$D^{\nu} t^k = \frac{\Gamma(k+1)}{\Gamma(k-\nu+1)} t^{k-\nu} , \quad -1 < k < \infty , \; 0 < \nu \leq k , \tag{5}$$

which is a generalization of the power formula.

§ 2. <u>The field extension of C^{∞}.</u> Let μ be a fixed real number greater than -1. Let C^{∞} denote the set of infinitely differentiable complex functions on $[0, \infty)$ for which the following operation is defined: for every $\phi, \psi \in C^{\infty}$,

$$\phi * \psi = \frac{1}{\Gamma(\mu+1)} Dt^{1-\mu} D^{\mu+1} \int_0^t \xi^{\mu} (t-\xi)^{\mu} \int_0^1 \phi(\xi x) \psi[(1-x)(t-\xi)] \, dx d\xi . \tag{6}$$

Following Dimovski [6], we call the operation $*$ a <u>convolution for B_{μ} in C^{∞}</u> if C^{∞} is closed under it and if it is bilinear, commutative, associative, distributive with respect to addition and satisfies the relation

$$B_{\mu} (\phi * \psi) = (B_{\mu} \phi) * \psi . \tag{7}$$

Moreover, we call a convolution <u>constant-preserving</u> if

$$\alpha * \phi(t) = \alpha \phi(t) \qquad \forall \alpha \in \mathbb{C} . \tag{8}$$

<u>Theorem 1.</u> The operation defined by equation (6) is a constant-preserving convolution for B_{μ} in C^{∞}.

Proof: C^∞ is clearly closed under the operation. Bilinearity, commutativity and distributivity are also clear. To prove associativity we have, on using (5),

$$t^p * (t^q * t^r) = \frac{p!q!r!\Gamma(\mu+p+1)\Gamma(\mu+q+1)\Gamma(\mu+r+1)}{\Gamma^2(\mu+1)(p+q+r)!\Gamma(\mu+p+q+r+1)} t^{p+q+r}$$

$$= (t^p * t^q) * t^r . \tag{9}$$

Due to the bilinearity, equation (9) still holds for polynomials. That it holds for elements of C^∞ follows from Weierstrass's Approximation Theorem. This same argument may be used to prove equation (7). Thus we only need to show that $B_\mu(t^q * t^r) = B_\mu(t^q) * t^r$. Indeed, $B_\mu(t^k) = t^{-\mu} D\, t^{\mu+1}\, D\{t^k\} = k(\mu+k)t^{k-1}$. Hence

$$B_\mu(t^q * t^r) = \frac{q!r!\Gamma(\mu+q+1)\Gamma(\mu+r+1)}{\Gamma(\mu+1)\Gamma(q+r)\Gamma(\mu+q+r)} t^{q+r-1}$$

$$= q(\mu+q)t^{q-1} * t^r = B_\mu(t^q) * t^r .$$

Finally, for $\alpha \in \mathbb{C}$, it follows from (3) and (4) that

$$\alpha * \phi = \frac{1}{\Gamma(\mu+1)} Dt^{1-\mu} D^{\mu+1} \int_0^t \xi^\mu(t-\xi)^\mu \int_0^t \alpha\phi(\xi x)dx d\xi$$

$$= \alpha\, Dt^{1-\mu} D^{n+1} I^{n-\mu} I^{\mu+1}\left[t^{\mu-1}\int_0^t \phi(u)du\right] = \alpha\phi(t)$$

where n = least integer greater than μ.

Theorem 2. C^∞ has no zero divisors.

Proof: The case when $\mu = 0, 1, 2, \ldots$ is proved in [1]. For $\mu \neq 0, 1, 2, \ldots$ it is easy to show that $\phi(t) * \psi(t) = 0$ implies that

$$\int_0^t (t-\xi)^{n-\mu-1} d\xi \int_0^\xi \eta^\mu(\xi-\eta)^\mu d\eta \int_0^1 \phi(\eta x)\psi\left[(1-x)(\xi-\eta)\right]dx = 0 .$$

From Titchmarsh's Theorem [4] and the fact that $t^{n-\mu-1} \neq 0$ on $(0, \infty)$ we have

$$\int_0^t \eta^\mu(t-\eta)^\mu d\eta \int_0^1 \phi(\eta x)\psi\left[(1-x)(t-\eta)\right]dx = 0 .$$

A change of variables enables us to use a theorem of Mikusinski and

Ryll-Nardzewski [7] to conclude that either $\phi = 0$ or $\psi = 0$.

Under the operations of addition and convolution as multiplication, C^∞ is a commutative ring without zero divisors. Since the convolution is constant-preserving, the identity element in C^∞ is the number 1. We may extend C^∞ into the quotient field, F, consisting of equivalence classes of ordered pairs (ϕ, ψ) of elements in C^∞ with nonzero second elements. Two pairs (ϕ_1, ψ_1) and (ϕ_2, ψ_2) are said to be equivalent if $\phi_1 * \psi_2 = \phi_2 * \psi_1$. Following Mikusinski, we call the elements of F operators and denote them by $\frac{\phi}{\psi}$. Addition, multiplication and scalar multiplication in F are defined by

$$\frac{\phi_1}{\psi_1} + \frac{\phi_2}{\psi_2} = \frac{\phi_1 * \psi_2 + \phi_2 * \psi_1}{\psi_1 * \psi_2} , \tag{10}$$

$$\frac{\phi_1}{\psi_1} \cdot \frac{\phi_2}{\psi_2} = \frac{\phi_1 * \phi_2}{\psi_1 * \psi_2} , \tag{11}$$

and

$$\alpha \frac{\phi}{\psi} = \frac{\alpha \phi}{\psi} . \tag{12}$$

These definitions are independent of the choice of representants. With these operations, F is in fact an algebra with the zero element given by $\frac{0}{\psi}$, $\psi \neq 0$ and the unit element by $\frac{\psi}{\psi}$, $\psi \neq 0$.

Operators of the form $\frac{\phi(t)}{1}$ constitute a subring of F isomorphic to C^∞ through the canonical map $\frac{\phi(t)}{1} \leftrightarrow \phi(t)$. F also contains certain locally integrable functions. Let

$$\Lambda \phi = \int_0^t \xi^{-\mu-1} \int_0^\xi \eta^\mu \phi(\eta) d\eta d\xi .$$

<u>Theorem 3.</u> For any $\phi(t) \in L_{loc}[0, \infty)$, $\frac{t}{\mu+1} * \phi = \Lambda \phi(t)$.
Hence, operators of the form $\frac{\phi(t)}{t}$ with $\phi(0) = 0$ may be identified with $f(t) \in L_{loc}[0, \infty)$ such that $\Lambda f < \infty \quad \forall t > 0$.

<u>Proof:</u> $\frac{t}{\mu+1} * \phi(t) = \frac{1}{\Gamma(\mu+1)} D\, t^{1-\mu} D^{\mu+1} \int_0^t \xi^\mu (t-\xi)^\mu \int_0^1 \phi(\xi x) \frac{(1-x)(t-\xi)}{\mu+1} dx d\xi$

$$= Dt^{1-\mu} D^{n+1} I^{n-\mu} I^{\mu+2} \{t^{\mu-2} \int_0^t \phi(\eta)(t-\eta)d\eta\}$$

$$= (1-\mu)t^{-\mu} \int_0^t \xi^{\mu-2} \int_0^\xi \phi(\eta)(\xi-\eta)d\eta d\xi + t^{-1} \int_0^t \phi(\eta)(t-\eta)d\eta$$

$$= (1-\mu)t^{-\mu} \int_0^t \phi(\eta) \left\{\frac{(\mu-1)t^\mu - \mu t^{\mu-1}\eta + \eta^\mu}{\mu(\mu-1)}\right\} d\eta + t^{-1} \int_0^t \phi(\eta)(t-\eta)d\eta$$

$$= \int_0^t \eta^\mu \phi(\eta) \int_\eta^t \xi^{-\mu-1} d\xi d\eta = \Lambda\, \phi(t)\ .$$

The second statement follows from the observation that for $\phi \in C^\infty$ with $\phi(0) = 0$,

$$\frac{\phi(t)}{t} = f(t) \in L_{loc}[0, \infty) \iff \phi(t) = t * f = (\mu+1)\Lambda f < \infty,\ \forall\ t > 0.$$

From theorem 3 and by induction, we have

<u>Theorem 4.</u> Let k be a positive integer; then for any $\phi(t) \in C^\infty$, $\frac{\Gamma(\mu+1)t^k}{\Gamma(k+\mu+1)k!} * \phi(t) = \Lambda^k \phi(t)$ where Λ^k = k-times application of Λ .

By virtue of the embedding of C^∞ in F , theorems 3 and 4 state that the integral operators Λ and Λ^k are in F and may be represented by

$$\Lambda = \frac{t}{\mu+1} \tag{13}$$

$$\Lambda^k = \frac{\Gamma(\mu+1)t^k}{\Gamma(k+\mu+1)k!} \tag{14}$$

Let V be the operator $\frac{\mu+1}{t}$ and V^k the k-times application of V .

<u>Lemma 1:</u> For any 2k times differentiable function $\phi(t)$,

$$V^k \phi(t) = B_\mu^k \phi(t) + \sum_{j=1}^k B_\mu^{k-j} \phi(0)V^j \tag{15}$$

where

$$B_\mu^{k-j} \phi(0) = B_\mu^{k-j} \phi(t)\Big|_{t\to 0^+}\ .$$

Proof: $\phi(t) = \Lambda B_\mu\phi(t) + \phi(0) = \frac{t}{\mu+1} * B_\mu\phi(t) + \phi(0)$

Thus $V\phi(t) = B_\mu\phi(t) + \phi(0)V$ and (15) is proved for $k = 1$.
The lemma follows by induction.

We now apply (15) to generate a number of operational formulas. The differential equations $B_\mu\ \phi(t) = \pm\ a\phi(t),\ \phi(0) < \infty$ are solved by $(2\sqrt{at})^{-\mu} I_\mu(2\sqrt{at})$ for the plus sign and $(2\sqrt{at})^{-\mu}J_\mu(2\sqrt{at})$ for the minus sign. Here $I_\mu(z)$ and $J_\mu(z)$ are Bessel functions of order μ. Since $\lim_{t\to 0^+}(2\sqrt{at})^{-\mu} I_\mu(2\sqrt{at}) = \lim_{t\to 0^+}(2\sqrt{at})^{-\mu} J_\mu(2\sqrt{at})$
$= \dfrac{1}{2^\mu\Gamma(\mu+1)}$, we have on using (15)

Theorem 5.
$$\frac{V}{V-a} = \Gamma(\mu+1)(at)^{-\frac{\mu}{2}} I_\mu(2\sqrt{at}) \tag{16}$$

$$\frac{V}{V+a} = \Gamma(\mu+1)(at)^{-\frac{\mu}{2}} J_\mu(2\sqrt{at}) \tag{17}$$

Through some simple calculations, the following formulas are derived from theorem 5.

Theorem 6.

$$\frac{V^2}{V^2-a^2} = \frac{\Gamma(\mu+1)}{2}(at)^{-\frac{\mu}{2}}\left[I_\mu(2\sqrt{at}) + J_\mu(2\sqrt{at})\right] \tag{18}$$

$$\frac{aV}{V^2-a^2} = \frac{\Gamma(\mu+1)}{2}(at)^{-\frac{\mu}{2}}\left[I_\mu(2\sqrt{at}) - J_\mu(2\sqrt{at})\right] \tag{19}$$

$$\frac{V^2}{V^2+a^2} = \Gamma(\mu+1)(at)^{-\frac{\mu}{2}}\left[\cos\frac{3\mu\pi}{4}\mathrm{ber}_\mu(2\sqrt{at}) + \sin\frac{3\mu\pi}{4}\mathrm{bei}_\mu(2\sqrt{at})\right] \tag{20}$$

$$\frac{aV}{V^2+a^2} = \Gamma(\mu+1)(at)^{-\frac{\mu}{2}}\left[\cos\frac{3\mu\pi}{4}\mathrm{bei}_\mu(2\sqrt{at}) - \sin\frac{3\mu\pi}{4}\mathrm{ber}_\mu(2\sqrt{at})\right] \tag{21}$$

$$\frac{V^2}{V^2+a^2} = (-1)^{\frac{n}{4}}\, n!\, (at)^{-\frac{n}{2}}\, \mathrm{ber}_n(2\sqrt{at}) \qquad \text{if } n \equiv 0 \pmod 4 \tag{22}$$

$$= (-1)^{\frac{n+2}{4}}\, n!\, (at)^{-\frac{n}{2}}\, \mathrm{bei}_n(2\sqrt{at}) \qquad \text{if } n \equiv 2 \pmod 4 \tag{23}$$

$$\frac{aV}{V^2+a^2} = (-1)^{\frac{n}{4}}\, n!\, (at)^{-\frac{n}{2}}\, \mathrm{bei}_n(2\sqrt{at}) \qquad \text{if } n \equiv 0 \pmod 4 \tag{24}$$

$$= (-1)^{\frac{n-2}{4}}\, n!\, (at)^{-\frac{n}{2}}\, \mathrm{ber}_n(2\sqrt{at}) \qquad \text{if } n \equiv 2 \pmod 4 \tag{25}$$

where $\mathrm{ber}_n(z)$ and $\mathrm{bei}_n(z)$ are Kelvin functions.

Formulas (18) to (21) are similar to those given by Meller [3] for the case $-1 < \mu < 1$. Formulas (22) to (25) are new. An interesting consequence of (22) - (25) is that the Kelvin functions bear the same relationship to the operator B_n as the circular functions to the operator D . For n, a nonnegative even integer, we have

$$B_n\{(at)^{-\frac{n}{2}}\, \mathrm{ber}_n(2\sqrt{at})\} = -\, a(at)^{-\frac{n}{2}}\, \mathrm{bei}_n(2\sqrt{at}) \tag{26}$$

$$B_n\{(at)^{-\frac{n}{2}}\, \mathrm{bei}_n(2\sqrt{at})\} = a(at)^{-\frac{n}{2}}\, \mathrm{ber}_n(2\sqrt{at}) \tag{27}$$

Following Mikusinski's theory, we may enlarge the table of formulas by considering sequences and series of operators, operational functions, their derivatives and their integrals. For example, it can be shown by parametric differentiation of (16) that

$$\frac{V}{(V-a)^{m+1}} = \Gamma(\mu+1)t^m(at)^{-\frac{\mu+m}{2}}\, I_{\mu+m}(2\sqrt{at}) \ . \tag{28}$$

§ 3. Relation to integral transformation and other calculi

The operational calculus can also be generated by the Meijer transformation [3] :

$$(k_\mu f)(p) = \frac{2p^{\frac{\mu}{2}+1}}{\Gamma(\mu+1)} \int_0^\infty f(t)\, t^{\frac{\mu}{2}} K_\mu(2\sqrt{pt})dt . \qquad (29)$$

If f is Lebesgue integrable on $(0, \infty)$ and it satisfies the bound $|f| < Ce^{2\gamma\sqrt{t}}\, t$, then the integral (29) converges in $\mathrm{Re}\sqrt{p} > \gamma$ and is analytic there. By virtue of the asymptotic and series expansions of $K_\mu(2\sqrt{pt})$ we can show by integration by parts twice that

$$k_\mu(B_\mu f) = p\, k_\mu f . \qquad (30)$$

Moreover, we conjecture that a convolution theorem can be proved for k_μ using our convolution process (6), i. e.

$$k_\mu(f_1 * f_2) = (k_\mu f_1)(k_\mu f_2) . \qquad (31)$$

For $\mu = 0, 1, 2, \ldots,$ and $f_i(t)$ of rapid descent as $t \to 0$ and $t \to \infty$, this is the case.

The relation between our calculus and the Mikusinski calculus for the operator D is given by

<u>Theorem 7 (Meller).</u> If $F(V) = f(t)$ and $F(D) = g(t)$, then

$$g(t) = \frac{1}{\Gamma(\mu+1)} \int_0^\infty \xi^\mu f(t\xi)e^{-\xi}\, d\xi \qquad (32)$$

$$\text{and } f(t) = \frac{\Gamma(\mu+1)}{t^\mu} \frac{1}{2\pi i} \int_{a-i\infty}^{a+i\infty} e^{pt} g(\tfrac{1}{p})(\tfrac{1}{p})^{\mu+1} dp \qquad (33) .$$

<u>Outline of proof:</u> From the Meijer and the Laplace-Carson transforms, we have

$$\frac{2p^{\frac{\mu}{2}+1}}{\Gamma(\mu+1)} \int_0^\infty f(t)\, t^{\frac{\mu}{2}} K_\mu(2\sqrt{pt})dt = p \int_0^\infty g(t)e^{-pt}dt .$$

Equation (32) follows by substituting for $K_\mu(2\sqrt{pt})$ its integral representation (see eq. (23) p. 82 of [8]). Equation (33) comes from the inversion of Laplace transforms.

Similarly, we have the following relation between our calculus and Ditkin's:

Theorem 8. If $F(V) = f(t)$ and $F(B_o) = h(t)$, then

$$f(t) = t^{-\mu} \frac{d}{dt} \int_o^t h(\xi)(t-\xi)^\mu \, d\xi \tag{34}$$

and $$h(t) = \frac{1}{\Gamma(1+\mu)\Gamma(n-\mu)} D^n \int_o^t f(\xi)\xi^\mu (t-\xi)^{n-\mu-1} \, d\xi \tag{35}$$

Meller's result [3] is a special case of this.

REFERENCES

[1] Koh, E. L., T. H. Darmstadt, preprint No. 240, 1975.

[2] Ditkin, V. A. and A. P. Prudnikov, Integral Transforms and Operational Calculus, Pergamon, 1965.

[3] Meller, N. A., Vichis. Matem. 6 (1960) 161 - 168.

[4] Mikusinski, J., Operational Calculus, Pergamon, 1959.

[5] Ross, B. (Ed.) Fractional Calculus and its Applications, Springer-Verlag, 1975.

[6] Dimovski, I. H., Compt. Rend. Acad. Bulg. Sci. 26 (1973) 1579 - 1582.

[7] Mikusinski, J. and C. R. Nardzewski, Studia Math. 13 (1) (1953) 62 - 68.

[8] Erdélyi, A. et. al. Higher Transcendental Functions, Vol. 2, McGraw-Hill, 1954.

An Oscillation Theory for Fourth Order Differential Equations

Kurt Kreith

1. Introduction. The study of higher order differential equations is frequently facilitated by representing such equations as systems of equations of lower order. In particular, if an nth order linear equation is represented as a first order system $\underline{x}' = P(t)\underline{x}$, then the nature of the vector field determined by $P(t)$ reflects some of the basic properties of the original equation.

For differential equations of even order one can also consider second order system representations of the form $\underline{x}'' = P(t)\underline{x}$ and seek to relate properties of the original equation to the vector field determined by $P(t)$. The advantage of the latter approach is that a second order representation leads to a lower dimensional and generally simpler vector field; the disadvantage is that whereas in a first order system the vector field determines the direction of solution curves, in a second order system it merely influences their direction. Thus in a second order representation one is likely to be concerned not only with the topological character of the field but also with its strength.

Second order systems representations were first used to study self-adjoint fourth order equations by Whyburn [9]. They were used by the author [4] as a means of obtaining oscillation properties of nonself-adjoint fourth order equations. Subsequent work by S. Cheng, A. Edelson and the author shows how such systems can be used to establish other basic properties of fourth order equations.

2. The System Representation. The general real linear nonselfadjoint fourth order differential equation

$$(2.1)\quad \ell[u] \equiv (p_2(t)u''-q_2(t)u')'' - (p_1(t)u'-q_1(t)u)' + p_0(t)u = 0$$

allows a second order system representation

$$(2.2)\quad \begin{aligned} y'' &= a(t)y + b(t)z \\ z'' &= c(t)y + d(t)z \end{aligned}$$

by means of the reduction and transformation given in [4]. The only aspect of the transformation required here is that simple, double, and triple zeros of u coincide with zeros of y, y', and z, respectively. We assume that the coefficients of (2.1) are sufficiently regular so that the coefficients of (2.2) are continuous and that $b(t) = \frac{1}{p_2(t)} > 0$ in $[\alpha,\infty)$.

For the special case

$$(2.3)\quad (p_2(t)u'')'' + p_0(t)u = 0$$

Leighton and Nehari [8] show that the oscillatory behavior of solutions depends in a fundamental way on whether $p_0(t) > 0$ or $p_0(t) < 0$ in $[\alpha,\infty)$. One of the advantages of the system representation (2.2) is that it allows one to extend this classification to the more general equation (2.1). Specifically, the case $p_0(t) > 0$ corresponds to the situation in which the vector force field $(ay+bz,\ cy+dz)$ has a spiral or center at $(0,0)$ while the case $p_0(t) < 0$ corresponds to a saddle point.

3. Rotation Theory. To formulate these generalizations we first introduce polar coordinates

$$r^2 = y^2 + z^2 \quad ; \quad \theta = \arctan \frac{y}{z}$$

in (2.2), obtaining as in [4]

$$r'' = r(\theta')^2 + \frac{1}{r} Q_1(y,z)$$
$$(r^2\theta')' = Q_2(y,z)$$

where Q_1 and Q_2 are quadratic forms defined by

$$Q_1(y,z) = ay^2 + (b+c)yz + dz^2$$
$$Q_2(y,z) = -cy^2 + (a-d)yz + bz^2.$$

The assumption that $Q_2(y,z)$ be positive definite in $[\alpha,\infty)$ corresponds to the case $p_0(t) > 0$ and allows one to extend to equation (2.1) many of the results which Leighton and Nehari establish in Part II of [8] for (2.3). For example, since $y(\alpha) = y'(\alpha) = 0$ implies that $(r^2\theta')(\alpha) = 0$, it follows easily that any solution of (2.1) can have at most one double zero. Accordingly, conjugate points are defined in terms of the zeros of a principal solution of (2.1) satisfying $u(\alpha) = u'(\alpha) = u''(\alpha) = 0$; for the system (2.2) this corresponds to a solution satisfying $y(\alpha) = y'(\alpha) = z(\alpha) = 0$ -- i.e., a solution emanating from the origin parallel to the z-axis. By way of another example, the monotonicity of such conjugate points now takes on a more general form [5].

3.1 Theorem. *Let $y_1(t)$, $z_1(t)$ and $y_2(t)$, $z_2(t)$ represent principal solutions of (2.1) emanating from the origin at $t = \alpha_1$ and $t = \alpha_2$, respectively, where $\alpha_1 < \alpha_2$. If the corresponding phase functions satisfy*

$$k\pi \leq \theta_1(\alpha_2) - \theta_2(\alpha_2) \leq (k+1)\pi$$

for some integer k, then

$$k\pi < \theta_1(x) - \theta_2(x) < (k+2)\pi$$

for all $x > \alpha_2$.

If the underlying equation is selfadjoint, this fact is reflected in its system representation by the identity $a = d$; indeed the system representation of the adjoint equation $\ell^*u = 0$ is obtained by interchanging a and d. Fortunately one rarely has to invoke this assumption to generalize Part II of [8]; a rare exception is the proposition that "if one solution of (2.3) is oscillatory, then all solutions are oscillatory" which does not hold for nonselfadjoint equations, as evidenced by $u^{(iv)} \pm u' + p_0 u = 0$ for sufficiently small constants $p_0 > 0$.

This phenomenon does, however, allow extension to the nonselfadjoint case by means of the system representation (2.2). In particular, it can be shown [5] that if one solution of (2.1) is oscillatory because the corresponding phase function $\theta(t)$ satisfies $\lim_{t \to \infty} \theta(t) = \infty$, then all solutions for which $\tilde{\theta}'(\beta) > 0$ for some $\beta > \alpha$ also satisfy $\lim_{t \to \infty} \tilde{\theta}(t) = \infty$. The difficulty is that there may also be solutions whose phase functions satisfy $\varphi'(t) < 0$ for all $t > \alpha$ but $\lim_{t \to \infty} \varphi(t) = \varphi_0 > -\infty$. It is this phenomenon which gives rise to nonoscillatory solutions. We do, however, have the following result which corresponds to the simpler proposition for (2.3) [7].

3.2 Theorem. Suppose (2.1) has an oscillatory solution $u(t)$ whose phase function satisfies $\lim_{t \to \infty} \theta(t) = \infty$. If $v(t)$ is a solution of $\ell^*v = 0$ whose

<u>phase function satisfies</u> $\varphi'(t) < 0$ <u>for all</u> $t > \alpha$, <u>then</u> $\lim_{t \to \infty} \varphi(t) = -\infty$.

<u>4. Saddle Points</u>. The case where $p_0(t) < 0$ in (2.3) is marked by the fact that solutions with triple zeros at $t = \alpha$ will be of constant sign for $t > \alpha$. In this case conjugate points are attained by solutions having double zeros at $t = \alpha$ and at $\beta = \eta(\alpha)$. The existence of such 2-2 conjugate points for nonselfadjoint fourth order equations is therefore also of interest.

Criteria for the existence of $\eta(\alpha)$ can be formulated qualitatively in terms of the force field of (2.2). Specifically in [6] trajectories $y(t)$, $z(t)$ are studied satisfying $y(\alpha) = y'(\alpha) = 0$; $z(\alpha) = 1$; $z'(\alpha) = v_0$. Denoting such a trajectory by $C(v_0)$, the question becomes whether one can choose a value of the parameter v_0 which will assure that $y(\beta) = y'(\beta) = 0$ for some $\beta > \alpha$. Sufficient conditions for such behavior (reflecting a sufficiently strong saddle point at (0,0)) can be shown to be as follows:

(A) If for some $t_0 \geq \alpha$ the quantities $y(t_0)$, $y'(t_0)$, $z(t_0)$, $z'(t_0)$ are all nonnegative (but not all zero), then $y(t)$, $y'(t)$, $z(t)$, $z'(t)$ remain positive for all $t > t_0$.

(B) No trajectory can satisfy $y(t) > 0$ and $z(t) < 0$ for arbitrarily large values of t.

(C) No trajectory has asymptotes of the form

(i) $z(t) \downarrow z_0 \geq 0$ and $y(t) \uparrow \infty$ as $t \to \infty$, or

(ii) $y(t) \downarrow y_0 \geq 0$ and $z(t) \uparrow \infty$ as $t \to \infty$,

nor can any trajectory tend to a finite limit point as $t \to \infty$.

(D) If $y(t) > 0$ for $t_1 \leq t \leq t_2$, then $z(t)$ can change sign at most once in $[t_1,t_2]$.

Specific conditions under which these qualitative conditions are satisfied are given by the following.

4.1 Theorem. *If the coefficients of* (2.2) *satisfy*

(i) $c(t) \geq a(t) > 0; \quad b(t) \geq d(t) > 0,$

(ii) $u'' + \min\{b(t)-d(t),\ c(t)-a(t)\}u = 0$ *is oscillatory at* $t = \infty$,

(iii) $\int^{\infty} tb(t)\, dt = \int^{\infty} tc(t)\, dt = \infty,$

then conditions (A) - (D) *above are satisfied.*

While this theorem only shows the existence of $\eta_1(\alpha)$, specific estimates for $\eta_1(\alpha) - \alpha$ can also be obtained using these techniques [3].

5. Systems Conjugate Points. Another application of the second order system approach to fourth order differential equations is due to Cheng [2]. Barrett [1] had studied selfadjoint fourth order differential equations and established conditions which assure the existence and non-existence of a systems conjugate point $\beta = \hat{\eta}(\alpha)$ defined by

$$u(\alpha) = u''(\alpha) = 0 = u(\beta) = u''(\beta).$$

By representing (2.1) in the form

$$\begin{aligned} z'' + A(t)z + B(t)y &= 0 \\ y'' + C(t)z + D(t)y &= 0 \end{aligned} \tag{5.1}$$

and assuming $A(t) \geq 0$, $D(t) \geq 0$, $B(t) > 0$, $C(t) > 0$ on $[\alpha,\infty)$, Cheng was able to formulate dynamical criteria for the existence of systems conjugate points in the nonselfadjoint case.

In terms of (5.1), systems conjugate points correspond to trajectories satisfying $z(\alpha) = y(\alpha) = 0 = z(\beta) = y(\beta)$. The basic qualitative criterion for their existence is that for every trajectory $z(t)$, $y(t)$ satisfying $z(\alpha) = y(\alpha) = 0$, $z'(\alpha) > 0$, $y'(\alpha) > 0$, it follows that $z(t)$ or $y(t)$ must eventually change sign. An example of the specific criteria which assure such behavior is contained in the following [2].

5.1 Theorem. *The existence of a systems conjugate point trajectory is assured by any of the following:*

(i) $u'' + A(t)u = 0$ *is oscillatory at* $t = \infty$.

(ii) $v'' + D(t)v = 0$ *is oscillatory at* $t = \infty$.

(iii) $\int^{\infty} B(t)\, dt = \infty$.

(iv) $\int^{\infty} C(t)\, dt = \infty$.

Cheng also shows that many other aspects of this problem are subject to analysis by such techniques. For example, if the systems focal point $\hat{\mu}(\alpha)$ is defined by $y(\alpha) = z(\alpha) = 0 = y'(\beta) = z'(\beta)$, then Cheng shows that $\hat{\mu}(\alpha)$ exists if and only if $\hat{\eta}(\alpha)$ exists but one necessarily has $\hat{\mu}(\alpha) < \hat{\eta}(\alpha)$. While satisfactory Sturm-type comparison theorems are notably lacking for 2-2 conjugate points, such theorems do exist for systems conjugate and focal points. Specifically, an increase in the coefficients A, B, C, or D will decrease $\hat{\mu}(\alpha)$ and $\hat{\eta}(\alpha)$.

A basic question is whether such techniques can be useful in studying equations of order 2n when $n > 2$. For example $y^{(vi)} + p(t)y = 0$ can readily be represented as a system of the form $\underline{x}'' = P(t)\underline{x}$ in E^3. While some of the techniques used to study systems conjugate points can be extended to this setting, the additional degrees of freedom severly complicate the theory based on rotation and saddle points. Thus the problem of establishing criteria for the existence of n-n conjugate points for nonselfadjoint equations of order 2n remains an open and challenging one.

References

1. J. Barrett, Systems-disconjugacy of a fourth-order differential equation, Proc. Amer. Math. Soc. 12(1961), 205-213.

2. S. Cheng, Systems-conjugate and focal points of fourth order nonselfadjoint differential equations, to appear.

3. A. Edelson and K. Kreith, Upper bounds for conjugate points of nonselfadjoint fourth order differential equations, to appear.

4. K. Kreith, A nonselfadjoint dynamical system, Proc. Edinburgh Math. Soc. 19(1974), 77-87.

5. ________, Rotation properties of a class of second order differential systems, J. Differential Eq. 17(1975), 395-405.

6. ________, Nonselfadjoint fourth order differential equations with conjugate points, Bull. Amer. Math. Soc. 80(1974), 1190-1192.

7. ________, Rotation properties of adjoint pairs of differential systems, to appear.

8. W. Leighton and Z. Nehari, On the oscillation of solutions of selfadjoint linear differential equations of fourth order, Trans. Amer. Math. Soc. 89(1958), 325-377.

9. W. Whyburn, On selfadjoint ordinary differential equations of the fourth order, Amer. J. Math. 52(1930), 171-196.

POINTWISE ERROR BOUNDS FOR THE EIGENFUNCTIONS OF ONE-DIMENSIONAL SCHRÖDINGER OPERATORS

Tassilo Küpper

1. Introduction

Let L be a one-dimensional Schrödinger operator given by $Lu := -u'' + Q\cdot u$ for all u in its domain of definition $D_L \subseteq L^2(-\infty,\infty)$. We assume that L is bounded from below and that its initial spectrum consists of simple eigenvalues $\lambda_1 < \lambda_2 < \dots$ with corresponding normalized eigenfunctions denoted by $u_1, u_2, \dots$. Additional assumptions on the potential Q will be required below.

Error estimates for the eigenfunctions of operators over compact intervals have been proved in [5]. Here we show how these methods can be applied to Schrödinger operators on the unbounded interval $]-\infty,\infty[$; similar results hold for semi-infinite intervals.

We shall use the fact that the eigenfunction u_j is almost concentrated within a compact interval $[a,b]$. In addition we use the fact that outside $[a,b]$ error bounds can often be given in terms of the potential Q (see Titchmarsh [8], Bazley-Fox [3]). Schröder's [6] "Monotoniesatz" is the main tool used to prove the results. Since $L-\lambda_j$ is not positive definite, there does not exist a function ψ with $(\psi-u_j)(x) \geq 0$, $(L-\lambda_j)\psi(x) > 0$ $(-\infty < x < \infty)$. Hence this theorem cannot be applied immediately. Therefore the interval $]-\infty,\infty[$ will be divided into (at least $j+2$) subintervals $]-\infty,a]$, $[a,t_1]$, $[t_1,t_2],\dots$, $[t_m,b]$, $[b,\infty[$ such that (for each of these subintervals I) the restriction of $L-\lambda_j$ to functions having support in I is positive definite. Consequently the "Monotoniesatz" can be applied to each of these restrictions.

Let v_j be a normalized approximation to u_j. Our error bound will be given as an upper bound $\psi \geq |u_j - v_j|$ *). The function $\psi|[t_i, t_{i+1}]$ $(i=0,\dots,m+1;\ t_o = a,\ t_{m+1} = b)$ is calculated as the solution of a positive definite boundary value problem involving the operator $L - \lambda_j$ as well as upper bounds for the residual $|(L-\lambda_j)v_j|$ and for $|u_j - v_j|$ at t_i, t_{i+1}. These terms can be evaluated numerically using upper and lower bounds to the eigenvalues. Outside $[a,b]$ we use Titchmarsh's estimates. We note, however, that our method also works in the infinite intervals $]-\infty, a]$, $[b, \infty[$, if the asymptotic behaviour of the eigenfunctions is a priori known.

2. The main result

Assume that the following assumptions hold; methods for their verification will be discussed later. Let j be a fixed positive integer.

(A1) Let $\underline{\lambda}_i, \overline{\lambda}_i$ be lower and upper bounds to λ_i $(i=1,\dots,j+1)$.

(A2) Let there be functions f_1, f_2, f_3, f_4 such that

$$\begin{aligned} u_j(a)\, f_1(x) &\leq u_j(x) \leq u_j(a) f_2(x) \qquad (x \leq a) \\ u_j(b)\, f_3(x) &\leq u_j(x) \leq u_j(b) f_4(x) \qquad (x \geq b)\ . \end{aligned} \tag{2.1}$$

(A3) Let $t_o = a < t_1 < \dots < t_m < t_{m+1} = b$ be a partition of $[a,b]$ such that all operators M_i $(i=0,\dots,m)$ are inverse positive; that is $M_i u \geq 0 \Rightarrow u \geq 0$. Here $M_i : C_2[t_i, t_{i+1}] \rightarrow$ $\rightarrow \mathbb{R}[t_i, t_{i+1}] \cap C_o]t_i, t_{i+1}[$ is defined by

$$(M_i u)(t) := \begin{cases} u(t_i) & t = t_i \\ (L - \overline{\lambda}_j) u(t) & t_i < t < t_{i+1} \\ u(t_{i+1}) & t = t_{i+1} \end{cases}$$

*) Here $\geq$ denotes the pointwise order relation between real functions.

(A4) Let there be constants ε_i such that

$$|u_j - v_j|(t_i) \le \varepsilon_i \qquad (i=0,\dots,m+1) .$$

<u>Theorem:</u> Suppose that (A1),...,(A4) hold. Let ψ' be a solution of

$$M_i\psi \ge \hat{\Psi} \qquad (i=0,\dots,m) \quad ;$$

where $\hat{\Psi}$ is given by:

$$\hat{\Psi}(t) := \begin{cases} \varepsilon_i & t = t_i \\ |(L-\overline{\lambda}_j)v_j|(t) + |\overline{\lambda}_j - \underline{\lambda}_j||v_j|(t) & t_i < t < t_{i+1} \\ \varepsilon_{i+1} & t = t_{i+1} \end{cases}$$

Then the error estimate

$$|u_j - v_j|(t) \le \psi(x) \qquad (a \le t \le b) \tag{2.2}$$

holds.

<u>Remark:</u> The error estimate (2.2) together with (2.1) provides an error bound for $|u_j - v_j|$ in the whole interval $]-\infty,\infty[$.

3. Proof and Comments

The residual can be estimated as follows:

$$\begin{aligned} |(L-\lambda_j)(v_j - u_j)| &= |(L-\lambda_j)v_j| \\ &\le |(L-\overline{\lambda}_j)v_j| + |\overline{\lambda}_j - \underline{\lambda}_j||v_j| . \end{aligned}$$

With regard to assumptions (A1),...,(A4), the proof is an immediate consequence of the "Monotoniesatz" [6]. Rather than carrying out the proof in detail I prefer to discuss methods for the verification of these assumptions.

<u>ad (A1):</u> Upper bounds $\overline{\lambda}_j$ as well as approximations v_j can be obtained by the Ritz-procedure. Methods to calculate lower bounds have been developed by Bazley and Fox [1].

<u>ad (A2):</u> If $\tilde{Q} := Q - \overline{\lambda}_j$ is positive and steadily increasing to infinity for $x \ge b$ and if $u_j(b)$ is positive, Titchmarsh [8] proves

(for $x \geq b$) :

$$\frac{2\,u_j(b)\exp[-(x-b)(\tilde{Q}(x+1))^{1/2}]}{1+[1+1/\tilde{Q}(x+1)]^{1/2}} \leq u_j(x) \leq u_j(b)\exp\left[-\int_b^x \tilde{Q}(y)^{1/2}\,dy\right]$$

A similar bound can be given for $x \leq a$.

In the case of potentials corresponding to one Electron Molecular Systems (which do not increase to infinity) Bazley and Fox [3] have shown how bounds for the semi-infinite intervals $]-\infty,a]$, $[b,\infty[$ can be calculated.

Such bounds for u_j in $]-\infty,a]$, $[b,\infty[$ can also be derived by an application of the "Monotoniesatz" if the asymptotic behaviour of u_j (say $\lim_{x\to\infty} u_j(x)/x^k = 0$) is a priori known. In that case, any positive solution ψ of $(L-\overline{\lambda}_j)\psi(x) > 0$ $(x \geq b)$ with $\psi(b) \geq |u_j|(b)$ and $\lim_{x\to\infty} \psi(x)/x^k > 0$ fulfills $|u_j|(x) \leq \psi(x)\ (x \geq b)$.
For detailed information concerning these techniques see Stoss [7].

ad (A3): Normally it will be possible to choose an equidistant partition in $[a,b]$. It has been shown in [5], that it suffices to select $m = j-1$ points t_i. Difficulties might occur if the eigenvalues are close together; in that case it will help to increase m.

ad (A4): For $h_1, h_2 \in C^2]-\infty,\infty[$ with $\gamma_i := [h_1h_2' - h_1'h_2](t_i) \neq 0$ we define

$$g_i(t) := \frac{1}{\gamma_i} \begin{cases} h_1(t)h_2(t_i) & t \leq t_i \\ h_1(t_i)h_2(t) & t \geq t_i \end{cases} \tag{3.1}$$

Assume that $g_i \in L^2(-\infty,\infty)$. Then a partial integration yields:

$$\begin{aligned} 0 &= \int_{-\infty}^{\infty}[g_i(L-\lambda_j)u_j](t)\,dt \\ &= \int_{-\infty}^{\infty}[g_i(L-\lambda_j)(u_j-v_j)](t)\,dt + \int_{\infty}^{\infty}[g_i(L-\lambda_j)v_j](t)\,dt \\ &= (u_j-v_j)(t_i) + \int_{-\infty}^{\infty}[(u_j-v_j)(L-\lambda_j)g_i](t)\,dt + \int_{-\infty}^{\infty}[g_i(L-\lambda_j)v_j](t)\,dt \end{aligned}$$

Consequently:

$$|u_j-v_j|(t_i) \le \|u_j-v_j\| \ \|(L-\lambda_j)g_i\| + \|g_i\| \ \|(L-\lambda_j)v_j\|$$

$$\le \|u_j-v_j\| \{ \|(L-\overline{\lambda}_j)g_i\| + |\overline{\lambda}_j-\underline{\lambda}_j| \|g_i\| \}$$

$$+ \|g_i\| \{ \|(L-\overline{\lambda}_j)v_j\| + |\overline{\lambda}_j-\underline{\lambda}_j| \}$$

Thus upper bounds to $|u_j-v_j|(t_i)$ are available, if upper bounds to $\|u_j-v_j\|$ are known. These have been given in terms of lower and upper bounds to the eigenvalues by Weinberger [9] in the case of operators with discrete spectrum:

$$\|u_j-v_j\| \le \sqrt{2}\sqrt{1-\sqrt{\eta_j}}$$

where η_j is defined as:

$$\eta_1 := 1 - (\overline{\lambda}_1-\underline{\lambda}_1)/(\underline{\lambda}_2-\underline{\lambda}_1)$$

$$\eta_j := \left\{ 1 - \frac{\overline{\lambda}_j - \underline{\lambda}_j}{\underline{\lambda}_{j+1} - \underline{\lambda}_j} \right\} \left\{ 1 - \frac{(\overline{\lambda}_j - \underline{\lambda}_j)(\overline{\lambda}_{j-1} - \underline{\lambda}_1)}{(\overline{\lambda}_j - \overline{\lambda}_{j-1})(\underline{\lambda}_j - \underline{\lambda}_1)} \right\} , \quad j > 1$$

Estimates of the overlap $(u_j,v_j) = 1 - \|u_j-v_j\|^2/2$ in the case of operators with a partly continuous spectrum have been derived, for example, by Hoffmann-Ostenhof [4].

The longest part in the estimates of $|u_j-v_j|(t_i)$ is usually given by $\|(L-\overline{\lambda}_j)g_i\|$. It is always possible to choose $g_i(t) := 1/2\exp(-|t-t_i|)$. On the other hand, since both $\|g_i\|$ and $\|(L-\overline{\lambda}_j)g_i\|$ should be small, a more careful choice of g_i is in general commendable. This can be achieved if $g_i(t) = G(t_i,t)$ where G is the Green's function of $L-\lambda$ for some suitable $\lambda < \lambda_j$. Here we recommend to calculate approximately the functions h_1,h_2 in (3.1) as L^2-solutions of

$$(L-\lambda)h_1(t) = 0 , \quad h_1(t_i) = 1 \qquad (t \le t_i)$$

$$(L-\lambda)h_2(t) = 0 , \quad h_2(t_i) = 1 \qquad (t \ge t_i)$$

for some $\lambda < \lambda_j$.

The functions h_1, h_2 can also be calculated by the Ritz procedure which leads to a nonhomogeneous system of linear equations. This gives $\| (L-\overline{\lambda}_j) g_i \| \approx |\lambda-\overline{\lambda}_j| \ \| g_i \|$.

We mention that (A1), (A2), (A3), (A4) can similarly be verified for slightly more general operators $Lu = -(pu')' + Qu$ and appropriate p.

For a symmetric potential Q the following special case with $\varepsilon_o = \varepsilon_1 = 0$ holds:

<u>Corollary:</u> Suppose that (A1), (A2) hold and choose $m = 0$. If ψ is a solution of

$$M_o\psi \geq \begin{cases} 0 & t = -b \\ (L-\overline{\lambda}_1)v|(t) + |\overline{\lambda}_1 - \underline{\lambda}_1| \ |v(t)| & -b < t < b \\ 0 & t = b \end{cases},$$

then there exists an eigenfunction u^* of L corresponding to λ_1 with $u^*(b) = v(b)$ and $|u^*-v|(t) \leq \psi(t)$ $(-b \leq t \leq b)$.

<u>Remarks:</u>

An analogous theorem holds for PDE's when L is a second order partial differential operator, defined on $D_L \subseteq L^2(\mathbb{R}^n)$. The $\mathbb{R}^n$ can be divided in a finite number of subdomains D_i such that $L-\lambda_j$ restricted to functions with support in D_i is positive definite. Further the assumptions (A1),(A2),(A3) can be fulfilled. Two difficulties, however, arise in the verification of the analogue to (A4). Again $|u_j-v_j|$ has to be estimated on the boundary of each D_i, which is no longer a finite set. This can be done if a suitable Green's function is known. On the other hand, there does not exist a similar procedure to construct an approximation to the Green's function as in the case of ordinary differential operators.

References:

[1] N.W. Bazley - D.W. Fox: Lower Bounds for Eigenvalues of Schrödinger's Equation.
The Physical Review 124, 483-492, 1961.

[2] N.W. Bazley - D.W. Fox: Error Bounds for Approximations to Expectation Values of Unbounded Operators.
J. Math. Phys. 7, 413-416, 1966.

[3] N.W. Bazley - D.W. Fox: Bounds for Eigenfunctions of One-Electron Molecular Systems.
Internat. J. of Quantum Chem., III, 581-586, 1969.

[4] T. Hoffmann-Ostenhof - M. Hoffmann-Ostenhof: Variational bounds to the overlap.
Chemical Physics Letters 31, 277-280, 1975.

[5] T. Küpper: Pointwise lower and upper Bounds for Eigenfunctions of ordinary differential Operators.
Battelle Report No. 96, 1975.

[6] J. Schröder: Monotonie-Eigenschaften bei Differentialgleichungen.
Lecture Notes 305, Springer 1973.

[7] H.J. Stoss: Monotonie-Eigenschaften bei Differentialgleichungen über nichtkompaktem Grundbereich.
Num. Math. 15, 61-73 (1970).

[8] E.C. Titchmarsh: Eigenfunction Expansions associated with second order differential Equations.
Clarendon Press Oxford 1946.

[9] H.F. Weinberger: Error Bounds in the Rayleigh-Ritz Approximation of Eigenvectors.
Journal of Research, Nat. Bureau of Standards, 64B, 217-225, 1960.

STABILITY OF SHOCK WAVES

Lorenzo Lara-Carrero

A magnificent example of stability problems of shock waves is provided by the theory of a single convex Hyperbolic Conservation Law in one space variable. This theory is due mainly to the work of Peter D. Lax. The point of view taken in this paper was started by David G. Schaeffer, Marty Golubitsky and John Guckenheimer (see references at end). The author himself developed the case with periodic initial conditions which is exposed here (more extense version will appear in Advances in Mathematics).

A single hyperbolic conservation law is a first order partial differential equation of the form:

$$u_t + f(u)_x = 0. \tag{1}$$

Here f is a smooth uniformly convex function. We usually think of u as a density and of f as a flux. We want to solve the Cauchy problem with initial data ϕ, smooth and periodic. It is well known that discontinuities arise for finite time, due to the intersection of contradictory signals coming from different initial points. This is a fact of life for non-linear equations. Nature makes a selection of this signals giving rise to shock waves. Mathematically, a shock wave is a discontinuity curve $x = s(t)$ along which a solution $u(x,t)$ has only a jump discontinuity and satisfies:

$$\frac{ds}{dt} = \frac{f(u+) - f(u-)}{u+ - u-} \qquad \text{(Shock Condition)}$$

$$u+ < u- \qquad \text{(Entropy Condition)}$$

We have denoted by u+ and u- the corresponding right and left limit of $u(x,t)$ at the point $x = s(t)$. As the names of this conditions suggest they have a physical and very intuitive meaning (see Lax, Monthly).

Regularity Theorem (3): For initial conditions in a subset of $C^\infty(S^1)$ which is a countable intersection of open dense sets, the Cauchy problem has a unique piecewise smooth solution with discontinuities only along shock waves. Moreover shocks collide only two at a time, and there is only a finite number of them starting in each period.

We remark that by a discontinous solution we mean a measurable function which is a weak solution in the sense of distributions. This regularity theorem, whose proof we omit, is taken as background to the ensuing discussion.

1. Singularity Theory

In 1957 Peter D. Lax generalized the following minimization scheme, first introduced by E. Hopf (1950) in the case where $f(u) = u^2 / 2$, for solving explicitly the Cauchy problem for any Conservation Law with convex smooth flux $f(u)$. If ϕ is the initial condition define:

$$G(x,t,y) = t.g(\frac{x-y}{t}) + \int_0^y \phi(s)ds$$

with $g(s) = sb(s) - f(b(s))$ and $b(s)$ the inverse function of $a(u)$. Now let $y_0(x,t)$ be a minimizing point of $G(x,t,.)$. The following theorem was proved by Lax.

Theorem 1.1: For fixed t, the funtion $G(x,t,.)$ has a unique minimum for all x except in a denumerable set of values of x. We also have $y_0(x,t) \leq y_0(z,t)$ if $x < z$. More importantly the function

$$u(x,t) = b(\frac{x-y_0(x,t)}{t}) \tag{2}$$

which is defined almost surely is a weak solution of the conservation law with initial value ϕ, a bounded measurable function.

We remark that weak solutions which are piecewise smooth satisfy necessarily the shock and entropy conditions along its discontinuity curves. Thus our regularity theorem above is about the same class of solutions as theorem 1.1. Indeed it is through the use of this scheme that the regularity theorem was proved (Schaeffer 1973 and Lara 1975). A classification of points in the x,t plane (t positive) was used which will be useful.

Case 1: Unique minima.

We single out the following two possibilities

$U = \{(x,t) | G(x,t,.)$ has a unique absolute nondegenerate minimum, i.e. $\frac{\partial^2 G}{\partial y^2} \neq 0\}$

$\Gamma_0^{(f)} = \{(x,t) | G(x,t,.)$ has a unique absolute degenerate minimum but $\frac{\partial^4 G}{\partial y^4} \neq 0\}$.

The set U is open and on it the minimization defines a function $y_0(x,t)$ and through (2) a function $u(x,t)$ which satisfied equation (1.1) in the classical sense (Schaeffer 1973, Lemma 1.1). On the other hand $\Gamma_0^{(f)}$ consists of isolated points and the superscript (f) is used to suggest that shocks are formed at these points (Schaeffer 1973, Lemma 1.4).

In this first case there is one possibility left and that is that $\frac{\partial^2 G}{\partial y^2} = 0$ and at the same time $\frac{\partial^4 G}{\partial y^4} = 0$.

Case 2: There are two or more minima.

We again single out two possibilities.

$\Gamma_1 = \{(x,t)|G(x,t,.)$ has precisely two minimas and at both $\frac{\partial^2 G}{\partial y^2} \neq 0\}$.

$\Gamma_0^{(c)} = \{(x,t)|G(x,t,.)$ has precisely three minima at all of them $\frac{\partial^2 G}{\partial y^2} \neq 0\}$.

Schaeffer (Lemma 1.2) shows that Γ_1 is a union of smooth curves across which the minimizing function has a jump discontinuity. Curves $\gamma(t)$ starting at points of $\Gamma_0^{(f)}$ continuing through points of Γ_1 and possibly colliding with another such curve at points of $\Gamma_0^{(c)}$ (Schaeffer 1973, Lemma 1.3) are shock waves. Points of collision are isolated.

To prove the regularity theorem all other possibilities left out of this classification are shown impossible. That is the half plane H is decomposed in the following way:

$$H = U \cup \Gamma_1 \cup \Gamma_0^{(f)} \cup \Gamma_0^{(c)}$$

for all initial data outside of a set of first category (Note that each of the sets defined in our classification depends on the initial data).

2. Asymptotic Behaviour of Shocks.

In order to understand which properties of solutions are stable, it is best to start by an analysis of the asymptotic behaviour of the shock waves. This will lead us naturally to the pertinent definition of stability (Definition 4.1). We introduce the following notation. For smooth periodic functions ϕ define:

$$m = \frac{1}{2\pi}\int_0^{2\pi} \phi(s)ds$$

$$\tilde{\phi}(x) = \int_0^x \phi\,(s)ds - mx$$

Note that $\tilde{\phi}$ is again periodic and smooth.

Theorem 2.1: If $\tilde{\Phi}$ has a unique absolute non-degenerate minimum x_0 in the interval $[0,2\pi]$ then as t goes to infinity there is precisely one shock s(t) such that every other shock is equal to it modulo 2π and such that

$$s(t) = x_0 + \pi + f'(m).t + o(1)$$

(Here as usual o(1) is a function which goes to zero as t goes to infinity).

Proof: We do the proof in three steps.

Reduction to the case $m = f(m) = f'(m) = 0$: Let u be a weak solution of (1) with Cauchy data ϕ. Then $v(x,t) = u(x-ct,t) - m$, with $c=f'(m)$, is a weak solution of

$$v_y + \tilde{f}(v)_x = 0$$

$$v(x,0) = \phi(x) - m;$$

where $\tilde{f}(v) = f(v+m) - f(m) - v.f'(m)$. By a simple computation one may check that weak solutions always satisfy the entropy and shock conditions along its discontinuities. Therefore a curve s(t) is a shock of u if and only if s(t) - c.t is a shock for v.

There is only one shock: The following can be made rigorous by a tedious $\epsilon - \delta$ argument. For large t the term $tg(\frac{x-y}{t})$ becomes very small thus leaving the integrated term as the predominant one. Then $y_0(x,t)$ tends to x_0 or to $x_0 + 2\pi$ as t goes to infinity. Using the non-degenracy of x_0 one may choose an appropiate neighborhood U of it and another U' of $x_0 + 2\pi$ such that auxiliary <u>smooth</u> functions $y_1(x,t)$ and $y_2(x,t)$ exist with the following properties: y_1 is the unique non-degenerate minimum of $G(x,t,.)$ in U, i.e. y_1 belong to U, similarly for y_2 in U'. These auxiliary functions exist only for t greater than or equal some bound. It is clear that for large t we will have $y_0 = y_1$ or $y_0 = y_2$. And in case $x = s(t)$ is a shock we will have both equalities valid at the same time. On the other hand we take advantage of the smoothness of the auxiliary functions to take the derivative of

$$L(x,t) = G(x,t,y_1(x,t)) - G(x,t,y_2(x,t)).$$

Note that for a fixed large t, a point x belongs to a shock if and only if it is a root of L, i.e. $L(x,t) = 0$. But we may compute:

$$\frac{\partial}{\partial x} L(x,t) = b(\frac{x - y_1}{t}) - b(\frac{x - y_2}{t}-) > 0$$

Thus for large t shocks are unique.

<u>Asymptotic Formula</u>: Let $x = s(t)$ denote the unique shock there exists for large t. Then for large t, $G(x,t,y_1) = G(x,t,y_2)$ where the y's denote the auxiliary functions defined above. Then using the Taylor ex pansion of G as a function of y we may write:

$$g''(o) \frac{(x - y_1)^2}{2t} + \Phi(y_1) = g''(o) \frac{(x - y_2)^2}{2} + \Phi(y_2) + 0(\frac{1}{t^2})$$

where Φ is the integral from 0 to y of ϕ.

Rearranging this equation we can write:

$$x = \frac{y_1 + y_2}{2} + \frac{\Phi(y_2) - \Phi(y_1)}{g''(o)(y_2 - y_1)} .t + 0(\frac{1}{t}).$$

Now substitute in the denominator $y_2 - y_1 = t.\ a(\phi(y_1)) - a(\phi(y_2))$ and compute the limit of the second term. This limit is $\frac{\Phi'(x_0)}{\phi'(x_0)}$, but recall that $\Phi'(x_0) = 0$ since x_0 is a minimum of Φ. To compute this limit we have used the already noted fact: y_0 tends to x_0 or to $x_0+2\pi$. Which is the same as to say y_1 tends to x_0 and y_2 tends to $x_0+2\pi$ as t goes to infinity.

Thus: $x = x_0 + \pi + 0(\frac{1}{t})$

which is the desired asymptotic formula in the case $m = f(m) = f'(m) = 0$. Now go back through the reductions done in the first step of this proof to get the complet asymptotic formula.

This finishes the proof.

3. What can go wrong?

From our asymptotic analysis of shock waves we can not expect to have stability in the metric sence. Even very small perturbations of the initial data will produce a change in the slope of the asymptotic limit of the shocks. Nevertheless uniqueness of the shock for large time is already a sort of weak asymptotic stability. If we are able to make an analysis of the geometry of shocks as they collide to form a unique one, we will be able to study their topological stability. This we do in the next section. Here we want to present an example of what can go wrong for small time.

Consider a conservation law with flow $f(u)$ such that $f(0) = 0$ and $f'(u)$ very small in the interval $(-1,1)$ being equal to 0 at $u = 0$ also.
Outside this interval $f(u)$ should grow quadratically. See figure 1. For simplicity assume f symmetric.

To give initial conditions showing non-stable behaviour, we make the following:

Definition: ϕ in $C^\infty(S^1)$ is a smooth step function if it is constant on a union of disjoint intervals and outside of them it has smooth steps, and at each interval where it is not constant it is monotone and has just one inflexion point (i.e. ϕ'' is strictly monotone and has only one zero).

Consider a smooth step function like the one in figure 2. To presserve symmetry suppose ϕ' satisfies $\phi'(y) = \phi'(y+\xi_2) = \phi'(y + \xi_3)$ for $\xi_1 - \varepsilon \leq y \leq \xi_1 + \varepsilon$, so that the smooth steps differ only by a const (1/3), and moreover suppose ϕ' is anti-symmetric in the ε-neighborhood of ξ_1 about ξ_1. The ξ_1 are inflexion points.

Suppose also that the smooth step occur in ε-neighborhoods of ξ_1, ξ_2, ξ_3, ξ_4 respectively, with ε small compared with the size of the intervals where ϕ is constant.

Now consider ϕ-m with m = -1/2. This makes the constant value of ϕ-m equal to $\frac{1}{2}$, $\frac{1}{6}$, $-\frac{1}{6}$, $-\frac{1}{2}$ respectively. The inflexion points corresponding to the decreasing smooth steps will give rise to three shocks $s_i(t)$, i=1,2,3. Due to the symmetry of $f(u)$ and the antisymmetry of ϕ' around each ξ_i we will get, using $\frac{ds}{dt} = \frac{f(u_+) - f(u_-)}{u_+ - u_-}$, the following slopes:

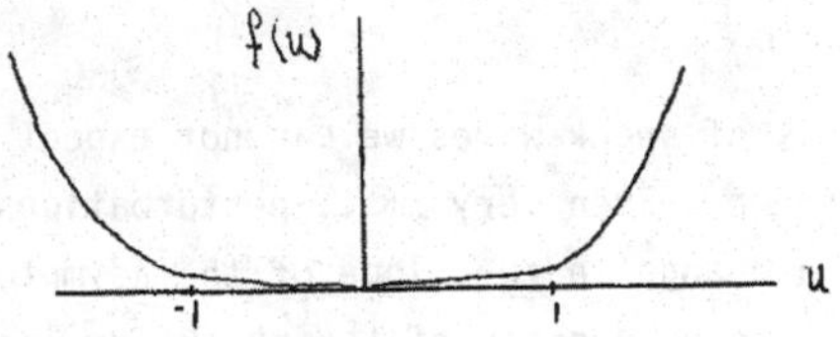

Figure 1

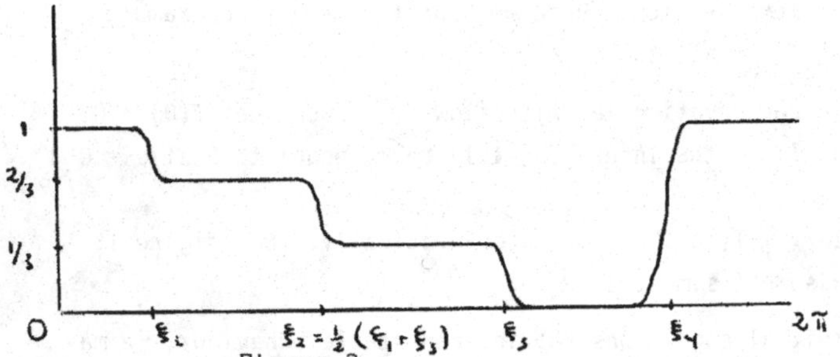

Figure 2

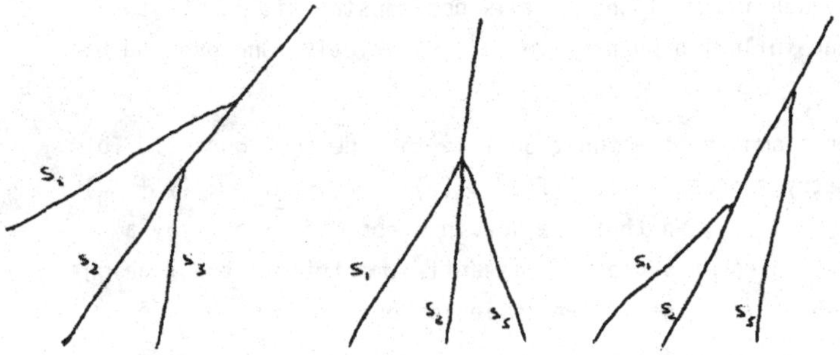

Collisions of Shock Waves

$$\frac{ds_2}{dt} = 0, \quad \frac{ds_3}{dt} = - \frac{ds_1}{dt}$$

for all time t for which s_1, s_2, s_3 are well defined (before collisions). Moreover we also have $[s_1(t^*) + s_3(t^*)]1/2 = \xi_2$ due to the symmetry of f and the relationship $\phi'(y) = \phi'(y + \xi_3)$ for y near ξ_1, where t* is the time at which both s_1 and s_3 start:

$$t^* = \frac{1}{1 + \phi'(\xi_1)a'(u_1)} = \frac{1}{1 + \phi'(\xi_3)a'(u_3)}$$

with $u_1 + m = \phi(\xi_1)$, $u_3 + m = \phi(\xi_3) = \phi(\xi_1) + \frac{2}{3}$ so $a'(u_1) = a'(u_3)$.

Then since $\xi_2 = \frac{1}{2}(\xi_1 + \xi_3)$ and since after a certain time shocks s_1 and s_3 must have non-zero slope (corresponding to the regions of constant value of ϕ-m) we will get a collision of all three shocks at the same point.

It is clear that this is a very unstable situation in the sense that changing m just a little we will not get a collision of the three shocks simultaneously.

If m = 1 - 2/3 we will get $\frac{ds_2}{dt} > \frac{ds_3}{dt}$ but both near zero, on the other hand $\frac{ds_1}{dt} > 1$. So for ϕ-m with m = 1-2/3 we get that s_1 collides with s_2 first then s_3. For m = -1-1/3 we get s_3 collides first with s_2 then s_1.

Note that these three ways in which collisions may occur are not diffeomorphically equivalent (the precise meaning of this equivalence is given in the next section). We shall show that for a large set of initial conditions shock waves are stable. Note that examples m = -1-1/3 and m = 1-2/3 will correspond to different equivalence classes and example m = -1/2 belongs to the exceptional set of initial conditions.

4. Infinitesimal Stability implies Stability.

The auxiliary function $G(x,t,y)$ introduced in section 1 was very useful in the asymptotic analysis of shocks. Now to study their stability it will be convenient to make the change of variables $y = x - a(u)t$, which is order preserving. Note that this change of variables comes out of rewriting equation 2. Thus, we define.

$$F(x,t,u) = G(x,t,x - a(u)t) = t(a(u)u - f(u) + \Phi(x - a(u)t)$$

where $\Phi(y)$ is the integral from 0 to y of ϕ.

It is clear that weak solutions as defined by equation 2 minimize $F(x,t,.)$. Conversely, if $u(x,t)$ minimizes $F(x,t,.)$ then it is a weak solution. Note that through the function F we obtain directly a solution of the Cauchy problem for the conservation law.

Definition 4.1: A solution u_1 of the conservation law corresponding to an initial condition ϕ_1 is called stable if and only if there is a neighborhood N of ϕ_1 in the space of smooth periodic functions such that for any ϕ in N there is a diffeomorphism h of $S^1 \times [0,\infty)$ onto itself mapping the shock set of u(the solution corresponding to ϕ) onto that of u_1. We also say the shock waves of u_1 are stable, to mean u_1 is stable.

With this definition of stability we can now state our main theorem.

Stability Theorem: Shock waves are stable for all smooth periodic initial conditions, except for a subset of these initial conditions which is of first category.

We shall sketch the proof of this theorem. A complete version is a part of the authors doctoral dissertation (M.I.T. 1975). We note in passing that an stochastic version of the stability theorem is also true. Defining a Gaussian probability measure on the space of smooth periodic functions we were able to show: shock waves are stable with probability one.

The main idea of the proof is to reduce stability of solutions u to stability of the associated function $F(x,t,u)$. Then the theory of stability of maps as developed by J. Mather and his followers is used. To use this theory we have to consider the map

$$\bar{F}(x,t,u) = (x,t,F(x,t,u)).$$

This is helpful in two ways: we get a map from $S^1 x[0,\infty)xR$ into itself (only in the case where ϕ has mean value 0) and the following sufficient condition for stability of solutions may be stated. Consider the class of smooth maps form $S^1 x[0,\infty)xR$ into itself of the form

$$G(x,t,u) = (h(x,t), H(x,t,u)).$$

Note that $\bar{F}$ belongs to this class of functions. Then stability of $\bar{F}$ in this class of functions implies stability of the solution u, and we reduce our proof to checking stability of $\bar{F}$. To do this we verify the conditions for infinitesimal stability and use a version of the famous theorem of Mather "infinitesimal stability implies stability" due to Latour which is specially adapted to our class of maps.

An important technical detail can not be ommited. The map ϕ goes into $\bar{F}$ is not continuous in the Whitney C^∞ topology due to non-compactness of their domains. This difficulty is overcome by making Mather's argument for the restriction of these maps to bounded t and u. Boundedness of u is no restriction at all, for the range of u's which is of interest is anyway bounded: bounded initial conditions give rise to bounded solutions. The restriction to bounded time gives us stability of the shock waves for bounded time and then we "paste" these shocks with the unique one existing for large time. This gives us stability for all time.

The general theory described above can not be applied directly to the case where the initial condition has mean different from 0, for then the map $\bar{F}$ does not map S^1 into itself as a function of x. A special argument has to be made which is interesting in itself as an introduction to the Mather theory of stability.

Let $Z = S^1 \times [0,\infty)$ and suppose we can find smooth functions $\Psi_m(p,w)$, $\Lambda_m(p,u)$, $g_m(p)$, with p in Z and u,w any real numbers, such that for m small.

1. $F_0(p,u) = \Psi_m(p, F_m(g_m(p), \Lambda_m(p,u))$

2. Both $\Psi_m(p,.)$, $\Lambda_m(p,.)$ are strictly increasing.

3. The function $g_m(.)$ is a diffeomorphism of Z onto itself.

Here $F_m(x,t,u) = F(x,t,u) + m(x - a(u)t$; m is a real number.

It is easy to check that if all these condtions are satisfied then g_m maps shocks corresponding to ϕ onto shocks corresponding to $\phi + m$ and we would have proved stability of the solution u corresponding to ϕ with respect to small changes in the mean value of ϕ.

We shall now construct such functions but only satisfying conditions 1 through 3 for p restricted to $S^1 \times [0,T]$, for any $T > 0$. The map g_m that we get will be a diffeomorphism of $S' \times [0,T]$ onto its image, mapping shocks which lie in its domain into shocks. Moreover g_m, for small m, will be near the identity in the following norms $g \to \|g\|_\infty, \|\frac{\partial g}{\partial t}\|_\infty, \|\frac{\partial g}{\partial x}\|_\infty$. This fact allows the extension of g_m to a diffeomorphism of $S^1 \times [0,\infty)$ onto itself preserving shocks (by using the asymptotic stability).

To construct Ψ_m, Λ_m and g_m we now get differential equations they must satisfy. We treat m as a smooth parameter.

From the first condition, relating all three unknown functions with F_0 and F_m, we get:

$$0 = \frac{\partial \Psi_m}{\partial m} + \frac{\partial \psi_m}{\partial w} \left\{ \frac{\partial F_m}{\partial m} + \frac{\partial F_m}{\partial x} \frac{\partial g_m^1}{\partial m} + \frac{\partial F_m}{\partial t} \frac{\partial g_m^2}{\partial m} + \frac{\partial F_m}{\partial u} \frac{\partial \Lambda_m}{\partial m} \right\}. \tag{4.1}$$

Now the function $\frac{\partial F_m}{\partial m} = x - a(u)t$ is known to us. <u>Suppose</u> we could find smoooth coefficients $a_m(p)$, $b_m(p)$, $c_m(p,u)$ and a function $\eta_m(p,w)$ such that

$$-\frac{\partial F_m}{\partial m} = a_m \frac{\partial F_m}{\partial x} + b_m \frac{\partial F_m}{\partial t} + c_m \frac{\partial F_m}{\partial u} + \eta_m(p, F_m) \tag{4.2}$$

<u>This assumption is called infinitesimal stability</u>. We shall show in this special case that infinitesimal stability implies stability. This will provide an example of the more general theory due mainly to J.N. Mather.

It may be verified that indeed infinitesimal stability holds for F_m, m small, for ϕ in $C^\infty(S^1)$ outside of a set of first category and in the stochastic case with probability one (Lara-Carrero 1975)

So assume (4.2) holds. Then from (4.1), dividing by $\frac{\partial \Psi_m}{\partial w}$ (which we want to be positive), we get:

$$\frac{\partial g_m^1}{\partial m}(p) = a_m(g_m(p)) \tag{4.3}$$

$$\frac{\partial g_m^2}{\partial m}(p) = b_m(g_m(p)) \tag{4.4}$$

$$\frac{\partial \Lambda_m}{\partial m}(p,u) = c_m(g_{m},(p),\ \Lambda_m(p,u)) \tag{4.5}$$

$$\frac{\partial \Psi_m}{\partial m}(p,w) = \eta_m(p,w)\,\frac{\partial \Psi_m}{\partial w}(p,w) \tag{4.6}$$

We have as initial conditions: $g_0(p) = p$, $\Lambda_0(p,u) = u$, $\Psi_0(p,w) = w$. Equations 4.3 to 4.6 form a system of ordinary and partial differential equations depending smoothly on the parameter p. Thus for (p,t) in $S^1 \times [0,T]$, with T finite we may indeed find a local solution of this system by the usual theory of differential equations. This shows that infinitesimal stability implies stability. This finishes our sketch of the proof of our main theorem.

Acknowledgements: This work is part of the author's Ph.D. dissertation done at M.I.T. under the helpful guidance of Professor David G. Schaeffer. A fellowship from the Instituto Venezolano de Investigaciones Científicas, Caracas, supported the author during that time.

References

1. Golubitsky, M and Schaeffer, D.G. (1975), Stability of Shock Waves for a Single Conservation Law, Advances in Math. 16, 65-71.

2. Latour, F. (1969), Stabilité des champs d'applications differentiables, généralization d'un theoreme de J.Mather. C.R. Acad. Sci. Paris 268 Ser A, 1331-1334.

3. Lara-Carrero, L. (1975), Hyperbolic Conservation Laws: Generic and Stochastic Regularity and Stability, PhD dissertation, Massachusetts Institute of Technology.

4. Lax, P.D.(1954), Weak solutions of nonlinear hyperbolic equations and their numerical computation, Comm.Pure Appl.Math. 7, 159-193.

5. Lax, P.D.(1957), Hyperbolic Systems of Conservation Laws II, Comm.Pure Appl.Math. 10, 537-566.

6. Lax, P.D.(1972), The Formation and Decay of Shock Waves, Amer. Math. Monthly 79, 227-241.

7. Schaeffer, D.G. (1973), A regularity theorem for conservation laws, Advances in Math. 11, 368-386.

REMARKS ON L_2 SOLUTIONS

N.G. Lloyd

Much work has been done on the classification of linear differential equations according to the number of independent square integrable solutions which they possess. Here we consider the class $\mathcal{S}$ of equations

$$y^{(n)} + a_{n-1}(t)\,y^{(n-1)} + \ldots + a_o(t)y = 0, \qquad (1)$$

where $a_i(t)$ $(i=0,1,\ldots,n-1)$ are complex-valued continuous functions of the real variable t. Equation (1) is identified with the point $a = (a_{n-1},\ldots,a_o)$ of the linear space of n-tuples of continuous functions $\mathbb{R} \to \mathbb{C}$.

Let C_k be the subset of $\mathcal{S}$ consisting of equations with k independent L_2 solutions (and no more); let S_k be the subset with at least k such solutions. It is natural to ask if $\mathcal{S}$ can be topologised in such a way that the C_k are open sets. Our first remark is that if we impose the natural requirement that $\mathcal{S}$ must be connected, then the C_k cannot all be open (nor can they all be closed!); that is, we can ensure that the C_k are open only at the expense of a disconnected space.

Let $\mathcal{S}$ be made into a topological space as follows. Define, for m=1,2,3,...,

$$p_m(a) = \max_{\substack{i=0,\ldots,n-1 \\ |t| \leq m}} |a_i(t)| .$$

These are seminorms on $\mathcal{S}$, and so generate a locally convex topology τ_1 on $\mathcal{S}$. In fact $\mathcal{S}$ is then metrisable and complete (that is, $\mathcal{S}$ is a Fréchet space).

A neighbourhood of $a\varepsilon\mathcal{S}$ contains a set of the form

$$\{b\varepsilon\mathcal{S}\,;|b_i(t)-a_i(t)| < \varepsilon,\ i=0,\dots,n-1,\ |t| \leq M\}$$

for some $\varepsilon>0$ and M.

The L_2 character of solutions is determined by the functions a_i in a neighbourhood of infinity.

<u>Proposition 1</u>. <u>For each</u> k, C_k <u>is dense in</u> $\mathcal{S}$, <u>and so is unbounded</u>.

Suppose that $a\varepsilon C_k$ has independent solutions $\phi_1,\dots,\phi_n$, of which $\phi_1,\dots,\phi_k$ are square integrable. With λ real and non-zero, let

$$a(\lambda) = (\lambda a_{n-1}(\lambda t),\dots,\lambda^n a_o(\lambda t)).$$

The solutions ψ of $a(\lambda)$ are related to the solutions ϕ of a by

$$\psi(t) = \phi(\lambda t).$$

The Wronskian $W(\psi_1,\dots,\psi_n)$ of $\psi_1,\dots,\psi_n$ satisfies

$$W(\psi_1,\dots,\psi_n)(t) = \lambda^{\frac{1}{2}n(n-1)}\, W(\phi_1,\dots,\phi_n)(\lambda t).$$

So $\psi_1,\dots,\psi_n$ are independent solutions of $a(\lambda)$. Clearly there is a one to one correspondence between the L_2 solutions of a and those of $a(\lambda)$; this means that $a(\lambda)\varepsilon C_k$ for $0<\lambda\leq 1$.

<u>Proposition 2</u>. <u>If</u> $a\varepsilon C_k$, <u>there is a path in</u> C_k <u>joining</u> a <u>to any given neighbourhood of</u> 0.

<u>Corollary</u> (1) C_o <u>is connected</u>. (2) <u>For each</u> k, $C_k\cup\{0\}$ <u>is connected</u>.

Let $\mathcal{A}$ be the collection of self-adjoint equations in $\mathcal{S}$ (with the a_i suitably smooth). It can be shown, using a standard representation of a self-adjoint differential operator, that if $a\,\varepsilon\mathcal{A}$, then the path of Proposition 2 remains in $\mathcal{A}$.

Now let $\mathcal{P}$ be the subset of $\mathcal{S}$ consisting of the equations with periodic coefficients. Floquet theory is now available to us; $a \in \mathcal{P} \cap C_k$ if and only if exactly k of its characteristic exponents have negative real parts (counting multiplicity).

Proposition 3. For each k, $\mathcal{P} \cap S_k$ is an open subset of $\mathcal{P}$.

Define $\mathcal{S}_1$ to be $\{a \in \mathcal{S};\ a_i$ bounded$\}$. Instead of the topology of compact convergence (τ_1) on $\mathcal{S}_1$, we could consider the topology of uniform convergence. Proposition 2 is unchanged, and the C_k are still unbounded; the density of C_k, however, does not now follow as before.

Finally we consider another topology on $\mathcal{S}$; it is defined by a neighbourhood system $\mathcal{N}$ for the origin O. A set U belongs to $\mathcal{N}$ if it contains one of the sets $U(\varepsilon,T)$:

$$U(\varepsilon,T) = \{b \in \mathcal{S};\ b = 0 \text{ for } |t| \geq T,\ |b| < \varepsilon \text{ for } |t| < T\}.$$

It may be checked that $\mathcal{N}$ does define a topology on the vector space $\mathcal{S}$, and that in this topology $\{a+U(\varepsilon,T);\ \varepsilon > 0,\ T > 0\}$ is a base of neighbourhoods of a. If the topology is denoted by τ_3, $(\mathcal{S},\tau_3)$ is certainly Hausdorff.

Proposition 4. In the topology τ_3, the sets C_k are open.

$(\mathcal{S},\tau_3)$ is clearly not connected; neither is it locally connected. The families $a(\lambda)$ of Proposition 2 are no longer paths.

References

1. J. L. Kelley, General topology (Van Nostrand, New York, 1955)

2. M. K. Kwong, 'L^p-perturbation of second order linear differential equations', Math. Ann. 215 (1975) 23-24.

3. A. P. Robertson and W. Robertson, Topological vector spaces (Cambridge University Press, 1964).

REGULARLY VARYING FUNCTIONS AND DIFFERENTIAL EQUATIONS

V. Marić and M. Tomić

1. The aim of this paper is to point out the possibility of the applications of a certain class of functions - which generally might be described as regularly varying functions in the sense of Karamata - in deriving asymptotics of solutions of some classes of ordinary differential equations. These functions are of frequent occurence in various branches of analysis and of probability theory but the authors feel that their usefulness in ordinary differential equations has been somewhat overlooked.

2. *Definition 1. A positive and measurable function* ρ *defined on* $[0,\infty)$ *is an o-regularly varying (o-RV) function if for all* $\lambda > 0$

$$\lim_{x\to\infty} \frac{\rho(\lambda x)}{\rho(x)} = \phi(\lambda)$$

where $0 < \phi(\lambda) < \infty$.

It is known that $\phi(\lambda) = \lambda^{\sigma}$. The number σ is the *index of regular variation of* ρ.

Definition 2. A positive and measurable function L defined on $[0,\infty)$ *is a slowly varying (SV) function if for all* $\lambda > 0$

$$\lim_{x\to\infty} \frac{L(\lambda x)}{L(x)} = 1.$$

It follows that $\rho(x) = x^{\sigma}L(x)$ and that a SV function is an o-RV function of index $\sigma = o$.

The basic result in the theory of o-RV functions is the following:

Representation theorem. For any o-RV function ρ *of index* σ *there exist a positive number B and bounded, measurable func-*

tions η *and* ε *converging to a finite number and zero respectively, such that for all* $x \geq B$

$$(1) \qquad \rho(x) = x^{\sigma}\exp\{\eta(x) + \int_B^x \frac{\varepsilon(t)}{t}\,dt\}.$$

The o-RV functions naturally fill in the orders of growth between those of any of two powers and of any two of the logarithmic scale generalizing thus both classes.

The o-regularly varying functions, including the slowly varying ones were introduced by J. Karamata in 1930 [1] (the term regularly increasing - "a croissance réguliere" was used for the former). He also obtained the basic results for the continuous case. These were later generalized to the measurable functions by T. Van Aardene-Ehrenfest, N.G. de Bruijn and J.Korevaar [2],[3].

For more complete information about the theory of the functions in question, see [4], [5], [6].

One extends the class of o-RV functions by the following

Definition 3. A positive measurable function g defined on (o,∞) *is O-regularly varying (O-RV) if for all* $o<x\leq x' \leq \lambda x$ $(\lambda > 1)$

$$o < m(\lambda) \leq \frac{g(x')}{g(x)} < M(\lambda) \leq +\infty.$$

This class was introduced by V.G. Avakumović in 1935,[7] and by J. Karamata who established the fundamental properties of such functions [8], e.g. a representation formula of the type (1) in which conditions $\eta(x) \to \infty$, $\varepsilon(x) \to 0, x \to \infty$ are replaced by the boundedness of these functions. Sometime O-RV functions are referred to as regularly bounded.

It turns out that for applications in differential equations the following definition equivalent to Def. 3 is more suitable [41]:

Definition 3'. A positive measurable function g defined on (o,∞) *is O-RV if there exist real numbers p,q such that* $x^p g(x)$ *is almost increasing and* $x^q g(x)$ *is almost decreasing.*

In that, we say (following S. Bernstein) that g is almost increasing if $x_1 < x_2$ implies $g(x_1) < Mg(x_2)$, $M > 1$; almost decreasing functions are defined likewise.

Both o-RV functions (including SV ones) and O-RV functions we shall call *regularly varying functions*.

These functions, mainly the o-RV and in particular the SV ones, are shown to be very useful in proving various results where precise asymptotic estimates of the involved functions are needed. Originally the use of o-RV functions was in Tauberian theorems (J. Karamata 1930, [9]), [10],[11],[12]. Later it has been extended to various problems of summability theory [13], [14],[15], cf. [16], Fourier series [17], [18],[19], cf. [20], analytic functions [21],[22], asymptotic behavior of integrals [23],[24], cf. [25], theory of numbers [26], functional analysis [27], functional, integral and difference equations [28],[29], [30],[31], Mercerian theorems [32], theory of probability and related fields (stohastic processes, renewal theory etc.) [33], [34],[35],[36], cf. [37].

This bibliography is only indicative and by no means complete. For, to the best knowledge of the authors, it should cover at least 150 entries already.

3. In the field of ordinary differential equations the regularly varying functions were first used by V.G. Avakumović who obtained sharp estimates of solutions of the generalized Thomas-Fermi equation [38]. We quote the following result as a typical one:

Theorem 1. *Let $\rho(x)$ be an o-RV function with the index of regularity $\sigma > -2$, and let $y(x)$ be a positive solution, tending to zero, of the equation*

$$y'' = f(x)y^{\lambda}, \qquad \lambda > 1. \tag{2}$$

If

$$f(x) \sim \rho(x), \quad x \to \infty,$$

then

$$y(x) \sim \{(1+\lambda+\sigma)(2+\sigma)(1-\lambda)^{-2}\}^{\frac{1}{\lambda-1}} \{x^2\rho(x)\}^{\frac{1}{1-\lambda}}, \quad x \to \infty.$$

This result considerably generalizes the case $f(x) = x^{\sigma}$ which was treated previously (cf. [39]). Th. 1 was extended in [40] to a wider class of equations of the type (2) in which $f(x) = \rho(\exp x^{\mu})$, $\mu < 1$ and y is replaced by $y^{\lambda}L_n(y^{-1})$, where L_n is a finite product of powers of iterated logarithms.

Recently, the following more general, but somewhat less precise, result is proved by the authors [41]:

Theorem 2. *Let $f(t)$ and $\phi(t)$ be positive and continuous for $t > o$ and let for large x*

A_1) $x^p f(x)$ *be almost increasing for some* $p < 2$

A_2) $x^{-q} f(x)$ *be almost decreasing for some* $q \geq 0$;

and let for small y i.e. for $y \to 0$,

B_1) $y^{-r} \phi(y)$ *be almost decreasing for some* $r > 1$

B_2) $y^{-s} \phi(y)$ *be almost increasing for some* $s > r$.

Then, there exist two constants $m > 0$, $M > m$ such that for any positive solution $y(x)$, tending to zero, of the equation

$$y'' = f(x)\phi(y) \tag{3}$$

there holds for $x \geq x_o$

$$m\{x^2 f(x)\}^{-1} \leq \frac{\phi\{y(x)\}}{y(x)} \leq M\{x^2 f(x)\}^{-1}. \tag{4}$$

At this point we may mention that the existence of solutions in question has been studied for the more general equation $y'' = f(x,y)$ under various assumptions on $f(x,y)$. We quote a result of P.K. Wong [42] which, specialized to our case, read as follows: Let $y_1 < y_2$ implies $y_1^{-1}\phi(y_1) < y_1^{-1}\phi(y_2)$; then the equation (3) has solutions which tend to zero if and only if

$$\int_a^x xf(x)dx \quad \text{diverges.} \tag{5}$$

It is known [39] that for the equation $y'' = x^{-2}y^\lambda, \lambda > 1$ one has $y(x) \sim c(\ln x)^{1/(1-\lambda)}$, $x \to \infty$. Hence, an approximation different from (4) is needed for the case

$$y'' = x^{-2}\psi(x)\phi(y) \tag{6}$$

where, roughly speaking, $\psi(x)$ is asymptotically smaller than any power x^ε, $\varepsilon > o$ (and larger than $x^{-\varepsilon}$). In fact, in this exeptional case there holds the following result, (to be proved elsewhere):

Theorem 3. Let (5) hold; if $\psi(x)$ is a SV function $L(x)$ and $\phi(y)$ satisfies conditions B_1) and B_2) from Th. 2, then, there exist two constants $m_1 > o$, $M_1 > m_1$ such that for any positive solution $y(x)$, tending to zero, of the equation (6) there holds for $x \geqslant x_o$

$$m_1\{\int_a^x t^{-1}L(t)dt\}^{-1} \leqslant \frac{\phi\{y(x)\}}{y(x)} \leqslant M_1\{\int_a^x t^{-1}L(t)dt\}^{-1} .$$

Observe that the function $L_1(x) = \{\int_a^x t^{-1}L(t)dt\}^{-1}$ is a SV one which tends to zero for $x\to\infty$ since the occuring integral diverges because of (5). Furthermore, the examples $L(t) = \ln t$ and $L(t) = (\ln t)^{-1}$ when

$$\int_a^x t^{-1}L(t)dt \sim 1/2 \ln^2 x, \quad \text{and} \quad \int_a^x t^{-1}L(t)dt \sim \ln\ln x$$

respectively, show that it is not possible to express the behavior of the occuring integral by a single formula for any $L(t)$.

The equation (3) has an interesting closure property expressed by

Theorem 4. If $f(x)$ and $\phi(y)$ are both O-RV functions (specified as in Th. 2.) then, any positive solution $y(x)$ of the equation (3), tending to zero, is also an O-RV function.

This follows, in fact, as a corollary to Th. 3. by using for the proof the Def. 3 of O-RV functions.

Needless to say, basic properties of regularly varying functions are indispensable in proving all mentioned results.

The authors are indebted to Professors S. Aljančić,Univ. of Belgrade, for several valuable remarks and R. Bojanić, Ohio State Univ. Columbus, Oh. for the help in completing the bibliography of RV functions.

References.

1. J. Karamata, Sur un mode de croissance régulière des fonction. *Matematica (Cluj) 4 (1930), 38-53.*

2. T. van Aardenne-Ehrenfest, N.G. de Bruijn, J. Korevaar, A note on slowly oscillating functions. *Nieuw. Arch. Wisk. 23 (1949), 77-86.*

3. N.G. de Bruijn, Pairs of slowly oscillating functions occuring in asymptotic problems concerning the Laplace transform. *Nieuw. Arch. Wisk. 7 (1959), 20-26.*

4. J. Karamata, Sur un mode de croissance régulière, Théoremes foundamentaux. *Bull. Soc. Math. France 61 (1933), 55-62.*

5. D. Adamović, Sur quelques properiétés des fonctions à croissance lente de Karamata I, II. *Mat. vesnik 3 (18) (1966), 123-136, 161-172.*

6. R. Bojanić, E. Seneta, Slowly varying functions and asymptotic relations. *J. Math. Analysis Appl. 34 (1971), 303-315.*

7. V. G. Avakumović, Sur une extension de la condition de convergence des théoremes inverses de sommabilité. *C.R. Acad. Sci. Paris 200 (1935), 1515-1517.*

8. J. Karamata, Remark on the proceding paper by V.G.Avakumović, with the study of a class of functions occuring in the inverse theorems of the summability theory (Serbo-Croatian). *Rad Jugoslav. Akad. Znan. Umjet. 254 (1936), 187-200.*

9. J. Karamata, Sur certains "Tauberian theorems" de M.M.Hardy et Littlewood. *Mathematica (Cluj) 3 (1930), 33-48.*

10. J. Karamata, Neuer Beweis und Verallgemeinerung der Taubershen Sätze, welche die Laplacesche und Stieltjessche Transformation betreffen. *J. Reine Angew. Math. 164 (1931),27-39.*

11. W. Feller, On the classical Tauberian theorems. *Arch.Math.14 (1963), 317-322.*

12. D. Drasin, Tauberian theorems and slowly varying functions. *Trans.Amer.Math. Soc. 133 (1968), 333-356.*

13. K. Knopp, Über eine Erweiterung des Äquivalenzsatzes der C- und H- Verfahren und eine klasse regular waschsenden Funktionen. *Math. Z. 49 (1943), 219-255.*

14. K. Knoop, Zwei Abelsche Sätze. *Publ. Inst. Math. (Beograd) 4 (1952), 89-94.*

15. S. Aljančić, J. Karamata, Regularly varying functions and Frullani's integral (Serbo-Croatian). *Zbornik radova Mat. Inst. SAN 5 (1956), 239-248.*

16. G. Doetsch, Handbuch der Laplace-Transformation, Band I. *Birkhäuser, Basel 1950.*

17. G. H. Hardy, W.W. Rogosinski, Notes on Fourier Series (III). *Quart. J. Math. 16 (1945), 49-58.*

18. S. Aljančić, R. Bojanić, M. Tomić, Sur l'integrabilité de certains séries trigonometrique. *Publ. Inst. Math. (Beograd) 8 (1955), 67-84.*

19. S. Aljančić, R. Bojanić, M. Tomić, Sur le comportement asymptotique au voisinage de zéro des séries trigonométrique de sinus a coefficient monotones. *Publ. Inst. Math. (Beograd) 10 (1956), 101-120.*

20. A. Zygmund, Trigonometric series, 2nd. ed. *Cambridge Univ. Press 1959.*

21. S.M. Shah, One entire functions of infinite order. *Arch.Math. 14 (1963), 323-327.*

22. A.A. Gol'dberg, J.V. Ostrovskii, Value distribution of meromorphic functions (Russian). *Nauka, Moscow 1970.*

23. S. Aljančić, R. Bojanić, M. Tomić, Sur la valeur asymptotique d'une classe des intégrales défines. *Publ. Inst. Math. (Beograd) 7 (1954), 81-94.*

24. A. Békéssy, Eine Verallgemeinerung der Laplaceshen Methode. *Publ. Math. Inst. Hung. Acad. Sci. 2 (1957), 105-125.*

25. E. Ja. Riekstinš, Asymptotic expansions of integrals, I. (Russian). *Ed. Zinatne, Riga 1974.*

26. J. Tull, A theorem in asymptotic number theory. *J. Austral. Math. Soc. 5 (1965), 196-206.*

27. R. Bojanić and M. Vulleumier, Asymptotic properties of linear operators. *Enseign. Math. 19 (283-308).*

28. R.R. Coifman, Sur l'équation fonctionelle d'Abel-Schroeder et l'iteration continue. *C. R. Acad. Sci. Paris, 258 (1964), 1976-1977.*

29. B. Stanković, On a class of singular integral equations (Serbo-Croatian). *Zbornik radova Mat. Inst. SAN 4 (1955), 81-130.*

30. S. Aljančić, Über den Perronschen Satz in der Theorie der Differenzengleichungen. *Publ. Inst. Math. (Beograd) 13 (1959), 47-56.*

31. T. Ganelius, Regularly varying functions and Poincare's theorem on difference equations. *Symp. Theor. Phys. Math., New York, 10 (1970), 7-17.*

32. S. Aljančić, Asymptotische Mercersätze für Hölder- und Cesaro-Mittel. *Publ. Inst.Math. (Beograd) 17 (1974), 5-16.*

33. J. Lamperti, Some limit theorems for stohastic processes. *J. Math. Mech. 7 (1958), 443-450.*

34. W. L. Smith, A note on the renewal function when the mean renewal lifetime is infinite. *J. Roy. Statist.Soc. Ser. B 23 (1961), 230-237.*

35. C. R. Heathcote, E. Seneta, D. Vere-Jones, A refinement of two theorems in the theory of branching processes. *Teor.Verojatnost. i Primenen. 12 (1967), 341-346.*

36. L. de Haan, On regular variation and its application to the weak convergence of sample extremes. *Mathematisch Centrum, Amsterdam 1970.*

37. W. Feller, An introduction to probability theory and its applications 2. *J. Wiley, New York 1966.*

38. V. G. Avakumović, Sur l'équation différentielle de Thomas-Fermi. *Publ. Inst. Math. (Beograd) 1 (1947), 101-113.*

39. R. Bellman, Stability theory of differential equations. *Mc. Graw-Hill, New York 1953.*

40. V. Marić, On asymptotic behavior of solutions of a class of second order nonlinear differential equations (Serbo-Croatian). *Zbornik radova Mat. Inst. SAN 4 (1955), 27-40.*

41. V. Marić, M. Tomić, Asymptotic properties of solutions of the equation $y'' = f(x)\phi(y)$. *(To appear).*

42. P.K. Wong, Existence and asymptotic, behavior of proper solutions of a class of second order nonlinear differential equations. *Pacific. J. Math. 13 (1963), 737-760.*

PROJECTION METHODS FOR LINEAR AND NONLINEAR SYSTEMS OF PARTIAL DIFFERENTIAL EQUATIONS

J. W. Neuberger

1. Introduction. The main material of this note is illustrated by a very simple example.

Suppose $\Omega = (-1,1)\times(-1,1)$. A vector field on Ω is a continuous function from Ω to E_2. The set of all vector fields is denoted by F. A conservative vector field $\binom{f}{g}$ is a vector field for which there is $u \in C^{(1)}(\Omega)$ such that $u_1 = f$, $u_2 = g$ ($u_1 = \partial f/\partial x$, etc.). Denote the set of conservative vector fields by K. Consider the simplest partial differential equation of all on Ω: the problem of finding $u \in C^{(1)}(\Omega)$ so that

$$(1) \qquad u_1 = 0 .$$

For this partial differential equation call a member $\binom{f}{g}$ of F a solution vector field if $f \equiv 0$. Denote by S the set of solution vector fields. The problem of solving (1) may then be recast as the problem of determining the intersection of S and K.

Define $L: F \to S$ so that $L\binom{f}{g} = \binom{0}{g}$ for all $\binom{f}{g} \in F$. Then L takes a member of F into what might be thought the nearest member of S. L is clearly an idempotent and hence a projection. If one has also a projection P from F onto K, then one might hope to solve (1) by taking limiting values of the iteration:

$$(PL)^n\binom{f}{g}, \quad n = 1, 2, \ldots .$$

For example in a Hilbert space setting, if L and P are orthogonal projections, $\{(PL)^n\}_{n=1}^{\infty}$ converges strongly to the orthogonal projection onto the intersection of the ranges of L and P (cf [2], problem 96).

Different approaches to choosing P are given in the following two paragraphs. Paragraph (i) is suggested by work in [3], [4], [6] and serves as an example for section 2 of this note in which higher order nonlinear analytic partial differential equations are considered. Paragraph (ii) serves as an example for section 3 in which L_2 solutions to nonlinear systems are obtained. The work of section 3 extends [7] considerably. Section 4 contains some global solvability conditions for linear systems and section 5 contains an application to a functional partial differential equation and to the fully nonlinear problem of finding u such that $f(u,u') = 0$.

(i) Denote by F_a, K_a, S_a the members of F, K, S respectively which have absolutely convergent power series in Ω about the origin. For $\binom{f}{g} \in F_a$, define first

$z(x,y) = \int_0^1 (f(xj,yj)x + g(xj,yj)y)$, $(x,y) \in \Omega$, $(j(t) = t$ for all t in $R)$ and then denote $\binom{z_1}{z_2}$ by $P\binom{f}{g}$. One may check that P is idempotent. One may also check that if $\binom{f}{g} \in F_a$, then $\{(PL)^n\binom{f}{g}\}_{n=1}^{\infty}$ converges to the member $\binom{r}{s}$ of F_a so that $r(x,y) = 0$, $s(x,y) = g(0,y)$, $(x,y) \in \Omega$. Hence $\binom{r}{s} \in K_a \cap S_a$.

(ii) Denote by F_2, K_2, S_2 the square integrable members of F, K, S respectively and denote $(\int_{\Omega}(f^2 + g^2))^{1/2}$ by $\|\binom{f}{g}\|$, $\binom{f}{g} \in F_2$. Denote by $\overline{F}_2$ a completion of F_2 relative to this norm and denote by $\overline{K}_2$, $\overline{S}_2$ the respective closures of K_2 and S_2 in $\overline{F}_2$. Extend L to $\overline{F}_2$ by continuity and define P to be the orthogonal projection of $\overline{F}_2$ onto $\overline{K}_2$. Here if $\binom{f}{g} \in F_2$ and has continuous extension to Ω, then $\{(PL)^n\binom{f}{g}\}_{n=1}^{\infty}$ converges in norm to $\binom{r}{s} \in F_2$ so that $r(x,y) = 0$,

$s(x,y) = 2^{-1}\int_{-1}^{1} g(j,y)$, $(x,y) \in \Omega$, so that $\binom{r}{s} \in K_2 \cap S_2$.

Some notation and definitions are now established. Denote by each of m, k and n a positive integer and by E a finite dimensional inner product space. Denote by $S(m,i;k)$ the space of E_k-valued, symmetric i-linear functions on E_m, $i = 1, 2, \ldots$. $S(m,0;k)$ denotes E_k. Denote $S(m,0;k) \times S(m,1;k) \times \cdots \times S(m,i;k)$ by S_i, $i = 1, 2, \ldots$. If i is a positive integer, then $S(m,i;k)$ is an inner product space with norm determined as follows: pick any orthonormal basis $e_1, \ldots, e_m$ of E_m. If $w \in S(m,i;k)$, define

$$\|w\| = (\Sigma_{p_1=1}^{m} \cdots \Sigma_{p_i=1}^{m} \|w(e_{p_1}, \ldots, e_{p_i})\|^2)^{1/2}$$

where the second norm is taken in E_k. For $Z = (z_0, a_1, \ldots, z_i) \in S_i$, $\|Z\| = (\Sigma_{q=0}^{i} \|z_q\|^2)^{1/2}$. The square of the norm of a member of $L(S_n,E)$ is calculated by taking a matrix obtained from an orthonormal basis of S_n and one for E and then taking the sum of the squares of the entries.

Denote by Ω an open subset of E_m. Denote by Γ_i the collection of continuous functions from Ω to S_i, $i = 1, 2, \ldots$ and suppose that A is a function with domain Γ_n and range a collection of continuous functions from Ω to $L(S_n,E)$. If $U \in \Gamma_n$, then the value of A at U is denoted by A_U. Denote by Γ_n' the subset of Γ_n consisting of those $U \in \Gamma_n$ for which there is $z \in C^{(n)}(\Omega)$ such that $U = (z, z', \ldots, z^{(n)})$ where $z^{(i)}$, $i = 1, \ldots, n$ denote the first n Fréchet derivatives of Z. General reference for Fréchet derivatives is [8]. If $Y = (y_0, \ldots, y_n) \in \Gamma_n$, then $\tau Y \in \Gamma_{n-1}$ denotes $(y_0, \ldots, y_{n-1})$. Suppose that for each $Z \in \Gamma_{n-1}$, F_Z is a continuous function from Ω to E. A very general functional partial differential equation may now be formulated:

Find $U \in \Gamma'_n$ (or in a suitable completion of a subset of Γ'_n) so that

(2) $A_U U = F_{\tau U}$, i.e., $A_U(p)U(p) = F_{\tau U}(p)$, $p \in \Omega$,

in perhaps a generalized sense.

Solutions to (2) are found under various restrictions on A and F but throughout the following is required:

(3) If $U \in \Gamma_n$, $A_U(p)A_U(p)^* = I$, the identity on E, for all $p \in \Omega$.

Problem (2) under restriction (3) includes a vast class of quasi-linear systems of partial differential equations. When neither of A_U and $F_{\tau U}$ depends upon U and $n = 1$, equation (2) represents a linear system of k first order inhomogeneous partial differential equations. Condition (3) is a consistency condition which may be achieved for many problems by a normalization.

Common to all the work is a function L on Γ_n so that if $U, W \in \Gamma_n$, then $L_U W = W - A_U^* A_U W + A_U^* F_{\tau U}$, i.e., $(L_U W)(p) = W(p) - A_U(p)^* A_U(p) W(p) + A_U(p)^* F_{\tau U}(p)$, $p \in \Omega$. Observe that $L_U^2 = L_U$ and that $A_U(L_U W) = F_{\tau U}$ so that if $L_U U = U$, $U \in \Gamma'_n$ then $A_U U = F_{\tau U}$, i.e., (2) is satisfied.

2. Analytic higher order problems. For this section take $k = 1$, $E = R$, keeping m and n arbitrary.

Suppose $B \in S(m,n;1)$, $\|B\| = 1$. A function u from a subset of E_m to a finite dimensional inner product space is called analytic at a point p provided $D(u)$ is open and there is $r > 0$ such that $u(x) = \Sigma_{q=0}^{\infty}(1/q!)u^{(q)}(p)(x-p)^q$ and $\Sigma_{q=0}^{\infty}(1/q!)\,\|u^{(q)}(p)\|\,\|x-p\|^q$ converges, $\|x-p\| < r$. Define $\|u\|_\delta = \Sigma_{q=0}^{\infty}(1/q!)\,\|u^{(q)}(0)\|\,\delta^q$ if u is analytic at 0 and the right hand side exists.

Suppose $r > 0$ and $\Omega_r = \{x \in E_m \mid \|x\| < r\}$. Denote by H_r the set of all real-valued functions u on Ω_r so that $\|u\|_\delta$ exists if $0 < \delta < r$. Denote by $H_r^{(n)}$ the set of all $S(m,n;1)$-valued functions w on Ω_r which are (1) analytic on Ω_r and (2) are such that $\|w\|_\delta$ exists for $0 < \delta < r$. Note that if $u \in H_r$, then $u^{(n)} \in H_r^{(n)}$. Suppose $g \in H_r$ and define $L : H_r^{(n)} \to H_r^{(n)}$ so that $LW = W - \langle B,W\rangle B + gB$ if $W \in H_r^{(n)}$. Define $P : H_r^{(n)} \to H_r^{(n)}$ as follows: If $W \in H_r^{(n)}$ first define $z(x) = \int_0^1 ((1-j)^{n-1}/(n-1)!)\, W(jx)x^n$, $\|x\| < r$, and then $PW = z^{(n)}$. Note $P^2 = P$.

Theorem 1. If $y, g \in H_r$, there is $u \in H_r$ such that if $0 < \delta < r$, then $\|(PL)^i y^{(n)} - u^{(n)}\|_\delta \to 0$ as $i \to \infty$. Moreover $\langle B,u^{(n)}\rangle = g$.

To prove this, note that $u(x) = \Sigma_{q=0}^{\infty}(1/q!)u^{(q)}(0)x^q$, $g(x) = \Sigma_{q=0}^{\infty}(1/q!)g^{(q)}(0)x^q$, $u^{(n)}(x) = \Sigma_{q=n}^{\infty}(1/(q-n)!)u^{(q)}(0)x^{q-n}$, $(Lu)(x) = \Sigma_{q=n}^{\infty}(1/(q-n)!)\{u^{(q)}(0)x^{q-n}$

$- [B(u^{(q)}(0)x^{q-n}) - g^{(q-n)}(0)x^{q-n}]B\}$ and so

$$((PL)u^{(n)})(x) = \Sigma_{q=n}^{\infty}(1/(q-n)!)\{u^{(q)}(0) - [(Bu^{(q)}(0))\cdot B - B\cdot g^{(q-n)}(0)]\}x^{q-n},$$

$\|x\| < r$, where if $C \in S(m,s;1)$, $D \in S(m,t;1)$, $s \geq t$, then CD denotes the element of $S(m,s-t;1)$ such that $(CD)(x_1, \ldots, x_{s-t}) = \langle C(x_1, \ldots, x_{s-t}), D\rangle$ for all $x_1, \ldots, x_{s-t} \in E_m$ and $C\cdot D$ denotes the symmetric product of C and D (see [3]). Note that $\langle CD,E\rangle = \langle C,D\cdot E\rangle$ for $C \in S(m,s;1)$, $D \in S(m,t;1)$, $E \in S(m,s-t;1)$. For $x \in E_m$, x^s denotes the member of $S(m,s;1)$ such that $x^s(y_1, \ldots, y_s) = \langle x,y_1\rangle\cdots\langle x,y_s\rangle$ for all $y_1, \ldots y_s \in E_m$. Define $Q_q w = (Bw)\cdot B$, $w \in S(m,q;1)$, $q \geq n$, $M_q w = (B\cdot w)B$, $w \in S(m,q;1)$, $q = 0, 1, \ldots$. By induction, $((PL)^i u^{(n)})(x) = \Sigma_{q=n}^{\infty}(1/(q-n)!)\{Q_q^i u^{(q)}(0) + [Q_q^{i-1}(B\cdot g^{(q-n)}(0)) + \cdots + Q_q(B\cdot g^{(q-n)}(0)) + B\cdot g^{(q-n)}(0)]\}x^{q-n} = \Sigma_{q-n}^{\infty}(1/(q-n)!)\{Q_q^i u^{(q)}(0) + B\cdot[((I-M_{q-n})^{i-1} + \cdots (I-M_{q-n}) + I)g^{(q-n)}(0)]\}x^{q-n}$. As in ([4], Th. 2), this converges to

$$\Sigma_{q=n}^{\infty}(1/(q-n)!)[u^{(q)}(0) - B\cdot(M_{q-n}^{-1}(Bu^{(q)}(0)))]x^{q-n}$$
$$+ \Sigma_{q=n}^{\infty}(1/(q-n)!)[B\cdot(M_{q-n}^{-1}g^{(q-n)}(0))]x^{q-n} \quad \text{as } i \to \infty,\ \|x\| < r,$$

and $\{\|(PL)^i u^{(n)}\|_\delta\}_{i=1}^{\infty}$ is bounded if $0 < \delta < r$. The conclusion to the theorem follows.

Denote by H the space of all real-analytic functions u on a connected open subset of E_m containing 0 so that the domain of u is maximal relative to this property. Suppose also $B \in S(m,n;1)$, $\|B\| = 1$, h is a positive integer $\leq n-1$ and for some $\alpha = \alpha_0, \alpha_1, \ldots, \alpha_{n-1} \in S_{n-1}$, f is an analytic function on an open subset of $E_m \times S_h$ containing $0, \alpha_0, \alpha_1, \ldots, \alpha_h$. Denote by $H(\alpha)$ the subset of H consisting of all $u \in H$ such that $u^{(i)}(0) = \alpha_i$, $i = 0, 1, \ldots, n-1$. If $u \in H(\alpha)$, then f_u denotes the member of H so that $f_u(x) = f(x,u(x),u'(x), \ldots,u^{(h)}(x))$ for all x in some region of E_m containing 0. Define $T: H(\alpha) \to H(\alpha)$ by

$$(Tu)(x) = \Sigma_{q=0}^{n-1}(1/q!)u^{(q)}(0)x^q + \int_0^1 ((1-j)^{n-1}/(n-1)!)[u^{(n)}(jx) - \langle B,u^{(n)}(jx)\rangle B + f_u(jx)B]x^n$$

for all x in some region of E_m containing 0.

Theorem 2. If $h \leq n/2$ and $u \in H(\alpha)$, then there is $y \in H(\alpha)$ and $r > 0$ so that $\langle B,y^{(n)}\rangle = f_y$ and $\|T^i u - y\|_\delta \to 0$ as $i \to \infty$, if $0 < \delta < r$.

This follows from ([4], Th. 1). For B with the additional property that if x is a nonzero member of E_m, then $Bx \neq 0$, it follows from [9] that the condition $h \leq n/2$ may be relaxed in some instances to $h \leq 3n/4$. It is conjectured that the theorem remains true if $h \leq n/2$ is replaced by $h \leq n-1$.

It is remarked that the proof of Theorem 2 and the improvement in [9] depend heavily on a tensor identity in [5].

3. Nonlinear L_2 results. First there is an iteration lemma for general real Hilbert space H. Denote by $L(H)$ the space of continuous linear transformations from H to H and by Γ a function from H to $L(H)$ so that $\Gamma(x)$ is symmetric and has numerical range in $[0,1]$ for all x in H. Suppose moreover that P is an orthogonal projection on H and Γ is strongly continuous, i.e., if $\{x_i\}_{i=1}^{\infty}$ converges to $x \in H$ and $w \in H$, then $\Gamma(x_i)w \to \Gamma(x)w$ as $i \to \infty$. If T is a symmetric nonnegative element of $L(H)$, then $T^{1/2}$ denotes the unique nonnegative symmetric member of $L(H)$ so that $(T^{1/2})^2 = T$.

Theorem 3. Suppose $w \in H$, $Q_0 = P$ and $Q_{n+1} = Q_n^{1/2}\Gamma(Q_n^{1/2}w)Q_n^{1/2}$, $n = 0, 1, \ldots$. Then $Q_1, Q_2 \ldots$ converges strongly to a member Q of $L(H)$. Moreover $\{Q_i^{1/2}w\}_{i=1}^{\infty}$ converges to $z \in H$ so that $Pz = z$ and $\Gamma(z)z = z$.

Proof. First note that Q_0 is symmetric and nonnegative. Using the fact that the range of Γ contains only symmetric nonnegative members of $L(H)$, one has by induction that each of $Q_0, Q_1, \ldots$ is symmetric and nonnegative. Moreover, if i is a nonnegative integer and $x \in H$, $\langle Q_{i+1}x,x\rangle = \langle Q_i^{1/2}\Gamma(Q_i^{1/2}w)Q_i^{1/2}x,x\rangle$ $= \langle\Gamma(Q_i^{1/2}w)Q_i^{1/2}x,Q_i^{1/2}x\rangle \leq \langle Q_i^{1/2}x,Q_i^{1/2}x\rangle = \langle Q_ix,x\rangle$, so that $Q_i \geq Q_{i+1} \geq 0$, $i = 0, 1, \ldots$. Hence $Q_0, Q_1, \ldots$ converges strongly on H to a symmetric nonnegative transformation Q. Also $Q_0^{1/2}, Q_1^{1/2}, \ldots$ converges strongly to $Q^{1/2}$ and hence $Q_0^{1/2}w, Q_1^{1/2}w, \ldots$ converges to $z \equiv Q^{1/2}w$. Therefore $\{\Gamma(Q_i^{1/2}w)\}_{i=1}^{\infty}$ converges strongly to $\Gamma(z)$ and hence $\{\Gamma(Q_i^{1/2}w)Q_i^{1/2}w\}_{i=1}^{\infty}$ converges to $\Gamma(z)z$. Note also that $PQ_0^{1/2} = Q_0^{1/2}$. Since if i is a positive integer, then $Q_i^{1/2}$ is the strong limit of a sequence of polynomials in Q_i, it follows by induction that $PQ_i^{1/2} = Q_i^{1/2}$. Since z is the limit of $\{Q_i^{1/2}w\}_{i=1}^{\infty}$, it follows that $Pz = z$. Since for each positive integer i and each $x \in H$, $\langle Q_{i+1}x,x\rangle = \langle\Gamma(Q_i^{1/2}w)Q_i^{1/2}x,Q_i^{1/2}x\rangle$, it follows that $\langle Qx,x\rangle = \langle Q^{1/2}x,Q^{1/2}x\rangle = \langle\Gamma(z)Q^{1/2}x,Q^{1/2}x\rangle$ and hence $\langle(I-\Gamma(z))Q^{1/2}x,Q^{1/2}x\rangle = 0$. Therefore $(I-(\Gamma(z))Q^{1/2}x = 0$ since $I-\Gamma(z)$ is symmetric and nonnegative. In particular $(I-\Gamma(z))z = 0$, i.e., $\Gamma(z)z = z$. This, together with the already established fact that $Pz = z$, is what was to be shown.

Theorem 3 is applied to nonlinear systems of partial differential equations. Suppose Ω is an open subset of E_m and for each $u \in \Gamma_n$, A_U is as in the introduction. Denote by F_2 the square integrable members of Γ_n and by $\overline{F}_2$ a completion of F_2 relative to the norm $\|W\| = (\int_\Omega(\|w_0\|^2 + \cdots + \|w_n\|^2)^{1/2}$, if $W = (w_0, \ldots, w_n) \in F_2$. Denote $\Gamma_n' \cap F_2$ by K_2 and denote the closure of K_2 in $\overline{F}_2$ by $\overline{K}_2$. Denote by P the orthogonal projection of $\overline{F}_2$ onto $\overline{K}_2$. In addition to (3) suppose that if $\{U_i\}_{i=1}^{\infty}$ is a Cauchy sequence in F_2, convergent to $U \in \overline{F}_2$, and $W \in F_2$, then

(4) $$\{W - A_{U_i}^*A_{U_i}W\}_{i=1}^{\infty} \text{ is a Cauchy sequence in } F_2.$$

Denote this limit by L_UW. Extend L_U continuously to all $W \in \overline{F}_2$.

Theorem 4. If $W \in \overline{F}_2$, $Q_0 = P$ and $Q_{i+1} = Q_i^{1/2} L_{Q_i^{1/2}W} Q_i^{1/2}$, $i = 0, 1, \ldots,$ then $\{Q_i^{1/2}W\}_{i=1}^{\infty}$ converges to a member Z of $\overline{K}_2$ such that

(5) $$A_Z Z = 0 .$$

Proof. The convergence is a direct consequence of Theorem 3 as is the fact that $L_Z Z = Z$. But this implies that $Z - A_Z^* A_Z Z = Z$, i.e., $A_Z^* A_Z Z = 0$. But then $0 = A_Z A_Z^* A_Z Z = A_Z Z$ and so (5) holds.

A natural question to raise about the preceding is whether the limit Z is always zero. The answer is no. For a wide class of linear systems, the process of Theorem 3 gives a "good" answer as can be seen from the following: Take $n = 1$. For an open subset Ω of E_m choose a linear transformation B from S_1 to E_k so that BB^* is invertible. Choose A so that if $U \in \Gamma_n$ and $W \in S$, then $A_U(p)W = (BB^*)^{1/2}BW$ -- that is A is independent of $U \in \Gamma$ and $p \in \Omega$. Write $C = (BB^*)^{1/2}B$ in place of $A_U(p)$, $U \in \Gamma$, $p \in \Omega$, so that (5) becomes

(6) $$CZ = 0$$

and $CC^* = I$ holds. Now (6) is equivalent to

(7) $$BZ = 0$$

and this represents a wide class of consistent first order constant coefficient systems of k partial differential equations in k unknowns. The solution obtained by the iteration is the nearest (in terms of the norm on $\overline{F}_2$) to the initial estimate W. The transformation T which takes initial estimates W into solutions Z is just the orthogonal projection of $\overline{F}_2$ onto the closure of the set of L_2 solutions to (7). Members of this closure which are not $C^{(1)}$ functions may be thought of as L_2 generalized solutions (cf [7]). The above considerations also apply to nonconstant coefficient problems.

The situation for nonlinear problems seems much more complicated. Extensive study of examples seems in order and numerical calculations should shed considerable light. It is remarked that the iteration in Theorem 4 does suggest a numerical procedure: L is given explicitly and the problem of computing values of P seems essentially a tractable potential theoretic problem. Details will have to be given elsewhere.

The process seems to have rather immediate application: Systems (5) include a wide variety of conservation systems (cf [1]). In practice, approximate solutions to (5) are sometimes found by picking $W_0 \in K_2$ and then solving

(8) $$A_{W_0} W = 0$$

for $W \in K_2$ -- picking a particular solution W according to physical requirements. It is proposed to use such a solution W to (8) as the starting point of an iteration as in Theorem 4. Limited experience with Theorems 1, 2, 4 seems to indicate that certain distinctive features of initial estimates W are retained in the limiting solution Z. Even the first few iterates seem to show nonlinear effects that appear to be lost in solutions to the linearized problem.

4. Solvability conditions for linear system. Suppose Ω is an open subset of E_m and C is a continuous function from Ω to $L(S_n,E)$ so that $C(p)C(p)^* = I$. Denote by H_2 the collection of all continuous square integrable functions from Ω to E and denote by $\overline{H}_2$ a completion. Consider the problem of finding $U \in \overline{K}_2$ such that

$$CU = g. \tag{9}$$

Denote by P the orthogonal projection of $\overline{F}_2$ onto $\overline{K}_2$. There is the following nonexistence result.

Theorem 5. If $g \in \overline{H}_2$, $g \neq 0$ and $PC^*g = 0$, then there is not $U \in \overline{K}_2$ such that

$$CU = g.$$

Proof. Suppose otherwise. Then there is $U \in \overline{K}_2$ such that $CU = g$ and so $\|g\|^2 = \langle g,g\rangle = \langle CU,g\rangle = \langle CPU,g\rangle = \langle U,PC^*g\rangle = 0$, a contradiction.

For an example, take Ω to be an open annulus in E_2 centered at the origin. Then there is not a $C^{(1)}$ function u on Ω such that $-yu_1(x,y) + xu_2(x,y) = 1$, $(x,y) \in \Omega$. Denote by L and T the transformations on $\overline{F}_2$ and by M the transformations on $\overline{H}_2$ so that $LW = W - C^*CW + C^*g$, $TW = W - C^*CW$, $W \in \overline{F}_2$, $Mf = CPC^*f$, $f \in \overline{H}_2$.

Theorem 6. If i is a positive integer and $W \in \overline{F}_2$, then $(PLP)^iW = (PTP)^iW + PC^*[I + (I-M) + \cdots + (I-M)^{i-1}]g$.

This is proved by induction using the fact that $(PTP)PC^* = PC^*(I-M)$.

Observe that M is symmetric and has numerical range in $[0,1]$. Denote by Π a spectral family for M on R (reference for spectral theory is ([10], chapter VII). Then $M = \int_{0-}^{1} j\, d\Pi$.

The following theorem gives a way to construct solutions U to $AU = f$, $U \in \overline{K}_2$, for f in a dense subset of $N(PC^*)^\perp$.

Theorem 7. If $g \in N(PC^*)^\perp$ and $\varepsilon > 0$, then there is $U \in \overline{K}_2$ such that $AU = g - \Pi(\varepsilon)g$. Moreover $g - \Pi(\varepsilon)g \to g$ as $\varepsilon \to 0$.

Proof. Suppose g and ε are as in the hypothesis and $g_\varepsilon \equiv g - \Pi(\varepsilon)g$. Then by Theorem 6, if i is a positive integer and $W \in \overline{F}_2$, $(PLP)^iW = (PTP)^iW + PC^*[I + (I-M) + \cdots + (I-M)^{i-1}]g_\varepsilon = (PTP)^iW + PC^*\int_{0-}^{1} j^{-1}(1-j)^n)\, d(\Pi g_\varepsilon)$

$$= (PTP)^iW + PC^*\int_{\varepsilon}^{1} j^{-1}(1-(1-j)^i)d(\Pi g) \to Y + PC^*\int_{\varepsilon}^{1} j^{-1}d(\Pi g) = Y + PC^*\int_{0-}^{1} j^{-1}d(\Pi g_\varepsilon) \equiv Z_\varepsilon$$

as $i \to \infty$ since $\Pi(t)g_\varepsilon = \Pi(t)(I - \Pi(\varepsilon))g = \begin{cases} (\Pi(t) - \Pi(\varepsilon))g & \text{if } \varepsilon \le t \le 1 \\ 0 & \text{if } 0 \le t < \varepsilon \end{cases}$,

where $(PTP)^i W \to Y$ as $i \to \infty$. Hence $CZ_\varepsilon = CY + CPC^* \int_{0-}^{1} j^{-1} d(\Pi g_\varepsilon) = \int_{0-}^{1} 1\, d(\Pi g_\varepsilon) = g_\varepsilon$.

Now $\Pi(\varepsilon)g \to \Pi(0)g$ as $\varepsilon \to 0$ due to the right strong continuity of Π. Since $g \in N(PC^*)^\perp$, $(I-M)\Pi(0)g = (I-M)\lim_{i\to\infty}(I-M)^i g = \Pi(0)g$. Hence $M\Pi(0)g = 0$ and so $0 = \langle M\Pi(0)g, \Pi(0)g\rangle = \langle CPC^*\Pi(0)g, \Pi(0)g\rangle = \|PC^*\Pi(0)g\|^2$. Therefore $\Pi(0)g \in N(PC^*)$ and hence $\|\Pi(0)g\|^2 = \langle \Pi(0)g, g\rangle = 0$ since $g \in N(PC^*)^\perp$. Therefore $\Pi(0)g = 0$ and so $\Pi(\varepsilon)g \to 0$ as $\varepsilon \to 0$ and the proof is complete.

5. Further examples. We first apply Theorem 4 to a nonlinear functional partial differential equation. Suppose Ω is a bounded open convex subset of E_m and B is a continuous function from $E_m \times E_k \times E_k$ to $L(S, E_k)$ so that the normalization $B(x,y,z)B(x,y,z)^* = I$ holds. If α and β are two continuous functions from Ω to Ω, then consider the problem of finding u from Ω to E_k so that

$$B(x, u(\alpha(x)), u(\beta(x))) \begin{pmatrix} u(x) \\ u'(x) \end{pmatrix} = 0, \quad x \in \Omega.$$

If $A_{(u,w)}(x) = B(x, u(\alpha(x)), u(\beta(x)))$, $x \in \Omega$, $(u,w) \in F_2$, then it is claimed that (4) holds and hence the iteration of Theorem 4 applies.

For a final example suppose Ω is an open subset of E_m and f is an E_k-valued $C^{(1)}$ function from all of $E_k \times S(m,1;k)$ so that $f_1(x,z)f_1(x,z)^* + f_2(x,z)f_2(x,z)^*$ is invertible, $x \in E_m$, $z \in S(m,1;k)$. Then the problem of finding an E_k-valued $C^{(1)}$ function u on Ω so that $f(u,u') = 0$ leads to the problem of finding such a function u so that $f_1(u,u')u' + f_2(u,u')u'' = 0$. Define $B : E_k \times S(m,1;k) \to L(S_2, S(m,1;k))$ so that $B(x,z)(r,s,t) = f_1(x,z)s + f_2(x,z)t$ for all $(r,s,t) \in E_k \times S(m,1;k) \times S(m,2;k) = S_2$. Now for $x, z \in E_k \times S(m,1;k)$, $B(x,z)B(x,z)^* = f_1(x,z)f_1(x,z)^* + f_2(x,z)f_2(x,z)^*$ is invertible and so if $C(x,z) \equiv (B(x,z)B(x,z)^*)^{1/2}B(x,z)$, $(x,z) \in E_k \times S(m,1;k)$, and $A_{(u,w,y)}(p) = C(u(p), w(p))$, $p \in \Omega$, $(u,w,y) \in \Gamma_2$, then (3) holds. It is claimed that (4) holds so that Theorem 4 applies.

REFERENCES

1. K. O. Friedrichs, On the laws of relativistic electro-magneto-fluid dynamics, Comm. Pure Appl. Math. 27 (1974), 749-808.

2. P. R. Halmos, A Hilbert space problem book, D. Van Nostrand (1967).

3. J. W. Neuberger, Tensor products and successive approximations for partial differential equations, Israel J. Math. 6 (1968), 121-132.

4. ____________, An iterative method for solving nonlinear partial differential equations, Advances Math. (to appear).

5. ____________, Norm of symmetric product compared with norm of tensor product, Linear Multilinear Algebra 2 (1974), 115-121.

6. ____________, A resolvent for an iteration method for nonlinear partial differential equations, Trans. Amer. Math. Soc. (to appear).

7. ____________, Partial differential equations and projections of non-conservative vector fields onto conservative vector fields.

8. F. und R. Nevanlinna, Absolute Analysis, Springer-Verlag, Berlin, 1959.

9. T. H. Pate, Lower bounds for the norm of the symmetric product, Linear Algebra and Its Applications (to appear).

10. F. Riesz and B. Sz.-Nagy, Functional Analysis, Ungar, New York (1955).

New results on strongly coupled systems of parabolic differential equations

Karl Nickel

Abstract

The boundary value problem for a nonlinear, strongly coupled, parabolic system of differential equations is considered. Error bounds for an approximation are given, a uniqueness theorem is proved. The results are derived with the aid of the theory of differential inequalities and by using an inequality of Bernstein. The results are applied to the 3-dimensional Prandtl boundary layer equations.

1. The problem

The following system of parabolic differential equations is considered:

(1a) $P(u,v) := u_t - f(t,x,y,u,v,u_y,\underline{v_y},u_x,u_{xx}) = 0$,

(1b) $Q(u,v) := v_t - g(t,x,y,u,v,\underline{u_y},v_y,v_x,v_{xx}) = 0$.

The functions f and g in (1) are in general nonlinear. The variable t is the "time"-variable, $x=(x_1,x_2,\dots,x_m)$ is the vector of the "space"-variables, y is a coupling variable. Letters as indices always denote partial derivatives.

The equations (1) are to be solved together with the corresponding boundary conditions:

(2a) $$Ru := u - h(t,x,y,u_n) = 0 \; ,$$

(2b) $$Su := v - k(t,x,y,v_n) = 0 \; .$$

In (2) the subscript n denotes the normal derivative.

Let $\tilde{u},\tilde{v}$ be an approximation to the solution u,v of (1),(2) in the following sense: Assume the existence of bounds $\rho_1(t)$, $\rho_2(t)$, $\sigma_1(t)$, $\sigma_2(t)$ for the following residual errors

$$|P(\tilde{u},\tilde{v})| \leqq \rho_1 \; , \quad |Q(\tilde{u},\tilde{v})| \leqq \rho_2 \; ,$$

$$|R\tilde{u}| \leqq \sigma_2 \; , \qquad |S\tilde{v}| \leqq \sigma_2 \; .$$

If ρ_1, ρ_2, σ_1, σ_2 are "small" then $\tilde{u},\tilde{v}$ is supposed to be "close" to the solution u,v. The goal of the following paper is therefore to construct errorbounds of the form

$$|u-\tilde{u}| \leqq \varepsilon_1 \; , \quad |v-\tilde{v}| \leqq \varepsilon_2 \; .$$

The error functions $\varepsilon_1(t)$, $\varepsilon_2(t)$ will be given explicitely; they depend linearly on the bounds ρ_1, ρ_2, σ_1, σ_2. Hence this result also contains a uniqueness theorem.

The equations (1a) and (1b) are coupled by the terms v, v_y and u, u_y. There is an extensive theory for the special case of <u>weakly</u>

coupled systems. These are such systems (1) where the underlined terms do not appear, which means that f does not depend on v_y and g depends not on u_y. For such systems the theory of differential inequalities can be applied, see V. Lakshmikantham - S. Leela [3], J. Szarski [8] and W. Walter [9]. With this theory much can be said about the behaviour of the solution of (1).

However the above equations (1) are strongly coupled systems. Unfortunately up to date there is no similar theory far-reaching enough for such equations or inequalities. The essential idea in the following paper is, to get bounds in v and u for the "trouble terms" v_y and u_y. Hence the strongly coupled system (1) will be replaced by a weakly coupled system. This can be done with the aid of an inequality of Bernstein [2].

There is however a high price to be paid for these results: opposite to weakly coupled systems strong regularity requirements have to be fulfilled for the functions u,v in y-direction. For weakly coupled systems the existence of u_y and v_y is already enough while here for strongly coupled systems the existence and continuity of $\partial^\mu u/\partial y^\mu$ and $\partial^\mu v/\partial y^\mu$ has to be assumed. Furthermore some global boundedness for these terms has to be true for all $\mu = 1,2,\ldots$!

One may ask if such assumptions are realistic at all. In section 5 therefore a criterion for one single (linear) parabolic differential equation will be given. If the assumptions of this criterion are satisfied, any solution has the required properties !

The new theory can be applied to much more general systems than (1), (2). This will be done in the future. Special cases have already been treated in [6] and [7].

2. Assumptions and definitions

2.1 The domain G

Denote $x = (x_1,x_2,\ldots,x_m)$, $m \in \mathbb{N}$, $x \in \mathbb{R}^m$ and $|x|^2 := \sum_{\mu=1}^{m} x_\mu^2$.

Let $B \subseteq \mathbb{R}^{m+1}$ be an open and bounded set with $B \subseteq \{0<t<T,\ x \in \mathbb{R}^m\}$ and with the closure $\overline{B}$. The boundary ∂B of B is divided into two disjoint subsets by $\partial B = \partial_p B \cup \partial_b B$. A point $(\overline{t},\overline{x}) \in \partial B$ belongs to the "backward" boundary $\partial_b B$ if there exist to $(\overline{t},\overline{x})$ both a half neighbourhood $H_t := \{\overline{t}-\delta<t<\overline{t}\}$ in the negative t-direction and a neighbourhood $N_x := \{\overline{t}, |x-\overline{x}|<\delta\}$ in the x-direction such that $H_t \cup N_x \subseteq \overline{B}$. The remaining points $\partial_p B$ of B are called the "parabolic" boundary (see figure).

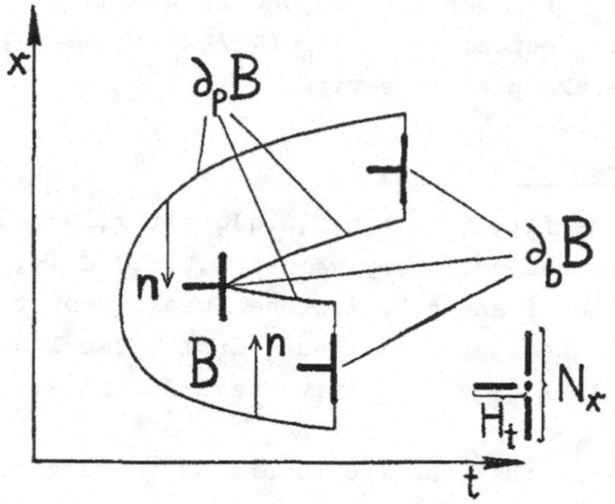

Figure. The basic domain.

The set $B_p := B \cup \partial_b B$ is called the "parabolic interior" of B.

Let $\partial_n B \subseteq \partial_p B$ be a subset of the parabolic boundary, where an (interior) normal n is defined: A point $(\overline{t},\overline{x}) \in \partial_p B$ belongs to $\partial_n B$ if there is a sequence of points $(\overline{t},x^{(\mu)}) \in B_p$ for $\mu \in \mathbb{N}$ with $\lim_{\mu\to\infty} x^{(\mu)} = \overline{x}$. The normal derivative ϕ_n of a function $\phi(t,x)$ relative to this normal n is then defined by

$$\phi_n(\overline{t},\overline{x}) := \lim_{\mu\to\infty} \frac{\phi(\overline{t},x^{(\mu)})-\phi(\overline{t},\overline{x})}{|x^{(\mu)} - \overline{x}|} \quad .$$

Define furthermore $G := B \times \mathbb{R}$, $G_p := B_p \times \mathbb{R}$, $\partial_p G := \partial_p B \times \mathbb{R}$, $\partial_n G := \partial_n B \times \mathbb{R}$.

In G_p the validity of the parabolic equations (1) is assumed. On $\partial_p G$ boundary values will be prescribed. The boundary conditions (2) are of the first or the third kind. This convenient way of writing the boundary operators was brought to the authors attention by R. Redheffer.

Sometimes it is not necessary to prescribe boundary values on a subset $\partial_o G \subseteq \partial_p G$. This is true if the functions f (and/or g) do not depend on u_{xx} for certain components x and if f(g) is monotone at $\partial_o G$ with respect to $\partial u/\partial x_\mu$ ($\partial v/\partial x_\mu$); see [4]. This will be the case in the example of section 6.

2.2 The right hand sides f,g,h,k

Let the functions f(t,x,y,a,b,c,d,p,q), g(t,x,...,q) be defined for $(t,x,y) \in G_p$ and arbitrary values $a,b,c,d \in \mathbb{R}$, $p \in \mathbb{R}^m$, $q \in \mathbb{R}^m \times \mathbb{R}^m$. Let f and g be isotone (weakly monotone increasing) with respect to the last variables q in the usual sense: one defines $q \leqq \bar{q}$ if the matrix $\bar{q}-q$ is positive semi-definite.

Assume furthermore that the two functions f and g satisfy Lipschitz conditions of the following kind:

$$(3)\quad \begin{cases} f(t,x,y,\bar{a},\bar{b},\bar{c},\bar{d},p,q) - f(t,x,y,a,b,c,d,p,q) \leqq \\ \leqq L_1(\bar{a}-a) + M_1|\bar{b}-b| + H_1|\bar{c}-c| + N_1|\bar{d}-d|, \\ g(t,x,y,\bar{b},\bar{a},\bar{d},\bar{c},p,q) - g(t,x,y,b,a,d,c,p,q) \leqq \\ \leqq L_2(\bar{a}-a) + M_2|\bar{b}-b| + H_2|\bar{c}-c| + N_2|\bar{d}-d| \end{cases}$$

for $(t,x,y) \in G_p$; $\bar{a} \geqq a, b, \bar{b}, c, \bar{c}, d, \bar{d} \in \mathbb{R}$, $p \in \mathbb{R}^m$, $q \in \mathbb{R}^m \times \mathbb{R}^m$. Kindly note that (3) is a one-sided Lipschitz condition with respect to the variable a but that (3) is two-sided with respect to b,c,d (compare with W. Walter [9], p. 103). It follows from this fact that the Lipschitz constants L_1, L_2 may be negative while

$M_1 \geq 0,\ldots,N_2 \geq 0$. If L_1 and L_2 are both negative then favorable error bounds are provided by the theorems of sections 3 and 4.

Let the functions $h(t,x,y,a)$, $k(t,x,y,a)$ be defined for $(t,x,y) \in \partial_n G$ and $a \in \mathbb{R}$. For the points $(t,x,y) \notin \partial_n G$ both functions h,k shall not depend on a. Assume that h and k are both isotone with respect to a.

2.3 The function class $\mathfrak{Z}$

A function $u(t,x,y)$ belongs to the class $\mathfrak{Z}$ if the following is true: u is defined and continuous on $\bar{G}$ with $\sup|u| < \infty$ on $\bar{G}$. Furthermore the derivative u_t exists in G_p, the derivatives u_x and u_{xx} exist and are contiuous in G_p and the derivatives u_n exist on $\partial_n G$.

2.4 The Bernstein class $\mathcal{B}(K,y)$

Let $0 \leq K \in \mathbb{R}$ be a real constant. A function $\phi(y)$ belongs to the class $\mathcal{B}(K,y)$ if the following is true: $\phi \in C^\infty(\mathbb{R})$, $\sup|\phi(y)| < \infty$ for $y \in \mathbb{R}$ and there exists a constant $c \in \mathbb{R}$ such that $|\partial^\mu \phi/\partial y^\mu| \leq cK^n$ on $\mathbb{R}$ for all $\mu \in \mathbb{N}$.

This class is essentially the class of the entire analytic functions of exponential type with an exponent not exceeding K, which are real and bounded on the real axis. See for this N.I. Achieser [1], pp. 131-139.

Define $\|\phi\| := \sup_{y \in \mathbb{R}} |\phi(y)|$. In this class $\mathcal{B}(K,y)$ the following holds:

Theorem (S.N. Bernstein [2]): Let $\phi,\psi \in \mathcal{B}(K,y)$. Then

$$\|\phi'-\psi'\| \leq K\|\phi-\psi\|. \tag{4}$$

3. *Lemma:* Let $v \in \mathfrak{Z}$ be a fixed function. Assume that the function $u \in \mathfrak{Z}$ is a solution of the problem

$$(1a) \qquad P(u,v) = 0 \qquad \text{in } G_p \ ,$$

$$(2a) \qquad Ru = 0 \qquad \text{on } \partial_p G \ .$$

Let $\tilde{u} \in \mathfrak{Z}$ be an approximation to the problem (1a), (2a) in the sense that

$$|P(\tilde{u},v)| \leq \rho(t) \quad \text{in } G_p \ ,$$

$$|R\tilde{u}| \leq \bar{\varepsilon}\, e^{L_1 t} \quad \text{on } \partial_p G \ ,$$

where $\bar{\varepsilon} \in \mathbb{R}$ and $\rho(t)$ is isotone and continuous on $0 \leq t \leq T$.

Then

$$|u-\tilde{u}| \leq \varepsilon(x) := \bar{\varepsilon}\, e^{L_1 t} + \int_0^t e^{L_1(t-s)} \rho(s)ds$$

on $\bar{G}$.

The proof is standard in the theory of parabolic inequalities, see W. Walter [9].

4. A posteriori bounds and uniqueness

Theorem: Assume $0 \leq K \in \mathbb{R}$. Let the pair of functions $u,v \in \mathfrak{Z} \cap \mathfrak{L}(K,y)$ be a solution of (1) in G_p and (2) on $\partial_p G$. Assume that the pair of functions $\tilde{u},\tilde{v} \in \mathfrak{Z} \cap \mathfrak{L}(K,y)$ satisfy the following conditions with two isotone and continuous functions $\rho_1(t)$, $\rho_2(t)$:

$$(5) \qquad |P(\tilde{u},\tilde{v})| \leq \rho_1(t), \quad |Q(\tilde{u},\tilde{v})| \leq \rho_2(t) \ \text{in } G_p.$$

Assume furthermore

$$|R\tilde{u}| \leq \bar{\varepsilon}_1 e^{L_1 t}, \quad |S\tilde{v}| \leq \bar{\varepsilon}_2 e^{L_2 t} \quad \text{on } \partial_p G$$

with the two Lipschitz constants L_1, L_2 *from (3) and with two real numbers* $\overline{\varepsilon}_1$, $\overline{\varepsilon}_2$. *Define*

$$R_1 := M_1 + KN_1,\ R_2 := M_2 + KN_2,\ \lambda := \sqrt{(L_1-L_2)^2 + R_1R_2}$$

($\lambda > 0$ w.l.o.g.) and herewith

$$\varepsilon_1(t) := \int_0^t \rho_1(s)\, e^{\frac{L_1+L_2}{2}(t-s)} \left[\cosh\tfrac{\lambda}{2}(t-s) + \frac{L_1-L_2}{\lambda}\sinh\tfrac{\lambda}{2}(t-s)\right] ds$$

$$+ \frac{2}{\lambda} R_1 \int_0^t \rho_2(s)\, e^{\frac{L_1+L_2}{2}(t-s)} \sinh\frac{\lambda}{2}(t-s)\, ds$$

$$+ \overline{\varepsilon}_1\, e^{\frac{L_1+L_2}{2}t}\left[\cosh\frac{\lambda}{2}t + \frac{L_1-L_2}{\lambda}\sinh\frac{\lambda}{2}t\right]$$

$$+ \frac{2}{\lambda} R_1 \overline{\varepsilon}_2\, e^{\frac{L_1+L_2}{2}t} \sinh\frac{\lambda}{2}t\,,$$

$$\varepsilon_2(t) := \frac{2}{\lambda} R_2 \int_0^t \rho_1(s)\, e^{\frac{L_1+L_2}{2}(t-s)} \sinh\frac{\lambda}{2}(t-s)\, ds$$

$$+ \int_0^t \rho_2(s)\, e^{\frac{L_1+L_2}{2}(t-s)} \left[\cosh\frac{\lambda}{2}(t-s) + \frac{L_2-L_1}{\lambda}\sinh\tfrac{\lambda}{2}(t-s)\right] ds$$

$$+ \frac{2}{\lambda} R_2 \overline{\varepsilon}_1\, e^{\frac{L_1+L_2}{2}t} \sinh\frac{\lambda}{2}t$$

$$+ \overline{\varepsilon}_2\, e^{\frac{L_1+L_2}{2}t}\left[\cosh\frac{\lambda}{2}t + \frac{L_2-L_1}{\lambda}\sinh\frac{\lambda}{2}t\right].$$

Then

$$(6) \qquad |u-\tilde{u}| \leqq \varepsilon_1(t)\,, \quad |v-\tilde{v}| \leqq \varepsilon_2(t) \qquad \text{on } \overline{G}.$$

An important <u>conclusion</u> from this theorem is: Any solution of the boundary value problem (1), (2) in the class of functions $\mathfrak{Z} \cap \mathfrak{L}(K,y)$ does depend continuously on the boundary values on $\partial_p G$ and on the residual error in G_p.

For $\bar{\varepsilon}_1 = \bar{\varepsilon}_2 = 0$, $\rho_1(x) \equiv \rho_2(x) \equiv 0$ one gets the

<u>*Uniqueness theorem:*</u> *Let f,g satisfy the Lipschitz condition (3). Assume $0 \leq K \in \mathbb{R}$. Then there is at most one solution u,v to the boundary value problem (1), (2) in the class of functions $\mathfrak{Z} \cap \mathfrak{L}(K,y)$.*

<u>Proof:</u> By (5) and (3) the following inequality holds in G_p:

$$|P(\tilde{u},v)| \leq |P(\tilde{u},\tilde{v})| + \rho_1 + M_1|v-\tilde{v}| + N_1|v_y-\tilde{v}_y| \; .$$

Hence by the Lemma of section 3 it follows that on $\bar{G}$ the following inequality is true:

$$(7) \quad |u-\tilde{u}| \leq \bar{\varepsilon}_1 \, e^{L_1 t} + \int_0^t e^{L_1(t-s)} [\rho_1(s) + M_1|v-\tilde{v}| + N_1|v_y-\tilde{v}_y|]ds.$$

Define for fixed $t \in [0,T]$ and for $(t,x,y) \in \bar{G}$:

$$d := u-\tilde{u} \; , \qquad D(t) := \sup|d| \; ,$$

$$\delta := v-\tilde{v} \; , \qquad \Delta(t) := \sup|\delta| \; .$$

Since $u,\tilde{u},v,\tilde{v} \in \mathfrak{L}(K,y)$ one gets from the inequality of Bernstein (4) for fixed t,x that

$$|dy| \leq K\, D(t) \quad \text{on} \quad \bar{G} \, ,$$

and that $$|\delta y| \leq K\, \Delta(t) \quad \text{on} \quad \bar{G}.$$

By inserting this in (7) one finds on $\bar{G}$ that

$$|u-\tilde{u}| = |d| \leq D(t) \leq \bar{\varepsilon}_1 e^{L_1 t} + \int_0^t e^{L_1(t-s)} [\rho_1(s)+R_1\Delta(s)]\, ds$$

and analogously.

$$|v-\tilde{v}| = |\delta| \leqq \Delta(t) \leqq \bar{\varepsilon}_2 e^{L_2 t} + \int_0^t e^{L_2(t-s)} [\rho_2(s) + R_2 D(s)]\, ds.$$

If one inserts the second inequality in the first one gets the linear Volterra integral inequality

$$D(t) \leqq R_1 R_2 \int_0^t e^{L_1(t-s)} \int_0^s e^{L_2(s-r)} D(r)dr\, ds$$

$$+ R_1 \int_0^t e^{L_1(t-s)} \int_0^s e^{L_2(s-r)} \rho_2(r)dr\, ds$$

$$+ \int_0^t e^{L_1(t-s)} [\rho_1(s) + R_1 \bar{\varepsilon}_2 e^{L_2 s}]\, ds$$

$$+ \bar{\varepsilon}_1 e^{L_1 t} .$$

This inequality has $\varepsilon_1(t)$ as maximal solution. This can be shown by means of the theory of integral inequalities (see W. Walter [9]). Hence $D(t) \leqq \varepsilon_1(t)$ on $0 \leqq t \leqq T$ and therefore the first inequality (6) is true. The second inequality (6) is proven in the same manner.

5. A criterion for $u \in \mathcal{L}(K,y)$

In the above theory no existence proof is given. The estimates (6) are derived under the assumption that there exists (at least) one solution u,v in both classes $\mathfrak{Z}$ and $\mathcal{L}(K,y)$. The first assumption $u,v \in \mathfrak{Z}$ for a solution of (1), (2) is more or less trivial. All the attributes of the class $\mathfrak{Z}$ are "natural" properties of a solution. The second condition $u,v \in \mathcal{L}(K,y)$ however seems to be very farefetched. Obviously there are very simple cases with $u,v \notin \mathcal{L}(K,y)$ for any $K \in \mathbb{R}$.

A trivial example for this statement can be found for $m = 1$, $\bar{B} := \{0 \leqq t \leqq T,\ 0 \leqq x \leqq 1\}$, $u := tx(1-x)e^{-y^2}$. The boundary condition on $\partial_p G$ is $u=h(t,x,y) := 0$ with $h \in \mathcal{L}(K,y)$ for any

$K \geq 0$. However $u \notin \mathcal{L}(K,y)$ for <u>all</u> $K \geq 0$. The function u satisfies for example the "parabolic" differential equation $u_t = u/t$ (where however no Lipschitz condition is satisfied for the right hand side) or $u_t = x(1-x)e^{-y^2}$ or ... .

It is very interesting to see that sometimes the property $u \in \mathcal{L}(K,y)$ can be concluded already a priori by inspecting the differential equation and/or the boundary conditions - without any existence theory ! For simplicity this will be shown not for the system (1), (2) but for a single parabolic differential equation of the special linear form

$$(8) \qquad u_t = a(t,x,y) + b(t,x)\, u + c(t,x)\, u_x + d(t,x)u_{xx} \quad \text{in } G_p.$$

Assume furthermore m=1 and the special boundary conditions of the first kind

$$(9) \qquad u = h(t,x,y) \qquad \text{on } \partial_p G.$$

<u>*Criterion:*</u> *Let the functions a,b,c,d be defined on G_p, let $d \geq 0$ in G_p and assume that $b(t,x)$ is integrable on $\overline{G}$ with respect to t. Assume $a \in \mathcal{L}(K,y)$ with $0 \leq K \in \mathbb{R}$ and let $\partial^\mu a/\partial y^\mu$ be continous and bounded for all $\mu \in \mathbb{N}$ in $\overline{G}$. Assume furthermore that $h \in \mathcal{L}(K,y)$ and that $\partial^\mu h/\partial y^\mu$ is bounded on $\overline{G}$ for all $\mu \in \mathbb{N}$. Let $u \in \mathfrak{Z}$ be a solution of the boundary value problem (8), (9) and assume that $\partial^\mu u/\partial y^\mu$ exist with $\partial^\mu u/\partial y^\mu \in \mathfrak{Z}$ for all $\mu \in \mathbb{N}$. Then $u \in \mathfrak{Z} \cap \mathcal{L}(K,y)$.*

<u>Proof:</u> The function $w := \partial^\mu u/\partial y^\mu$ solves the boundary value problem

$$w_t = a_\mu + bw + cw_x + dw_{xx} \quad \text{in } G_p \;,$$

$$w = h_\mu \text{ on } \partial_p G$$

for all $\mu=0,1,\ldots$. Here $a_\mu := \partial^\mu a/\partial y^\mu$ and $h_\mu := \partial^\mu h/\partial y^\mu$. Define

Define $B(t,x) := \exp - \int_0^t b(s,x)ds$. The function

$$(10)\quad v(t,x,y) := B(t,x)w(t,x,y) + \int_0^t B(t,x)B^{-1}(s,x)a_\mu(s,x,y)ds$$

solves the boundary value problem

$$v_t = c\, v_x + d\, v_{xx} \quad \text{in } G_p ,$$

$$v = V(t,x,y) := B(t,x)\, h_\mu(t,x,y) + + \int_0^t B(t,x)B^{-1}(s,x)a_\mu(s,x,y)ds \text{ on } \partial_p G.$$

By the maximum-minimum-principle we have $|v| \leq \sup|V|$ on $\bar{B}$. Since $V \in \mathcal{L}(K,y)$ it follows that there exists a constant $\bar{c}$ such that $|v| \leqq \bar{c}\, K^\mu$. By using (10) one sees then the existence of a constant c such that $|w| \leqq c\, K^\mu$. Hence $u \in \mathcal{L}(K,y)$.

6. Application to boundary layer theory.

The Prandtl boundary layer equations for a three dimensional unsteady fluid flow can be written in the Crocco form as

$$(11)\quad \begin{cases} u_t = -z\, u_x - v\, u_y - u\, v_y + a\, u_z + \nu\, u^2 u_{zz} =: f , \\ v_t = -z\, v_x - v\, v_y + a\, v_z - b \quad + \nu\, u^2 v_{zz} =: g . \end{cases}$$

The physical meaning of the variables will not be given here, see for details [5]. The 4 independend variables are denoted as t,x,y,z; in the above notation $m=2$ and $x_1=x$, $x_2=z$. The functions $a=a(t,x,y)$, $b=b(t,x,y)$ are given; $\nu > 0$ is a constant. The equations (11) are clearly of the type considered in the previous sections of this paper.

The region G is defined by

$$G := \{0 < t < T,\ 0 < x < X,\ -\infty < y < \infty,\ 0 < z < U(t,x,y)\}$$

with a given function $U(t,x,y)$.

The boundary conditions for (11) are:

$$
(12)\begin{cases}
\underline{t = 0:} & u(0,x,y,z),\ v(0,x,y,z) \text{ given.} \\
\underline{x = 0:} & u(t,0,y,z),\ v(t,0,y,z) \text{ given.} \\
\underline{x = X:} & \text{no boundary condition necessary (see [4]).} \\
\underline{z = 0:} & u(t,x,y,0) = \dfrac{a(t,x,y)/\nu}{u_z(t,x,y,0)},\ v(t,x,y,0) = 0. \\
\underline{z=U(t,x,y):} & u(t,x,y,U) = 0,\ v(t,x,y,U) = V(t,x,y).
\end{cases}
$$

From these boundary conditions one sees that z=0 is the subset $\partial_n G \subseteq \partial_p G$, obviously $\partial/\partial n = \partial/\partial z$ on $\partial_n G$. In section 2.2 it was assumed that $h(t,x,y,u_n)$ is isotone with respect to u_n. Therefore only the case $a(t,x,y) \leqq 0$ will be considered (favorable pressure gradient in the fluid flow).

One finds the following partial derivatives:

$$\frac{\partial f}{\partial u} = -v_y + 2\nu u u_{zz},\ \frac{\partial f}{\partial v} = -u_y,\ \frac{\partial f}{\partial u_y} = -v,\ \frac{\partial f}{\partial v_y} = -u\ ,$$

$$\frac{\partial g}{\partial u} = 2\nu u v_{zz}\qquad ,\ \frac{\partial g}{\partial v} = -v_y,\ \frac{\partial g}{\partial u_y} = 0,\ \frac{\partial g}{\partial v_y} = -v.$$

Therefore Lipschitz constants $L_1,\dots,N_2$ in the Lipschitz conditions (3) can be given only if a priori bounds of the solution u,v and its derivatives are known. Unless such bounds are explicitely known the application of the theorem of section 4 to the Prandtl equations (11) is not possible. An uniqueness proof can however be given if only the existence of such constants is assumed, while their value need not to be known:

Uniqueness theorem: Let $u,v \in \mathfrak{Z}$ and $a(t,x,y) \leqq 0$ in G_p. Let the following derivatives be bounded in $\overline{B}$: u_y, v_y, u_{zz}, v_{zz}. Then the boundary layer equations (11) have at most one solution u,v $u,v \in \mathfrak{Z} \cap \mathfrak{L}(K,y)$ for $0 \leqq K \in \mathbb{R}$ which satisfy the boundary conditions (12).

Acknowledgement

I wish to thank very much professor Raymond M. Redheffer, UCLA/USA for many fruitful discussions and suggestions.

References

[1] Achieser, N.I.: Vorlesungen über Approximationstheorie. Akademie-Verlag, Berlin, 1953.

[2] Bernstein, S.N.: Sur une propriété des fonctions entières. Comptes rendus, 176 (1923).

[3] Lakshmikantham, V. and S. Leela: Differential and Integral Inequalities. Vol. I and II. Academic Press, New York and London, 1969.

[4] Nickel, K.: Ein Eindeutigkeitssatz für instationäre Grenzschichten. Math. Z. 74 (1960), 209-220.

[5] Nickel, K.: The Crocco-transformation for the three-dimensional Prandtl boundary layer equations. MRC Technical Summary Report No. 1594. Madison / Wisc. 1975.

[6] Nickel, K.: Errorbounds and Uniqueness for the Solutions of nonlinear, strongly coupled, parabolic systems of differential equations. MRC Technical Summary Report No. 1596. Madison / Wisc. 1975.

[7] Nickel, K.: On strongly coupled systems of nonlinear parabolic differential equations. To appear in russian in the proceedings of the conference in honor of the 75th birthday of I.G. Petrovski in Moscow, January 27 to 31, 1976.

[8] Szarski, J.: Differential Inequalities. Monografie Matematiczne, Tom 43. Warszawa 1965.

[9] Walter, W.: Differential and Integral Inequalities. Springer-Verlag 1970.

Addendum: In the meantime R. Redheffer and W. Walter extended the above results to systems of arbitrary many equations. In addition to this, they proved the first existence theorem (!) for special linear systems of strongly coupled parabolic equations and for the above class of solutions $\mathfrak{Z} \cap \mathfrak{L}(K,y)$.

On a nonlinear diffusion equation arising in population genetics

L.A. Peletier

1. Introduction

In this paper we report on some joint work with P.C. Fife of the University of Arizona. A detailed account will appear elsewhere.

Let u be a measure for the genetic composition of a population distributed linearly, say along a river bank. We assume that u takes values in the interval [0,1] and that it depends on the location x in the habitat $\Omega \subset \mathbb{R}^1$ and on the time t. A simple model, discribing the effects of dispersal and selective advantages of the genotypes leads to the nonlinear parabolic equation.

$$u_t = Du_{xx} + f(x,u) \qquad x \in \Omega,\ t > 0 \tag{1}$$

in which D is a positive constant related to the rate of dispersal and f a source term related to the relative survival fitnesses of the genotypes. The function f depends on u and, if the environment is inhomogeneous, on x as well [4].

One possible choice for f is

$$f(x,u) = s(x)\, u(1-u) \qquad x \in \Omega,\ u \in [0,1].$$

This case was considered by Conley [2], who established, under suitable conditions on s, the existence of a monotonic equilibrium solution in an infinite habitat: $\Omega = \mathbb{R}$. It was also considered by Fleming [5], who obtained results about the existence and the stability properties of equilibrium solutions when $\Omega = (-1,1)$ and u satisfies zero Neumann boundary conditions at $x = \pm 1$.

Another appropriate choice for f is:

$$f(x,u) = u(1-u)[u-a(x)] \qquad x \in \Omega,\ u \in [0,1]. \tag{2}$$

in which $a \in C'(\Omega)$ and $a(x) \in (0,1)$ for all $x \in \Omega$. It has been shown by Fife and Peletier [3] that if $\Omega = \mathbb{R}$, $a' < 0$ and $a(-\infty) > \frac{1}{2} > a(+\infty)$, there exists a monotonically increasing equilibrium solution which, moreover, is stable.

In this paper we shall assume that f is given by (2) and that $\Omega = (-1,1)$. At the boundaries $x = \pm 1$ we impose zero Neumann boundary conditions. Thus, writing $\lambda = D^{-1}$ and rescaling the time variable, we arrive at the following problem:

$$\text{(I)} \begin{cases} u_t = u_{xx} + \lambda f(x,u) & -1 < x < 1,\ t > 0 \\ u_x(-1,t) = u_x(+1,t) = 0 & t > 0 \\ u(x,0) = \psi(x) & -1 \le x \le 1 \end{cases}$$

in which $\psi \in C(\bar{\Omega})$. We shall be interested in questions of existence and stability for the equilibrium solutions of this problem.

2. Equilibrium solutions

Clearly any equilibrium solution u of problem I is a solution of the problem

$$\text{(II)} \begin{cases} u'' + \lambda f(x,u) = 0 & -1 < x < 1 \\ u'(-1) = u'(+1) = 0 & \end{cases}$$

in which primes denote differentiation.

We begin with a few preliminary observations:

(i) For any $\lambda \geq 0$, the functions $u = 0$ and $u = 1$ are solutions of problem II. This is evident from the function f. In what follows we shall call $u = 0$ and $u = 1$ the trivial solutions of problem II.

(ii) For any $\lambda \geq 0$ there exists at least one nontrivial solution of problem II, i.e. a solution u such that $0 < u(x) < 1$ for $-1 \leq x \leq 1$. This is a consequence of the fact that $f_u(x,0) < 0$ and $f_u(x,1) < 0$ for $-1 \leq x \leq 1$ and a result due to Amann [1].

(iii) Let $\lambda > 0$, and let u(x) be a solution of problem II. Then

$$\int_{-1}^{1} f(x,u(x))\, dx = 0. \tag{3}$$

This follows when we integrate the equation and use the boundary conditions.

For small values of λ we obtain the following result.

THEOREM 1. There exists a number $\lambda_0 > 0$ such that for $\lambda \in [0,\lambda_0]$ there is a continuous branch of nontrivial solutions $u(x,\lambda)$ of problem II in $C([-1,1])$. Moreover

$$||u(.,\lambda) - \alpha|| \to 0 \quad \text{as } \lambda \to 0,$$

where

$$\alpha = \tfrac{1}{2} \int_{-1}^{1} a(x)\, dx.$$

Here $||.||$ denotes the supremum norm on $C([-1,1])$.

Proof. It follows from an elementary estimate that

$$||u(.,\lambda) - c|| \to 0 \quad \text{as } \lambda \to 0$$

for a suitable constant $c \in [0,1]$.

By (3) and the continuity of f, c must satisfy the equation

$$\int_{-1}^{1} f(x,c)\,dx = 0.$$

In view of our assumption about f this means that $c = 0$, $c = \alpha$ or $c = 1$.

It follows by means of a standard argument involving the contraction mapping theorem that the branches $u(x,\lambda)$ are continuous and unique. Those converging to $c = 0$ and $c = 1$ are clearly the trivial solutions $u = 0$ and $u = 1$. Hence there remains a unique branch converging to $u = \alpha$, its existence being insured by the second preliminary observation.

Next we consider large values of λ. We now make the following assumption.

Assumption. $a \in C^1([-1,1])$, $a' < 0$ and $a(0) = \frac{1}{2}$.

THEOREM 2. There exists a number $\lambda_1 > 0$ such that if $\lambda \geq \lambda_1$, problem II has at least three nontrivial solutions.

Proof Because $a(1) \in (0,\frac{1}{2})$, there exists a number $b \in (0,1)$ such that

$$\int_{0}^{b} f(1,u)\,du = 0.$$

Choose $c \in (b,1)$. Then it can be shown that for λ large enough the solution v of the initial value problem

$$\text{(III)}\begin{cases} v'' + \lambda f(x,v) = 0 \\ v(1) = c,\ v'(1) = 0 \end{cases}$$

vanishes at a point $\xi \in (-1,1)$, whilst being positive on $(\xi,1]$. Clearly the composite function

$$w^{-}(x) = \begin{cases} 0 & -1 \le x \le \xi \\ v(x) & \xi < x \le 1 \end{cases}$$

is a subsolution of problem II. Similarly we can construct a supersolution w^{+}. Since for λ large enough, $w^{-} < w^{+}$, this implies that there exists a solution u_0 of problem II such that $w^{-} < u_0 < w^{+}$ [6].

By means of a shooting technique, varying c in problem III, we can construct a solution u_1 of problem II such that $0 < u_1 < u_0$. In the same manner we can construct a solution u_2 of problem II such that $u_0 < u_2 < 1$.

3. Stability

Let $u(x)$ be an equilibrium solution. Then we can define the quadratic functional

$$Q(v) = \int_{-1}^{1} \{v'^2 - \lambda f_u(x,u(x))v^2\}\, dx$$

on the Sobolev space $H^1(-1,1)$.

LEMMA [5]. If u is an isolated equilibrium solution of problem I and there exists an element $v \in H^1(-1,1)$ such that $Q(v) < 0$. Then u is unstable.

This criterion enables us to establish the following result.

THEOREM 3. For small values of $\lambda > 0$, any nontrivial equilibrium solution of problem I is unstable.

<u>Proof</u>. It follows from our choice of f that

$$\int_{-1}^{1} f_u(x,\alpha)\, dx > 0.$$

For small values of λ, $||u-\alpha||$ is small and hence

$$\int_{-1}^{1} f_u(x,u(x))\, dx > 0$$

as well. Thus, if we choose $v(x) \equiv 1$ we obtain

$$Q(1) = -\lambda \int_{-1}^{1} f_u(x,u(x))\, dx < 0.$$

By the Lemma this implies that u is unstable.

Finally we consider large values of λ. The following result is an immediate consequence of the proof of Theorem 2.

THEOREM 4. As regards the stability of the solution u_0 we can distinguish three possibilities:

(i) u_0 is stable from below, and there exists another equilibrium solution u_0^* of problem I such that $u_0 < u_0^* < w^+$, which is stable from above;

(ii) u_0 is stable from above, and there exists another equilibrium solution u_0^* of problem I such that $w^- < u_0^* < u_0$ which is stable from below;

(iii) u_0 is stable from above and from below.

REFERENCES

1. Amann, H., Existence of multiple solutions for nonlinear elliptic boundary value problems, Indiana Univ. Math.J. 21 (1972) 925 - 935.
2. Conley, C., An application of Wazewski's method to a nonlinear boundary value problem which arises in population genetics, Univ. of Wisconsin Math. Research Center Tech. Summary Report No. 1444, 1975.
3. Fife P.C., and L.A. Peletier, Nonlinear diffusion in population genetics. To appear
4. Fisher, R.A., Gene frequencies in a cline determined by selection and diffusion, Biometrics 6 (1950) 353 - 361.

5. Fleming, W.H., A selection-migration model in population genetics, Journal Math. Biology 2 (1975) 219 - 233.

6. Sattinger, D.H., Topics in stability and bifurcation theory, Lecture Notes in Mathematics, Vol. 309, Springer, New York (1973).

A Generalization of the Flaschka-Leitman Theorem.

Andrew T. Plant

The purpose of this paper is to give a simple proof of the 'Flaschka-Leitman Property' for the nonlinear nonautonomous functional differential equation

$$x'(t) = F(t,x(t),x_t) \qquad x_s = \phi \in C \tag{1}$$

in Banach space X. Here $C = C(I;X)$, $I = [-r,0]$ or $(-\infty,0]$ and $x_t \in C$, $x_t(\theta) = x(t+\theta)$.

If this problem has a unique continuous solution then we may define the evolution operator $U(t,s)$ at ϕ by $U(t,s)\phi = x_t$. Notice that

$$x(t) = \begin{cases} U(s,s)\phi(t-s) & t \leq s \\ U(t,s)\phi(0) & t > s \end{cases} \tag{2}$$

The generators $A(t)$ of U are formally given by

$$D(A(t)) = \{ \phi \mid \phi \text{ continuously differentiable}, \; \phi'(0) = F(t,\phi(0),\phi) \} \tag{3}$$

$$A(t)\phi = -\phi'$$

For a demonstration of this we refer the reader to [3] and [5].

In [3],[7],[8] sufficient conditions on F are given to ensure that the generators A(t) satisfy the assumptions of the Crandall-Pazy Theorem [2]. Hence we may define the productintegral

$$U(t,s)\phi = \lim_{\substack{n\to\infty \\ m/n\to t-s}} \prod_{i=1}^{m} [I + 1/nA(s + i/n)]^{-1}\phi \tag{4}$$

and in passing note that this limit is uniform on compact t-sets, and $U(t,s)\phi$ is continuous in t.

The interesting and important question now arises as whether or not the segments $U(t,s)\phi$ 'fit together' to form a continuous curve in X. More precisely: Is it true that $U(t,s)\phi = x_t$ where $x(t)$ is defined by (2)? An equivalent assertion is that $\psi(\theta,\sigma) = U(\sigma,s)\phi(\theta)$ is constant on lines $\theta + \sigma$ = constant.

An affirmative answer to this problem is given by Flaschka and Leitman in [4] for the case $F(t,\phi(0),\phi) = G(\phi)$, G Lipschitz continuous from $C(I;X)$ to X and I compact. In this case $U(t,s)$ reduces to a semigroup $T(t) = U(t+s,s)$ and convergence of (4) is a consequence of the Crandall-Liggett Theorem [1]. The proof given by Flaschka and Leitman of this property uses a probabilistic argument. We prove the following generalization using an elementary method.

THEOREM. *For* $t \geq 0$ *let* $D(A(t)) \subset C^1(I;X)$ *and* $A(t)\phi = -\phi'$. *Moreover suppose for some* $\phi \in C(I;X)$ *the productintegral* (4) *converges uniformly on compact* t-*sets to* $U(t,s)\phi$, *and* $t \to U(t,s)\phi$ *is continuous. Then* $\psi(\theta,\sigma) = U(\sigma,s)\phi(\theta)$ *is constant on lines* $\theta + \sigma$ = *constant*.

We point out that if $A(t)$ is defined by (3) then the above theorem makes no explicit assumptions on F.

However the existence and convergence of the productintegral (4) will require further restrictions on $D(A(t))$ and hence on F.

PROOF OF THEOREM. First note that there is no loss in generality in assuming $s = 0$. Since $U(t,0)\phi$ is continuous in t, $\psi(\theta,\sigma)$ is continuous on $I \times [0,\infty)$. Now for sufficiently large positive integer n define the functions ψ_n on $I \times [0,\infty)$ by

$$\psi_n(\theta,\sigma) = \prod_{i=1}^{m} [I + 1/nA(i/n)]^{-1}\phi(\theta) \qquad \frac{m-1}{n} < \sigma \leq \frac{m}{n}$$

Then $\psi_n \to \psi$ uniformly on compact subsets of $I \times [0,\infty)$. Moreover

$$(I + 1/nA(m/n))\psi_n(\theta,m/n) = \psi_n(\theta,(m-1)/n)$$

and so

$$\psi_n(\theta,m/n) - \psi_n(\theta,(m-1)/n) = \frac{1}{n}\frac{\partial}{\partial\theta}\psi_n(\theta,m/n)$$

Integrating this expression with respect to θ from α to β

$$\int_\alpha^\beta \psi_n(\theta,m/n) - \psi_n(\theta,(m-1)/n)\, d\theta = \frac{1}{n}[\psi_n(\beta, m/n) - \psi_n(\alpha,m/n)]$$

$$= \int_{\frac{m-1}{n}}^{m/n} \psi_n(\beta,\sigma) - \psi_n(\alpha,\sigma)\, d\sigma$$

Now add for $m = p + 1, \ldots\ldots, q$

$$\int_\alpha^\beta \psi_n(\theta,q/n) - \psi_n(\theta,p/n)\,d\theta = \int_{p/n}^{q/n} \psi_n(\beta,\sigma) - \psi_n(\alpha,\sigma)\,d\sigma$$

Finally let $n \to \infty$, $p/n \to s$, $q/n \to t$ to obtain

$$\int_\alpha^\beta \psi(\theta,t) - \psi(\theta,s)\,d\theta = \int_s^t \psi(\beta,\sigma) - \psi(\alpha,\sigma)\,d\sigma$$

Hence $H(\alpha,\beta;s,t) = \int_s^t \int_\alpha^\beta \psi(\theta,\sigma)\,d\theta\,d\sigma$ has continuous first partial derivatives satisfying $H_t + H_s = H_\beta + H_\alpha$. Consequently

$$G_\delta(\tau) = \delta^{-2} H(\ \theta-\tau\ ,\ \theta+\delta-\tau\ ;\ \sigma+\tau\ ,\ \sigma+\delta+\tau\ )$$

is continuously differentiable and $G'_\delta(\tau) = 0$. But $G_\delta(\tau) \to \psi(\theta - \tau\ ,\ \sigma + \tau)$ as $\delta \downarrow 0$. Hence $\psi(\theta,\sigma)$ is constant on lines $\theta + \sigma = \text{constant}$.

We point out that we have generalized the main result in [4] and Proposition 1 of [3]. Further generalizations may be found in [6] where the interesting case of L^p initial data is also considered.

As remarked at the end of [4], formally one has $\partial/\partial\sigma U(\sigma,s)\phi(\theta) = -A(\sigma)U(\sigma,s)\phi(\theta) = \partial/\partial\theta U(\sigma,s)\phi(\theta)$. That is $\psi_\sigma - \psi_\theta = 0$ with boundary data $\psi(\theta,s) = \phi(\theta)$. The boundary data on the σ-axis is nonclassical : It has functional form. However the property that ψ is constant on the characteristics $\theta + \sigma = \text{constant}$ is, as we have shown, preserved from the classical case.

REFERENCES

[1] M.G.Crandall & T.M.Liggett, Generation of semigroups of nonlinear transformations on general Banach spaces, Amer. J. Math., 93 (1971), 265-298.

[2] ___________ & A.Pazy, Nonlinear evolution equations in Banach spaces, Israel J. Math. 11 (1972), 57-94.

[3] J.Dyson & R.V.Bressan, Functional differential equations and nonlinear evolution operators, Edinburgh J.Math. (To appear).

[4] H.Flaschka & M.J.Leitman, On semigroups of nonlinear operators and the solution of the functional differential equation $\dot{x}(t) = F(x_t)$, J. Math. Anal. and Appl., 49 (1975), 649-658.

[5] J.Hale, Functional differential equations, Appl. Math. Series, Vol.3, Springer-Verlag, New York, 1971.

[6] A.T.Plant, Nonlinear semigroups of translations in Banach space generated by functional differential equations, J. Math. Anal. and Appl. (Submitted).

[7] G.F.Webb, Autonomous nonlinear functional differential equations and nonlinear semigroups, J. Math. Anal. and Appl., 46 (1974) 1-12.

[8] _______ , Asymptotic stability for abstract nonlinear functional differential equations, Proc. Amer. Math. Soc. (To appear).

Hilbert's Projective Metric Applied to a Class of Positive Operators

A. J. B. Potter

1. Introduction

This paper is concerned with the study of a class of positive operators defined on a cone in a Banach space. Operators in this class have been called strictly (-1)-convex. There are several examples of such operators.

It has been known for some time that positive homogeneous operators are contractions in Hilbert's projective metric. In [4] it is shown that much larger classes of positive operators are contractions in Hilbert's metric and therefore existence theorems for such can be proved by mere applications of the contraction mapping principle. Strictly (-1)-convex operators are a generalization of a class of operators considered in [4]; we show in this paper that they have very similar properties.

We apply our theorems to the study of non-linear Hammerstein equations. It is interesting to note that we can allow the non-linear term in our equations to have a singularity at the origin.

Much of the work in this paper is motivated by results which appear in Krasnoselskii's book (Chapter 6 [2]). Many of our proofs are motivated by proofs in that chapter.

2. Preliminaries

Throughout this paper X denotes a real Banach space with a closed solid cone K, int(K) denotes the interior of K and U the unit sphere in X. The relations $\leq$ and $<$ are defined in the usual way (i.e.: $x \leq y$ and $x < y$ if and only if $y - x \in K$ and $y - x \in \text{int}(K)$ respectively). $x \lneqq y$ means $y - x \notin K$. We assume that the norm is monotonic on int(K), that is if $0 < x \leq y$ then $\|x\| \leq \|y\|$. $\mathbb{R}^+$ denotes the non-negative reals and $\mathbb{R}^{++}$ the positive reals. For $r \in \mathbb{R}^+$ we define $E_r = \{x \in \text{int}(K): \|x\| = r\}$.

2.1 Definition: For $x,y \in \text{int}(K)$ define

$$M(x/y) = \inf\{\lambda: x \leq \lambda y\},$$

$$m(x/y) = \sup\{\mu: \mu x \leq y\},$$

and $$d(x/y) = \log(M(x/y)/m(x/y)).$$

(d is called Hilbert's projective metric.)

2.2 Proposition: For $x,y \in \text{int}(K)$

(i) $0 < m(x/y)y \leq x \leq M(x/y)y$,

(ii) $d(\lambda x,\mu y) = d(x,y)$ for all $\lambda,\mu \in \mathbb{R}^{++}$,

(iii) if $x,y \in E_r$ then $0 < m(x/y) \leq 1 \leq M(x,y) < \infty$,

(iv) (E_r,d) is a metric space for all $r \in \mathbb{R}^{++}$.

Proof: A proof of these statements can be found in [1].

2.3 Definition: Let A be a mapping from int(K) into int(K) (we say A is positive). Then

(i) A is decreasing if $x \leq y$ implies $A(x) \geq A(y)$,

(ii) for $E \subset \text{int}(K)$, the contraction ratio, $K(A,E)$, of A on E is defined by

$$K(A,E) = \inf\{\lambda: d(A(x),A(y)) \leq \lambda d(x,y) \text{ for all } x,y \in E\}.$$

The main consideration of this paper is the study of a class of positive mappings which satisfy a certain "convexity" type condition. To be precise we make the following definition.

2.4 Definition: Let A be a positive operator defined on int(K). We say A is strictly (-1)-convex if for each $x \in \text{int}(K)$ and for each t such that $0 < t < 1$ there exists a number $\eta(x,t) > 0$ such that

$$A(tx) \leq (1-\eta)t^{-1}A(x). \quad \text{(A)}$$

2.5 Remarks: (i) A is strictly (-1)-convex if and only if for each $x \in \text{int}(K)$ and s such that $s > 1$ there exists a number $\eta'(x,t) > 0$ such that

$$A(sx) \geq (1+\eta')s^{-1}A(x). \quad \text{(B)}$$

(ii) An elementary example of a strictly (-1)-convex operator is the following. Take $X = \mathbb{R}$, $K = \mathbb{R}^+$ and so $\mathrm{int}(K) = \mathbb{R}^{++}$. Define $A: \mathrm{int}(K) \to \mathrm{int}(K)$ by

$$A(u) = 1/(u^\alpha + u^\beta) \text{ where } \alpha > \beta > 0.$$

It is easy to show that A is positive strictly (-1)-convex. It is also decreasing. Note the behaviour of A as $u \to 0$.

(iii) It is not difficult to show that if A is strictly (-1)-convex and decreasing then A^2 is u_0-concave in the sense of Krasnoselskii (for any $u_0 \in \mathrm{int}(K)$) (see page 187 [2] for the definition of u_0-concave).

3. Eigenvalue Problems

In this section we consider the solution set to the eigenvalue problem

$$A(x) = \lambda x. \tag{C}$$

We assume A is a positive operator defined on int(K) which is strictly (-1)-convex and decreasing. A positive solution to (C) is an element $(x,\lambda) \in \mathrm{int}(K) \times \mathbb{R}^{++}$ which satisfies (C).

3.1 Lemma: Suppose (x_1,λ_1) and (x_2,λ_2) are positive solutions of (C). If $\lambda_1 \geq \lambda_2$ then $x_1 \leq x_2$.

Proof: Suppose $x_1 \not\leq x_2$. Then $M(x_1/x_2) > 1$ (by definition of M). Put $M = M(x_1/x_2)$. Then

$$\begin{aligned} x_1 &= (1/\lambda_1)A(x_1) \geq (1/\lambda_1)A(Mx_2) \\ &\geq (1/\lambda_1)(1+\eta')M^{-1}A(x_2) \\ &= (\lambda_2/\lambda_1)(1+\eta')M^{-1}x_2, \end{aligned}$$

where $\eta' > 0$.

Suppose $\xi = (\lambda_2/\lambda_1)(1+\eta')M^{-1} \leq 1$. Then

$$\lambda_1 x_1 = A(x_1) \leq A(\xi x_2) \leq (1-\eta)\xi^{-1}A(x_2) = (1-\eta)\xi^{-1}\lambda_2 x_2$$

where $\eta \geq 0$. Hence $\lambda_1 x_1 \leq \xi^{-1}\lambda_2 x_2$. Thus

$$x_1 \leq (1/(1+\eta'))Mx_2,$$

and so by definition of M

$$M \le (1/(1+\eta'))M,$$

which is impossible since $M > 1$. Hence $\xi > 1$, and so $(\lambda_2/\lambda_1) > M/(1+\eta')$. But

$$A(x_2) \ge A(Mx_2) \ge (1+\eta')M^{-1}A(x_2)$$

(since $M > 1$ and since A is decreasing). Therefore

$$M/(1+\eta') \ge 1,$$

so $\lambda_2 > \lambda_1$. This proves the lemma.

<u>3.2 Corollary</u>: For each $\lambda > 0$ there is at most one positive solution to (C).

We now consider the existence of solutions to equation (C). It is not difficult to show that if (x_1,λ_1) and (x_2,λ_2) are two solutions then there exists a solution to $A(x) = \lambda x$ for all λ such that $\lambda_2 \le \lambda \le \lambda_1$ under only slight further assumptions on A (see Theorem 4.8 [3]). This requires merely an application of well known fixed point theorems to the operator A_λ^2 (where $A_\lambda = (1/\lambda)A$). (It is easy to deduce that A_λ^2 has a unique fixed point in int(K) and thus A_λ has a fixed point.) But this leaves undecided whether there actually exists a solution of (C). This is the point at which Hilbert's projective metric is used.

<u>3.3 Lemma</u>: Let A be a positive strictly (-1)-convex decreasing operator on int(K). Then

$$d(A(x),A(y)) < d(x,y) \text{ for all } x,y \in E_r.$$

(Thus $K(A,E_r) \le 1$).

<u>Proof</u>: Let $x,y \in E_r$. Then

$$0 < m(x/y)y \le x \le M(x/y)y$$

(by 2.2(i)). Since A is decreasing

$$A(My) \le A(x) \le A(my)$$

where $M = M(x/y)$ and $m = m(x/y)$. But by 2.2(iii) $0 < m \le 1 \le M$ and so by definition of strictly (-1)-convex and 2.5(i) there exist η and $\eta' > 0$ such that

$$(1+\eta')M^{-1}A(y) \le A(x) \le (1-\eta)m^{-1}A(y).$$

Thus $$M(A(x)/A(y)) \le (1-\eta)m^{-1}$$

and $$m(A(x)/A(y)) \ge (1+\eta')M^{-1}.$$

Therefore $$d(A(x),A(y)) \le \log(M(1-\eta)/m(1+\eta))$$
$$< \log(M/m) = d(x,y).$$

3.4 Remark: This lemma is useful in proving existence of solutions to (C). Consider the operator $T: E_r \to E_r$ defined by $T(x) = rA(x)/\|A(x)\|$. Note that the existence of a fixed point of T is equivalent to the existence of a solution of (C). Also

$$\begin{aligned} d(T(x),T(y)) &= d(A(x)/\|Ax\|, A(y)/\|Ay\|) \\ &= d(A(x),A(y)) \qquad \text{(see 2.2(ii))} \\ &< d(x,y) \qquad \text{for all } x,y \in E_r. \end{aligned}$$

It is suggestive that the solution of a fixed point of T could be proved by the contraction mapping principle. This idea is applied to integral equations in the next section.

4. An Integral Equation

We consider the positive solutions of the eigenvalue problem

$$\lambda u(x) = \int_\Omega k(x,y)f(y,u(y))dy, \quad x \in \Omega \tag{D}$$

where $\Omega \subset \mathbb{R}^n$ is compact, $k: \Omega \times \Omega \to \mathbb{R}^{++}$ is continuous and $f: \Omega \times \mathbb{R}^{++} \to \mathbb{R}^{++}$ is also continuous. [Note we can allow the possibility that $f(x,u) \to \infty$ as $u \to 0$.]

Let $X = C(\Omega)$ endowed with supremum norm and let $K = \{u \in X: u(x) \ge 0 \text{ for all } x \in \Omega\}$. Then $\text{int}(K) = \{u \in X: u(x) > 0 \text{ for all } x \in \Omega\}$. Define $C: \text{int}(K) \to \text{int}(K)$ by

$$Cv(x) = \int_\Omega k(x,y)v(y)dy, \quad x \in \Omega.$$

It can be shown (see [1]) that if

$$a = \min\{k(x,y): (x,y) \in \Omega \times \Omega\} \text{ and}$$
$$b = \max\{k(x,y): (x,y) \in \Omega \times \Omega\}$$

then $$d(C(x),C(y)) \le \alpha d(x,y) \text{ for all } x,y \in \text{int}(K)$$

where $\alpha = \tanh((1/2)\log(b/a)) < 1$. Define $F: \text{int}(K) \to \text{int}(K)$ by $F(u)(x) = f(x,u(x))$, we assume that for each x, $f(x,u)$ is strictly

(-1)-convex and decreasing as a function of u (see 2.5(ii)). It follows that F is strictly (-1)-convex and decreasing. Since C is positive and linear it follows that A = CF is positive decreasing and for $x,y \in E_r$.

$$d(A(x),A(y)) = d(CF(x),CF(y)) \leq \alpha d(F(x),F(y)) \leq \alpha d(x,y)$$

(the last inequality follows from Lemma (3.3)).

Now E_r is complete in this special case (see [1]) and so the operator $T: E_r \to E_r$ defined by $T(x) = A(x)/\|A(x)\|$ has a unique fixed point (see (3.4)) in E_r. That is for each $r > 0$ there exists $\lambda_r > 0$ and $u_r \in E_r$ such that $A(u_r) = \lambda_r u_r$. We have solved equation (D). Moreover for each r, u_r is unique. In view of Lemma (3.1) and the remarks made after it, it would seem reasonable to suppose the following diagram is a true image of the solution set.

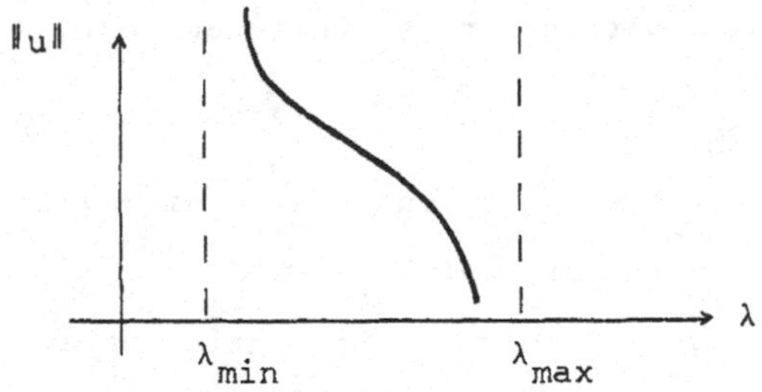

where λ_{min} = inf{λ: there exists a positive solution to (D)}

λ_{max} = sup{λ: there exists a positive solution to (D)}.

Bounds for λ_{min} and λ_{max} would have to be achieved by a closer scrutiny of the equation involved.

5. References

1. Bushell, P.J., Hilbert's metric and positive contraction mappings in a Banach space. Arch. Rat. Mech. Anal. 52, 4, 330-338 (1973).

2. Krasnoselskii, M.A., Positive solutions of operator equations. Noordoff (Groningen) (1964).

3. Potter, A.J.B., Existence theorem for a non-linear integral equation. J. Lond. Math. Soc. (2), 11, 7-10 (1975).

4. Potter, A.J.B., Applications of Hilbert's projective metric to certain classes of non-homogeneous operators (to appear).

A Limit-point Criterion for $-(py')' + qy$

Thomas T. Read

We shall establish a limit-point criterion for the formally symmetric second order differential expression

$$My = -(py')' + qy \tag{1}$$

on the interval $[a,\infty)$. Here p and q are real-valued functions with p locally absolutely continuous and q locally integrable. The criterion includes several known criteria, including both interval criteria and "criteria of Levinson type", and is particularly effective in dealing with expressions in which q is oscillatory but $|\int_a^x q|$ increases relatively slowly.

A simple example of the type of oscillatory behavior that can be handled is

$$-y'' - (xe^x \sin(e^x))y \tag{2}$$

or, in fact,

$$-y'' - x\,G'(x)y \tag{3}$$

where G is any bounded differentiable function. We shall return to (3) following the proof of Theorem 1.

More generally, it suffices for q to be decomposable into the sum of two functions, one of which is as described above and the other of which is bounded below by a suitable function on a part at least of $[a,\infty)$. It is, for instance, a special case of Theorem 3 that, with $p = 1$, if the negative part, q^-, of q satisfies $\int_a^x q^- \le Kx^3$ then q can be so decomposed.

Our main result is as follows.

<u>Theorem 1</u>. If there is a nonnegative locally absolutely continuous function w and a decomposition $q = q_1 + q_2$ of q such that

(i) $pw'^2 \le K_1$,

(ii) $-q_1 w^2 \le K_2$,

(iii) $p^{-1/2} w^{\alpha}(x) |\int_a^x q_2 w^{1-\alpha}| \le K_3$ for some constant α, $0 \le \alpha \le 1$,

(iv) $\int_a^{\infty} wp^{-1/2} = \infty$,

then (1) is limit-point.

Remarks 1. The coefficient q may be complex-valued if the conclusion is rephrased to the statement that $My = 0$ has a solution not in $L_2(a,\infty)$. This is done in [6]. In this situation the decomposition is of the real part of q; the imaginary part does not affect matters.

Remarks 2. When $q_2 = 0$ and $w > 0$ the change of notation $w^{-2} = M$ transforms Theorem 1 into the well-known limit-point criterion of Levinson [5].

Remarks 3. The case $p = 1, \alpha = 0$ is closely related to a theorem of Knowles [4]. Details are given in [6].

Remarks 4. Hypotheses (i) and (ii) may be combined into $pw'^2 - q_1 w^2 \le K$ provided $\alpha = 1$ in (iii). This version may be proved in a way very close to Theorem 1 of Atkinson and Evans [2] which contains a similar hypothesis.

Proof: It will be shown that $py'^2 w^2$ is in $L_1(a,\infty)$ whenever $My = 0$ and y is in $L_2(a,\infty)$. Assuming this, suppose that u and v are solutions with $p(u'v-uv') = 1$. Then

$$(p^{1/2}u'w)v - (p^{1/2}v'w)u = wp^{-1/2} \tag{4}$$

so that if both u and v are in $L_2(a,\infty)$, then the left side of (4) is in $L_1(a,\infty)$ by Schwarz's inequality. But the right side of (4) is not in $L_1(a,\infty)$ by (iv), so $My = 0$ must have a solution not in $L_2(a,\infty)$.

To establish the assertion concerning py'^2w^2, suppose that $My = 0$ with y in $L_2(a,\infty)$. Then

$$(py')'yw^2 - q_1y^2w^2 - q_2y^2w^2 = 0. \tag{5}$$

We integrate (5) from a to x and investigate the terms separately. First,

$$\int_a^x (py')'yw^2 = py'yw^2]_a^x - \int_a^x py'^2w^2 - 2\int_a^x pyy'ww'.$$

Set $H(x) = \int_a^x py'^2w^2$. Then by (i) and Schwarz's inequality,

$$2|\int_a^x pyy'ww'| \leq 2K_1|\int_a^x p^{1/2}yy'w| \leq 2K_1||y||H^{1/2}(x).$$

Next, with $Q(x) = \int_a^x q_2w^{1-\alpha}$,

$$\int_a^x q_2y^2w^2 = y^2w^{1+\alpha}Q(x) - \int_a^x Q[2yy'w^{1+\alpha} + (1+\alpha)y^2w^\alpha w'].$$

Now $|y^2w^{1+\alpha}Q| \leq K_3p^{1/2}y^2w$. Also, by Schwarz's inequality,

$$|\int_a^x Qyy'w^{1+\alpha}| \leq K_3 \int_a^x |p^{1/2}yy'w| \leq K_3||y||H^{1/2}(x),$$

and

$$|\int_a^x Qy^2w^\alpha w'| \leq K_3 \int_a^x |p^{1/2}y^2w'| \leq K_1K_3||y||^2.$$

Finally, $-\int_a^x q_1y^2w^2 \leq K_2||y||^2$.

Thus, from (5),

$$A(x) = py'yw^2(x) + K_3 p^{1/2}y^2w(x) \geq H(x) + C_1 H^{1/2}(x) + C_2. \tag{6}$$

The right side of (6) is either bounded (if H is) or approaches infinity with x (if H does). Our assertion is precisely that the first alternative must occur. Thus the proof will be complete if we can show that the left side of (6) cannot be eventually bounded away from 0.

To see this, suppose $A(x) \geq c > 0$ for all $x \geq x_0$. Set

$$N = \{x \geq x_0 \colon yy'(x) < 0\}.$$

Then on N, $K_3p^{1/2}y^2w > |py'yw^2|$. Also $K_3p^{1/2}y^2w > c$ there so that $(K_3/c)y^2 > (p^{1/2}w)^{-1}$. Thus

$$-\int_N y'/y < K_3 \int_N (p^{1/2}w)^{-1} < (K_3{}^2/c) \int_N y^2 < \infty.$$

But then $\log(y(x)/y(x_0))$ is bounded below, that is, y is bounded away from 0 on $[x_0,\infty)$. This is impossible since y is in $L_2(a,\infty)$ and the proof is complete.

To return now to the example (3) mentioned above, note that the hypotheses of Theorem 1 are satisfied with $q_1 = 0$, $q_2(x) = -xG'(x)$, $w(x) = x^{-1}$, and $\alpha = 0$ since then $|\int_a^x q_2 w| \le 2||G||_\infty$.

As a more involved illustration of this type of choice of w we derive a limit-point criterion due to J. Walter [7].

<u>Theorem 2</u>. Let $p \in C^2(a,\infty)$. If there is a positive $C^2(a,\infty)$ function g such that

(i) $|(g^2p^{1/2})'p^{1/2}| \le K$,

(ii) $-q \le 1/pg^4 - (pg')'/g$,

(iii) $\int_a^\infty g^2 = \infty$,

then (1) is limit-point.

<u>Proof</u>: Make the change of variable $s = \int_a^x p^{-1/2}$. If $h(s) = x$, then $h'(s) = p^{1/2}(h(s))$. If $My = 0$ and $z(s) = y(h(s))p^{1/4}(h(s))$, then

$$-z''(s) + Q(h(s))z(s) = 0 \tag{7}$$

where $Q = q - (p(p^{-1/4})')'/p^{-1/4}$. Moreover $\int_0^\infty z^2 ds = \int_a^\infty y^2 dx$ so that (1) is limit-point if and only if (7) is.

Set $w = g^2p^{1/2}$ and $W(s) = w(h(s))$. Then $|W'(s)| = |(g^2p^{1/2})'p^{1/2}(h(s))| \le K$ and $\int_0^\infty W\,ds = \int_a^\infty w\,p^{-1/2}\,dx = \int_a g^2 dx = \infty$.

Now write $Q = Q_1 + Q_2$ where $Q_2 = (1/2)(p^{1/2}w')'/wp^{-1/2}$. It is straightforward to verify that

$$(pg')'/g = (p(p^{-1/4})')'/p^{-1/4} + (1/2)(p^{1/2}w')'/wp^{-1/2} - (1/4)pw'^2/w^2.$$

Thus $-Q_1 = -Q + Q_2 = -q + (p(p^{-1/4})')'/p^{-1/4} + (1/2)(p^{1/2}w')'/wp^{-1/2}$

$$\leq (pg^4)^{-1} + (1/4)pw'^2/w^2 \quad \text{by (ii)}$$

$$\leq W^{-2}(1+W'^2/4) \leq W^{-2}(1+K^2/4).$$

Also $2\int_0^s Q_2 W\, ds = \int_a^{h(s)} (p^{1/2}w')'dx = (g^2p^{1/2})'p^{1/2}]_a^{h(s)}$ is bounded by (i). Thus, Theorem 2 follows from Theorem 1 with $\alpha = 0$.

If $w = 0$ on subintervals of $[a,\infty)$, then Theorem 1 becomes in effect an interval criterion--the intervals of interest being those on which $w > 0$. Consider, for instance,

$$-y'' - x^3[(1+\cos x)^2 + \cos(x^4)]y. \tag{8}$$

Here we shall take $q_1(x) = -x^3(1+\cos x)^2$ and $q_2(x) = -x^3 \cos(x^4)$. Now for all sufficiently large integers n, $x^3(1+\cos x)^2 < Kn$ when $|x - (2n+1)\pi| < n^{-1/2}$. Thus if for such n we define w on $[a_n,b_n] = [(2n+1)\pi-n^{-1/2},(2n+1)\pi+n^{-1/2}]$ by $w(x) = (x-a_n)(b_n-x)/(b_n-a_n)$, then $-q_1w^2 \leq K/4$ on $[a_n,b_n]$. If we set $w = 0$ on the remainder of $[a,\infty)$, then this inequality will be valid on the entire interval. Moreover, $|w'| \leq 1$ and $\int_{a_n}^{b_n} w = 2n^{-1}/3$ so that $\int_a^\infty w = \infty$. Finally, $w(x) \leq Kx^{-1/2}$ on $[a,\infty)$ so that $w(x)\int_a^x q_2 \to 0$ as $x \to \infty$. Thus it follows from Theorem 1 with $\alpha = 1$ that (8) is limit-point.

As an interval criterion Theorem 1 contains the following result which is due, for leading coefficient one, to Eastham [3].

Theorem 3. Suppose that there is a sequence $\{I_n\}_{n=1}^\infty$, $I_n = [a_n,b_n]$, of pairwise disjoint intervals in $[a,\infty)$ and a sequence $\{v_n\}_{n=1}^\infty$ of positive numbers such that for each n,

(i) $v_nP_n^2 \geq K > 0$ where $P_n = \int_{a_n}^{b_n} p^{-1/2}$,

(ii) $\sum_{n=1}^{\infty} v_n^{-1} = \infty$,

(iii) $\int_{a_n}^{b_n} q^- \, dt \le Cv_n^2 P_n^3 \min_{I_n} p^{1/2}$ where q^- is the negative part of q .

Then $-(py)' + qy$ is limit-point.

Remarks 1. Eastham [3] has shown that the hypotheses of Theorem 3 are satisfied for $-y'' + qy$ if $\int_a^x q^- \le Kx^3$. Thus in particular it will follow from the proof that any such function can be decomposed as in Theorem 1.

Remarks 2. A recent result of Atkinson [1, Theorem 11] can be established, for real coefficients, by an argument very similar to the proof of Theorem 3.

Proof. We may assume that $K \ge 1$. Divide each I_n into $[v_n P_n^2] + 1$ subintervals ([...] is the greatest integer function) so that on each subinterval J,

$$(2v_n P_n)^{-1} \le \int_J p^{-1/2} \le (v_n P_n)^{-1}.$$

On at least half of these,

$$\int_J q^- \le 2Cv_n P_n \min_{I_n} p^{1/2} \tag{9}$$

For each subinterval $J = [c,d] \subset I_n$ on which (9) is valid, choose $e \in (c,d)$ so that $\int_c^e p^{-1/2} = P_n/2$. Define w on each such subinterval by $w(x) = \int_c^x p^{-1/2}$ for $c \le x \le e$ and $w(x) = w(e) - \int_e^x p^{-1/2}$ for $e \le x \le d$. Then on J,

$$\max w = (1/2) \int_J p^{-1/2} \le (2v_n P_n)^{-1},$$

and

$$\int_J wp^{-1/2} = (1/4)(\int_J p^{-1/2})^2 \ge (4v_n P_n)^{-2}.$$

Set $w = 0$ on the remaining subintervals of each I_n and on the complement of the union of the I_n's.

Then for each n, since (9) holds on at least $(1/2)([v_n P_n^2] + 1)$ subintervals of I_n,

$$\int_{I_n} wp^{-1/2} \geq (4v_n P_n)^{-2}(1/2)v_n P_n^2 = (1/8)v_n^{-1}.$$

Hence by (ii), $\int_a^\infty wp^{-1/2} = \infty$. Also w is clearly absolutely continuous with $p(w')^2 \leq 1$ a.e.. Thus (i) and (iv) of Theorem 1 are satisfied. It remains to construct a suitable decomposition of q.

Define a step function q_0 which is constant on each of the subintervals $J = [c,d]$ constructed above (whether (9) holds or not) by

$$q_0 = \int_c^d q^-/(d-c)$$

on $[c,d]$.

Similarly, on the interval $[b_n, a_{n+1}]$ between I_n and I_{n+1}, set $q_0 = \int_{b_n}^{a_{n+1}} q^-/(a_{n+1}-b_n)$. Thus $[a,\infty)$ is the union of subintervals on each of which $\int (q^- - q_0) = 0$. Note that $q_0 w^2$ is bounded above on $[a,\infty)$. This is clear outside the union of the subintervals where (9) holds, for then $w = 0$. On a subinterval $J = [c,d] \subset I_n$ for which (9) is valid,

$$q_0 w^2 \leq 2Cv_n P_n \min_{I_n} p^{1/2}/(2v_n P_n^2)^2 (d-c) \leq C$$

since $\min p^{1/2}/(d-c) \leq (\int_J p^{-1/2})^{-1} \leq 2v_n P_n$.

Now decompose q by setting $q_1 = q^+ - q_0$, and $q_2 = -q^- + q_0$. From the previous paragraph, $-q_1 w^2 \leq C$ on $[a,\infty)$. Also $wp^{-1/2}(x) \int_a^x q_2$ is nonzero only in the subintervals on which (9) holds, and on such a subinterval $[c,d]$, $\int_a^x q_2 = \int_c^x q_2$ since $\int_a^c q_2 = 0$. Hence on such a subinterval

$$wp^{-1/2}(x)\left|\int_a^x q_2\right| \leq 2Cv_nP_n \min_{I_n} p^{1/2}/2v_nP_np^{1/2}(x) \leq C.$$

Thus (ii) and (iii) of Theorem 1 are also satisfied for this decomposition and Theorem 3 now follows from that result.

References

1. F. V. Atkinson, Limit-n criteria of integral type, Proc. Roy. Soc. Edinburgh(A), 73 (1974/75), 167-199.
2. F. V. Atkinson and W. D. Evans, On solutions of a differential equation which are not of integrable square, Math. Z., 127 (1972), 323-332.
3. M. S. P. Eastham, On a limit-point method of Hartman, Bull. London Math. Soc., 4 (1972), 340-344.
4. I. Knowles, A limit-point criterion for a second-order linear differential operator, J. London Math. Soc. 8 (1974), 719-727.
5. N. Levinson, Criteria for the limit-point case for second order differential operators, Časopis Pešt.Mat., (1949), 17-20.
6. T. T. Read, A limit-point criterion for expressions with oscillatory coefficients, to appear.
7. J. Walter, Bemerkungen Zu dem Grenzpunktfallkriterium von N. Levinson, Math Z. (1968), 345-350.

NONSYMMETRIC DIRAC DISTRIBUTIONS IN SCATTERING THEORY

Elemer E. Rosinger

1. INTRODUCTION

We consider the following Schroedinger equation, with the potential any positive power of the Dirac δ distribution

(1) $$(D^2+k^2+\alpha(\delta(x))^m)u(x) = 0, \quad x \in R^1,$$

and the initial conditions

(2) $$u(x_0) = y_0, \quad Du(x_0) = y_1$$

where $k \in R^1$, $m \in (0,\infty)$, $\alpha \in R^1$, $x_0 \in (-\infty,0)$, $y_0, y_1 \in R^1$.

It is shown, that function solutions of the form

(3) $$u(x) = u_-(x) + (u_+(x)-u_-(x)) \cdot H(x), \quad x \in R^1,$$

with $u_-, u_+ \in C^\infty(R^1)$ and H the Heaviside function, can be constructed.

More precisely, let $u_- \in C^\infty(R^1)$ be the unique solution of

(4) $$(D^2+k^2)u(x) = 0, \quad x \in R^1,$$

with the initial conditions (2), and suppose for $x \in R^1$

$$u_-(x) = \begin{cases} c_1 \cos kx + c_2 \sin kx & \text{if } k \neq 0 \\ c_1 + c_2 x & \text{if } k = 0 \end{cases}$$

If $m \in (0,1)$ and $\alpha \in R^1$, then $u_+ = u_-$.

If $m = 1$ and $\alpha \in R^1$, then, for $x \in R^1$

$$u_+(x) = \begin{cases} c_1 \cos kx + (c_2 - \alpha c_1/k) \sin kx & \text{if } k \neq 0 \\ c_1 + (c_2 - \alpha c_1)x & \text{if } k = 0 \end{cases}$$

If $m = 2$, then u_+, therefore u in (3), exists only for $\alpha = (n\pi)^2$, with $n = 0,1,2,\ldots$, and in this case

$$u_+ = \begin{cases} u_- & \text{if } \alpha = (2n\pi)^2, \text{ with } n = 0,1,2,\ldots \\ -u_- & \text{if } \alpha = ((2n+1)\pi)^2, \text{ with } n = 0,1,2,\ldots \end{cases}$$

If $m \in (2,\infty)$ and $\alpha \in (0,\infty)$, then u_+ is not unique and for $x \in R^1$

$$u_+(x) = \begin{cases} \sigma c_1 \cos kx + (\sigma c_2 + K c_1/k) \sin kx & \text{if } k \neq 0 \\ \sigma c_1 + (\sigma c_2 + K c_1)x & \text{if } k = 0 \end{cases}$$

for any $\sigma \in \{-1,1\}$ and $K \in R^1$ given. In particular, for $K = 0$ we obtain $u_+ = \pm u_-$.

For m positive integer, the equation (1), (2) containing the power $(\delta(x))^m$ of the Dirac δ distribution and the solution (3) are considered within the associative and commutative algebras with unit element and containing the distributions in $D'(R^1)$, constructed in [7], [8], [9] and [10]. In these algebras the Dirac δ distribution and its derivatives are necessarily nonsymmetric (see [9]) since relations as

$$D^q\delta(-x) \neq cD^q\delta(x), \quad \forall \; q = 0,1,2,\ldots, \quad c \in C^1$$

hold.

In the case of $m \in (0,\infty)$ *arbitrary, the equation (1), (2), the power* $(\delta(x))^m$ *and the solution (3) are considered in the usual "weak" sense.*

2. THE WEAK SOLUTION u^*

For $\omega > 0$, $K \in R^1$ *we define* $V(\omega,K,x) = \begin{cases} 0 \text{ if } x \in R^1 \setminus (0,\omega) \\ K \text{ if } x \in (0,\omega) \end{cases}$

and we consider $\alpha(\delta(x))^m$ *being represented by*

$$(5) \qquad \lim_{\omega \to 0} V(\omega,\alpha/\omega^m,\cdot)$$

assuming for $\delta(x)$ *the nonsymmetric representation given by* $\lim_{\omega\to 0} V(\omega,1/\omega,\cdot)$. *It is important to mention that in the case of* m *positive integer, by considering (1), (2) and (3) within the mentioned algebras containing* $D'(R^1)$, *the solutions (3) will be independent of the particular "weak" process (5).*

The equation (1) is replaced by

$$(6) \qquad (D^2+k^2+V(\omega,\alpha/\omega^m,x))u(x) = 0, \quad x \in R^1.$$

Denote by $u(\omega,\cdot) \in C^1(R^1) \cap C^\infty(R^1 \setminus \{0,\omega\})$ *the unique solution of (6), (2) on* R^1.

Denote by A *the set of all* $(m,\alpha) \in (0,\infty) \times R^1$ *such that* $\forall\ y_0, y_1 \in R^1$: $\exists\ z_0, z_1 \in R^1$:

$$(7) \qquad \lim_{\omega\to 0} u(\omega,\omega) = z_0, \quad \lim_{\omega\to 0} Du(\omega,\omega) = z_1$$

It results that (for some of the proofs see the next section) A *does not depend on* $k \in R^1$ *and* $x_0 \in (-\infty,0)$. *For* $(m,\alpha) \in A$ *given, the initial conditions (2) define uniquely* $z_0, z_1 \in R^1$ *by (7).*

Denote by $u_+ \in C^\infty(R^1)$ the unique solution of (4) with the initial conditions

(8) $\qquad u(0) = z_o, \quad Du(0) = z_1$

and define

(9) $\qquad u^*(x) = u_-(x) + (u_+(x)-u_-(x)) \cdot H(x), \quad x \in R^1.$

It can easily be seen that

$$A = ((0,1] \times R^1) \cup ((0,\infty) \times \{0\}) \cup (\{2\} \times \{(n\pi)^2 \mid n = 0,1,2,\ldots\})$$

and for any $(m,\alpha) \in A$ we obtain $\lim_{\omega\to 0} u(\omega,\cdot) = u^*$ in the following sense

*) $\qquad u(\omega,\cdot) = u^*$ on $(-\infty,0]$, $\forall\ \omega > 0$

**) $\qquad$ for each $p = 0,1,2,\ldots$, $\lim_{\omega\to 0} D^p u(\omega,\cdot) = D^p u^*$ uniformly on every $[a,\infty)$, with $a > 0$, except the case $k = p = 0$, when the above limit is uniform only on every $[a,b]$, with $0 < a < b < \infty$.

We can now extend the set A and still obtain a weak solution u^*. Indeed, denote by B the set of all $(m,\alpha) \in (0,\infty) \times R^1$ such that

$\exists\ (\omega_\nu \mid \nu = 0,1,\ldots) \subset (0,\infty)$:

***) $\qquad \lim_{\nu\to\infty} \omega_\nu = 0$

****) $\qquad \forall\ y_o,y_1\ R^1 : \exists\ z_o,z_1 \in R^1$:

(10) $\qquad \lim_{\nu\to\infty} u(\omega_\nu,\omega_\nu) = z_o, \quad \lim_{\nu\to\infty} Du(\omega_\nu,\omega_\nu) = z_1$

It results again that B does not depend on $k \in R^1$ and $x_o \in (-\infty,0)$. Moreover, for $(m,\alpha) \in B$ given, the initial conditions (2) will result due to (10), in $z_o,z_1 \in R^1$.

Now, we can proceed as before and obtain u^* by (9).

It can be seen that $B = A \cup ((2,\infty) \times (0,\infty))$. Suppose given

$(m,\alpha) \in B \setminus A = (2,\infty) \times (0,\infty)$ and the initial conditions (2). The way we obtain z_o, z_1 through (10), depends on the sequence $(\omega_\nu \mid \nu = 0,1,\ldots)$. The weak solutions u^* obtained in this case, will still satisfy *) and **) above.

3. PROOFS

Consider for $h \in R^1$, the differential equation

$$(11) \qquad (D^2+h)y(x) = 0, \quad x \in R^1,$$

and for $x \in R^1$, the 2×2 matrix $W(h,x) = \exp(xA_h)$ where

$$A_h = \begin{pmatrix} 0 & 1 \\ -h & 0 \end{pmatrix}$$

If $v \in C^\infty(R^1)$ is the unique solution of (11) with the initial conditions $v(a) = b$, $Dv(a) = c$, where $a,b,c \in R^1$, then

$$(12) \qquad \begin{pmatrix} D^{2p}v(x) \\ D^{2p+1}v(x) \end{pmatrix} = (-1)^p h^p W(h,x)W(h,a)^{-1} \begin{pmatrix} b \\ c \end{pmatrix}, \quad \forall\ p = 0,1,\ldots, \quad x \in R^1$$

Assume now $(m,\alpha) \in (0,\infty) \times R^1$, $y_o, y_1 \in R^1$ given.
Then, for $\omega > 0$, (12) results in

$$(13) \qquad \begin{pmatrix} u(\omega,\omega) \\ Du(\omega,\omega) \end{pmatrix} = W(k^2+\alpha/\omega^m, \omega) W(k^2, x_o)^{-1} \begin{pmatrix} y_o \\ y_1 \end{pmatrix}$$

therefore $(m,\alpha) \in A$ only if $\lim\limits_{\omega \to 0} W(k^2+\alpha/\omega^m, \omega)$ exists and it is finite.
Suppose $k \neq 0$.
If $\alpha > 0$, then $k^2 + \alpha/\omega^m > 0$

hence

$$(14) \qquad W(k^2+\alpha/\omega^m,\omega) = \begin{pmatrix} \cos L & \frac{1}{H}\sin L \\ -H\sin L & \cos L \end{pmatrix}$$

with

$$(15) \qquad H = \sqrt{k^2 + \alpha/\omega^m}, \quad L = \omega H$$

Suppose now, $\alpha < 0$. Since we consider $\omega \to 0$, the assumption $k^2 + \alpha/\omega^m < 0$ can be made, hence

$$(16) \qquad W(k^2 + \alpha/\omega^m,\omega) = \frac{1}{2}\begin{pmatrix} \exp(L) + \exp(-L) & \frac{1}{2}(\exp(L) - \exp(-L)) \\ H(\exp(L) - \exp(-L)) & \exp(L) + \exp(-L) \end{pmatrix}$$

with

$$(17) \qquad H = \sqrt{-k^2 - \alpha/\omega^m}, \quad L = \omega H$$

In both cases of α,a simple computation of limits will give the required expression

$$A = ((0,1] \times R^1) \cup ((0,\infty) \times \{0\}) \cup (\{2\} \times \{(n\pi)^2 \mid n = 0,1,2,\ldots\}).$$

For $k = 0$, the above result follows easily.
The property *) of the weak solution u^* is immediate, while the property **) results from (12).

Now, we show that $B = A \cup ((2,\infty) \times (0,\infty))$. First, the inclusion $(2,\infty) \times (0,\infty) \subset B$. Suppose $(m,\alpha) \in (2,\infty) \times (0,\infty)$ given and $k \neq 0$. Then, (15) results in $\lim_{\omega\to 0} H = \lim_{\omega\to 0} L = +\infty$.
Therefore, given any sequence $(\omega_\nu \mid \nu = 0,1,\ldots) \subset (0,\infty)$, with $\lim_{\nu\to\infty} \omega_\nu = 0$, a necessary condition for the existence and finiteness of $\lim_{\nu\to\infty} W(k^2+\alpha/\omega_\nu{}^m, \omega_\nu)$ is, due to (14), that $\lim_{\nu\to\infty} \sin L_\nu = 0$, where

$$(18) \qquad L_\nu = \omega_\nu \sqrt{k^2+\alpha/\omega_\nu{}^m}.$$

This remark suggests the construction of sequences $(\omega_\nu \mid \nu = 0,1,\ldots)$ sa-

*tisfying ***) and ****). Indeed, define* $\theta : (0,\infty) \to (0,\infty)$ *by* $\theta(\omega) = \omega\sqrt{k^2+\alpha/\omega^m}$, *then, there exists* $A > 0$ *such that for each* $a > A$, *the equation* $\theta(\omega) = a$ *has exactly two solutions* $0 < \bar{\omega}_1(a) < \bar{\omega}_2(a)$.
Moreover

$$\lim_{a\to\infty} \bar{\omega}_1(a) = \infty. \tag{19}$$

Suppose now $(n_\nu \mid \nu = 0,1,\ldots)$ *is a sequence of positive integers and* $(e_\nu \mid \nu = 0,1,\ldots)$ *is a sequence of nonzero real numbers, such that* $\lim_{\nu\to\infty} n_\nu = \infty$, $\lim_{\nu\to\infty} e_\nu = 0$ *and* $n_\nu\pi + e_\nu > A$, *for* $\nu = 0,1,\ldots$
Define $(\omega_\nu \mid \nu = 0,1,\ldots)$ *with* $\omega_\nu = \bar{\omega}_1(n_\nu\pi + e_\nu)$, *then* $\lim_{\nu\to\infty} \omega_\nu = 0$, *due to (19). Taking into account (18), it results*
$L_\nu = \theta(\omega_\nu) = n_\nu\pi + e_\nu$, *hence*

$$\cos L_\nu = (-1)^{n_\nu} \cos e_\nu \tag{20}$$

Denoting

$$H_\nu = \sqrt{k^2 + \alpha/\omega_\nu^{\,m}} \tag{21}$$

it results $-H_\nu \sin L_\nu = (-1)^{n_\nu+1} e_\nu H_\nu \dfrac{\sin e_\nu}{e_\nu}$, *hence*

$$\lim_{\nu\to\infty} (-H_\nu \sin L_\nu) = \lim_{\nu\to\infty} (-1)^{n_\nu+1} e_\nu \sqrt{k^2+\alpha/(\bar{\omega}_1(n_\nu\pi+e_\nu))^m}. \tag{22}$$

We notice that $\bar{\omega}_1 \in C^1((A,\infty))$ *and* $\lim_{a\to\infty} \bar{\omega}_1^{-1}(a) = 0$, *thus*

$$\lim_{\nu\to\infty} \frac{e_\nu^{\,2}}{(\bar{\omega}_1(n_\nu\pi+e_\nu))^m} = \lim_{\nu\to\infty} \frac{1}{\left(\dfrac{\bar{\omega}_1(n_\nu\pi)}{|e_\nu|^{2/m}} + |e_\nu|^{1-\frac{2}{m}} \bar{\omega}_1^{-1}(n_\nu\pi+\xi_\nu e_\nu)\right)^m}$$

where $\xi_\nu \in (0,1)$, *hence*

$$\lim_{\nu\to\infty} |\, e_\nu \,|^{1-\frac{2}{m}} \bar{\omega}_1^{-1}(n_\nu\pi + \xi_\nu e_\nu) = 0.$$

Obviously, the limit

$$\lim_{\nu\to\infty} \frac{\bar{\omega}_1(n_\nu\pi)}{|e_\nu|^{2/m}}$$

can assume any value from 0 *up to and including* $+\infty$*, depending on a proper choice of* $n_\nu \to \infty$ *and* $e_\nu \to 0$.

Therefore, due to (22) and (20), (21), it results that for any $\sigma \in \{-1,1\}$ *and* $K \in R^1$*, there exists a sequence* $(\omega_\nu|\nu = 0,1,\ldots) \subset (0,\infty)$*, with* $\lim_{\nu\to\infty} \omega_\nu = 0$*, and such that*

$$\lim_{\nu\to\infty} W(k^2+\alpha/\omega_\nu^{\ m}, \omega_\nu) = \begin{pmatrix} \sigma & 0 \\ K & \sigma \end{pmatrix}$$

The inclusion $B \subset A \cup ((2,\infty) \times (0,\infty))$ *results from the relation* $B \cap ((2,\infty) \times (-\infty,0)) = \emptyset$ *which follows easily from (16) and (17). For* $k = 0$ *the proof follows in the same way. The inclusion* $A \subset B$ *is obvious.*

REFERENCE

1. Braunss G., Liese R.: *Canonical products of distributions and causal solutions of nonlinear wave equations.* J.Diff.Eq. 16,3,1974,399-412

2. Fuchssteiner B.: *Eine assoziative Algebra ueber einem Unterraum der Distributionen.* Math. Ann. 178,1968,302-314

3. Guettinger W.: *Generalized functions in elementary particle physics and passive system theory: recent trend and problems.* SIAM J.Appl.Math. 15,4,1967,964-1000

4. Mikusinski J.: *On the square of the Dirac delta distribution.* Bull.Acad.Pol.Sci. 14,9,1966,511-513

5. Rosinger E.: *Embedding of the $D'(R^n)$ distributions in pseudo-topological algebras.* *Stud.Cerc.Mat. 18,5,1966,687-729*

6. *Pseudotopological spaces. Embedding of the $D'(R^n)$ distributions into algebras.* *Stud.Cerc.Mat. 20,4,1968,553-582*

7. *A distribution multiplication theory.* *Haifa Technion's Preprint Series, AMT-31, October 1974*

8. *A modified distribution multiplication theory.* *Haifa Technion's Preprint Series, AMT-33, October 1974*

9. *The principle of nonsymmetry in the algebras containing the Schwartz distributions.* *Haifa Technion's Preprint Series, AMT-37, Januar 1975*

10. *Extensions of the distribution multiplication theory.* *Haifa Technion's Preprint Series, No. 50, June 1975*

11. Schwartz L.: *Sur l'impossibilite de la multiplication des distributions.* *C.R.Acad.Sci.Paris, 239,1954,847-848*

A Maximum Principle for a Class of Functionals in Nonlinear Dirichlet Problems

Philip W. Schaefer and René P. Sperb

1. Introduction

Recently, the Hopf maximum principles [5] for elliptic partial differential equations have been used to deduce inequalities for certain functionals which are defined on positive solutions of various linear and nonlinear elliptic boundary value problems. These inequalities then lead to upper and/or lower bounds for various important quantities in some physical problems of interest. Payne [2] used this procedure to compute bounds for the maximum stress in the Saint Venant torsion problem in terms of geometric properties of the cross section of the beam. In [3], Payne and Stakgold obtained bounds for the mean-to-peak neutron density ratio, a quantity of importance in a nuclear reactor operating at criticality. They extended their results to the nonlinear equation $\Delta u + w(u) = 0$ in [4]. More recently, Schaefer and Sperb [6] considered the Dirichlet and Robin problems for this nonlinear equation and extended and improved some of the earlier results by means of a functional which was optimal in a certain sense.

Here we shall extend the above mentioned procedure to the inhomogeneous, nonlinear Dirichlet problem in $n > 2$ dimensions. Specifically, we let D be a domain in Euclidean n-space with sufficiently smooth boundary ∂D and assume that u is a positive solution of

$$(1.1) \qquad \Delta u + \lambda\rho(x)f(u) = 0 \quad \text{in} \quad D$$

$$(1.2) \qquad u = 0 \quad \text{on} \quad \partial D ,$$

where Δ is the Laplace operator, λ is a positive parameter, $\rho(x)$ is a positive C^2 function in D, and $f(u)$ is a positive C^1 function of $u \geq 0$. The existence of positive solutions to (1.1), (1.2) has been discussed by several authors (see, for example, [1]). We shall define the functional and develop the maximum principle in section 2 and then remark about other problems, applications, and open questions in section 3.

2. The Maximum Principle

Let u be a positive solution of (1.1). We define the functional

$$\Phi = \frac{|\nabla u|^2}{\rho} g(u) + h(u) , \tag{2.1}$$

where g and h are arbitrary functions to be chosen so that Φ satisfies an elliptic differential inequality.

We shall use the comma notation for partial differentiation and the summation convention on repeated indices. Thus, we have

$$\Phi,_k = \frac{2gu,_i u,_{ik}}{\rho} + \frac{|\nabla u|^2 g' u,_k}{\rho} - \frac{|\nabla u|^2 g\rho,_k}{\rho^2} + h'u,_k \tag{2.2}$$

$$\begin{aligned}\Phi,_{kk} = {} & \frac{2gu,_{ik}u,_{ik}}{\rho} + \frac{2gu,_i u,_{ikk}}{\rho} + \frac{4u,_i u,_{ik} g' u,_k}{\rho} - \frac{4u,_i u,_{ik} g\rho,_k}{\rho^2} \\ & + \frac{|\nabla u|^4 g''}{\rho} + \frac{|\nabla u|^2 g' u,_{kk}}{\rho} - \frac{2|\nabla u|^2 g' u,_k \rho,_k}{\rho^2} - \frac{|\nabla u|^2 g\rho,_{kk}}{\rho^2} \\ & + \frac{2|\nabla u|^2 g\rho,_k \rho,_k}{\rho^3} + h''u,_k u,_k + h'u,_{kk} .\end{aligned} \tag{2.3}$$

From (2.2) it follows that

$$\begin{aligned}\frac{4u,_{ik}u,_{ik}u,_j u,_j g^2}{\rho^2} \geq {} & - H_k \Phi,_k + \frac{|\nabla u|^6 g'^2}{\rho^2} + \frac{|\nabla u|^4 g^2 |\nabla\rho|^2}{\rho^4} \\ & + |\nabla u|^2 h'^2 - \frac{2|\nabla u|^4 gg' u,_i \rho,_i}{\rho^3} + \frac{2|\nabla u|^4 g'h'}{\rho} \\ & - \frac{2|\nabla u|^2 gh' u,_i \rho,_i}{\rho^2} ,\end{aligned} \tag{2.4}$$

where $-H_k = \Phi,_k - \frac{2|\nabla u|^2 g' u,_k}{\rho} + \frac{2|\nabla u|^2 g\rho,_k}{\rho^2} - 2h'u,_k$.

We now use (1.1), (2.2), and (2.4) in (2.3) and collect terms so that

$$\Phi,_{kk} + \frac{L_k\Phi,_k}{|\nabla u|^2} \geq \frac{|\nabla u|^4}{\rho}\{g'' - \frac{3g'^2}{2g}\} + \frac{|\nabla u|^2 g|\nabla\rho|^2}{2\rho^3} + \frac{|\nabla u|^2 g' u,_i\rho,_i}{\rho^2}$$

$$(2.5) \qquad + \frac{u,_i\rho,_i}{\rho}\{h' - 2\lambda fg\} + |\nabla u|^2\{(h' - 2\lambda fg)' + \lambda fg' - \frac{h'g'}{g} - g\frac{(\Delta\rho)}{\rho^2}\}$$

$$+ \frac{\rho h'}{2g}\{h' - 2\lambda fg\} \quad ,$$

where $L_k = \frac{\rho}{2g} H_k$.

Since

$$(2.6) \qquad \frac{|\nabla u|^2 g' u,_i\rho,_i}{\rho^2} \leq \frac{|\nabla u|^2}{2}\{\frac{g'^2|\nabla u|^2}{g\rho} + \frac{g|\nabla\rho|^2}{\rho^3}\} \quad ,$$

we obtain

$$\Phi,_{kk} + \frac{L_k\Phi,_k}{|\nabla u|^2} \geq \frac{|\nabla u|^4}{\rho}\{g'' - \frac{2g'^2}{g}\} + \frac{u,_i\rho,_i}{\rho}\{h' - 2\lambda fg\}$$

$$(2.7) \qquad + |\nabla u|^2\{(h' - 2\lambda fg)' + \lambda fg' - \frac{h'g'}{g} - g\frac{(\Delta\rho)}{\rho^2}\}$$

$$+ \frac{\rho h'}{2g}\{h' - 2\lambda fg\} \ .$$

We now ask that g and h be chosen so that the right side of (2.7) is nonnegative. Clearly,

$$(2.8) \qquad g > 0, \ (g^{-1})'' \leq 0, \ h = 2\lambda\int_0^u f(\eta)g(\eta)d\eta, \ \Delta\rho \leq 0 \ \text{ in } D$$

will suffice. Thus we have,

THEOREM 1: If u is a positive solution of (1.1) and g, h, and ρ satisfy (2.8), then Φ takes its maximum either on ∂D or at a critical point of u.

We note that if we take $g \equiv 1$ at the outset, then there is no need for (2.6) and ρ must then satisfy

$$\frac{|\nabla\rho|^2}{2\rho} - \Delta\rho \geq 0 \quad \text{in } D,$$

i.e., $\Delta\rho^{1/2} \le 0$ in D. Thus for simplicity, we assume that

$$(2.9) \qquad g \equiv 1, \; h = 2\lambda\int_0^u f(\eta)d\eta, \; \Delta\rho^{1/2} \le 0, \text{ in } D.$$

In order to rule out the occurrence of the maximum at a point on ∂D, we consider $\frac{\partial\Phi}{\partial n}$ at an arbitrary point $P \in \partial D$. Since $|\nabla u| = -u_n$ on D, we have

$$(2.10) \qquad \frac{\partial\Phi}{\partial n} = \frac{2u_n u_{nn}}{\rho} - \frac{|\nabla u|^2}{\rho^2}\rho_n + 2\lambda f u_n,$$

where the subscript denotes the outer normal derivative. Moreover, on ∂D

$$\Delta u = u_{nn} + (n-1)Ku_n = -\lambda\rho f$$

for K the average curvature of ∂D, so that

$$(2.11) \qquad \frac{\partial\Phi}{\partial n} = -|\nabla u|^2\{\frac{2(n-1)K}{\rho} + \frac{1}{\rho^2}\frac{\partial\rho}{\partial n}\}.$$

Consequently, if

$$(2.12) \qquad 2(n-1)K + \frac{\partial}{\partial n}(\ell n\, \rho) \ge 0 \text{ on } \partial D$$

we conclude by Hopf's second maximum principle that Φ cannot take its maximum on ∂D. Hence we state

<u>THEOREM 2</u>: If u is a positive solution of (1.1), (1.2), where $\Delta\rho^{1/2} \le 0$ in D and $\frac{\partial}{\partial n}(\ln \rho) \ge -2(n-1)K$ on ∂D, then

$$\Phi = \frac{|\nabla u|^2}{\rho} + 2\lambda\int_0^u f(\eta)d\eta$$

takes its maximum at a critical point of u.

3. Remarks

Although the previous analysis is valid for n = 2 dimensions, certain improvements and extensions are possible when n = 2. Using the identity

$$u,_{ik}u,_{ik} = (\Delta u)^2 + 2(u,_{xy}^2 - u,_{xx}u,_{yy})$$

instead of (2.4), we obtain an elliptic equation that Φ satisfies; an equation which does not contain any $u,_i\rho,_i$ terms. Then choosing g, h, and ρ such that

$$g > 0,\ (\ell n\ g)'' \geq 0,\ h = 2\lambda\int_0^u f(\eta)g(\eta)d\eta,\ \Delta(\ell n\ \rho) \leq 0 \text{ in } D,$$

we arrive at the conclusion of Theorem 1. Moreover, when $g \equiv 1$, one obtains the boundary requirement (2.12) with $n = 2$ and K the curvature of ∂D. This boundary requirement is also encountered if u is subject to the mixed conditions

$$u = 0 \text{ on } \Gamma_1 \neq \emptyset,\ \frac{\partial u}{\partial n} = 0 \text{ on } \Gamma_2,\ \Gamma_1 \cup \Gamma_2 = \partial D.$$

In the event that (2.12) is not satisfied, one is able to include a term αu in the definition of Φ, where α is a positive constant to be chosen to satisfy the corresponding analog of (2.12). One is referred to [7] for more details.

As one application, we consider an extension of the results in [3]. In an inhomogeneous, monoenergetic, nuclear reactor operating at criticality, the neutron density ratio E defined by

$$E = \frac{\int_D \rho v dx}{v_m \int_D \rho dx} \tag{3.1}$$

plays a fundamental role. Here v is the first eigenfunction with associated eigenvalue λ of the inhomogeneous fixed membrane problem

$$\Delta v + \lambda\rho v = 0 \text{ in } D \tag{3.1}$$

$$v = 0 \text{ on } \partial D, \tag{3.2}$$

and v_m is the maximum value of v in $D \cup \partial D$. From Theorem 2 we deduce that

$$|\nabla v| \leq (\lambda\rho)^{1/2} v_m\ . \tag{3.3}$$

Now integrating over the boundary and observing that

(3.4) $$-\int_{\partial D} \frac{\partial v}{\partial n}\, ds = \lambda \int_D \rho v dx,$$

we obtain

(3.5) $$E \leq \frac{L}{M\lambda^{1/2}} ,$$

where

$$L = \int_{\partial D} \rho^{1/2} ds, \quad M = \int_D \rho dx .$$

As a second application, we consider the nonlinear eigenvalue problem ($\rho = 1$ for simplicity)

(3.6) $$\Delta u + \lambda u^p = 0 \quad \text{in} \quad D$$

(3.7) $$u = 0 \quad \text{on} \quad \partial D$$

where $p > 0$. It can be shown [7] that

(3.8) $$(u_{max})^{\frac{1-p}{2}} N(p) \leq d \sqrt{\frac{2\lambda}{p+1}}$$

where $N(p) = \sqrt{\pi}\; \Gamma(\frac{1}{p+1})[(p+1)\; \Gamma(\frac{p+3}{2(p+1)})]^{-1}$

and d is the radius of the largest inscribed circle in D with its center at a point in D at which u takes its maximum. We observe that if $0 < p < 1$, (3.8) gives an upper bound for u_{max}, if $p > 1$, (3.8) gives a lower bound for u_{max}, and if $p = 1$, (3.8) results in a lower bound for the first eigenvalue of the fixed membrane problem, namely,

$$\lambda \geq \frac{\pi^2}{4d^2} .$$

We close by noting that extensions to problems with boundary conditions of the third kind in $n \geq 2$ dimensions for the inhomogeneous problem and $n > 2$ dimensions for the homogeneous problem remain to be done. Extensions to uniformly elliptic operators have not been accomplished but would seem to be possible by reasoning analogous to that presented here.

REFERENCES

1. H. B. Keller and D. S. Cohen, Some positone problems suggested by nonlinear heat generation, J. Math Mech, 16 (1967), 1361-1376.
2. L. E. Payne, Bounds for the maximum stress in the Saint Venant torsion problem, Indian J. Mech. and Math., Special Issue, 1968, 51-59.
3. L. E. Payne and I. Stakgold, On the mean value of the fundamental mode in the fixed membrane problem, Appl. Anal., 3, 1973, 295-306.
4. L. E. Payne and I. Stakgold, Nonlinear problems in nuclear reactor analysis, Proc. Conf. on Nonlinear Problems in Physical Sciences and Biology, Springer Lecture Notes in Math. No. 322, 1972, 298-307.
5. M. H. Protter and H. F. Weinberger, Maximum principles in differential equations, Prentice-Hall, Inc., 1967.
6. P. W. Schaefer and R. P. Sperb, Maximum principles for some functionals associated with the solution of elliptic boundary value problems, to appear Arch. Rational Mech. Anal.
7. P. W. Schaefer and R. P. Sperb, Maximum principles and bounds in some inhomogeneous elliptic boundary value problems, submitted.

GLOBAL METHODS FOR THE CONSTRUCTION OF CONVERGENT SEQUENCES OF BOUNDS FOR SYSTEMS OF ORDINARY INITIAL VALUE PROBLEMS

G. SCHEU

Summary. Systems of ordinary nonlinear differential equations with initial conditions are considered. For the solution of such problems convergent sequences of bounds are constructed iteratively. Numerically it is possible to calculate the sequences of bounds on a grid only. To estimate the local truncation error, sequences of bounds are needed on the whole interval. Therefore, data on the grid are interpolated and the truncation error is estimated suitably. To account for the round-off error, all data are computed by use of interval mathematics. In this way it is guaranteed in the mesh points that the solution of the initial value problem is bracketed by the numerically computed values of the bounds and in addition, in the whole interval the solution is bracketed by the interpolation polynomials. The order of accuracy of the interpolating polynomials and the sequences of bounds is given. By use of this method, sequences of bounds are computed numerically for the solution of some sample problems.

1. Introduction. The well known error estimates of methods for the numerical solution of ordinary initial value problems have the disadvantage that in general a certain derivative of the unknown solution is needed. Therefore, they are inappropriate for quantitative computations unless an a priori estimate is known for the solution. The numerical approximation is computable in general on a grid only. Estimates are needed for the difference of the solution and the numerical approximation between the mesh points, too (global).

Therefore, by use of the theory of differential-inequalities, sequences of bounds for the solution are now constructed iteratively. Each iterate is approximated one-sidedly by an interpolation polynomial (HERMITE). Thereby, the interpolation error is taken into account suitably. This method yields explicit equations, because it is applied to the equations of the iteration method. The sequences of bounds, which are constructed as described, yield rigorously valid estimates, i.e.

the numerical data on the grid are the support values of the corresponding interpolation polynomials, in addition these polynomials are globally valid bounds of the solution of the initial value problem. The influence of the round-off errors is taken into account by use of interval mathematics, e.g. [3].

2. Problem. To the initial value problem

(1a) $y' = f(t,y), \quad t_o < t \leq T, \quad y(t_o) = \eta, \quad t_o, T \in \mathbb{R}_o^+,$

with $y(t), \eta \in Y \subset \mathbb{R}^M$, $M \in \mathbb{N}$, and Y convex and suitably chosen, the following equivalent integral equation is adjoined

(1b) $y(t) = \eta + \int_{t_o}^{t} f(\tau, y(\tau))\, d\tau, \quad t \in [t_o,T].$

Assume

(2) $y := (y_1,\ldots\ldots,y_M)^T \in C^{m+1}\,[t_o,T],$

(3) $f := (f_1,\ldots\ldots,f_M)^T \in C^m\,[[t_o,T] \times Y],$

with $m \in \mathbb{N}$ is chosen corresponding to the used interpolation formulae. The functions f_i, $i = 1(1)M$, are assumed quasi-monotone, i.e. to decrease as follows with respect to the second argument $y \in Y$:

(4) $f_i(t,u) \geq f_i(t,v)$ for $u \leq v, \quad u_i = v_i, \quad i = 1(1)M,$

$t \in [t_o,T], \quad u, v \in Y,$ e.g. [7, p.42].

Due to assumption (3), there exists one and only one solution of the problem (1), e.g. [6, p.48].

Remark 1: Integration, monotonicity, inequalities and so on are defined componentwise (natural partial order). ¤

Remark 2: By addition of suitable terms on both sides of the differential equation (1a) the quasi-monotonicity of the function f can always be achieved. If the function f is quasi-monotone increasing the equations (5) are uncoupled. ¤

3. Iteration scheme. By use of the theory of differential inequalities, bounds are computed iteratively [5, 6].

Theorem 1: Sequences of upper and lower bounds $(\overline{y}_n)$, $(\underline{y}_n)$, of the solution y of problem (1) are computed as follows:

(5a) $\overline{y}'_{n+1} := f(t,\underline{y}_n)$, $\quad t_o < t \leq T,\quad$ $\overline{y}_{n+1}(t_o) \geq \eta + \overline{\varepsilon}$, $\quad n \in \mathbb{N}_o$,

(5b) $\underline{y}'_{n+1} := f(t,\overline{y}_n)$, $\quad \underline{y}_{n+1}(t_o) \leq \eta - \underline{\varepsilon}$,

with $\overline{\varepsilon}, \underline{\varepsilon} \in (\mathbb{R}_o^+)^M$.

Under the conditions

(6a) $\overline{y}'_o \geq \overline{y}'_1 = f(t,\underline{y}_o)$, $\quad t_o < t \leq T,\quad$ $\overline{y}_o(t) \geq \eta + \overline{\varepsilon}$,

(6b) $\underline{y}'_o \leq \underline{y}'_1 = f(t,\overline{y}_o)$, $\quad \underline{y}_o(t) \leq \eta - \underline{\varepsilon}$,

the sequences $(\overline{y}_n)$, $(\underline{y}_n)$ are monotone and uniformly convergent towards continuous limit functions $\overline{y}$, $\underline{y}$ and it is

(7a) $\underline{y} := \lim\limits_{n\to\infty} \underline{y}_n \leq y \leq \overline{y} := \lim\limits_{n\to\infty} \overline{y}_n$, $\quad t \in [t_o,T]$.

Analogously, the sequences $(\overline{y}'_n)$, $(\underline{y}'_n)$ are monotone and uniformly convergent to continuous limit functions $\overline{y}'$, $\underline{y}'$ and

(8a) $\overline{y}' := \lim\limits_{n\to\infty} \overline{y}'_n \leq y' \leq \overline{y}' = \lim\limits_{n\to\infty} \overline{y}'_n$, $\quad t \in [t_o,T]$,

holds true.

Proof: Because of the conditions (5),(6), induction yields by use of theorem 12 X(β) in [7, p.85] the following inequalities

(7b) $\underline{y}_o \leq \underline{y}_1 \leq \ldots \leq \underline{y}_n \leq \ldots \leq \underline{y} \leq y \leq \overline{y} \leq \ldots \leq \overline{y}_n \leq \ldots \leq \overline{y}_1 \leq \overline{y}_o$, $\quad t \in [t_o,T]$,

and analogously

(8b) $\underline{y}'_o \leq y'_1 \leq \ldots \leq \underline{y}'_n \leq \ldots \leq \underline{y}' \leq y' \leq \overline{y}' \leq \ldots \leq \overline{y}'_n \leq \ldots \leq \overline{y}'_1 \leq \overline{y}'_o$,
$t \in [t_o,T]$.

The sequences $(\overline{y}_n)$, $(\underline{y}_n)$ and $(\overline{y}'_n)$, $(\underline{y}'_n)$, respectively, are monotone and bounded, and therefore they converge pointwise to the functions $\overline{y}$, $\underline{y}$ respectively $\overline{y}'$, $\underline{y}'$.
These functions are continuous, because the function f and the integral operator are continuous. Therefore the assumptions of the theorem of DINI are satisfied and the sequences of functions thus are seen to be uniformly convergent. []

Remark 3: The functions $\overline{y}$, $\underline{y}$ in general are not solutions of problem (1). Since the derivations $(\overline{y}'_n)$, $(\underline{y}'_n)$ are bounded, the sequences of functions $(\overline{y}_n)$, $(\underline{y}_n)$ are equicontinuous, e.g. [6, p.56]. ¤

Remark 4: For fixed $n \in \mathbb{N}_0$, the iteration problems (5) are of monotone kind (inverse isotone), e.g. [2, p.275]. ⌑

Remark 5: The advantages of employing the iteratively constructed bounds are as follows:

i) the point- and component-wise error estimates as given by the usual distance of the bounds obviously are better than global error estimates by use of norms in well known iteration schemes, and

ii) it is possible to estimate globally and rigorously the interpolation error, because the interpolation problem is defined by the iteration equations (5). This will be shown in the following. ⌑

4. *Construction of global bounds by means of interpolation.*

For the numerical computation of the sequences of bounds, the iteration (5) is used. Consider the mesh points

(9) (t_σ), $\sigma = 0(1)\, S+1$, $S \in \mathbb{N}$, $t_0 := t_0$, $t_{S+1} := T$,

which are determined a posteriori. For the starting functions $\overline{y}_0$, $\underline{y}_0$, a polynomial of degree one in the variable t is assumed for simplicity in each interval $[t_\sigma, t_{\sigma+1}]$, $\sigma = 0(1)S$, and beginning at the point t_σ the point $t_{\sigma+1}$, $\sigma = 0(1)S$, is determined in such a way that the conditions (6) are satisfied. Starting from the initial values $\eta + \overline{\varepsilon}$, $\eta - \underline{\varepsilon}$ at t_0 and the polynomial of degree one, sequences of bounds are computed iteratively on each interval $[t_\sigma, t_{\sigma+1}]$, $\sigma = 0(1)S$, which bracket the solution y of problem (1) because of theorem 1. This holds true only if computations are carried out in the field of real numbers and if the iterates $\overline{y}_n$, $\underline{y}_n$, $n \in \mathbb{N}_0$, are explicitly computable in each point of the interval $[t_\sigma, t_{\sigma+1}]$, $\sigma = 0(1)S$.

Since it is possible by use of equation (5) to compute numerically the bounds $\overline{y}_n$, $\underline{y}_n$, $n \in \mathbb{N}$, on a grid only, they have to be interpolated in order to obtain global bounds on the interval $[t_\sigma, t_{\sigma+1}]$, $\sigma = 0(1)S$. Since the interpolation error is taken into account suitably, the iteration scheme has to be changed as follows for the actual numerical applications:

$$
\begin{array}{lll}
\text{(10a)}\quad \overline{v}'_{n+1} := f(t\,,\, \underline{u}_n(t)), & & \overline{v}_{n+1}(t_\sigma) = \overline{v}_0(t_\sigma) = \overline{y}_0(t_\sigma), \\
 & t_\sigma < t \le t_{\sigma+1}, & \qquad\qquad n \in \mathbb{N}_0, \\
\text{(10b)}\quad \underline{v}'_{n+1} := f(t\,,\, \overline{u}_n(t)), & \sigma = 0(1)S, & \underline{v}_{n+1}(t_\sigma) = \underline{v}_0(t_\sigma) = \underline{y}_0(t_\sigma),
\end{array}
$$

$$\text{(11a)}\quad \overline{u}_n := \overline{S}_{2q,n} + \overline{R}_n, \qquad n \in \mathbb{N}, \quad t \in [t_\sigma, t_{\sigma+1}],$$

$$\text{(11b)}\quad \underline{u}_n := \underline{S}_{2q,n} + \underline{R}_n,$$

where

$$\text{(12a)}\quad \overline{u}_o := \overline{v}_o := \overline{y}_o, \qquad t \in [t_\sigma, t_{\sigma+1}],$$

$$\text{(12b)}\quad \underline{u}_o := \underline{v}_o := \underline{y}_o,$$

$\overline{S}_{2q,n}$, $\underline{S}_{2q,n}$ denote the interpolation polynomials of the functions $\overline{v}_n$, $\underline{v}_n$, $t \in [t_\sigma, t_{\sigma+1}]$, and

$$\text{(13a)}\quad \overline{R}_n := \max_{t \in [t_\sigma, t_{\sigma+1}]} |R_n \{f(t, \underline{u}_{n-1}(t))\}| \qquad n \in \mathbb{N}_o,$$

$$\text{(13b)}\quad \underline{R}_n := \max_{t \in [t_\sigma, t_{\sigma+1}]} |R_n \{f(t, \overline{u}_{n-1}(t))\}|,$$

represent an estimate of the truncation error, by use of the interpolation error, compare the equations (19).

The equations (10, (11) yield numerically computable bounds $\overline{u}_n$, $\underline{u}_n$ with $\overline{u}_n \geq \overline{y}_n$, $\underline{u}_n \leq \underline{y}_n$, $t \in [t_\sigma, t_{\sigma+1}]$, and this for every iteration step $n \in \mathbb{N}$ and on the whole interval $[t_\sigma, t_{\sigma+1}]$, i.e. they are globally valid numerically computed bounds which bracket the solution of problem (1). These iterations are carried out up to an $N_\sigma \in \mathbb{N}$, $\sigma = 0(1)S$, and then the iteration is carried out analogously in the subsequent interval $[t_{\sigma+1}, t_{\sigma+2}]$. Here the finally computed values of the bounds are used as initial values for the next interval $[t_{\sigma+1}, t_{\sigma+2}]$, $\sigma = 0(1)S-1$.

Starting from the known functions (12) and the initial values of (10) in each step of iteration the following 2q data are computed by use of the differentiated differential equations (10), e.g. [1, p.259]

$$\text{(14a)}\quad \overline{v}^{(k)}_{n+1}(t_\sigma) = D^{k-1} f(t_\sigma, \underline{u}_n(t_\sigma)), \quad \overline{v}^{(k)}_{n+1}(t_{\sigma+1}) = D^{k-1} f(t_{\sigma+1}, \underline{u}_n(t_{\sigma+1})),$$

and

$$\text{(14b)}\quad \underline{v}^{(k)}_{n+1}(t_\sigma) = D^{k-1} f(t_\sigma, \overline{u}_n(t_\sigma)), \quad \underline{v}^{(k)}_{n+1}(t_{\sigma+1}) = D^{k-1} f(t_{\sigma+1}, \overline{u}_n(t_{\sigma+1})),$$

$k = 1(1)q$, $n \in \mathbb{N}_o$, where $D^{k-1} f$ denotes the total derivation of the function f. The pertinent interpolation polynomials are

$$\text{(15)}\quad \overline{S}_{2q,n+1} \quad \text{and} \quad \underline{S}_{2q,n+1}, \quad t \in [t_\sigma, t_{\sigma+1}], \quad n \in \mathbb{N}_o.$$

The values of the functions $\overline{v}_{n+1}(t_{\sigma+1})$, $\underline{v}_{n+1}(t_{\sigma+1})$ are computed by use of

(16a) $$\overline{v}_{n+1}(t_{\sigma+1}) := \overline{S}_{2q,n+1}(t_{\sigma+1}) = \overline{v}_n(t_\sigma) + \sum_{k=1}^{q} C_{kq}h^k[D^{k-1}f(t_\sigma,\underline{u}_n(t_\sigma))+ +(-1)^k D^{k-1}f(t_{\sigma+1},\underline{u}_n(t_{\sigma+1}))],$$

and

(16b) $$\underline{v}_{n+1}(t_{\sigma+1}) := \underline{S}_{2q,n+1}(t_{\sigma+1}) = \underline{v}_n(t_\sigma) + \sum_{k=1}^{q} C_{kq}h^k[D^{k-1}f(t_\sigma,\overline{u}_n(t_\sigma)) + + (-1)^k D^{k-1}f(t_{\sigma+1},\overline{u}_n(t_{\sigma+1}))],$$

with $q \in \mathbb{N}$, $n \in \mathbb{N}_0$,

(17) $$C_{kq} := \frac{1}{k!}\,\frac{q(q-1)\dots(q-k+1)}{2q(2q-1)\dots(2q-k+1)}, \quad k = 1(1)q,$$

and the step size

(18) $h_\sigma := t_{\sigma+1} - t_\sigma$, $\sigma = 0(1)S$.

The following estimate, with fixed $t \in [t_\sigma,t_{\sigma+1}]$, pertains to the interpolation error

(19a) $$|R_{n+1}\,\{f(t,\underline{u}_n(t))\}| \leq |C_q\, h^{2q+1}D^{2q}\, f(\tau,\underline{u}_n(\tau))|,$$

$$\tau \in [t_\sigma,t_{\sigma+1}], m:=2q,$$

(19b) $$|R_{n+1}\,\{f(t,\overline{u}_n(t))\}| \leq |C_q\, h^{2q+1}D^{2q}\, f(\tau,\overline{u}_n(\tau))|,$$

with

(20 $$C_q := (-1)^q\,\frac{q!\,q!}{(2q)!}\,\frac{1}{(2q+1)!}\,.$$

Remark 6: The equations (16) are well known quadrature formulae, which are obtained by HERMITE interpolation of the integrand, e.g. [1,p.258]. ◻

Remark 7: It is possible also to use other interpolation schemes. ◻

Theorem 2: If the function f is monotonically decreasing for $y \in Y$ and if the total derivative D^{2q} f is numerically computable or estimatable in the whole interval $[t_\sigma,t_{\sigma+1}]$, then it is possible to determine as follows numerically computable sequences of bounds $\overline{u}_n$, $\underline{u}_n$, $n \in \mathbb{N}$, in the interval $[t_\sigma,t_{\sigma+1}]$, $\sigma = 0(1)S$:

Assume:

$$(10a)\quad \overline{v}'_{n+1} := f(t,\underline{u}_n(t)),\quad t_\sigma<t\le t_{\sigma+1},\ \sigma=0(1)S,\quad \overline{v}_{n+1}(t_\sigma) = \overline{v}_o(t_\sigma) = \overline{y}_o(t_\sigma),$$

$$(10b)\quad \underline{v}'_{n+1} := f(t,\overline{u}_n(t)),\quad n\in\mathbb{N}_o,\quad \underline{v}_{n+1}(t_\sigma) = \underline{v}_o(t_\sigma) = \underline{y}_o(t_\sigma),$$

$$(11a)\quad \overline{u}_n := \overline{S}_{2q,n} + \overline{R}_n,$$

$$n\in\mathbb{N},\quad t\in[t_\sigma,t_{\sigma+1}],$$

$$(11b)\quad \underline{u}_n := \underline{S}_{2q,n} + \underline{R}_n,$$

where

$$(12a)\quad \overline{u}_o := \overline{v}_o := \overline{y}_o,$$

$$t\in[t_\sigma,t_{\sigma+1}],$$

$$(12b)\quad \underline{u}_o := \underline{v}_o := \underline{y}_o,$$

with the interpolation polynomials $\overline{S}_{2q,n}$, $\underline{S}_{2q,n}$ of equation (15) and the truncation errors $\overline{R}_n$, $\underline{R}_n$ of equation (19). Further the starting functions $\overline{u}_o$, $\overline{u}_1$ and $\underline{u}_o$, $\underline{u}_1$ have to satisfy the conditions

$$(21a)\quad \overline{u}_1 \le \overline{u}_o,$$

$$t\in[t_\sigma,t_{\sigma+1}],$$

$$(21b)\quad \underline{u}_o \le \underline{u}_1,$$

analogously to condition (6). Then

$$(22a)\quad \underline{1.}\quad \overline{v}_1 = \overline{y}_1,\qquad t\in[t_\sigma,t_{\sigma+1}],\ \text{if}\qquad \overline{v}_1(t_\sigma) = \overline{y}_1(t_\sigma),$$

$$(22b)\quad \underline{v}_1 = \underline{y}_1,\qquad \underline{v}_1(t_\sigma) = \underline{y}_1(t_\sigma),$$

$$(23a)\quad \underline{2.}\quad \overline{y}_n \le \overline{u}_n,\qquad t\in[t_\sigma,t_{\sigma+1}],\quad n\in\mathbb{N},$$

$$(23b)\quad \underline{u}_n \le \underline{y}_n,$$

3. for fixed $n\in\mathbb{N}$ in each interval $[t_\sigma,t_{\sigma+1}]$, $\sigma = 0(1)S$, the numerically constructed bounds $\underline{u}_n$, $\overline{u}_n$ of the equation (10), (11) approach the iterates $\overline{y}_n$, $\underline{y}_n$ of equation (5) with the order of convergence $O(h^{2q+1})$.

Remark 8: If the function $D^{2q} f$ possesses special properties, for example monotonicity or absence of a change of sign etc., then it is possible to estimate the error in equation (13) in a simpler way. □

Proof: Equations (5), (10) and (12) yield

$$(24a)\quad \overline{v}'_1 = f(t,\underline{y}_o(t)) = \overline{y}'_1,\qquad t\in(t_\sigma,t_{\sigma+1}],\qquad \overline{v}_1(t_\sigma) = \overline{y}_1(t_\sigma),$$

$$(24b)\quad \underline{v}'_1 = f(t,\overline{y}_o(t)) = \underline{y}'_1,\qquad \underline{v}_1(t_\sigma) = \underline{y}_1(t_\sigma),$$

i.e. part one is correct.

Because of equations (10), (11) and (22) the relations

$$(25a)\quad 0 \le \overline{w}_1 := \overline{u}_1 - \overline{y}_1 \le 2\,\overline{R}_1,$$
$$t \in [t_\sigma, t_{\sigma+1}],$$
$$(25b)\quad 0 \le \underline{w}_1 := \underline{y}_1 - \underline{u}_1 \le 2\,\underline{R}_1,$$

holds true. Therefore the mean value theorem of differential calculus yield

$$(26a)\quad \overline{d}_2' := \overline{v}_2' - \overline{y}_2' = f_y(t, \underline{y}_1 - \delta \underline{w}_1)\,\underline{w}_1,\ \delta \in \mathbb{R}^M,\quad 0 < \delta < 1,\ t \in (t_\sigma, t_{\sigma+1}],$$

or

$$(27a)\quad \overline{d}_2(t) = \int_{t_\sigma}^{t} f_y(\tau, \underline{y}_1(\tau) - \delta \underline{w}_1(\tau))\,\underline{w}_1(\tau)\,d\tau \ge 0,\quad t \in [t_\sigma, t_{\sigma+1}],$$

because the function f is monotonically increasing for $y \in Y$. Analogously

$$(26b)\quad \underline{d}_2' := \underline{y}_2' - \underline{v}_2' = f_y(t, \overline{y}_1 + \delta \overline{w}_1)\,\overline{w}_1,\ \delta \in \mathbb{R}^M,\quad 0 < \delta < 1,\ t \in (t_\sigma, t_{\sigma+1}],$$

or

$$(27b)\quad \underline{d}_2(t) = \int_{t_\sigma}^{t} f_y(\tau, \overline{y}_1(\tau) + \delta \overline{w}_1(\tau))\,\overline{w}_1(\tau)\,d\tau \ge 0,\quad t \in [t_\sigma, t_{\sigma+1}].$$

Because of the equations (10), (11) and (13)

$$(28a)\quad 0 \le \overline{u}_2 - \overline{v}_2 \le 2\,\overline{R}_2,$$
$$t \in [t_\sigma, t_{\sigma+1}].$$
$$(28b)\quad 0 \le \underline{v}_2 - \underline{u}_2 \le 2\,\underline{R}_2,$$

By use of the estimates (27), (28)

$$(29a)\quad 0 \le \overline{u}_2 - \overline{y}_2 = \overline{u}_2 - \overline{v}_2 + \overline{v}_2 - \overline{y}_2 \le 2\,\overline{R}_2 + \overline{d}_2 =: \overline{D}_2,$$
$$t \in [t_\sigma, t_{\sigma+1}].$$
$$(29b)\quad 0 \le \underline{y}_2 - \underline{u}_2 = \underline{y}_2 - \underline{v}_2 + \underline{v}_2 - \underline{u}_2 \le 2\,\underline{R}_2 + \underline{d}_2 =: \underline{D}_2,$$

Because of the estimates

$$(30a)\quad 0 \le \overline{u}_n - \overline{v}_n \le 2\,\overline{R}_n,$$
$$t \in [t_\sigma, t_{\sigma+1}],\quad n \in \mathbb{N},$$
$$(30b)\quad 0 \le \underline{v}_n - \underline{u}_n \le 2\,\underline{R}_n,$$

induction yields

$$(31a)\quad 0 \le \overline{u}_n - \overline{y}_n \le \overline{D}_n,$$
$$t \in [t_\sigma, t_{\sigma+1}],\quad n \in \mathbb{N} \setminus \{1\},$$
$$(31b)\quad 0 \le \underline{y}_n - \underline{u}_n \le \underline{D}_n,$$

i.e. the inequalities (23) of part two.
By use of the mean value theorem of integral calculus as applied to the equations (27), the estimates (25) yield the order of convergence $O(h^{2q+1})$ for the functions $\overline{D}_2$, $\underline{D}_2$, because of the estimates (13), (19). Induction and the inequalities (31) yield part three. [ ]

Remark 9: At each further iteration, the data (14), the values of the functions in equation (16), and the truncation error of equation (13) have to be recomputed. ¤

Remark 10: The truncation error consists of the following contributions:

i) inequalities, ii) iterations,
iii) interpolations, and iv) round-off errors.

Even though this error theoretically can be made as small as desired, it grows exponentially in any actual numerical example. ¤

5. Numerically carried out examples. Several examples have been treated by use of the developed procedure, e.g. [4]. Here the following example is discussed

(32) $y' = 1 + y^2, \quad 0 < t \leq T < \pi/2, \quad y(0) = 0, \quad y(t) = \tan t.$

Here, the function f is monotonically increasing, compare remark 2. For this initial value problem, sequences of bounds are numerically computed

a) without consideration of round-off errors, and
b) with consideration of round-off errors, by use of TRIPLEX, i.e. all computations were carried out by use of interval mathematics, e.g.[3].

The table shows the numerically computed bounds

	$\overline{y}$ (1.5)	y (1.5)	$\underline{y}$ (1.5)	$\overline{y}(1.5)-\underline{y}(1.5)$
a)	*1.41014* 2106	*1.410141995*	*1.4101419* 27	1.7782,-06
b)	*1.4101* 62770		*1.4101* 05311	5.7459,-04

Here, the equidistant step size h = 1/100 and a method of order seven (q = 3) have been employed.
The number of iterations was:
in case a) 8 to 10 and in case b) 4 to 5.
The increase of computation time due to employing case b) instead of case a) is approximately a factor of ten (UNIVAC 1108 of the University of Karlsruhe).

References.

[1] BÖHMER, K.: Spline-Funktionen. Stuttgart: Teubner 1974.

[2] COLLATZ, L.: Funktionalanalysis und Numerische Mathematik. Berlin, Göttingen, Heidelberg: Springer 1964. (*)

[3] MOORE, R.E.: Intervallanalyse. München, Wien: Oldenbourg 1969. (*)

[4] SCHEU, G., ADAMS, E.: Zur numerischen Konstruktion konvergenter Schrankenfolgen für Systeme nichtlinearer gewöhnlicher Anfangswertaufgaben. In: Proceedings of the Interval Symposium, pp.279-287. Berlin, Heidelberg,New York: Springer 1975.

[5] SPREUER, H., ADAMS, E., SRIVASTAVA, U.N.: Monotone Schrankenfolgen für gewöhnliche Randwertaufgaben bei schwach gekoppelten nichtlinearen Systemen. ZAMM 55,211-218 (1975).

[6] WALTER, W.: Gewöhnliche Differentialgleichungen. Berlin, Heidelberg, New York: Springer 1972.

[7] WALTER, W.: Differential- und Integral-Ungleichungen. Berlin, Göttingen, Heidelberg, New York: Springer 1964. (*)

(*) These books are also available in English.

Bifurcation from a Multiple Eigenvalue

M. Shearer

Let X, Y be real Banach spaces. Consider the equation

$$F(\lambda, u) \equiv Lu + \lambda L_1 u + M[u] + R(\lambda, u) = 0 \tag{1.1}$$

for $(\lambda, u) \in E = \mathbb{R} \times X$ $\quad(\|(\lambda,u)\| = (|\lambda|^2 + \|u\|^2)^{\frac{1}{2}})$.

Here, $F : E \to Y$ is continuously Frechet differentiable and

$$F(\lambda, 0) = 0 \qquad \lambda \in \mathbb{R} \tag{1.2}$$

$L : X \to Y$ is a bounded linear Fredholm operator with index zero and nullity $s \geq 1$. $M[u] = M(u,..,u)$ where $M(u_1,..,u_{p+1})$ is a bounded $(p+1)$-linear operator from X to Y. $L_1 : X \to Y$ is a bounded linear operator, and $R : E \to Y$ represents higher order terms in λ and u:

$$\|R(\lambda,u) - R(\lambda,v)\| \leq C\|u-v\|\left\{|\lambda|^2 + |\lambda|(\|u\|^r + \|v\|^r) + \|u\|^{p+r} + \|v\|^{p+r}\right\}$$

for $(\lambda,u), (\lambda,v)$ in E, with norms less than $\eta > 0$. C is a positive constant, and r is a positive integer.

Let $\Gamma = \{(\lambda, 0) : \lambda \in \mathbb{R}\}$. Then $F(\Gamma) = 0$, and Γ is the set of trivial solutions to (1.1). If, in every E-neighbourhood of $(\lambda,u) = (0,0)$ there exists a non-trivial solution to (1.1), then $\lambda = 0$ is a bifurcation point for (1.1) with respect to Γ.

Let N, R be the null and range spaces of L. Then there exist a closed subspace X_o of X and an s-dimensional subspace Y_o of Y such that $X = X_o \oplus N$ and $Y = R \oplus Y_o$. Let $P : X \to X_o$ and $Q : Y \to R$ be projections defined by

$$Px = 0 \quad x \in N \; ; \; Px = x \quad x \in X_o \; ; \; Qy = 0 \quad y \in Y_o \; ; \; Qy = y \quad y \in R$$

Then (1.1) may be written, for $u = z + v$, $z \in N$, $v \in X_o$:

$$QF(\lambda, z+v) = 0 \tag{1.3}$$

$$(I-Q)F(\lambda, z+v) = 0 \tag{1.4}$$

Lemma 1.1

Under the above conditions on F, there exists $\delta > 0$ such that if $\|(\lambda, z)\| < \delta$ there is a unique solution $v = v(\lambda, z) \in X_0$ to (1.3) satisfying

(1) $v(\lambda, z)$ is continuously differentiable in (λ, z) for $\|(\lambda, z)\| < \delta$

(2) $\|v(\lambda, z)\| \leqslant K \|z\| (|\lambda| + \|z\|^p)$ if $\|(\lambda, z)\| < \delta$, for some positive constant K.

Proof (details in 3.)

The existence, uniqueness and differentiability properties of $v(\lambda, z)$ follow from the Implicit Function Theorem in Banach space. (2) is a consequence of the estimates on $F(\lambda, u) - Lu$.

Substituting $v = v(\lambda, z)$ into (1.4), one obtains the so-called branching equations:

$$(I - Q)F(\lambda, z + v(\lambda, z)) = 0 \qquad \|(\lambda, z)\| < \delta \tag{1.5}$$

Solving (1.1) for small (λ, u) in E is equivalent to solving (1.5) for small (λ, z) in $\mathbb{R} \times N$. In particular, to solve (1.5) for non-zero λ, it is convenient to introduce a scaling parameter ε as follows. Let $x_1, \ldots, x_s$ be a basis for N and set

$$\lambda = \nu \varepsilon^p \quad \text{where} \quad \nu = \operatorname{sgn} \lambda \;\; (p \text{ even}) \; ; \quad \nu = +1 \;\; (p \text{ odd})$$

$$z = \varepsilon x = \varepsilon \sum_{i=1}^{s} a_i x_i \quad ; \quad a = (a_1, \ldots, a_s) \in \mathbb{R}^s \tag{1.6}$$

Substituting into (1.5), and dividing by ε^{p+1}, one obtains a system in $\mathbb{R}^s$:

$$f_\nu(a) + \varepsilon^q g_\nu(a; \varepsilon) = 0 \tag{1.7}$$

Here, $f_\nu = (f_1, \ldots, f_s)$, $g_\nu = (g_1, \ldots, g_s)$, where for $j = 1, \ldots, s$, $f_j(a)$ is the j-th component in Y_0 of $(I - Q)(\nu L_1 x + M[x])$, $q = \min(p, r)$, and $\varepsilon^q g_j(a; \varepsilon)$ represents higher order ε terms.

Remark : The scaling (1.6) may be made without losing generality if there exists $\xi > 0$ such that $F(0,u) = 0$, $\|u\| < \xi$ implies $u = 0$, a sufficient condition for this being: (2.)

(A1) $(I-Q)M[z] = 0$, $z \in N$ implies $z = 0$.

Theorem 1.2

Suppose there exists x_0 in N such that

$$(I-Q)(\nu L_1 x_0 + M[x_0]) = 0 \qquad (1.8)$$

and : (A2) $(I-Q)(\nu L_1 z + M_u[x_0] z) = 0$, $z \in N$ implies $z = 0$.

Then there exist $\eta, \delta > 0$ and continuously differentiable functions $x:(-\eta,\eta) \to N$, $w:(-\eta,\eta) \to X_0$ such that

(1) $F(\nu\varepsilon^p, \varepsilon x(\varepsilon) + \varepsilon^{p+1} w(\varepsilon)) = 0 \qquad |\varepsilon| < \eta$

(2) $x(0) = x_0$, $w(0) = w_0$: $L w_0 = -\nu L_1 x_0 - M x_0$

(3) If $(\lambda,u) = (\nu\varepsilon^p, \varepsilon z + v)$ is a solution to (1.1) with $z \in N$, $v \in X_0$, $\|v\| + \|z - x_0\| < \delta$, $|\varepsilon| < \eta$ then $z = x(\varepsilon)$ and $v = \varepsilon^p w(\varepsilon)$.

Proof (see 1. or 2. for details)

If $x = x_0$ is given by (1.6) with $a = a_0$, then the conditions of the theorem state that $(a_0,0)$ is a solution to (1.7), and the Jacobian matrix $\frac{df_\nu}{da}(a_0)$ of (1.7) at $(a_0,0)$ is non-singular. The result now follows from the Implicit Function Theorem and Lemma 1.1 .

Remarks :

1) Assumption (A2) corresponds to Dancers non-degeneracy condition (2.), but (A1) is also called a non-degeneracy condition by some authors (e.g. Toland 5.) especially when using topological degree methods.

2) (See 1. or 2.) For each $x_0 \in N$ satisfying (1.8) and (A2), set

$$C(x_0) = \{(\nu\varepsilon^p, \varepsilon x(\varepsilon) + \varepsilon^{p+1} w(\varepsilon)) : |\varepsilon| < \eta\}$$

where x, w and η are given by the theorem.

Suppose the following assumption is satisfied.

(A3) (A1) holds and (A2) holds for every $x_0 \in N$ satisfying (1.8). Then the set $D = \{x_0 \in N : x_0 \text{ satisfies (1.8) for } \nu = \pm 1\}$ has a finite number of elements $x_0^1, \ldots, x_0^m$, $m \geq 1$. Moreover, there exists a neighbourhood U of $(0,0)$ in E such that if $(\lambda, u) \in U$ is a solution to (1.1), then $(\lambda, u) \in C(x_0)$ for some $x_0 \in D$. Thus, the solutions to (1.1) in U are entirely described by a finite number of curves $C(x_0^1) \cap U, \ldots, C(x_0^m) \cap U$ with $x_0^1 = 0$, $C(x_0^1) \subset \Gamma$.

2. A Degenerate Case

A solution $a = a_0$ of the equation

$$f_\nu(a) = 0 \qquad a \in \mathbb{R}^s \tag{2.1}$$

is isolated if there exists $\rho > 0$ such that, if $a = a_1$ is a solution to (2.1) with $\|a_0 - a_1\| < \rho$, then $a_1 = a_0$. The topological index, $\mathrm{ind}(a_0)$ may then be defined by

$$\mathrm{ind}(a_0) = \deg(f_\nu, 0, B_\theta(a_0))$$

where deg is the finite dimensional degree function, and $B_\theta(a_0)$ is the open ball, centre a_0, radius θ, $0 < \theta < \rho$ in $\mathbb{R}^s$.

A more general form of Theorem 1.2 is:

Theorem 2.1

Suppose $a_0 \in \mathbb{R}^s$ is an isolated solution to (2.1) and $\mathrm{ind}(a_0)$ is non-zero. Then there exist $\eta > 0$ and continuous functions $x : (-\eta, \eta) \longrightarrow N$, $w : (-\eta, \eta) \longrightarrow X_0$ satisfying (1), (2) of Theorem 1.2.

Proof (details in 4.)

By the homotopy property of degree,

$$\deg(f_\nu(\ ) + \varepsilon^2 g_\nu(\ ;\varepsilon), 0, B_\theta(a_0)) = \mathrm{ind}(a_0) \neq 0$$

for all small enough ε. Thus there is a solution $a(\varepsilon)$ to (1.7) for each small enough ε. The result now follows using (1.6) and Lemma 1.1.

If $\operatorname{ind}(a_o) = 0$, there may still be bifurcating solutions, but it is necessary to study the terms in (1.7) depending on ε. Suppose

(A4) The null space of $\frac{df_\nu}{da}(a_o)$ is one-dimensional.

Then the basis $x_1,..,x_s$ for N may be chosen so that

$$\frac{\partial f_i}{\partial a_1}(a_o) = 0 \qquad i = 1,..,s \tag{2.2}$$

Let $J_\nu(a_o)$ be the matrix $\frac{df_\nu}{da}(a_o)$ with $\frac{\partial f_i}{\partial a_1}(a_o)$ replaced by $g_i(a_o;0)$, $i = 1,..,s$.

Theorem 2.2

Let $\nu = +1$ or -1. Suppose $a_o = (a_1^o,..,a_s^o)$ is an isolated solution to (2.1) satisfying (A4), and that $x_1,..,x_s$ have been chosen so that (2.2) holds. Suppose $g_\nu(a;\varepsilon)$ is continuously differentiable in a neighbourhood of $(a_o,0)$ and

(A5) $\det J_\nu(a_o) \neq 0$.

Let q be odd. Then there exist $\eta, \delta > 0$ and real valued functions $t_2,..,t_s,b$, continuously differentiable on $(-\eta,\eta)$ such that

1) $t_i(0) = a_i^o$ $i = 2,..,s$; $b(0) = 0$

2) For each μ, $|\mu| < \eta$, $(a;\xi) = (a_1^o + \mu, t_2(\mu),..,t_s(\mu); b(\mu))$ is a solution to the equation

$$f_\nu(a) + \xi\, g_\nu(a;\xi^{\frac{1}{q}}) = 0 \tag{2.3}$$

3) If $(a;\xi) \in \mathbb{R}^{s+1}$ is a solution to (2.3) with $\|a - a_o\| + |\xi| < \delta$ and $|a_1 - a_1^o| < \eta$, then $a_1 = a_1^o + \mu$, $a_i = t_i(\mu)$ $i = 2,..,s$ and $\xi = b(\mu)$.

4) $b(\mu) = 0$, $|\mu| < \eta$ implies $\mu = 0$.

Let q be even. Then there exist $\eta, \delta > 0$ and, for $\sigma = +1$ and $\sigma = -1$, real valued functions $t_2^\sigma,..,t_s^\sigma, b^\sigma$, continuously differentiable on $(-\eta,\eta)$ such that

1)' $t_i^\sigma(0) = a_i^o$ $i = 1,..,s$; $b^\sigma(0) = 0$

2)' For each μ, $|\mu| < \eta$, $(a;\xi) = (a_1^o + \mu, t_2^\sigma(\mu),..,t_s^\sigma(\mu); b^\sigma(\mu))$

is a solution to the equation

$$f_\nu(a) + \xi\, g_\nu(a;\sigma|\xi|^{\frac{1}{2}}) = 0 \tag{2.4}$$

3)' If $(a;\xi) \in \mathbb{R}^{s+1}$ is a solution to (2.4) with $\|a-a_o\| + |\xi| < \delta$ and $|a_1 - a_1^o| < \eta$, then $a_1 = a_1^o + \mu$, $a_i = t_i^\sigma(\mu)$ $i = 2,..,s$ and $\xi = b^\sigma(\mu)$.

4)' $b^\sigma(\mu) = 0$, $|\mu < \eta$ implies $\mu = 0$.

Proof

$J_\nu(a_o)$ is the Frechet derivative of the left hand side of both (2.3) and (2.4) with respect to $(a_2,..,a_s;\xi)$ at $(a;\xi) = (a_o;0)$. 1) - 3) and 1)' - 3)' now follow from the Implicit Function Theorem.

If either 4) or 4)' does not hold for some $\eta > 0$, then a_o is not an isolated solution to (2.1), contrary to assumption.

When q is odd, Theorem 2.2 guarantees a curve of solutions to (1.1) in E passing through $(\lambda, u) = (0,0)$ of the form

$$(\lambda(\mu),u(\mu)) = (\nu(b(\mu))^{\frac{p}{2}}, (b(\mu))^{\frac{1}{2}}(x(\mu) + (b(\mu))^{\frac{1}{2}} w(\mu))) \tag{2.5}$$
$$(|\mu| < \eta)$$

where $x: (-\eta,\eta) \longrightarrow N$, $w: (-\eta,\eta) \longrightarrow X_o$ are continuously differentiable functions satisfying (2) of Theorem 1.2. When q is even, $\sigma = \pm 1$, a corresponding statement will be true if it can be shown that $b^\sigma(\mu)$ has a minimum at $\mu = 0$. If both b^+ and b^- have a minimum at $\mu = 0$, there are two curves of solutions to (1.1) of the form (2.5) which will coincide if $g_\nu(a;\varepsilon)$ is even in ε for ε near zero. If p and q are both odd, showing that $b(\mu)$ has a maximum (respectively, a minimum) will indicate, if $a_o \neq 0$, that the bifurcating curve (2.5) of solutions to (1.1) is subcritical (resp. supercritical).

Let $J_\nu^2(a_o)$ be the matrix $\frac{df_\nu}{da}(a_o)$ with $\frac{\partial f_i}{\partial a_1}(a_o)$ replaced $\frac{\partial^2 f_i}{\partial a_1^2}(a_o)$. Let ' denote ordinary differentiation with respect to μ.

Corollary 2.3

Let the conditions of Theorem 2.2 hold and suppose the left hand sides of (2.3),(2.4) are twice continuously differentiable with respect to $(a;\xi)$ in a neighbourhood of $(a_o;0)$. Then the implicit functions of the theorem are twice continuously differentiable in a neighbourhood of $\mu = 0$, and :

1) If q is odd

$$b''(0) = - \frac{\det J^2_\nu(a_o)}{\det J_\nu(a_o)}$$

2) If q is even

$$b^{\sigma\,\prime\prime}(0) = - \frac{\det J^2_\nu(a_o)}{\det J_\nu(a_o)} \qquad (\sigma = \pm 1)$$

Proof

The first statement is immediate from a corollary to the Implicit Function Theorem.

1) q odd. Substitute $(a;\xi) = (a^o_1 + \mu, t_2(\mu), \ldots, t_s(\mu); b(\mu))$ into (2.3). Differentiating with respect to μ and setting $\mu = 0$, (2.2) and (A5) imply

$$b_1'(0) = t_2'(0) = \ldots = t_s'(0) = 0 \qquad (2.6)$$

Differentiating again with respect to μ , setting $\mu = 0$ and using (2.6) :

$$J_\nu(a_o) \begin{bmatrix} b''(0) \\ t_2''(0) \\ \vdots \\ t_s''(0) \end{bmatrix} = - \begin{bmatrix} \frac{\partial^2 f_1}{\partial a_1^2}(a_o) \\ \vdots \\ \frac{\partial^2 f_s}{\partial a_1^2}(a_o) \end{bmatrix}$$

from which the result follows, by inverting $J_\nu(a_o)$.

2) may be proved similarly.

References

1. J.B. McLeod and D.H. Sattinger: Loss of Stability and Bifurcation at a Double Eigenvalue, J. Funct. Anal. 14 (1973).

2. E.N. Dancer: Bifurcation Theory in Real Banach Space, Proc. Lond. Math. Soc. 23 (1971).

3. D. Sather: Branching of Solutions of an Equation in Hilbert Space, Arch. Rat. Mech. Anal. 36 (1970).

4. D. Sather: Branching of Solutions of Nonlinear Equations in Hilbert Space, Rocky Mtn. J. Math. 3 (1973).

5. J.F. Toland: Global Bifurcation for k-set Contractions Without Multiplicity Assumptions, to appear.

Another approach to the Dirichlet problem for very strongly nonlinear elliptic equations.

C.G. Simader

Introduction

During the last ten years many mathematicians proved the existence of weak solutions of the Dirichlet problem for certain nonlinear elliptic equations in divergence form. Nearly all the existence proofs have been performed by means of the theory of monotone operators. Here we notice in particular two methods: 1) monotone operators in certain Orlicz-Sobolev spaces constructed with respect to the strong nonlinearities, e.g. [10], 2) monotone operators which may be not everywhere defined on the Sobolev spaces under consideration, e.g. [12]. But most of these results are restricted to bounded domains. There is a very simple reason for this restriction. In general, Rellich's compactness theorem need no longer hold in unbounded domains and the abstract operators constructed relative to the differential equation under consideration lose important compactness properties. But Berger-Schechter [1],[2] overcame this difficulty considering, roughly spoken, certain weighted Sobolev spaces and performed existence proofs in unbounded domains. This technique was refined and improved by several British mathematicians e.g. Edmunds, Evans and Webb, together with the Italian Moscatelli ([5] - [9], [18]). Recently, with quite another technique, Hess [13] succeeded in proving certain results in unbounded domains under assumptions similiar to the case of bounded domains.

It is the purpose of the present paper to report on some recent results ([14],[16],[17]) obtained by a quite different approach. Our method works in bounded as well as in unbounded domains, covers a considerably wider class of nonlinearities, which we want to call "very strong nonlinearities". A typical example is the equation

$$-\Delta u - \Sigma_1^N \partial_i(\partial_i u \exp|\partial_i u|) + u \exp|u| = f$$

Our method completely avoids the theory of monotone operators and uses from nonlinear functional analysis solely the classical fixed point theorem of Schauder, applied in a very obvious manner (Lemma 1). The remaining proofs only use very elementary facts on Sobolev spaces, linear functional analysis and integration theory. This method, developed by Leinfelder and the author [14], was first applied to the study of nonlinear perturbations of Schrödinger operators with strongly singular potentials.

Notations. In the sequel let $G \subset \mathbb{R}^N$ denote an open set. Assume that all functions under consideration are real valued. For (Lebesgue-) measurable $f,g : G \longrightarrow \mathbb{R}$ such that $fg \in L^1(G)$ let $(f,g)_o := \int_G fg\, dx$. For $1 \leq p < \infty$ and $f \in L^p(G)$ let $\|f\|_{o,p} := (\int_G |f|^p dx)^{1/p}$. If $u \in C_o^\infty(G)$, $m \in \mathbb{N}$, let $\|u\|_{m,p} := (\sum_{|\alpha|\leq m} \|D^\alpha u\|_{o,p}^p)^{1/p}$ and for $u,v \in C_o^\infty(G)$ let $(u,v)_m := \sum_{|\alpha|\leq m} (D^\alpha u, D^\alpha v)_o$. By $W_o^{m,p}(G)$ $(1 \leq p < \infty$, $m \in \mathbb{N})$ we denote the closure of $C_o^\infty(G) \subset W^{m,p}(G)$ with respect to the norm $\|\cdot\|_{m,p}$. If $p = 2$, $W_o^{m,2}(G)$ is a Hilbert space with inner product $(\cdot,\cdot)_m$. If $\alpha = (\alpha_1, \ldots, \alpha_N)$, where the α_i are nonnegative integers, let us denote as usual $|\alpha| = \Sigma_1^N \alpha_i$ and $D^\alpha = \Pi_1^N (\partial/\partial x_i)^{\alpha_i}$. For $m,N \in \mathbb{N}$ given, let $r = r(m,N)$ be the number of those α satisfying $|\alpha| \leqslant m-1$ and $s = s(m,N)$ the number of those α with $|\alpha| = m$. For $\eta \in \mathbb{R}^r$ we write $\eta = (\eta_\alpha)_{|\alpha|\leq m-1}$ and for $\zeta \in \mathbb{R}^s$ similiarly $\zeta = (\zeta_\alpha)_{|\alpha|=m}$. If $\xi \in \mathbb{R}^{r+s}$, let $\xi = (\eta,\zeta)$, where $\eta \in \mathbb{R}^r$, $\zeta \in \mathbb{R}^s$. For $u \in W_o^{m,p}(G)$, let us write $\eta(u)(x) := (D^\alpha u(x))_{|\alpha|\leq m-1}$ and similarily $\zeta(u)$, $\xi(u)$.

Let us state now the Dirichlet problem in the weak sense for nonlinear equations: Let $G \subset \mathbb{R}^N$ be an open set, and let for $|\alpha| \leqslant m$ $(m \in \mathbb{N})$ functions $N_\alpha : G \times \mathbb{R}^{r+s} \longrightarrow \mathbb{R}$ be given, each satisfying a Carathéodory condition (that is: $N_\alpha(x,\cdot) : \mathbb{R}^{r+s} \longrightarrow \mathbb{R}$ is continuous for almost all $x \in G$ and $N_\alpha(\cdot,\xi) : G \longrightarrow \mathbb{R}$ is measurable for all $\xi \in \mathbb{R}^{r+s}$). Then, for given $f \in L^p(G)$ $(1 \leqslant p < \infty)$ we call a function $u \in W_o^{m,q}(G)$ $(1 \leqslant q < \infty)$ such that $N_\alpha(\cdot,\xi(u)) \in L^1_{loc}(G)$ is satisfied for $|\alpha| \leqslant m$, a weak solution of the Dirichlet problem for the equation

$$\sum_{|\alpha|\leq m} (-1)^{|\alpha|} D^\alpha N_\alpha(\cdot,\xi(u)) = f \quad \text{in } G$$

$$D^\sigma u\big|_{\partial G} = 0 \quad \text{on } \partial G \quad \text{for } |\sigma| \leqslant m-1$$

if $N[u,\Phi] := \sum_{|\alpha|\leq m} (N_\alpha(\cdot,\xi(u)), D^\alpha \Phi)_o = (f,\Phi)_o$ for each $\Phi \in C_o^\infty(G)$.

We give now certain properties of the semilinear form $N[u,\Phi]$.

2. The basic Lemma.

First we consider the case of a nonlinear perturbed linear elliptic operator. For this purpose let us assume

$$(H.1)\quad \begin{cases} \text{Let } B: W_\circ^{m,2}(G) \times W_\circ^{m,2}(G) \longrightarrow \mathbb{R} \quad (m \in \mathbb{N}) \text{ be a continuous} \\ \text{bilinear form such that with a constant } C_\circ > 0 \\ B[u,u] \geq C_\circ \|u\|^2_{m,2} \quad \text{for } u \in W_\circ^{m,2}(G) \end{cases}$$

All our existence proofs go back to the following rather simple situation.

<u>Lemma 1.</u> Let (H.1) be satisfied. Assume that for $|\alpha| \leq m-1$ functions $N_\alpha : G \times \mathbb{R}^r \longrightarrow \mathbb{R}$ are given, each satisfying a Carathéodory-condition. Let $G_1 \subset G$ be a bounded open subset and $K \geq 0$ a constant such that

(1) $\quad |N_\alpha(x,\eta)| \leq K\,\chi_{G_1}(x)$ for $(x,\eta) \in G \times \mathbb{R}^r$ for $|\alpha| \leq m-1$,

where $\chi_{G_1}(x) = 1$ for $x \in G_1$ and $\chi_{g_1}(x) = 0$ otherwise.

Then, for every $f \in L^2(G)$ there is a $u \in W_\circ^{m,2}(G)$ such that $N_\alpha(u) := N_\alpha(.,\eta(u)) \in L^2(G)$ and

(2) $$B[u,\Phi] + \sum_{|\alpha|\leq m-1} (N_\alpha(u), D^\alpha\Phi)_\circ = (f,\Phi)_\circ \quad \text{for every } \Phi \in W_\circ^{m,2}(G)$$

Let us briefly sketch the proof in the case $m = 1$ and $B[u,\Phi] := (u,\Phi)_1$. Let $f \in L^2(G)$ be given and let for $u \in W_\circ^{1,2}(G)$

(3) $$L_u(\Phi) := (f,\Phi)_\circ - (N_\circ(u),\Phi)_\circ\,, \quad \phi \in W_\circ^{1,2}(G)$$

Since

(4) $$|L_u(\Phi)| \leq (\|f\|_{0,2} + K\mu(G_1)^{1/2})\|\Phi\|_{0,2} \leq (\|f\|_\circ + K\mu(G_1)^{1/2})\|\Phi\|_{1,2},$$

L_u is a continuous linear functional on $W_\circ^{1,2}(G)$ and, with a uniquely determined $w \in W_\circ^{1,2}(G)$ it admits by the Riesz representation theorem the discription $L_u(\Phi) = (w,\Phi)_1$. Since this may be done for every $u \in W_\circ^{1,2}(G)$, a (nonlinear) operator $M : W_\circ^{1,2}(G) \longrightarrow W_\circ^{1,2}(G)$ is defined such that by (3)

(5) $\quad (M(u),\Phi)_1 = (f,\Phi)_o - (N_o(u),\Phi)_o$ holds for $u,\Phi \in W_o^{1,2}(G)$.

Now our problem is equivalent to the fixed point equation $u = M(u)$ being solved in the sequel. By (4) we conclude for $\Phi := M(u)$

(6) $\quad \|M(u)\|_1 \leqslant \|f\|_{o,2} + K\,\mu(G_1)^{1/2} := C(f)$.

Let $u,v \in W_o^{1,2}(G)$, then we derive from (1) and (5)

(7) $\quad \|M(u) - M(v)\|_{1,2}^2 \leqslant \|N_o(v) - N_o(u)\|_{o,2}\|M(u) - M(v)\|_{L^2(G_1)}$

First we see from this $\|M(u) - M(v)\|_{1,2} \leqslant \|N_o(v) - N_o(u)\|_{o,2}$. Since the operator N_o is continuous (this is a well known fact for Nemytskii-operators), the operator M is continuous too. Further, let (u_n) be an arbitrary sequence from $W_o^{1,2}(G)$. Then, by (6), the sequence $v_n := M(u_n)$ is a bounded sequence in $W_o^{1,2}(G)$. Now take any $\psi \in C_o^\infty(\mathbb{R}^N)$ such that $\psi \equiv 1$ in a neighbourhood of G_1. Then, $(v_n\psi)$ is a bounded sequence in $W_o^{1,2}(G \cap \operatorname{supp}\psi)$. Therefore, by Rellich's theorem there is a $v \in W_o^{1,2}(G)$ such that $v_n\psi \longrightarrow v\psi$ in $L^2(G)$ and therefore $v_n|_{G_1} \longrightarrow v|_{G_1}$ in $L^2(G_1)$. But then we derive from (7) and (6)

$$\|M(u_n) - M(u_m)\|_{1,2}^2 \leqslant 2\,C(f)\,\|M(u_n) - M(u_m)\|_{L^2(G_1)} \longrightarrow 0 \ (n,m \to \infty).$$

So we see that M maps $W_0^{1,2}(G)$ completely continuously into $W_0^{1,2}(G)$. Hence, and, by Schauder's fixed point theorem, there is a $u \in W_o^{1,2}(G)$ such that $u = M(u)$ holds, which proves the Lemma in the special case considered.

3. Strongly nonlinear perturbed linear equations.

The proof above is very simple but the nonlinearities occurring from applications are not decent enough to satisfy the restrictive assumptions of Lemma 1, consider e.g. the problem of finding for a given $f \in L^2(G)$ (open $G \subset R^N$), a solution of the equation

(8) $\quad -\Delta u + \lambda u + p\,u \exp u = f$ in G, $u|_{\partial G} = 0$,

where we assume $p \in L^1_{loc}(G)$, $p \geqslant 0$ in G and $\lambda > 0$ in general, but $\lambda \geqq 0$ for bounded G or for those G being bounded in one direction (at least after a suitable rotation of coordinates). Equations of this figure have been the starting point of our considerations. But how does one bridge the gap between the situation considered in Lemma 1 and that occuring with equation (8) ? For brevity let us denote $B[u,u] := (u,\Phi)_1 + \lambda(u,\Phi)_o$ and $g(t) := t \exp t$. Observe that

(9) $B[u,u] > C_o\|u\|_1^2$ for $u \in W_o^{1,2}(G)$,

where $C_o = C_o(G,\lambda) > 0$ for $\lambda > 0$ if G is unbounded and $\lambda \geqslant 0$ if G is bounded in one direction.

For $n \in \mathbb{N}$ let $\Phi_n(t) := t$ if $|t| \leqslant n$ and $\Phi_n(t) := n\, t/|t|$ otherwise and let $g^{(n)} := g \circ \Phi_n$. Since $g(t)t \geqslant 0$ for $t \in \mathbb{R}$, the cut-off applied preserves this and we have $g^{(n)}(t)t \geqslant 0$ for $t \in \mathbb{R}$. Let further for $n \in \mathbb{N}$ $p^{(n)} := \Phi_n \circ p$, $B_n := \{x \in \mathbb{R}^N \mid |x| < n\}$, $\chi_n(x) = 1$ for $x \in B_n$ and $\chi_n(x) = 0$ otherwise (last cut-off is necessary only if G is unbounded). Then, for given $f \in L^2(G)$, the equation

(10) $B[u,\Phi] + (\chi_n p^{(n)} g^{(n)}(u),\Phi)_o = (f,\Phi)_o$ for $\Phi \in W_o^{1,2}(G)$

satisfies the assumptions of Lemma 1 and for every $n \in \mathbb{N}$ we get a solution $u_n \in W_o^{1,2}(G)$. Now we consider equation (10) as a suitable approximation of equation (8) in the following sense: Equation (10) is by Lemma 1 easily solvable and we get a sequence of solutions $(u_n) \in W_o^{1,2}(G)$. Question: Does this sequence converge in some sense to the weak solution of equation (8) being under consideration? It does: Taking $\Phi = u_n$, we conclude from (9) and (10)

$$C_o\|u_n\|_{1,2}^2 \leqslant B[u_n,u_n] + (\chi_n p^{(n)} g^{(n)}(u_n),u_n)_o \leqslant \|f\|_{o,2}\|u\|_{o,2} \leqslant$$

$$\leqslant C_o/2\, \|u_n\|_{1,2}^2 + C_o^{-1}/2\, \|f\|_{o,2}^2$$

and therefore

(11) $C_o/2\, \|u_n\|_{1,2}^2 + (\chi_n p^{(n)} g^{(n)}(u_n),u_n)_o \leqslant C_o^{-1}/2\, \|f\|_{o,2}^2$.

Since $\chi_n p^{(n)} g^{(n)}(u_n)u_n \geqslant 0$ we first conclude $\|u_n\|_{1,2}^2 \leqslant C(f)$ and therefore, there is a subsequence $(u_{n'}) \subset (u_n)$ and a $u \in W_o^{1,2}(G)$ such that $u_{n'} \longrightarrow u$ weakly in $W_o^{1,2}(G)$. As in the "proof" of Lemma 1 apply Rellich's theorem gradually to $G \cap B_k$ $(k \in \mathbb{N})$, which gives a sequence $(u_{n''}) \subset (u_{n'})$ such that $u_{n''} \longrightarrow u$ a.e. in G_1. Since $u_{n''} \chi_{n''} p^{(n'')} g^{(n'')}(u_{n''}) u_{n''} \longrightarrow p\, g(u)\, u$ a.e. in G and the integrals are uniformly bounded, we conclude by Fatou's theorem $p\, g(u) u \in L^1(G)$. An easy calculation (compare [16], Lemma 3) yields the fact that for any $G' \subset G$ such that G' is bounded and $\overline{G'} \subset G$ we have $p\, g(u) \in L^1(G')$ and $\int_{G'} |p\, g(u) - \chi_{n''} p^{(n'')} g^{(n'')}(u_{n''})|\, dx \longrightarrow 0$ For fixed $\Phi \in C_0^\infty(G)$ choose now some $G' \subset\subset G$ such that $\text{supp}\Phi \subset G' \subset\subset G$ and consider (10) for $u_{n''}$. Then, passing to the limit $n'' \longrightarrow \infty$ we find

$$B[u,\Phi] + \int_G p\, g(u)\, \Phi dx = (f,\Phi)_0 ;$$

that is, u is the desired solution.

These considerations are the prototype for all our proofs. Roughly speaking, we have to construct those nonlinearities to prove an estimate of type (11) in such a manner that we may conclude two facts: 1) $\|u_n\|_{m,2} \leqslant$ const (independent of $n \in \mathbb{N}$) and 2) integrals involving strong nonlinearities satisfying certain sign-conditions (e.g. $0 \leqslant \int_G \chi_n\, p^{(n)}\, g^{(n)}(u_n)\, u_n\, dx$) are bounded from above independently from $n \in \mathbb{N}$. From the first fact we conclude weak convergence of a subsequence to some limit function u (which turns out to be the desired solution); from the second we conclude by means of Fatou's theorem and the consequenc of Rellich's theorem mentioned above, that certain interesting limits belon belong to $L^1_{loc}(G)$. But then we conclude from the last property the $L^1_{loc}(G)$ convergence of the strong nonlinearites to the corresponding one of the solution. But if we consider "very strong nonlinearities" (see below), we have to obtain in addition the following information: 3) If $m \in \mathbb{N}$ is the order of the equation, we have to conclude that there is a subsequence $(u_{n''}) \subset (u_n)$ of the solutions of approximating equations, such that in addition $D^\alpha u_{n'''} \longrightarrow D^\alpha u$ a. e. n G, where u denotes the weak limit of the sequence $(u_{n'}) \subset (u_n)$ and $|\alpha| \leqslant m$. All the assumptions made in the following theorems and n [14], [16], [17] are of such a type to get 1) and 2) and in particular, if very strong nonlinearities are considered, to get property 3). So far it is a purely technical question to modify and improve the assumptions of [14], [16], [17] to make them applicable to other specific problems under consideration.

Along the line cited above we get the following results ([16]).

Assume that for $|\alpha| \leqslant m-1$ there are p_α and g_α be given, satisfying respectively

(H.2) $p_\alpha \in L^1_{loc}(G)$ and $p_\alpha \geqslant 0$

(H.3) $g_\alpha \in C^\circ(\mathbb{R})$ such that $g_\alpha(t)\, t \geqslant 0$ for $t \in \mathbb{R}$.

Then we conclude

Theorem 1. Let $G \subset \mathbb{R}^N$ be an open set and let (H.1), (H.2) and (H.3) be satisfied. Then, for any $f \in L^2(G)$ there is a $u \in W_\circ^{m,2}(G)$ such that $p_\alpha g_\alpha(D^\alpha u) \in L^1_{loc}(G)$ ($p_\alpha g_\alpha(D^\alpha u) \in L^1(G)$ if $p_\alpha \in L^1(G)$), $p_\alpha g_\alpha(D^\alpha u) D^\alpha u \in L^1(G)$ and

$$B[u,\Phi] + \sum_{|\alpha| \leq m-1} (p_\alpha g_\alpha(D^\alpha u), D^\alpha \Phi)_\circ = (f,\Phi)_\circ \quad \text{for } \Phi \in C_\circ^\infty(G).$$

It is very easy to add to the equation certain operators of Nemytskii-type [16]. We then split these operators such that one part is subordinate to a sign-condition and the other part satisfies a certain smallness-condition with respect to (H.1). In particular, Theorem 1 extends a famous result of Hess [12] to unbounded domains.

4. Very strongly nonlinear equations.

Considering also semilinear forms, we assume the following conditions of Leray-Lions (compare [15], p. 182):

Let $m \in \mathbb{N}$ and $p \in \mathbb{R}$, $1 < p < \infty$, $p' := p/(p-1)$. Assume that for $|\alpha| \leqslant m$ there are $A_\alpha : G \times \mathbb{R}^{r+s} \longrightarrow \mathbb{R}$ be given, each satisfying a Carathéodory condition, and that there is a constant $c > 0$ and a $k \in L^{p'}(G)$ such that

(H.4)
i) $|A_\alpha(x,\xi)| \leqslant c\ |\xi|^{p-1} + h(x)$ for $|\alpha| \leqslant m$ and all $(x,\xi) \in G \times \mathbb{R}^{r+s}$: Further, there is a constant $C_0 > 0$ and $h \in L^1(G)$ such that

$$\sum_{|\alpha|\leq m} A_\alpha(x,\xi)\xi_\alpha \geqslant C_0 \sum_{|\alpha|\leq m} |\xi_\alpha|^{\ p} - h(x) \quad \text{for} \quad (x,\xi) \in G \times \mathbb{R}^{r+s}.$$

ii) For $x \in G$, $\eta \in \mathbb{R}^r$ and $\zeta,\zeta' \in \mathbb{R}^s$ with $\zeta \neq \zeta'$ let $\sum_{|\alpha|\leq m} \left(A_\alpha(x,\eta,\zeta) - A_\alpha(x,\eta,\zeta')\right)(\zeta_\alpha - \zeta'_\alpha) > 0.$

If G is unbounded,wwe have to assume

ii') $\sum_{|\alpha|\leq m} (A_\alpha(x,\xi) - A_\alpha(x,\xi'))\cdot(\xi_\alpha - \xi'_\alpha) > 0$ for $x \in G$ and $\xi,\xi' \in \mathbb{R}^{r+s}$ with $\xi \neq \xi'$.

If (H.4) is satisfied, then let for $u,v \in \overset{\circ}{W}{}^{m,p}(G)$

$$B[u,v] := \sum_{|\alpha|\leq m} (A_\alpha(.,\xi(u),D^\alpha v)_0$$

To define "strong nonlinearities" and "very strong nonlinearites", we assume:

(H.5)
For $|\alpha| \leqslant m-1$ let $g_\alpha \in C^0(\mathbb{R})$ be given such that $g_\alpha(t)t \geqslant 0$ $(t \in \mathbb{R})$. Further assume that there is a $C > 0$ such that $g_\alpha(t)s \leqslant g_\alpha(t)t + C\ g_\alpha(s)s$ for $s,t \in \mathbb{R}$.

As is readily seen, if g_α is monotone nondecreasing, then (H.5) holds with $C = 1$. Assumption (H.5) is considerable weaker than the consitions of Browder [4] and Hess [12]. They had to assume that for any $\varepsilon > 0$ there is a constant $C(\varepsilon) > 0$ sucht that $g_\alpha(t)s \leqslant \varepsilon\ g_\alpha(t)t + C(\varepsilon)\ g_\alpha(s)s$. This condition was necessary for being able to apply Vitali's convergence theorem in a suitable situation. Our proof avoids this theorem.

(H.6)
For $|\alpha| = m$ let $g_\alpha \in C^0(\mathbb{R})$ be monotonous nondecreasing. Let $G_\alpha(t) := g_\alpha(t)t$ and assume that G_α is convex and satisfies $G_\alpha(t) \geqslant 0$ for $t \in \mathbb{R}$.

Then we prove easily

<u>Theorem 2.</u> Let $G \subset \mathbb{R}^N$ be open, $m \in \mathbb{N}$ and assume that (H.1) holds. For $|\alpha| \leqslant m$ let p_α satisfy (H.2) and assume that (H.5) and (H.6)

hold. Then, for every $f \in L^2(G)$ there is a $u \in W_\circ^{m,2}(G)$ such that for $|\alpha| \leqslant m$ $p_\alpha g_\alpha(D^\alpha u)D^\alpha u \in L^1(G)$, $p_\alpha g_\alpha(D^\alpha u) \in L^1_{loc}(G)$ (and $p_\alpha g_\alpha(D^\alpha u) \in$ $\in L^1(G)$ if $p_\alpha \in L^1(G)$) and

$$B[u,\Phi] + \sum_{|\alpha|\leq m} (p_\alpha g_\alpha(D^\alpha u), D^\alpha \Phi)_\circ = (f,\Phi)_\circ \quad \text{for} \quad \Phi \in C_\circ^\infty(G).$$

Also in the case when we consider only strong nonlinearities of order $\leqslant$ m - 1, this result is stronger than the corresponding one of Webb[18] (also for bounded domains), since we assume $p_\alpha \in L^1_{loc}(G)$ instead of $p_\alpha \in L^1(G)$.

In the case of a semilinear form we get

Theorem 3. Let $G \subset \mathbb{R}^N$ be open, let $m \in \mathbb{N}$ and $1 < p < \infty$, $p' := p/(p-1)$ and assume that (H.4) is satisfied. For $|\alpha| \leqslant m$ let p_α be given, satisfying (H.2), and assume that (H.5) and (H.6) hold. Then, for every $f \in L^{p'}(G)$ there is a $u \in W_\circ^{m,p}(G)$ such that for $|\alpha| \leqslant m$ $p_\alpha g_\alpha(D^\alpha u)D^\alpha u \in L^1(G)$, $p_\alpha g_\alpha(D^\alpha u) \in L^1_{loc}(G)$ (and $p_\alpha g_\alpha(D^\alpha u) \in$ $\in L^1(G)$ if $p_\alpha \in L^1(G)$) and

$$B[u,\Phi] + \sum_{|\alpha|\leq m} (p_\alpha g_\alpha(D^\alpha u), D^\alpha \Phi)_\circ = (f,\Phi)_\circ \quad \text{for} \quad \Phi \in C_\circ^\infty(G),$$

where the semilinear form is definied with respect to the A_α, as mentioned above.

These theorems extend numerous well known results, concerning the class nonlinearities being admitted as well as the underlying domains G. Again we could easily add certain Nemytskii - type nonlinearities (compare [16]). Also in that case the proof is based on Lemma 1 and similar to the proof of Theorem 1.

Now the question arises if we could drop the "regular" term $B[u,\Phi]$. This is possible :

Theorem 4. Let $G \subset \mathbb{R}^N$ be open, $m \in \mathbb{N}$ and let $1 < p < \infty$. Assume that for $|\alpha| \leqslant m$ there are strictly monotone increasing functions g_α such that with $G_\alpha(t) := g_\alpha(t)t$ we have $g_\alpha(t) \geqslant 0$ $(t \in \mathbb{R})$ and G_α is convex. Assume that there is a constant $C_\circ > 0$ such that $(|\alpha| \leq m)$ $|G_\alpha(t)| \geqslant C_\circ |t|^{p-1}$. Then, for every $f \in L^{p'}(G)$ there is a $u \in W_\circ^{m,p}(G)$

such that $g_\alpha(D^\alpha u)D^\alpha u \in L^1(G)$, $g_\alpha(D^\alpha u) \in L^1_{loc}(G)$ ($g_\alpha(D^\alpha u) \in L^1(G)$ if $\mu(G) < \infty$) and

$$\sum_{|\alpha|\leq m} (g_\alpha(D^\alpha u), D^\alpha \Phi)_\circ = (f,\Phi)_\circ \quad \text{for} \quad \Phi \in C_\circ^\infty(G).$$

The last Theorem (compare [17]) is intimitely connected with a famous result of Gossez [10], but in proving it we neither use monotone operators nor Orlicz space arguments. But if we want to drop the assumptions $|G_\alpha(t)| \geq c_\circ |t|^p$ with $p > 1$, we clearly have to use certain arguments of Orlicz space theory.

5. The special case m = 1.

Up till now, all our results do not depend on the order of the specific equation considered. But as one may expect, in the case $m = 1$ we can get sharper results. As Hess [11] observed, the assumptions on the non-linearities depending solely on u can be considerable relaxed. This result was extended to unbounded domains by Webb [18]. Both results are included in

Theorem 5. Assume that the assumptions of Theorem 2 or Theorem 3 respectively are satisfied, but replace for $|\alpha| = 0$ assumption (H.5) by

(H.5)' $\begin{cases} \text{Assume that } g_\circ \in C^\circ(\mathbb{R}) \text{ and } g_\circ(t)t \geq 0 \text{ for } t \in \mathbb{R}. \text{ Further,} \\ \text{let } p_\circ : G \longrightarrow \mathbb{R} \text{ be measurable, } p_\circ \geq 0 \text{ and let} \\ p_\circ|_{G'} \in L^1(G') \text{ for each measurable } G' \subset G \text{ such that} \\ \mu(G') < \infty. \end{cases}$

Then, the assertions of Theorem 2 and 3 respectively still holds true.

Further, under certain assumptions we can prove better quality for the nonlinearities either in case of Theorem 2 or Theorem 3. For example, If $g_\circ$ is monotone nondecreasing and if for $|\alpha| = m = 1$ (H.6) is satisfied, we can prove for the weak solution of the equation

$$-\Delta u - \Sigma_1^N \partial_i(p_i g_i(\partial_i u)) + g_\circ(u) = f$$

($f \in L^2(G)$ given, $u \in \overset{\circ}{W}{}^{1,2}(G)$) that $g_\circ(u) \in L^2(G)$ holds. This is done with the same method as used in [14], compare [16]. Moreover, it not too difficult to prove in the case m = 1 that weak solutions are strong solutions and, under suitable assumptions, we can prove classical differentiability properties.

6. Concluding remarks.

i) Here we have not investigated the question of uniqueness and stability of solutions. In the case m = 1 this may be done along the same line as in [14]. If m > 1, those questions seem to be rather delicate.

ii) The results mentioned above and proved in [14], [16], [17] admit a lot of generalizations. In section 3 we scetched the proof of gaining those results: It is a purely technical question how to generalize these results within the framework of the techniques applied here to technically more involved situations.

iii) The well-known results concerning weak solutions of strongly nonlinear equations (in bounded domains) caused considerable effort in developing the theory of monotone operators. But things grew more and more complicated in considering unbounded domains. Here we present a rather simple method of proof, applying to bounded and to unbounded domains as well as to so-called "very strong nonlinearities". Our proofs demand little knowledge of functional analysis and integration theory, but no deep results. On the other hand, modern results on monotone operators have been explained and justified by examples from the theory of nonlinear elliptic boundary value problems. It is the author's opinion that there seems to be a rather wide gap between the modern theory of monotone operators and their applications and the results achieved by more straightforward methods. The author hopes that his results may stimulate the colleagues working on the theory of monotone operators to develope this theory for being better applicable to problems arising from analysis.

References

[1] Berger, M.S. and Schechter, M.: L^p embedding and nonlinear eigenvalue problems for unbounded domains. Bull. Amer. Math. Soc. 76, 1299 - 1302 (1970)

[2] Berger, M.S. and Schechter, M.: Embedding theorems and quasilinear elliptic boundary value problems for unbounded domains. Trans. Amer. Math. Soc. 172, 261 -278 (1973)

[3] Browder, F.E.: Existence theorems for nonlinear partial differential equations. Proc. Sympos. Pure Math. 16, 1 - 60, Providence, R.I.: Amer. Math. Soc. 1970

[4] Browder, F.E.: Existence theory for boundary value problems for quasilinear elliptic systems with strongly nonlinear lower order terms. Proc. Sympos. Pure Math. 23, 269 - 286, Providence, R.I.: Amer. Math. Soc. 1973

[5] Edmunds, D.E. and Evans, W.D.: Elliptic and degenerate elliptic operators in unbounded domains. Ann. Scuola Norm. Sup. Pisa 27, 591 - 640 (1973)

[6] Edmunds, D.E. and Moscatelli, V.B.: Semi-coercive nonlinear problems. Bolletino Un. Mat. It. 11, 144 - 153 (1975)

[7] Edmunds, D.E., Moscatelli, V.B. and Webb, J.R.L.: Opérateurs elliptiques non linéaires dans des domaines non bornés. C.R.Acad. Sc. Paris 278, 1505 - 1508 (1974)

[8] Edmunds, D.E., Moscatelli, V.B. and Webb, J.R.L.: Strongly nonlinear elliptic operators in unbounded domains. Publ. Math. Bordeaux 4, 6 - 32 (1974)

[9] Edmunds, D.E. and Webb, J.R.L.: Quasilinear elliptic problems in unbounded domains. Proc. Royal Soc. London A, 334, 397 - 410 (1973)

[10] Gossez, J.-P.: Nonlinear elliptic boundary value problems for equations with rapidly (or slowly) increasing coefficients. Trans. Amer. Math. Soc. 190, 163 - 205 (1974)

[11] Hess, P.: A strongly nonlinear elliptic boundary value problem. J. Math. Anal. Appl. 43, 241 - 249 (1973)

[12] Hess, P.: On nonlinear mappings of monotone type with respect to two Banach spaces. J. Math. pures et appl. 52, 13 - 26 (1973)

[13] Hess, P.: Nonlinear elliptic problems in unbounded domains. To appear in Abhandlungen der Akademie der Wissenschaften der DDR

[14] Leinfelder, H. and Simader, C.G.: Bemerkungen über nichtlineare Störungen von Schrödinger-Operatoren. manuscripta math. 17, 187 - 204 (1975)

[15] Lions, J.L.: Quelques méthodes de résolution des problèmes aux limites non linéaires. Paris: Dunod, Gauthier-Villars 1969

[16] Simader, C.G.: Über schwache Lösungen des Dirichletproblems für streng nichtlineare elliptische Differentialgleichungen. To appear in Math. Z.

[17] Simader, C.G.: Remarks on certain stongly nonlinear elliptic differential operators. To appear.

[18] Webb, J.R.L.: On the Dirichlet problem for strongly non-linear elliptic operators in unbounded domains. J. London Math. Soc. 10, 163 - 170 (1975)

Global estimates for non-linear wave equations and linear wave equations with non-linear boundary constraints

B. D. Sleeman

§1 Introduction

In recent years considerable interest has been shown in questions of global existence of solutions to non-linear wave equations and non-linear parabolic equations. It is known that such equations do not possess global solutions in time for arbitrary choices of initial data. Since sufficient conditions on initial data for non-existence of global solutions correspond to necessary conditions for global existence it is important to develop as many techniques as possible to establish non-existence theorems. One of the most widely applicable techniques is the so-called "concavity method" developed by R. J. Knops, H. A Levine and L. E. Payne [see 4, 6, 7]. In its simplest form the "concavity method" depends on the following observation; if $F(t)$ is a concave function of t on $[0,T)$ such that $F(0) > 0$, $F'(0) < 0$ and $T \geq -F(0)/F'(0)$ then $F(t)$ has a zero in $[0,T)$. This follows from the fact that for such a function the graph of F lies below any tangent line. This implies that $F(t) \leq F(0) + t\,F'(0)$ and so $F(t)$ has a zero in $[0,-F(0)/F'(0))$ say at T_o. Consequently if $G(t) = F^{-1}(t)$ then $G(t)$ is unbounded on $[0,T_o]$.

This paper treats a modification of the concavity method to obtain non-existence theorems of a somewhat different character; the most striking of which is that positive solutions to non-linear wave equations may have compact support in time. The method also enlarges the class of non-linearities which may be treated.

To illustrate the idea two representative examples involving the wave equation will be considered

Problem A

Let $\Omega \subset R^m$ be a bounded domain with a smooth boundary and let $f : R^1 \to R^1$ be a given continuously differentiable function except possibly at the origin. Let $\underset{\sim}{n} = (n_1,\ldots,n_m)$ denote the outward directed normal to $\partial\Omega$. Consider u to be a real-valued classical solution to

$$\frac{\partial^2 u}{\partial t^2} = \Delta u + f(u) \quad \text{in} \quad \Omega \times [0,T),$$

$$u(x,0) = u_o, \qquad u_o \in C^2(\bar{\Omega})$$

$$u_t(x,0) = v_o, \qquad v_o \in C^1(\bar{\Omega})$$

$$\frac{\partial u}{\partial n} = 0 \quad \text{on} \quad \partial\Omega \times [0,T),$$

where Δ is the m-dimensional Laplacian, $\frac{\partial u}{\partial n} \equiv \sum_{i=1}^{m} n_i \frac{\partial u}{\partial x_i}$ denotes the outward directed normal derivative of u on $\partial\Omega$. Here u_o, v_o are prescribed on $\bar{\Omega}$ the closure of Ω and x designates a point in R^m.

Problem B

Adopt the same notation and conditions as described in problem A and consider u to be a real-valued classical solution to

$$\frac{\partial^2 u}{\partial t^2} = \Delta u \quad \text{in} \quad \Omega \times [0,T),$$

$$u(x,0) = u_o, \qquad u_o \in C^2(\bar{\Omega}),$$

$$u_t(x,0) = v_o, \qquad v_o \in C^1(\bar{\Omega}),$$

$$\frac{\partial u}{\partial n} = f(u) \quad \text{on} \quad \partial\Omega \times [0,T).$$

§2. Non-linear wave equations

Define

$$E(0) \equiv \int_\Omega (v_o^2 + |\nabla u_o|^2)dx - 2\int_{\partial\Omega}\left(\int^{u_o} f(z)dz\right) ds. \tag{2.1}$$

For the purposes of comparison we collect some results of Levine [6] together in the following theorem

Theorem 1.

Let $u(x,t) : \bar{\Omega} \times [0,T) \to R^1$ be a real valued classical solution to problem A, let $f(0) = 0$ and $\int_0^{u(x,t)} \{zf'(z) - (4\alpha + 1)f(z)\}dz > 0$ for some $\alpha > 0$. Then $u(x,t)$ can only exist on a bounded interval $[0,T)$ in the sense that there is a T_o, $0 < T_o < \infty$ such that if $T = T_o$ then

$$\lim_{t\to T_-} \int_\Omega u^2 dx = +\infty,$$

whenever the initial data satisfies any one of the following four sets of inequalities.

(α_1) $E(0) < 0$.

(β_1) $E(0) = 0, \quad \int_\Omega u_o v_o dx = \lambda \int_\Omega u_o^2 dx > 0.$

(γ_1) $E(0) > 0, \quad \int_\Omega u_o v_o dx = \lambda \int_\Omega u_o^2 dx > 0.$

$$E(0) \leq \frac{\alpha}{2(2\alpha + 1)} \left(\int_\Omega u_o v_o fx \right)^2 \left(\int_\Omega u_o^2 dx \right)^{-1} .$$

(δ_1) $\int_\Omega u_o v_o dx = \lambda \int_\Omega u_o^2 dx > 0$

$$\frac{\alpha}{2(2\alpha + 1)} \left(\int_\Omega u_o v_o dx \right)^2 \left(\int_\Omega u_o^2 dx \right)^{-2} < E(0) < \left(\int_\Omega u_o v_o dx \right)^2 \left(\int_\Omega u_o^2 dx \right)^{-1}$$

Remark

Levine gives upper bounds for T in each of the four cases.

We establish the following theorems.

Theorem 2.

Let $u(x,t) : \bar{\Omega} \times [0,T) \to R^1_+$ be a real valued classical solution to problem A

and let $\int_\infty^{u(x,t)} \{ zf'(z) + (4\alpha - 1)f(z)\}dz \leq 0$ (*)

for some $\alpha \geq 1$.

α_2 β_2 } If $E(0) \leq 0, \quad \int_\Omega u_o v_o dx < 0,$

then there is a T_o, $0 < T_o < \infty$ such that if $T = T_o$,

$$\lim_{t\to T_o^-} \int_\Omega u_t^2 dx = +\infty,$$

where $T_o \leq \frac{1}{2\alpha} \left(\int_\Omega u_o^2 dx \right) \left| \int_\Omega u_o v_o dx \right|^{-1}.$

On the interval of existence

$$\int_\Omega u^2 dx \le \int_\Omega u_o^2 dx \left[1 + 2\alpha \frac{\int_\Omega u_o v_o dx}{\int_\Omega u_o^2 dx} t \right]^{1/\alpha} .$$

Theorem 3.

Let $u(x,t) : \bar{\Omega} \times [0,T) \to R_+^1$ be a real valued classical solution to problem A and satisfying (*) for some $\alpha \ge 1$.

(γ_2) If $E(0) > 0$, $\int_\Omega u_o v_o dx < 0$

and $E(0) \le \frac{\alpha}{2(2\alpha - 1)} \left(\int_\Omega u_o v_o dx \right)^2 \left(\int_\Omega u_o^2 dx \right)^{-1}$

then there exists a T_o, $0 < T_o < \infty$ such that if $T = T_o$

$$\lim_{t \to T_o^-} \int_\Omega u^2 dx = 0.$$

Theorem 4.

Let $u(x,t) : \bar{\Omega} \times [0,T) \to R_+^1$ be a real valued classical solution to problem A and satisfying (*) for some $\alpha \ge 1$.

(δ_2) If $\int_\Omega u_o v_o dx < 0$

and

$$\frac{\alpha}{2(2\alpha - 1)} \left(\int_\Omega u_o v_o dx \right)^2 \left(\int_\Omega u_o^2 dx \right)^{-1} < E(0) < \left(\int_\Omega u_o v_o dx \right)^2 \left(\int_\Omega u_o^2 dx \right)^{-1},$$

then there exists a T_o, $0 < T_o < \infty$ such that if $T = T_o$

$$\lim_{t \to T_o^-} \int_\Omega u^2 dx = 0,$$

and $\int_\Omega u_t^2 dx > E(0), \quad t \in [0,T_o)$

where $T_o^2 = \int_\Omega u_o^2 dx \Big/ E(0).$

We make the following remarks regarding theorems 2-4.

(I) In each theorem the nonlinearity $f(u)$ may take, for example the general form

$$f(u) = u^{-(4\alpha-1)}\phi(u), \quad u > 0$$

where $\phi(u)$ is a monotone non-decreasing function of u.

(II) Theorem 2 is a non-existence theorem in the sense that the derivative u_t cannot exist pointwise globally in time. Theorem 3, 4, admit the possibility of positive solutions to non-linear wave equations having compact support in time.

(III) Similar theorems may be established if $\alpha \in (0,1]$. In this case we work with the easily derived fundamental inequality

$$FF'' + (\alpha - 1)F'^2 \leq 2F \left\{ \int_\Omega uu_{tt}dx + (3 - 2\alpha) \int_\Omega u_t^2 dx \right\}.$$

The results are much the same and are omitted.

(IV) As in the work of Levine [6] the method extends to more general problems, for example systems of non-linear wave equations. Weak solutions may also be considered provided they are suitably defined.

(V) Similar results may also be obtained for certain abstract non-linear wave equations of the form

$$Pu_{tt} = -Au + f(u),$$

where P, A are positive linear operators, f is a gradient operator and $u : [0,T) \to H$ (a Hilbert space), is a positive solution in an appropriately defined sense. This together with remarks (IV) will be taken up elsewhere.

(VI) The method is widely applicable in the sense that properties of existence and uniqueness are not required and so non-well posed problems may be discussed.

(VII) The proofs of the theorems show that if $\int_\Omega u_o v_o dx \geq 0$ then we obtain results which say that $\int_\Omega u^2 dx$ is bounded, grows at most algebraically or at most exponentially with time.

Proofs of Theorems 2, 3, 4

Proof of Theorem 2.

Let
$$F(t) = \int_\Omega u^2 dx, \tag{2.2}$$

then
$$F'(t) = 2\int_\Omega uu_t dx, \tag{2.3}$$

and
$$F''(t) = 2\int_\Omega u_t^2 dx + 2\int_\Omega uu_{tt} dx. \tag{2.4}$$

A computation then shows that

$$FF'' + (\alpha - 1)F'^2 = 4(1-\alpha)S^2 + 2F\left\{\int_\Omega uu_{tt} - (1-2\alpha)\int_\Omega u_t^2 dx\right\}, \tag{2.5}$$

where

$$S^2 = \left[\left(\int_\Omega u^2 dx\right)\left(\int_\Omega u_t^2 dx\right) - \left(\int_\Omega uu_t dx\right)^2\right] \geq 0$$

from Schwarz inequality. Thus

$$FF'' + (\alpha - 1)F'^2 \leq 2F\left\{\int_\Omega uu_{tt} dx - (1-2\alpha)\int_\Omega u_t^2 dx\right\}. \tag{2.6}$$

On using the differential equation in Problem A and invoking the divergence theorem we find

$$\int_\Omega uu_{tt} dx = -\int_\Omega |\nabla u|^2 dx + \int_\Omega uf(u)dx. \tag{2.7}$$

Let $H(t) = \int_\Omega u_t^2 \, dx$, then

$$H'(t) = 2\int_\Omega u_t u_{tt} dx = 2\int_\Omega u_t \Delta u dx + 2\int_\Omega u_t f(u)dx$$

$$= -\int_\Omega |\nabla u|_t^2 dx + 2\int_\Omega u_t f(u)dx, \tag{2.8}$$

after some manipulation. An integration now gives

$$H(t) = \int_\Omega v_o^2 dx + \int_\Omega |\nabla u_o|^2 dx - \int_\Omega |\nabla u|^2 dx + 2\int_\Omega\int_\infty^u f(z)dzdx$$

$$-2\int_\Omega\int_\infty^{u_o} f(z)dzdx$$

$$= 2 \int_\Omega \int_\infty^u f(z)dz - \int_\Omega |\nabla u|^2 dx + E(0). \tag{2.9}$$

Substituting (2.7), (2.9) in (2.6) we get

$$FF'' + (\alpha - 1)F'^2 \leq 2F \left\{ \int_\Omega uf(u)dx - 2\alpha \int_\Omega |\nabla u|^2 dx \right.$$

$$\left. - (1 - 2\alpha)E(0) - 2(1 - 2\alpha) \int_\Omega \int_\infty^u f(z)dz\, dx \right\}$$

$$= 2F \left\{ \int_\Omega \int_\infty^u \{zf'(z) + (4\alpha - 1)f(z)dz\}dx \right.$$

$$-2\alpha \int_\Omega |\nabla u|^2 dx$$

$$\left. - (1 - 2\alpha)E(0) \right\}$$

$$\leq 0. \tag{2.10}$$

Thus $(F^\alpha(t))'' \leq 0$ and so $F^\alpha(t)$ is concave and lies below any tangent line; in particular it lies below the tangent line at the origin.

That is as long as the solution exists

$$F^\alpha(t) \leq F^\alpha(0) + \alpha F'(0)F^{\alpha-1}(0)t$$

$$= F^\alpha(0)[1 + \frac{\alpha F'(0)}{F(0)} t],$$

or

$$F(t) = \int_\Omega u^2 dx \leq F(0)[1 + \frac{\alpha F'(0)}{F(0)} t]^{1/\alpha} . \tag{2.11}$$

In particular $\int_\Omega u^2 dx$ cannot exist as positiver longer than

$$t = T = - \frac{F(0)}{\alpha F'(0)} = \frac{1}{2\alpha} \left(\int_\Omega u_o^2 dx \right) \left| \int_\Omega u_o v_o dx \right|^{-1}.$$

Furthermore from (2.3) we have

$$F'(t) = 2\int_\Omega uu_t dx \geq -2\left(\int_\Omega u^2 dx\right)^{\frac{1}{2}}\left(\int_\Omega u_t^2 dx\right)^{\frac{1}{2}}$$

$$= -2\ F^{\frac{1}{2}}(t)\left(\int_\Omega u_t^2 dx\right)^{\frac{1}{2}}. \qquad (2.12)$$

Also since

$$(F^\alpha(t))'' \leq 0$$

then

$$(F^\alpha(t))' \leq (F^\alpha(0))'$$

i.e.,

$$F'(t) \leq F^{\alpha-1}(0)F'(0)F^{-\alpha+1}(t) \qquad (2.13)$$

which together with (2.12) gives

$$2\left(\int_\Omega u_t^2\right)^{\frac{1}{2}} \geq F^{\alpha-1}(0)|F'(0)|F^{-\alpha+\frac{1}{2}}(t)$$

or

$$\int_\Omega u_t^2 \geq \frac{1}{4}[F'(0)]^2[1 + \frac{\alpha F'(0)}{F(0)}t]^{\frac{1-2\alpha}{2}}.$$

Thus

$$\lim_{t\to T_o}\int_\Omega u_t^2 dx = +\infty$$

where

$$T_o \leq \frac{1}{2\alpha}\left(\int_\Omega u_o^2 dx\right)\left|\int_\Omega u_o v_o dx\right|^{-1}.$$

<u>Proof of Theorem 3.</u>

Let

$$F(t) = \int_\Omega u^2 dx + Q^2, \qquad (2.14)$$

where Q is a positive constant to be determined. Proceeding as in the proof of theorem 2 we find, as long as the solution exists, that

$$FF'' + (\alpha - 1)F'^2 = 4(1-\alpha)S^2 + 4(1-\alpha)Q^2\int_\Omega u_t^2 dx$$

$$+ 2F\left\{\int_\Omega uu_{tt} - (1-2\alpha)\int_\Omega u_t^2 dx\right\}$$

$$\leq 2F\left\{\int_\Omega uu_{tt}dx - (1-2\alpha)\int_\Omega u_t^2 dx\right\}. \qquad (2.15)$$

Use of (2.7) and (2.9) leads to the estimate

$$FF'' + (\alpha - 1)F'^2 \le 2(2\alpha - 1)E(0)F. \tag{2.16}$$

Now write $2(2\alpha - 1)E(0) = \mu^2 Q^2$, (2.17)

where μ (> 0) is to be chosen, then

$$FF'' + (\alpha - 1)F'^2 \le \mu^2 Q^2 F \le \mu^2 F^2. \tag{2.18}$$

If $G(t) = F^\alpha(t)$ then (2.18) may be written as

$$G''(t) \le \alpha\mu^2 G(t)$$

or

$$G''(t) + \mu\sqrt{\alpha}\, G'(t) \le \mu\sqrt{\alpha}\, (G'(t) + \mu\sqrt{\alpha}\, G(t)), \tag{2.19}$$

as long as the solution exists. Integration of this inequality leads to the estimate

$$G(t) \le \cosh \mu\sqrt{\alpha}\, t\,[G(0) + \frac{G'(0)}{\mu\sqrt{\alpha}} \tanh \mu\sqrt{\alpha}\, t\,] \tag{2.20}$$

Thus if we can find a μ such that

$$G'(0) \le -\mu\sqrt{\alpha}\, G(0), \tag{2.21}$$

then it follows from (2.20) that

$$\lim_{t\to T_-} G(t) = 0$$

where

$$T = \frac{1}{\mu\sqrt{\alpha}} \tanh^{-1}[\frac{-\mu\sqrt{\alpha}\, G(0)}{G'(0)}]. \tag{2.22}$$

Consequently $\int_\Omega u^2 dx$ cannot exist in the limit as $t \to T_-$. Indeed if a μ can be found such that (2.21) holds then from the definition of G(t) and (2.20) we see that

$$\lim_{t\to T_0} \int_\Omega u^2 dx = 0 \tag{2.23}$$

where T_0 is the unique positive solution to

$$Q^{2\alpha} = \cosh \mu\sqrt{\alpha}\, t\,[F(0) + \frac{\sqrt{\alpha}}{\mu} F'(0) \tanh \mu\sqrt{\alpha}\, t\,]F^{\alpha-1}(0). \tag{2.24}$$

Now (2.21) holds if and only if there exists a μ such that $\sqrt{\alpha}\, F'(0) + \mu F(0) \le 0$ or such that the polynomial

$$P(\mu) \equiv \mu^2 \int_\Omega u_o^2 + 2\sqrt{\alpha}\mu \int_\Omega u_o v_o + 2(2\alpha - 1)E(0) \tag{2.25}$$

is somewhere negative or has equal roots. The conditions of theorem 3 ensure that this is so; indeed (2.25) has two positive real roots. This completes the proof.

Proof of Theorem 4.

Proceeding as in theorem 3 but with $Q^2 = 0$ we find that

$$FF'' + (\alpha - 1)F'^2 \leq 2(2\alpha - 1)E(0)F. \tag{2.26}$$

Set $\nu^2 = 2(2\alpha - 1)E(0)$ and $G(t) = F^\alpha(t)$, then $G(t)$ satisfies the inequality

$$G''(t) \leq \alpha\nu^2 G^{\frac{\alpha-1}{\alpha}}(t). \tag{2.27}$$

Since $\int_\Omega u_o v_o dx < 0$ it follows that $F'(0) < 0$ so that $F'(t) < 0$ for $t \in [0,\eta)$ say. Thus multiplying both sides of (2.27) by $G'(t)$ gives

$$\tfrac{1}{2}(G'^2)' \geq \frac{\alpha^2\nu^2}{(2\alpha - 1)} \left(G^{\frac{2\alpha-1}{\alpha}}\right)', \qquad t \in [0,\eta). \tag{2.28}$$

and an integration shows that

$$\left(G'(t) - \mu\, G^{\frac{2\alpha-1}{2\alpha}}(t)\right)\left(G'(t) + \mu\, G^{\frac{2\alpha-1}{2\alpha}}(t)\right)$$

$$\geq \left(G'(0) - \mu\, G^{\frac{2\alpha-1}{2\alpha}}(0)\right)\left(G'(0) + \mu\, G^{\frac{2\alpha-1}{2\alpha}}(0)\right) \tag{2.29}$$

where $\mu^2 = \dfrac{2\alpha^2\nu^2}{2\alpha - 1}$.

Since $F'(0) < 0$ the right hand side of (2.29) will be positive provided

$$G'(0) + \mu\, G^{\frac{2\alpha-1}{2\alpha}}(0) < 0. \tag{2.30}$$

The assumed smoothness of u implies that the factor

$$G'(t) + \mu\, G^{\frac{2\alpha-1}{2\alpha}}(t) < 0$$

for all t.

That is

$$F'(t) + \frac{\mu}{\alpha} F^{\frac{1}{2}}(t) < 0.$$

An integration shows that

$$F(t) < (F^{\frac{1}{2}}(0) - \frac{\mu t}{2\alpha})^2,$$

i.e.
$$\int_\Omega u^2 dx \leq \int_\Omega u_0^2 dx \left(1 - \frac{\mu t}{2\alpha}\left(\int_\Omega u_0^2 dx\right)^{-\frac{1}{2}}\right)^2. \tag{2.31}$$

Arguing as in the proof of theorem 2 we find that as long as the solution exists

$$\int_\Omega u_t^2 \geq E(0).$$

From (2.31) we conclude

$$\lim_{t\to T_-} \int_\Omega u^2 dx = 0$$

where
$$T^2 = \int_\Omega u_0^2 dx \Big/ E(0).$$

After a little algebra we find that (2.30) implies

$$E(0) < \left(\int_\Omega u_0 v_0 dx\right)^2 \Big/ \int_\Omega u_0^2 dx.$$

and the theorem follows.

§3 Linear wave equations with non-linear boundary constraints.

In this section we consider problem B and once again for the purposes of comparison state the following theorem due to Levine and Payne [7].

Theorem 5.

Let u(x,t) be a classical solution to problem B. If f(0) = 0 and

$$2\int_{\partial\Omega}\left(\int_0^{u_0} f(z)dz\right) ds > \int_\Omega \left(|\nabla u_0|^2 + v_0^2\right) dx \tag{3.1}$$

where f(z) has the form $|z|^{(4\alpha+1)}\phi(z)$ $(\alpha > 0)$ and $\phi(z)$ is a monotone non-decreasing function of z then there is a $T < \infty$ such that

$$\lim_{t\to T_-} \int_\Omega u^2 dx = +\infty,$$

and hence u is pointwise unbounded in $\Omega \times (0,T)$.

We prove the following.

Theorem 6.

Let $u(x,t) > 0$ be a classical solution to problem B such that the initial data u_o, v_o satisfy $\int_\Omega u_o v_o dx < 0$. If in addition

$$2 \int_{\partial\Omega} \left(\int_\infty^{u_o} f(z)dz \right) ds > \int_\Omega (|\nabla u_o|^2 + v_o^2)dx \tag{3.2}$$

where $f(z)$ has the form $z^{-(4\alpha-1)}\phi(z)$ $(\alpha \geq 1)$ and $\phi(z)$ is a monotone non-decreasing function of z then there is a $T < \infty$ such that

$$\lim_{t \to T_-} \int_\Omega u_t^2 \, dx = \infty$$

and as long as the solution exists

$$\int_\Omega u^2 dx \leq \int_\Omega u_o^2 dx \left[1 + 2\alpha t \frac{\int_\Omega u_o v_o dx}{\int_\Omega u_o^2 dx} \right]^{1/\alpha} \tag{3.3}$$

Proof

Let
$$F(t) = \int_\Omega u^2 dx. \tag{3.4}$$

Proceeding as we did in the proof of theorem 3 we obtain the differential inequality

$$FF'' + (\alpha - 1)F'^2 \leq 2F \left\{ \int_\Omega uu_{tt} dx - (1 - 2\alpha) \int_\Omega u_t^2 \right\}. \tag{3.5}$$

In analogy with (2.7) and (2.9) we find

$$\int_\Omega uu_{tt} dx = \int_{\partial\Omega} uf(u)ds - \int_\Omega |\nabla u|^2 dx, \tag{3.6}$$

and

$$\int_\Omega u_t^2 dx = \int_\Omega (|\nabla u_o|^2 + v_o^2)dx - 2 \int_{\partial\Omega} \int_\infty^{u_o} f(z)dz$$
$$+ 2 \int_{\partial\Omega} \int_\infty^{u} f(z)dz - \int_\Omega |\nabla u|^2 dx. \tag{3.7}$$

Using (3.6), (3.7) in (3.5) and the conditions of the theorem we find

$$FF'' + (\alpha - 1)F'^2 \leq 0. \tag{3.8}$$

This implies that as long as the solution exists

$$F(t) = \int_\Omega u^2 dx \leq \int_\Omega u_o^2 dx \left[1 + 2\alpha t \frac{\int_\Omega u_o v_o dx}{\int_\Omega u_o^2 dx} \right]^{1/\alpha} . \tag{3.9}$$

Again arguing as in the proof of theorem 3 we find

$$\lim_{t \to T_-} \int_\Omega u_t^2 dx = +\infty ,$$

where $T \leq \frac{1}{2\alpha} \left(\int_\Omega u_o^2 dx \right) \left| \int_\Omega u_o \, v_o dx \right|^{-1}$. This completes the proof.

Remarks similar to those pertaining to theorems 2-4 are relevant here also.

§4 Concluding Remarks.

The methods used in this paper for the treatment of wave equations may also be applied to similar classes of abstract non-linear parabolic equations of the form $Pu_t = -Au + f(u)$. Some aspects of this have been considered by McKay [9]. Certain stability and non-existence problems in continuum mechanics have recently been discussed using concavity arguments based on the functional $G(t) = F^{\frac{1}{2}}(t) = \|u\|$, (i.e. the case in which $\alpha = \frac{1}{2}$). See Hills and Knops [2] and Knops and Straughan [5].

The study of non-linear equations having solutions with compact support in time is attracting much interest at the present time. This is mainly due to the fact that such phenomena are not present in linear equations in general. However most of this work seems to be directed towards non-linear evolutionary or parabolic equations. For reference to this and other aspects of non-linear problems we cite Kalashnikov [3] and the survey article by Brezis [1].

Finally, the disappearance of solutions to dissipative wave equation has been recently considered by A Majda [8].

Acknowledgement

The author would like to thank Professor Levine, who first aroused the author's interest in the concavity method, for stimulating discussions and correspondence concerning this work. Thanks are also due to Professor Knops for communicating some of the results of his investigations prior to publication.

References

[1] H. Brezis , Monotone operators, nonlinear semigroups and applications. Proceedings of the International Congress of Mathematicians, Vancouver 1974 Vol 2 p 249-255.

[2] R. N. Hills, R. J Knops, Qualitative results for a class of general materials. S.I.A.M. J. Applied Math (to appear).

[3] A. S. Kalashnikov, Propagation of disturbances in problems of non-linear heat conduction with absorption. U.S.S.R. Comp. Maths. and Math. Phys 14. 4. (1974).

[4] R. J. Knops, H. A. Levine, L. E. Payne, Nonexistence instability and growth theorems for solutions to an abstract nonlinear equation with applications to elastodynamics. Arch. Rat. Mech. Anal. 55 (1974) 52-72.

[5] R. J. Knops, B. Straughan, Nonexistence of global solutions to nonlinear Cauchy problems arising in mechanics. Univ. of Lecce (1975) symposium on trends of applications of pure mathematics to mechanics (to appear).

[6] H. A. Levine, Instability and nonexistence of global solutions to nonlinear wave equations of the form $P\dot{u}_{tt} = -Au + \mathfrak{F}(u)$. Trans. Amer. Math. Soc. 192 (1974) 1-21.

[7] H. A. Levine, L. E. Payne, Nonexistence theorems for the heat equation with nonlinear boundary conditions and for the porous medium equation backward in time. J. Diff. Equations 16 (1974) 319-334.

[8] A. Majda, Disappearing solutions for the dissipative wave equation. Indiana Univ. Math. J. 24 1119-1133, (1975).

[9] T. M. McKay, Review of "concavity" as applied to nonlinear evolutionary equations. M.Sc. thesis University of Dundee (1975).

Difference approximations for some pseudodifferential equations in R^n

Ernst Stephan

Introduction

In this paper pseudodifferential equations of the form

$$(0.1) \qquad (Op(a)u)(x) \equiv (2\pi)^{-n} \int_{R^n} e^{i<x,\xi>} a(x;\xi)\left(\int_{R^n} e^{-i<\xi,y>} u(y)dy\right)d\xi = v(x), \quad x \in R^n$$

are approximated by suitable difference approximations. The function $a(x;\xi)$ is the symbol of the pseudodifferential operator $Op(a)$ (hereafter ψdo). Equations of the form (0.1) contain differential equations, regular and singular integral equations as well as integro-differential equations. If the ψdo in (0.1) is elliptic, then a slight modification enables us to examine equation (0.1) not only locally (see Seely [5]) but even on the whole space R^n. Since the embedding of the Sobolev space $H^s(R^n)$ into $H^t(R^n)$ for real s,t with $s > t$ is not compact as for finite domains, the a-priori estimate for an elliptic ψdo $Op(a)$ in (0.1) does not guarantee its decomposition into a definite operator and a compact operator. Extra conditions for the symbol $a(x;\xi)$ are needed so that the perturbation of the definite term becomes compact even in the whole space.

The approximation schemes are given by the following families

$$(0.2) \qquad (Op_h(a_h)u_h)(x) \equiv (2\pi)^{-n} \int_{Q_h^n} e^{i<x,\xi>} a_h(x;\zeta(\xi),\bar{\zeta}(\xi))\tilde{u}_h(\xi)d\xi = v_h(x)$$

where $x \in R_h^n$ and $h \in (0,h_o] \equiv I$. These families are defined on a sequence of equidistant, infinite grid points R_h^n of an n-dimensional grid with mesh widths $h=(h_1,\ldots,h_n)$. In (0.2) $\tilde{u}_h$ denotes the discrete Fourier transform of a grid function u_h. In the following $\tilde{u}_h$ is at least square integrable on the cube Q_h^n.
$a_h(x;\zeta(\xi),\bar{\zeta}(\xi))$ denotes a suitable approximation of $a(x;\xi)$.

The purpose of this paper is the investigation of the solvability of the discretized equations (0.2) and the convergence of the solutions u_h to u as $h \to o$. The underlying frame work for this investigation is the theory of the discrete convergence of operators in suitable approximation schemes. Here, F. Stummel's version of this theory is used following the presentation in [7].

The crucial point in this approach is to find suitable approximations of the symbol $a(x;\xi)$ by symbols a_h such that an a-priori estimate holds; moreover, the approximating operators in (0.2) are consistent and their adjoints can be decomposed into sums of positive definite and weakly discretely compact sequences.If these approximations to the symbols are found, then a theorem by Stummel secures the inverse stability of the family $\{Op_h(a_h)\}_{h\in I}$, the unique solvability of (0.2), and the discrete convergence $Op_h(a_h)^{-1} \to Op(a)^{-1}$ as $h \to o$ provided that (0.1) is uniquely solvable.

In § 1 pseudodifference operators $Op(a_h)$ are introduced as parametric families $\{Op_h(a_h)\}_{h\in I}$ of operators of the form (0.2). In § 2 it is shown that for certain strongly elliptic ψdo's the approximation problems (0.2) are uniquely solvable for any $h \in I$, and the solutions u_h of (0.2) converge strongly (in the discrete sense [7]) to the solution u of (0.1) as $h \to o$. For the special case of the inverse Bessel potential operator $\Lambda^m \equiv (1+(\frac{\partial}{\partial x})^2)^{m/2}$, $m \in R$, in (0.1) with $n = 1$ the infinite systems of equations corresponding to (0.2) are explicitely formulated, and in this case, they can be uniquely solved independent of the mesh width h . Here, Hilbert's cut off method in ℓ^2 is an appropriate method for constructing the discrete solutions.

The results of this paper present parts of the author's dissertation [6].

§ 1 Pseudodifferential and pseudodifference operators

In order to solve equation (0.1) in the whole space R^n, we consider ψdo's with symbols which provide a suitable behaviour at infinity. To this end let us introduce the following notation:

We denote by S^m, $m \in R$, the set consisting of all functions $a(x;\xi) \in C^\infty(R^n \times R^n)$ such that

(1.1) $\lim_{|x|\to\infty} a(x;\xi) = a(\infty;\xi) \neq o$ uniformly in $\xi \neq o$ and in $\frac{x}{|x|}$,

(1.2) $\sup_{x,\,\xi\in R^n} (1+|\xi|)^{|\gamma|-m} |x^\alpha D_x^\beta D_\xi^\gamma a'(x;\xi)| < \infty$ for any multi-index α,β,γ

where $a'(x;\xi) \equiv a(x;\xi) - a(\infty;\xi)$,

(1.3) $\sup_{\xi\in R^n} (1+|\xi|)^{|\gamma|-m} |D_\xi^\gamma a(\infty;\xi)| < \infty$ for any multi-index γ.

Here the usual abbreviations are used:

$$x^\alpha = x_1^{\alpha_1} \cdots x_n^{\alpha_n}, \quad D_x^\beta = (-i\frac{\partial}{\partial x_1})^{\beta_1} \cdots (-i\frac{\partial}{\partial x_n})^{\beta_n}, \quad D_\xi^\gamma = (-i\frac{\partial}{\partial \xi_1})^{\gamma_1} \cdots (-i\frac{\partial}{\partial \xi_n})^{\gamma_n}.$$

Remark 1.1: For $a \in S^m$ the ψdo $Op(a)$ is continuous from $\mathcal{S}$ into $\mathcal{S}$ and from $H^s(R^n)$ into $H^{s-m}(R^n)$ for any real s, where $\mathcal{S}$ denotes the Schwarz space of tempered functions. Furthermore there exists a symbol $a^* \in S^m$ with $Op(a^*) = Op(a)^*$ where $Op(a)^*$ is the formal adjoint operator with respect to the L^2 scalar product. In addition, if $b \in S^p$, then there exists a symbol $c \in S^{m+p}$ such that $Op(c) = Op(a)Op(b)$ (see Seeley [5]).

Remark 1.2: The ψdo $Op(a)$ is called elliptic, if there exist constants k, $C > o$ such that $|a_m(x;\xi)| > C|\xi|^m$ for all $\xi \in R^n$ with $|\xi| > k$, where a_m denotes the principal symbol of a. If $a \in S^m$ has a suitable asymptotic expansion in terms of homogeneous functions, then in the elliptic case an a-priori estimate holds, and in this estimate the perturbation term is an operator of order $-\infty$, which is, in general, not compact (see [5]).

Remark 1.3: Using a fundamental lemma by Kohn and Nirenberg [4, p. 289] we obtain a result similar to the one given by Volevich and Kagan [10, p. 252]: For $a \in S^m$, $b \in S^p$ and for any positive integer N, the operators $Op(c_\alpha)$ and T_N given by

$$Op(a)Op(b) = \sum_{|\alpha|<N} Op(c_\alpha) + T_N$$

are completely continuous from H^s into $H^{s-m-p+|\alpha|-\lambda}$ and $H^{s-m-p+N-\lambda}$, respectively, for $\lambda > o$ and any non-vanishing multi-index α. Here

$$c_\alpha(x;\xi) \equiv (iD_\xi)^\alpha a(x;\xi)\, D_x^\alpha b(x;\xi)/\alpha\,! \quad \text{(see [6,pp.25-27])}.$$

To obtain difference approximations for (0.1) where the ψdo $Op(a)$ has a symbol of class S^m, $m \in R$, let us define one parametric families $Op(a_h) = \{Op_h(a_h)\}_{h \in I}$ of operators $Op_h(a_h)$ on the grid functions $u_h = r_h u$ with $u \in \mathcal{S}$ and for simplicity we write $u_h \in \mathcal{S}_I$ (see [2]). Here it is understood that r_h is the pointwise restriction to the grid.

We denote by $^*S_I^m$, $m \in R$, the set consisting of all families of complex valued functions $a_h(x;\xi)$ defined for all $x \in R_h^n$, $\xi \in Q_h^n$ and $h \in I$ such that

(1.4) $\lim_{|x|\to\infty} a_h(x;\xi) = a_h(\infty;\xi) \neq o$ uniformly for $\xi \neq o$, $\frac{x}{|x|}$ and $h \in I$, respectively,

(1.5) $$\sup_{h \in I} \sup_{\substack{x \in R_h^n \\ \xi \in Q_h^n}} (1+|\zeta(\xi)|)^{-m} |x^\alpha \partial_h^{\beta_1} \bar{\partial}_h^{\beta_2} a_h'(x;\xi)| < \infty \quad \text{for any multi-index } \alpha, \beta_1, \beta_2$$

where $a_h'(x;\xi) \equiv a_h(x;\xi) - a_h(\infty;\xi)$,

(1.6) $$\sup_{h \in I} \sup_{\xi \in Q_h^n} (1+|\zeta(\xi)|)^{-m} |a_h(\infty;\xi)| < \infty$$

with $\zeta(\xi) = \prod_{j=1}^n \zeta_j(\xi_j)$, $\zeta_j(\xi_j) = \dfrac{e^{ih_j\xi_j}-1}{ih_j}$.

In (1.5) ∂_h (and $\bar{\partial}_h$) denote the forward (and backward) directed difference operators multiplied by $(-i)$. Analogously to the ψdo $Op(a)$ (see [5]) we state

Definition 1.1: The pseudodifference operator $Op(a_h)$ with symbol $a_h \in {}^*S_I^m$ (hereafter

<u>ψdio) is defined on</u> $u_h \in S_I$ <u>by the family</u>

$$(1.7)\qquad (Op_h(a_h)u_h)(x) \equiv (2\pi)^{-n} \int_{Q_h^n} e^{i<x,\xi>} a_h(x;\xi)\, \tilde{u}_h(\xi)d\xi \;, \quad h \in I$$

<u>with</u> $\tilde{u}_h(\xi) = h_1 \ldots h_n \sum_{x \in R_h^n} u_h(x)\, e^{-i<x,\xi>}$ <u>and</u> $Q_h^n \equiv \{\xi = (\xi_1,\ldots,\xi_n) \in R^n : |\xi_j| \le \frac{\pi}{h_j}\}$.

Since the composition of two such ψdio's in general does not become a ψdio again, let us restrict the class of symbols, $^*S_I^m$, in the following way.
<u>We denote by</u> S_I^m, $m \in R$, <u>the set consisting of all</u> $a_h(x;\zeta(\xi),\bar{\zeta}(\xi)) \in {}^*S_I^m$ <u>such that</u>

$$(1.8)\qquad \sup_{h \in I}\ \sup_{\substack{x \in R_h^n \\ \zeta \in B_h^n}} (1+|\zeta|)^{|\gamma_1|+|\gamma_2|-m} |x^\alpha \partial_h^{\beta_1} \bar{\partial}_h^{\beta_2} D_\zeta^{\gamma_1} D_{\bar{\zeta}}^{\gamma_2} a_h'(x;\zeta,\bar{\zeta})| < \infty \quad .$$

$$(1.9)\qquad \sup_{h \in I}\ \sup_{\zeta \in B_h^n} (1+|\zeta|)^{|\gamma_1|+|\gamma_2|-m} |D_\zeta^{\gamma_1} D_{\bar{\zeta}}^{\gamma_2} a_h(\infty;\zeta,\bar{\zeta})| < \infty$$

<u>for any multi-index</u> $\alpha, \beta_1, \beta_2, \gamma_1, \gamma_2$,

<u>where</u> B_h^n <u>denotes the ball</u> $B_h^n \equiv \{\zeta = (\zeta_1,\ldots,\zeta_n) \in C^n \;:\; |1+ih_j\zeta_j| \le 1,\ 1 \le j \le n\}$.

In (1.8), (1.9) the derivatives $D_\zeta = (-i\frac{\partial}{\partial\zeta_1})\ldots(-i\frac{\partial}{\partial\zeta_n})$, $D_{\bar{\zeta}} = (-i\frac{\partial}{\partial\bar{\zeta}_1})\ldots(-i\frac{\partial}{\partial\bar{\zeta}_n})$ are defined with $\frac{\partial}{\partial\zeta_j}$ and $\frac{\partial}{\partial\bar{\zeta}_j}$ as the Pompeju derivatives (see e. g. Vekua [9,I §4]), which can be written as partial derivatives in the case of sufficiently smooth functions. Therefore (1.8), (1.9) do not imply the holomorphy of the symbols.

<u>Remark 1.4:</u> With the discrete Fourier transform any family of linear combinations of difference operators can be written as a ψdio (1.7); e. g. the family $A_h \equiv \sum_{|\alpha| \le k} \sum_{\beta+\gamma=\alpha} a_h^{\beta,\gamma}(x)\, \partial_h^\beta \bar{\partial}_h^\gamma$, $h \in I$, with $a_h^{\beta,\gamma} \in S_I$ has the symbol

$$\sum_{|\alpha| \le k} \sum_{\beta+\gamma=\alpha} a_h(x)\, \zeta(\xi)^\beta\, \bar{\zeta}(\xi)^\gamma \in S_I^m, \quad \text{where } \bar{\zeta}(\xi) = \prod_{j=1}^n \bar{\zeta}_j(\xi_j),\ \bar{\zeta}_j(\xi_j) = \frac{1-e^{-ih_j\xi_j}}{ih_j} .$$

Similar results as for ψdo's are obtained for ψdio's:

<u>Theorem 1.1: If</u> $a_h \in S_I^m$, <u>then</u> $Op(a_h)$ <u>is continuous from</u> $H_I^s(R_h^n)$ <u>into</u> $H_I^{s-m}(R_h^n)$ <u>for any real</u> s, <u>where</u> $H_I^s(R_h^n)$ <u>is the scale of discrete Sobolev spaces analogous to</u> $H^s(R^n)$ <u>introduced by Frank</u> [2].

<u>Proof:</u> The continuity of $Op(a_h)$ is shown by the representation

$$(1.10)\qquad \widetilde{Op_h(a_h)u_h}(\xi) = a_h(\infty;\xi)\, \tilde{u}_h(\xi) + (2\pi)^{-n} \int_{Q_h^n} \tilde{a}_n'(\xi-\eta;\eta)\tilde{u}_h(\eta)d\eta$$

using the smoothness condition at infinity (1.5) implying $a_h'(\cdot,\xi) \in S_I$ uniformly in $\xi \neq o$, the decay of the discrete Fourier transform of a function in S_I and the growth conditions (1.5), (1.6) (see [6, pp. 32,33]).

Theorem 1.2: If $a_h \in S_I^m$ and $b \in S_I^p$, then there exists a family $c_h \in {}^*S_I^{m+p}$ with $Op(c_h) = Op(b_h)Op(a_h)$. Furthermore $c_h(x;\xi)$ has the asymptotic expansion

$$(1.11) \quad c_h \sim \sum_{k=o}^{\infty} \sum_{|\alpha_1|+|\alpha_2|=k} \frac{(1+ih\zeta)^{\alpha_1}}{\alpha_1!} \frac{(1-ih\bar{\zeta})^{\alpha_2}}{\alpha_2!} (iD_\zeta)^{\alpha_1} (iD_{\bar{\zeta}})^{\alpha_2} b_h(x;\zeta,\bar{\zeta}) \times$$

$$\times \partial_h^{\alpha_1} \bar{\partial}_h^{\alpha_2} a_h(x;\zeta,\bar{\zeta})$$

$$= \sum_{k=o}^{\infty} c_h^{m+p-k} \quad \text{such that} \quad c_h - \sum_{k=o}^{N-1} c_h^{m+p-k} \in {}^*S_I^{m+p-N} \quad \text{holds.}$$

To a_h there exists also a family $a_h^* \in {}^*S_I^m$ with $Op(a_h^*) = Op(a_h)^*$ and $Op(a_h)^* \equiv \{Op_h(a_h)^*\}_{h \in I}$ where $Op_h(a_h)^*$ is the formal adjoint operator to $Op_h(a_h)$ with respect to the ℓ^2 scalar product.

Proof: With the representation (1.10), the periodicity properties of the discrete Fourier transform and (1.5), (1.6) we obtain $Op(c_h) = Op(b_h)Op(a_h)$ with $c \in {}^*S_I^{m+p}$. By estimating with (1.8), (1.9) the remainder in the Taylor expansion for $b_h(x;\zeta,\bar{\zeta})$ we arrive at (1.11). The other statements are proved similarly (see [6, pp. 37-49]).

For ψdo's it is known that the ellipticity implies the existence of a parametrix and a-priori estimates. Similarly let us state the following

Definition 1. 2: The ψdio $Op(a_h)$ is called strongly elliptic, if $a_h \in S_I^m$ and there are constants $k, c, \hat{c} > o$ such that

$$(1.12) \quad c|\zeta|^m \leq |a_h(x;\zeta,\bar{\zeta})| \leq \hat{c}\,|\zeta|^m \qquad \text{for any} \quad h \in I,\ x \in R_h^n, \text{ and } \zeta \in B_h^n \text{ with } |\zeta| > k\ .$$

For strongly elliptic ψdio's we have the following theorem, which is the basis of the solvability of the approximation equations (0.2).

Theorem 1.3: Let $a_h \in S_I^m$ satisfy the growth condition (1.12) and fulfill

$$(1.13) \quad a_h(x;\zeta,\bar{\zeta}) \neq o \qquad \text{for any} \quad h \in I,\ x \in R_h^n \text{ and } \zeta \in B_h^n, \text{ respectively,}$$

then for any given number N there exist $b_h^1, b_h^2 \in S_I^{-m}$ with

$$(1.14) \quad b_h^e = \frac{1}{a_h} + \sum_{j=1}^{N} b_h^{e(j)}\ , \quad b_h^{e(j)} \in S_I^{-m-j}, \quad e = 1,2, \quad \text{such that}$$

$$T^1 \equiv Op(b_h^1)Op(a_h)\text{-identity} \quad \text{and} \quad T^2 \equiv Op(a_h)Op(b_h^2) - \text{identity}$$

have symbols $t_h^e \in {}^*S_I^{-N-1}$, $e = 1,2$, vanishing as $|x| \to \infty$.

Proof: According to Theorem 1.2 $t_h^e(x;\xi)$ has an asymptotic expansion (1.11). Using the set-up (1.14) the unknown discrete symbols $b_h^{e(j)}$, $e = 1,2$, $j = 1,\dots,N$, are constructed by recursion. The strong ellipticity (1.12) implies the boundedness of the ψdio's T^e, $e = 1,2$, from $H_I^s(R_h^n)$ into $H_I^{s+N+1}(R_h^n)$ for any real s. Finally, (1.13) and (1.14) secure in connection with Theorem 1.2 that their symbols $t_h^e(x;\xi)$ vanish as $|x| \to \infty$ (see [6, pp. 82-83]).

In [1] for another class of elliptic operators in (0.2) Frank constructed a-priori estimates. However, there neither the conditions of the solvability to the equations (0.2) nor the convergences of the approximated solutions of (0.2) are given.

§2 Discrete approximation of the solution of a strongly elliptic pseudodifferential equation

For any Sobolev space $H^s(R^n)$, $s \in R$, we can show ([6, pp. 66-71]) the existence of a discrete approximation $A(H^s, \prod_{h \in I} H_h^s, R^s)$ with the discrete Sobolev space $H_h^s(R_h^n)$ introduced by Frank [2]. Therefore the solvability of the approximation problems (0.2) can be characterized by the following equivalence theorem proved by Stummel [7, p. 55].

<u>Theorem 2.1:</u> <u>Let</u> $Op(a)$ <u>be surjective and let</u> $Op_h(a_h)$, $h \in I$, <u>form a stable sequence of continuously invertible, bijective operators. The inverse stability of the approximating operators</u> $Op_h(a_h)$ <u>and the consistency of the pair</u> $(Op(a), \{Op_h(a_h)\}_{h \in I})$ <u>are necessary and sufficient for the existence of</u> $Op(a)^{-1}$ <u>and the discrete convergence of the inverse operators.</u>

Now let us formulate conditions on the symbols $a \in S^m$ in (0.1) and $a_h \in S_I^m$ in (0.2) such that Theorem 2.1 can be applied. That leads to

<u>Theorem 2.2:</u> <u>Let</u> $Op(a)$ <u>be a bijective</u> ψ<u>do with</u> $a \in S^m$; <u>let the symbols</u> $a_h \in S_I^m$ <u>of the approximating operators</u> $Op_h(a_h)$, $h \in I$, <u>admit the decomposition</u>

$$(2.1) \qquad a_h(x;\zeta(\xi),\bar\zeta(\xi)) = a_h^o(x;\zeta(\xi),\bar\zeta(\xi)) + a_h^1(x;\zeta(\xi),\bar\zeta(\xi))$$

<u>where</u> $a_h^o \in S_I^m$ <u>satisfy</u> (1.12), (1.13) <u>and</u> $a_h^1 \in S_I^k$ <u>for</u> $k < m - \frac{n}{2}$ <u>fulfill</u>

$$(2.2) \qquad a_h^1(\infty;\zeta,\bar\zeta) = o \quad \text{for any } h \in I \text{ and } \zeta \in B_h^n,$$

<u>then the condition of consistency</u>

$$(2.3) \qquad \lim_{h \to o} |a_h(x;\zeta(\xi),\bar\zeta(\xi)) - a(x;\xi)| = o$$

<u>implies the unique solvability of the approximation problems</u> (0.2) <u>for any</u> $h \in I$ <u>and the solutions</u> u_h <u>of</u> (0.2) <u>converge strongly in the discrete sense to the solution</u> u <u>of</u> (0.1) <u>as</u> $h \to o$.

In what follows, an outline of the proof which is based on Theorem 2.1 will be given.

1.: The conditions (1.12), (1.13) secure the existence of sequences of regularizers

$Op_h(b_h^1)$, $Op_h(b_h^2)$, $h \in I$, such that

$$(2.4) \qquad T_h^1 \equiv Op_h(b_h^1)Op_h(a_h^o)\text{-identity}, \quad T_h^2 \equiv Op_h(a_h^o)Op_h(b_h^2)\text{-identity}, \quad h \in I$$

with $t_h^e \in {}^*S_I^{-N-1}$, $e = 1,2$, for $N > \frac{n}{2} - 1$ define discretely compact operator families from the cartesian product $\prod_{h \in I} H_h^s$ into itself for any positive real s.

Furthermore, there exists a constant c such that for any subsequence $\Lambda \subset I$, for any $h \in \Lambda$ and for any sequence $u_h \in \prod_{h \in I} H_h^s$ the following a-priori estimate holds:

$$(2.5) \qquad c\,|u_h|_{s,h} \leq |Op_h(a_h^o)u_h|_{s-m,h} + |T_h^1 u_h|_{s,h}$$

with

$$|u_h|_{s,h}^2 = (2\pi)^{-n} \int_{Q_h^n} (1+|\zeta(\xi)|^2)^s\,|\tilde{u}_h(\xi)|^2 d\xi \,.$$

2.: For the approximation of problems in elliptic differential equations Stummel proved in [7, p. 71] the inverse stability of the approximating operators under the assumption that an a-priori inequality holds involving weakly discretely compact perturbation terms.

3.: Due to Theorem 1.2 the adjoint operators in (2.4) satisfy a corresponding inequality (2.5) with weakly discretely compact perturbation terms (see [7, p. 66]). Hence, the sequence of the adjoint operators is inverse stable. In the same way the inverse stability of $\{Op_h(a_h^o)\}_{h \in I}$ is proved since the adjoint ψdio $Op(a_h^o)^*$ fulfills (1.12), (1.13) (see [6 , pp. 88-91]).

4.: According to (2.2) a_h^1 define only discretely compact perturbations, which do not destroy the existence of the inverses to the approximating operators.

5.: The desired stability of $\{Op_h(a_h)\}_{h \in I}$ follows by Theorem 1.1 by the choice $a_h \in S_I^m$.

6.: The condition (2.3) guarantees the consistency of the pair $(Op(a), \{Op_h(a_h)\}_{h \in I})$:

$$(2.6) \qquad |Op_h(a_h)r_h u - r_h Op(a)u|_{s-m,h} \to o \quad \text{as} \quad h \to o \quad \text{for any} \quad u \in \mathcal{S}$$

Using Lebesgue's theorem the relation (2.6) follows by the representation (1.10) and the special choice of the symbol classes S^m and ${}^*S_I^m$ (see [6 , pp. 72-75]).

Remark 2.1: The discrete compactness ([7, p. 66], [8, pp. 5, 15]) of the operator families in 1 is shown by carrying over Remark 1.3 to the situation in the theory of discrete convergence. This follows by the discrete compactness of the embedding in $\prod_{h \in I} H_h^t$ of suitable subsets of the sequence space $\prod_{h \in I} H_h^s$ for real s, t with $s > \frac{n}{2}$,

$s - \frac{n}{2} > t$ (see [6, p. 76-81, 83-85]).

Remark 2.2: Due to condition (2.3) the symbol a of the ψdo Op(a) has a decomposition analogous to (2.1) satisfying conditions similar to (1.12), (1.13). Hence, in Theorem 2.2 the ψdo Op(a) is strongly elliptic and belongs to a class which was characterized by Kohn and Nirenberg [4, p. 283].

2.1. Difference approximations of equations with inverse Bessel potential operators

For the approximation of

$$(2.7) \qquad (\Lambda^m u)(x) = v(x), \quad x \in R^1, \quad u \in H^m$$

where the ψdo $\Lambda^m \equiv (1+(\frac{\partial}{\partial x})^2)^{m/2}$ has the symbol $(1+|\xi|^2)^{m/2} \in S^m$, $m \in R$, let us define by

$$(2.8) \qquad a_h^e(x;\zeta(\xi),\bar{\zeta}(\xi)) \equiv \begin{cases} (1+|\zeta(\xi)|^2)^{m/2} & , e = 1 \\ [(\zeta(\xi)+i)(\bar{\zeta}(\xi)-i)]^{m/2} & , e = 2 \end{cases}$$

families (o.2). For any fixed $h > o$ the series expansion of (2.8) in terms of $e^{ih\xi}$ converges absolutely and uniformly for $\xi \in Q_h$. Inserting this expansion in (o.2), the equation becomes the following infinite system of linear equations:

$$(2.9) \qquad f_j(h)^{-m/2} \sum_{\nu=o}^{\infty} \sum_{\mu=o}^{\nu} a_{\mu\nu} (2+f_j(h))^{(m/2)-\nu} u_h(x+(\nu-2\mu)h) = v_h(x), \quad x \in R_h^1$$

with

$$a_{\mu\nu} \equiv (-1)^\nu \binom{m/2}{\nu} \binom{\nu}{\mu}, \quad f_j(h) \equiv \begin{cases} h^2 & , j = 1 \\ \frac{h^2}{1+h} & , j = 2 \end{cases}$$

It is easily shown that for this example all the assumptions of Theorem 2.2 are satisfied; hence, (2.9) is uniquely solvable. The solutions u_h converge discretely to the solution $u \in H^m$ of (2.7). Furthermore, the symbols (2.8) satisfy a positivity condition. Consequently, the corresponding approximation operator $Op_h(a_h)$ can be decomposed into the sum of a definite and a compact operator on the Hilbert space $\ell^2(R_h^n)$ for any fixed $h > o$. Hence, by a result by Hildebrandt and Wienholtz [3, p. 371] the cut off method converges independently of the mesh width providing a construction for finding the solutions u_h (see [6, p. 96-101]).

The numerical results for (2.7) with $m = 1$ and $m = \frac{8}{3}$ show the effectiveness of our approximation method (see [6, pp. 106-113]).

References

[1] Frank, L. S., Difference operators in convolutions
Soviet Math. Dokl. 9 (1968), 831-834

[2] Frank, L. S., Spaces of network functions
Math. USSR Sbornik, 15 (1971), 183-226

[3] Hildebrandt, S. and Wienholtz, E., Constructive Proofs of Representation Theorems in Separable Hilbert Space
Comm. Pure Appl. Math. 17 (1964), 369-373

[4] Kohn, J. J. and Nirenberg, L., An algebra of pseudo-differential operators, Comm. Pure Appl. Math. 18 (1965), 269-305

[5] Seeley, R., Topics in pseudo-differential operators
C.I.M.E. - II Ciclo, Rom 1969, 169-305

[6] Stephan, E., Differenzenapproximationen von Pseudo-Differentialoperatoren
Dissertation, Darmstadt 1975

[7] Stummel, F., Diskrete Konvergenz linearer Operatoren I
Math. Ann. 190 (1970), 45-92

[8] Stummel, F., Discrete Convergence of Mappings
Proceedings of the Conference on Numerical Analysis, Dublin, August 1972

[9] Vekua, I. N., Verallgemeinerte analytische Funktionen
Akademie-Verlag, Berlin, 1963

[10] Volevich, L. R. and Kagan, V. M., Hypoelliptic pseudo-differential operators in the theory of functional spaces
Trans. Moscow Math. Soc. 20 (1969), 243-283

Remarks to Galerkin and least squares methods with finite elements for general elliptic problems

Ernst Stephan and Wolfgang Wendland

Introduction:

Finite element methods have succeeded in many practical treatments e.g. for elliptic partial differential boundary value problems, for second kind integral equations and even for certain first kind integral equations. In all these cases the approximate solutions defined by the Galerkin or by the Gauss equations converge to the desired solution with optimal order. In these problems all the operators belong to the wider class of pseudodifferential operators (hereafter ψdo's). The purpose of this paper is to shed some light on the applicability of the Galerkin method or the least squares method to linear equations involving elliptic ψdo's.

According to Vainikko's result [35] one can expect the stability of the usual Galerkin method only for coercive problems. In the latter case the stability follows from the convergence proof by Hildebrandt and Wienholtz [18] (see also [26]). The coerciveness is well known for strongly elliptic partial differential boundary value problems (from Gårdings inequality), and for Riesz-Schauder operators as in integral equations of the second kind. Hence, we consider the wider class of strongly elliptic ψdo's in §2 after formulating the general problem in §1. First we investigate problems on compact manifolds. Here, strong ellipticity implies coerciveness for ψdo's according to Kohn and Nirenberg [21; p. 283].

For boundary value problems we adapt smoothness conditions imposed by Višik, Eskin [40] and Dikanskii [12, 13]. For such a smooth ψdo of order α the factorization index along the boundary becomes an integer and the strong ellipticity yields $\kappa = \frac{\alpha}{2} = L-M$ where L is the number of boundary conditions and M is the number of coboundary operators. The latter were introduced by Višik and Eskin in order to obtain Fredholm mappings. Assuming the corresponding Šapiro-Lopatinski condition and using Dikanskii's generalized Green formula [13] we arrive again at bilinear equations. Here we consider only simple coercive variational problems and not the general case (see e.g. [25,p.200]).

For the coercive problems it is shown that the Galerkin method with finite elements as trial functions leads to an optimal rate of convergence. In the case of compact manifolds this generalizes the results for first kind equations in [20, 22, 23, 24, 28, 29] and is also applicable to strongly elliptic differential operators, to certain classes of singular integral operators and to the classical integral equations of the second kind. In the case of boundary value problems this generalizes the results by

Schultz [33] and others for differential operators to the ψdo problems.
In § 3 we consider least squares methods. For the stability of Gaussian equations it is appropriate that the mapping defined by the original problem is an isomorphism between the space of definition and the image space. A well known example is the regular elliptic boundary value problem in the sense of Agmon-Douglis-Nirenberg [1] . In the case of ψdo's the corresponding a-priori estimates are those by Seeley [34 , p.239] on compact manifolds and by Dikanskii [13] for boundary value problems. But here the L_2-norm over the domain leads necessarily to interpolation spaces on the boundary [25, p. 188] whereas only L_2-products are realizable. According to Nitsche [31], Bramble and Schatz [10] and Aubin [4] one has to modify the least squares method by imposing weighted L_2-boundary norms. It turns out that the results by Bramble and Schatz remain valid for the ψdo problems on compact manifolds as well as for the ψdo boundary value problems namely, the least squares method with finite elements leads to an optimal rate of convergence.

§ 1 Formulation of the problem and general assumptions

Let M be a sufficiently smooth n-dimensional closed compact manifold with a domain $G \subseteq M$. Then the following two cases (i), (ii) require different approaches:

(i) $G = M$, $n \geq 1$, and
(ii) G is bounded by a (n-1)-dimensional sufficiently smooth manifold Γ, $n \geq 2$.

We consider equations

$$(1.1) \qquad \tilde{\tilde{A}}u(x) \equiv A(x;D)u + \sum_{k=1}^{M} F_k \rho_k = g(x) \qquad \text{for } x \in G ,$$

where A is a given elliptic ψdo of order α on M and the unknown u is always extended by o for $x \notin \bar{G} \equiv G \cup \Gamma$. [1)]

In case (i), α can be any real number and the terms F_k disappear $(M = o)$.

In case (ii) for simplicity we assume that α is an integer. Here the $\rho_k(x')$ are unknown densities for $x' \in \Gamma$ generating generalized potentials

$$(1.2) \qquad F_k\rho_k \equiv G_k(x;D)\ (\rho_k(x') \otimes \delta_\Gamma(x_n))$$

where the G_k are the coboundary operators given by ψdo's of orders α_k on M, operating on distributions $\rho_k \otimes \delta_\Gamma$ which are concentrated on Γ.
The boundary conditions are given by

$$(1.3) \qquad \tilde{B}_j\tilde{u} \equiv \gamma' B_j(x;D)u + \sum_{k=1}^{M} E_{jk}(x';D')\rho_k = g_j(x') \quad \text{for} \quad x' \in \Gamma ,\ j=1,\dots,L$$

1) It is understood that u and g can also be vector valued functions with ν components in which case we assume that all the single orders in (1.1) equal α .

where the B_j are ψdo's of order β_j on M; $\gamma' \equiv \gamma p$ where γ is the trace operator and p denotes the operator of restriction to G : $pH^s(M) = H^s(G)$; and the E_{jk} are ψdo's of order $a_k+\beta_j-\alpha+1$ on Γ. The collections of unknowns and data are denoted by $\tilde{u}$ and $\tilde{g}$ respectively belonging to the following function spaces:

$$(1.4) \qquad \tilde{u} \equiv (u,\rho_1,\ldots,\rho_M) \in X^s \equiv H^s(G) \times \prod_{k=1}^{M} H^{s-\alpha+\alpha_k+1/2}(\Gamma),$$

$$(1.5) \qquad \tilde{g} \equiv (g,g_1,\ldots,g_L) \in Y^{s-\alpha} \equiv H^{s-\alpha}(G) \times \prod_{j=1}^{L} H^{s-\beta_j-1/2}(\Gamma),$$

where $H^s(G)$ and $H^t(\Gamma), s,t \in R$ denote the Sobolev-spaces with the L_2 scalar products $(\ ,\ )_o$ on G and $<\ ,\ >_o$ on Γ, respectively. The norm in the product space X^s is defined by

$$|||\tilde{u}|||_s \equiv ||u||_s + \sum_{k=1}^{M} |\rho_k|_{s-\alpha+\alpha_k+\frac{1}{2}}$$

and, correspondingly, in $Y^{s-\alpha}$.

In case (i), the spaces simplify to $X^s = H^s(M)$ and $Y^{s-\alpha} = H^{s-\alpha}(M)$.

In case (ii), since a-priori estimates are needed, we assume (according to Dikanskii [13]) $\alpha > \beta_j + \frac{1}{2}$ and that the principal symbols of A, G_k and B_j belong to the classes D_α, D_{α_k}, D_{β_j}, respectively. Similarly, the lower order terms shall be smooth operators in the sense of [40, p. 101]. Further, let us assume that (1.1), (1.3) satisfy the Šapiro-Lopatinskii condition.

In both cases ((i) [34], (ii) [13]), the mapping $\tilde{u} \to \tilde{g}$ defined by (1.1), (1.3) is Fredholm. For simplicity, we assume that the deficiency of this mapping is zero. For a unique solution of (1.1), (1.3) let us require

$$(1.6) \qquad \Lambda_\ell u = c_\ell\ ,\ \ell = 1,\ldots,N \equiv \text{nullity of } ((1.1),\ (1.3))$$

where the Λ_ℓ are suitably choosen continuous linear functionals on X^s. Hence, the extended problem (1.1), (1.3), (1.6) is uniquely solvable.

For the approximation of (1.1), (1.3), (1.6) by Galerkin's or by least squares methods we use regular finite element spaces $\tilde{H} \subset H^m(G)$ and $\tilde{H}_k \subset H^{m_k}(\Gamma)$, consisting of regular (t_k,m_k)-systems $S_h^{(t,m)}(G)$ and $S_h^{(t_k,m_k)}(\Gamma)$ in the sense of Babuška and Aziz [7], $t_k \geq m_k + 1$ $(m_o \equiv m)$ and m_k integers $\geq o$. They are characterized by the following

Convergence property: For $-(m+1) \leq \hat{t} \leq s \leq (m+1), -m \leq s; \hat{t} \leq m$ the following approximation condition holds: For every $u \in H^s$ exists a $\chi \in \tilde{H}$ such that

$$(1.7) \qquad ||u - \chi||_{\hat{t}} \leq c\, h^{s-\hat{t}}\ ||u||_s$$

where c is independent of u.

Let us assume the corresponding convergence property for each of the boundary spaces $\tilde{H}_k(\Gamma)$, too.

To get estimates of optimal order we further assume the following fundamental stability property:

Inverse assumption: For $t \le s$; $|t|$, $|s| \le m$, there exist constants $M = M(t,s,m)$ such that

(1.8) $$||\chi||_s \le M\, h^{t-s} ||\chi||_t \quad \text{for all } \chi \in \tilde{H}\,.$$

For splines corresponding to regular triangulations condition (1.8) was proved by Nitsche [30], see also [4, pp. 125, 137].

For simultaneous approximations one often uses $\tilde{H}_k \equiv \gamma\tilde{H}$. Then (1.7) for $\tilde{H}_k(\Gamma)$ follows from (1.7) for $\tilde{H}(G)$ with $m_k = m - \frac{1}{2}$ and $t_k = t - \frac{1}{2}$ [7, p. 101]; the inverse assumption (1.8) for $\tilde{H}_k(\Gamma)$ requires additional regularity properties of the triangulations on Γ defined by the traces of the triangulations on M (see Babuška [6]).

In the case (ii), the functions in G and on Γ will be approximated simultaneously by elements of the product space

(1.9) $$\tilde{\mathcal{H}} \equiv \tilde{H}(G) \times \tilde{H}_1(\Gamma) \times \ldots \times \tilde{H}_M(\Gamma)\,.$$

§ 2 Galerkins's method and strongly elliptic problems

Since Vainikko's result [36] implies that Galerkin's methods converge only for the sum of a definite and a compact operator we restrict us here to strongly elliptic problems.

A is strongly elliptic if there exists a C^∞ function $\theta(x)$ on M such that the inequality

(2.1) $$\operatorname{Re}\ \theta(x)a'(x,\xi) \ge \gamma > 0 \quad \text{holds for all } \xi \in R^n \text{ with } |\xi| = 1$$

where γ is some constant independent of x and ξ and a' is the principle symbol of A.

Multiplication of (1.1) by θ leads to $\theta \equiv 1$ without loss of generality.

2.1 Strongly elliptic operators on closed manifolds without boundary

As Kohn and Nirenberg proved in [21, p. 283], Garding's inequality is valid for strongly elliptic operators with $\theta \equiv 1$ in (2.1). Consequently one has the following

Coerciveness property: A has the form

(2.2) $$A = D + K$$

where D is definite

(2.3) $$\operatorname{Re}(Dv,v)_0 \ge \gamma' ||v||^2_{\frac{\alpha}{2}}$$

and $K : H^{\frac{\alpha}{2}} \to H^{-\frac{\alpha}{2}}$ is completely continuous.

The Galerkin equations to (1.1), (1.6) for $u^* \in \tilde{H}$ are

$$(2.4) \qquad A[u^*,\phi] \equiv (Au^*,\phi)_o = (g,\phi)_o = (Au,\phi)_o \qquad \text{for all } \phi \in \tilde{H} ,$$

$$(2.5) \qquad \Lambda_\ell u^* = c_\ell = \Lambda_\ell u \qquad \text{for } \ell = 1,\dots,N.$$

For sufficiently small h these equations are uniquely solvable and u^* converges to u as $h \to o$ with optimal order:

Theorem 2.1: If A is strongly elliptic and the Λ_ℓ are continuous in $H^{\frac{\alpha}{2}}(G)$ and $-m \leq \frac{\alpha}{2} \leq t \leq s \leq m+1$; $t \leq m$, then there exist constants $h_o > o$ and $c \geq o$ such that the Galerkin equations (2.4), (2.5), are uniquely solvable for all $o < h \leq h_o$ and

$$(2.6) \qquad ||u-u^*||_t \leq c\, h^{s-t}\, ||u||_s$$

Proof: If we consider the bilinear equations in the Hilbert-space $H^{\frac{\alpha}{2}}(G) \times \mathbb{C}^N$ then the results of Hildebrandt and Wienholtz [18] imply solvability and stability of the Galerkin equations for $h \leq h_o$ and the mapping $\underset{\sim}{G} : u \to u^*$ defines Galerkin's operator with the properties

$$(2.7) \qquad ||\underset{\sim}{G}u||_{\frac{\alpha}{2}} \leq C||g||_{-\frac{\alpha}{2}} \leq C'||u||_{\frac{\alpha}{2}} ,$$

$$(2.8) \qquad \underset{\sim}{G}\,\phi = \phi \qquad \text{for all } \phi \in \tilde{H} .$$

By Nitsche's technique [31], (2.7) and (2.8) imply with (1.8) the inequality

$$||u-u^*||_t \leq ||u-\chi||_t + M\, h^{\frac{\alpha}{2}-t}\, ||\underset{\sim}{G}(u-\chi)||_{\frac{\alpha}{2}} \leq ||u-\chi||_t + C'Mh^{\frac{\alpha}{2}-t}\, ||u-\chi||_{\frac{\alpha}{2}}$$

for all $\chi \in \tilde{H}$. Hence (1.7) leads to the desired estimate (2.6)

Applications and corresponding results:

2.1.1: $A(x;D)$ is an even order strongly elliptic differential operator.

Then the characteristic polynomial satisfies (2.1) and Theorem 2.1 corresponds to the results on quasioptimal convergence (see e. g. [33]).

2.1.2: Singular integral equations of Mikhlin's type [27].

Here A is of order zero and (2.1) is a criterion for the convergence in any space dimension n .

In the special case $n = 1$ Gohberg and Fel'dman investigated in [16] also the Galerkin method by another approach (and modified the method even for nonvanishing Fredholm index). For $n = 1$, A is defined by

$$(2.11) \qquad Au(\zeta) \equiv a(\zeta)u(\zeta) + \frac{1}{\pi i} \oint_{M_1} \frac{u(z)b(z,\zeta)}{z-\zeta}\, dz + T_1u(\zeta)$$

where M_1 is a simply closed smooth curve in the complex plane and T_1 is an integral operator with smooth kernel. The principal symbol is

$$(2.12)\qquad a'(z,\xi) = \begin{cases} c(z) \equiv a(z) + b(z,z) & \text{for } \xi \geq 1, \\ d(z) \equiv a(z) - b(z,z) & \text{for } \xi \leq -1. \end{cases}$$

Here strong ellipticity (2.1) is equivalent to convergence of (2.4), (2.5) with every $\tilde{H}$ [16, p. 62 ff.]. Gohberg and Fel'dman proved that under additional assumptions on $\tilde{H}$,

$$(2.13)\qquad c \neq o,\quad d \neq o \quad \text{and index } c = \text{index } d = o\,,$$

is necessary and sufficient for the convergence of (2.4), (2.5) [16, pp. 141, 142, 152]. In this case, (1.1), (1.4) can be formulated as systems in projected spaces. With

$$(2.14)\qquad a' \equiv \begin{pmatrix} c & o \\ o & d \end{pmatrix} \quad \text{and} \quad \theta \equiv \begin{pmatrix} \bar{c} & o \\ o & \bar{d} \end{pmatrix}$$

(2.13) is equivalent to the strong ellipticity.

The convergence of optimal order (2.6) in the special case $t = o$ and (2.1) can also be obtained from a result by K. Atkinson [3] in connection with the stability of the Galerkin equations.

2.1.3: Fredholm integral equations of the second kind

For these equations with a weakly singular kernel, the operator A has the form

$$(2.15)\qquad A = I + K$$

where I is the identity, K is compact, $\alpha = o$ and $a'(x;\xi) \equiv 1$. Here our result (2.6) can also be obtained from the convergence of Galerkin's method [2,11] and is related to Ben Nobles more general results on colocation methods [9].

2.1.4: Fredholm integral equations of the first kind

Theorem 2.1 is also valid for operators as

$$(2.16)\qquad A(x;D)u \equiv -\int_{M_1} \log |x-y|\, u(y)\, ds_y + Ku$$

where K is of order less than -1. Equations with such A arise in conformal mapping, Gaier [15] and in the single layer method for solving interior and exterior boundary value problems for strongly elliptic differential equations in the plane. The latter method was developed by Fichera [14], Hsiao and MacCamy [19] and Ricci [32]. Here A is strongly elliptic of order $\alpha = -1$ with the principal symbol

$$(2.17)\qquad a'(x,\xi) = \frac{\text{const.}}{|\xi|} \qquad \text{for} \quad |\xi| \geq 1\,.$$

The convergence (2.6) was proved for (2.16) by Hsiao and Wendland in [20] and by Mme. Le Roux [22, 23, 24] for $K = o$ and special finite elements. In the corresponding case $M_2 \subset R^3$ the symbol is still given by (2.17) and our result (2.6) contains the results by Nedelec [28] and Nedelec and Planchard [29].

2.1.5: Finite difference methods for ψdo's

E. Stephan investigated in [35] finite difference approximations to strongly elliptic ψdo's in R^n proving convergence by a different approach. His smoothness conditions at infinity allow an interpretation on the unit sphere $M_n \subset R^{n+1}$; and the use of special $\tilde{H}$ seems to yield his results from Theorem 2.1. But this is yet to be done.

2.2. Strongly elliptic operators in bounded domains

Since $a'(x,\xi)$ belongs to D_α, by following Višik and Eskin in [40 pp. 95-97] a straightforward elementary computation yields the following

Lemma 2.1: If A is a strongly elliptic pseudo differential operator with a' belonging to D_α then the factorization index κ equals $\frac{\alpha}{2}$.

Hence, the order α must be even:

(2.18) $\alpha = 2\kappa, \quad L - M = \kappa \quad (\kappa = o, \pm 1,\ldots).$

In the following we restrict our considerations only to the two simplest cases $M = o$ and $L = o$.

2.2.1 $L = \kappa \geq o$ and $M = o$:

Here (1.1) does not contain any coboundary operators and (1.3) reduces to boundary conditions for u only. The reduced problem with homogeneous boundary conditions $g_j = o$, $j = 1,\ldots,L$ defines the bilinear form

(2.19) $A[v,w] \equiv (Av,w)_o$ for $v,w \in H^{2p}$ with $B_j v = o, j = 1,\ldots, p \equiv \kappa$

For simplicity, let us assume that A becomes V-coercive with respect to

(2.20) $V \equiv$ closure of $\{v \in H^{2p} | B_j v = o, j=1,\ldots,p\}$ in $H^p(G)$ (see [25, p.200 ff.]).

Therefore, A can be decomposed into

(2.21) $A[\ ,\ ] \equiv D[\ ,\ ] + K[\ ,\ ]$

where D is a definite and K is a compact continuous bilinear form on $V \times V$:

(2.22) $\operatorname{Re} D[v,v] \geq \gamma' ||v||_p^2$ for all $v \in V$ with a constant $\gamma' > o$.

For Galerkin's method let us restrict $\tilde{H}$ to the subspace

(2.23) $\tilde{H}_V \equiv \{\chi \in \tilde{H} \mid \hat{B}_j \chi = o,\ j = 1,\ldots,L\}$

where the $\hat{B}_j$ are suitable approximations to the B_j. For the time being let us consider $\hat{B}_j = B_j$ although in practive $B_j \neq \hat{B}_j$ (see [33]). The Galerkin equations are again (2.4), (2.5) where $\tilde{H}$ has to be replaced by $\tilde{H}_V$. Theorem 2.1 is valid guaranteeing the optimal order of convergence. Examples are regular boundary value problems for strongly elliptic partial differential equations [33], singular integral equations of Mikhlin's type with definite symbols and Fredholm integral equations of the second kind in bounded regions [9], [2].

2.2.2 $M = p \equiv -\kappa > o,\ L = o$:

Here equation (1.1) remains the same whereas the boundary conditions (1.3) are cancelled. Let us identify $\tilde{w}$ with the special distribution

$$(2.24)\qquad \tilde{w} \equiv w + \sum_{k=1}^{p} \psi_k(x') \otimes \delta^{(k-1)}(x_n),\ w \in C^\infty(G),\ \psi_k \in C^\infty(\Gamma),\ \gamma G_j^* w = o$$

and let us compute the L_2-scalar product of $\tilde{\tilde{A}}u$ with (2.24) for $Au \in C_o^\infty$. Then we obtain the following bilinear form

$$(2.25)\qquad A[\tilde{u},\tilde{w}] \equiv (Au,w)_o + \sum_{j,k=1}^{p} \langle H_{jk}\,\rho_j,\ \psi_k \rangle_o = (g,w)_o + \sum_{k=1}^{p} \langle D_n^{(k-1)} g,\psi_k \rangle_o$$

where the H_{jk} are defined by

$$(2.26)\qquad H_{jk}\ \rho_j \equiv \gamma' D_n^{(k-1)}\ G_j(\rho_j \otimes \delta(x_n)).$$

Here D_n denotes the derivative to the local coordinate x_n in the direction of the inner normal to Γ in x' [13]. The H_{jk} turn out to be ψdo's of orders $\alpha_j + k$ on Γ. For obtaining a coercive bilinear form (2.25) let us define:

A, H_{jk} form a strongly elliptic system of order $\alpha = -2p$ if

$$(2.27)\qquad \alpha_k = \alpha + k - 1 = -2p + k - 1 \qquad \text{for} \quad k = 1,\dots,p$$

and if there exist p functions $\theta_1(x'),\dots,\theta_p(x')$ such that the matrix defined by the principal symbols $h'_{jk}(x',\xi')$,

$$(2.28)\qquad \mathrm{Re}\ ((\theta_j(x')h'_{jk}(x',\xi')|\xi'|^{1-\alpha-j-k})) \geq \gamma > o$$

becomes positive definite.

Using $\theta_k\rho_k$ in $\tilde{u}$ (1.4) instead of ρ_k leads to $\theta_k \equiv 1$ without loss of generality.

A can be extended by continuity to a continuous bilinear form on

$$(2.29)\qquad V \equiv Z \times \prod_{k=1}^{p} H^{k-p-\frac{1}{2}}(\Gamma) \quad \text{where}$$

$$Z \equiv \text{closure of } \{f \in L^2(M) \text{ with } \operatorname{supp} f \subset G\} \text{ in } H^{-p}(M)^{1)}$$

and let us assume that (2.25) on V is equivalent to (1.1).

The bilinear form A becomes V-coercive since Gårding's inequality hold's accord-

1) It can be shown that the following norms are equivalent on Z :
$'||g||_{-p} \simeq ||g||_{H^{-p}(G)} \simeq |||g||_{H^{-p}(M)}$ where $'||g||_{-p}$ denotes the norm used by Višik and Eskin [40, p. 92].

ing to Kohn and Nirenberg [21]. Using the bilinear form (2.25) in the Galerkin equations (2.4), (2.5) with $\tilde{\mathcal{H}}$ instead of $\tilde{H}$ we obtain again the validity of Theorem 2.1 under the additional conditions $-m_k \le k-p-\frac{1}{2} \le k+t-\frac{1}{2} \le k+s-\frac{1}{2} \le m_k+1$, $k+t-\frac{1}{2} \le m_k$ in the form

$$(2.30) \qquad |||\tilde{u}-\tilde{u}^*|||_t \le c\, h^{s-t}\, |||\tilde{u}|||_s .$$

Remark: The symbol (2.18) of the first kind integral operator (2.17) does not belong to D_α. Nevertheless, the investigations by Višik [38] and by Višik and Eskin [39] suggest that Theorem 2.1 might also be valid for operators with principal symbol $|\xi|^\alpha$ $\alpha \in R$.

§ 3 Least squares methods for general elliptic problems

By following Bramble and Schatz [10], let us consider the following least squares approximation: Find $\tilde{u}^* \in \tilde{\mathcal{H}}$ which minimizes

$$(3.1) \qquad ||\tilde{A}\tilde{\chi}-g||_o^2 + \sum_{j=1}^{L} h^{2\beta_j-2\alpha+1} |\tilde{B}_j\tilde{\chi}-g_j|_o^2 + \sum_{\ell=1}^{N} |\Lambda_\ell\tilde{\chi}-c_\ell|^2 \to \min.$$

over $\tilde{\chi} \in \tilde{\mathcal{H}}$. With the bilinear form

$$(3.2) \qquad [\tilde{v},\tilde{w}] \equiv (\tilde{A}\tilde{v},\tilde{A}\tilde{w})_o + \sum_{j=1}^{L} h^{2\beta_j-2\alpha+1} \langle \tilde{B}_j\tilde{v},\tilde{B}_j\tilde{w}\rangle_o + \sum_{\ell=1}^{N} \Lambda_\ell\tilde{v}\,\overline{\Lambda_\ell\tilde{w}}$$

the Gaussian equations to (3.1) are

$$(3.3) \qquad [\tilde{u}^*,\tilde{\chi}] = [\tilde{u},\tilde{\chi}] \qquad \text{for all } \tilde{\chi} \in \tilde{\mathcal{H}} .$$

The basis of the convergence proofs by Nitsche [31] and Bramble and Schatz [10] is the a-priori estimate

$$(3.4) \qquad ||u||_t + \sum_{k=1}^{M} |\rho_k|_{t-\alpha+\alpha_k+\frac{1}{2}} \le c\,\{||g||_{t-\alpha} + \sum_{j=1}^{L} |g_j|_{t-\beta_j-\frac{1}{2}} + \sum_{\ell=1}^{N} |c_\ell|\}$$

of the solution $\tilde{u}$ to (1.1), (1.3), (1.6) which is valid in case (i) for all $t \in R$ (see Seeley [34, p. 239]; $L = M = o$) and in case (ii) for $\beta_j + \frac{1}{2} < \alpha \le t$, $t \ge o$ (this follows from [13, p. 74]). For $\tilde{u} \in X^s$, $s \ge \max\{\alpha, o\}$, $\beta_j + \frac{1}{2} < \alpha$ and $\alpha - \max\{o,\alpha,\beta_j+1\} \le t$, the a-priori estimate (3.4) also holds (this follows from Theorems 1.1 and 2.3 in Dikanskii's work [13] and with interpolation).
The a-priori estimate implies that [,] is a scalar product for every $h > o$ and (3.3) is uniquely solvable. The least squares solutions converge to $\tilde{u}$ with optimal order. In particular, let be

$$(3.5) \qquad 2\alpha - m - 1 \le t \le s \le m+1,\ t \le m,\ \alpha \le s \quad \text{and}$$

$$(3.6) \qquad \begin{array}{l} \alpha-\max\{o,\alpha,\beta_j+1\} \le t\ , \quad \beta_j + 1 + \alpha_k - \alpha - m_k \le t\ , \\ s-\alpha+a_k \le m_k + \frac{1}{2}\ , \quad -m_k - \frac{1}{2} \le t-\alpha+\alpha_k\ , \quad |\alpha_k+1| \le m_k + \frac{1}{2} \end{array}$$

Theorem 3.1: In case (i) for (3.5) and in case (ii) for (3.5) and (3.6) there exists a constant c such that

$$(3.7) \qquad |||\tilde{u} - \tilde{u}^*|||_t \le c h^{s-t} \, |||\tilde{u}|||_s \, .$$

The proof can be obtained by a slight modification of Baker's proof in [8]. Baker gave a new proof of the results obtained by Bramble and Schatz for the special case of the Dirichlet problem. The proof is based on the a-priori estimate (3.4), on the inequality

$$(3.8) \qquad |\gamma v|_o \le c \, \{\varepsilon^{-1} \, ||v||_o + \varepsilon ||v||_1\} \quad \text{for every} \quad \varepsilon > o \quad \text{and} \quad v \in H^1$$

with a constant c independent of ε, the interpolation Lemma, the trace Theorem [25] and Nitsche's trick.

Applications and corresponding results:

For an even order elliptic differential operator A Theorem 3.1 repeats the well known results by Aubin [4], Babuška [5], Nitsche [31] and Bramble and Schatz [10].

But it should be pointed out that (3.7) also holds for general regular elliptic boundary value problems in the sense of Agmon-Douglis-Nirenberg [1]. As a special example let us formulate the standard Riemann-Hilbert problem for generalized analytic functions in a simply connected plane domain (I.N. Vekua [37], Haack-Wendland [17]):

$$\begin{aligned} \tilde{A}w &\equiv w_{\bar z} - aw - b\bar{w} = g && \text{in} \quad G \, , \\ \tilde{B}w &\equiv \mathrm{Re}\, e^{i\tau(s)} \, w = g_1 && \text{on} \quad \Gamma \\ \Lambda w &\equiv \oint_\Gamma \mathrm{Im}\, e^{i\tau(s)} w\sigma(s) ds = c \, , && \text{with} \ \oint d\tau = o \end{aligned}$$

where $\sigma > o$. Here is $\alpha = 1$, $\beta = o$, $L = N = 1$, $M = o$.

Obviously, Theorem 3.1 holds for any elliptic problem on compact closed manifolds. More general than in §2, Theorem 3.1 holds for singular integral equations and for integral equations of the first kind on compact manifolds as well as on bounded domains.

References:

[1] Agmon, S., Douglis, A. and Nirenberg, L., Estimates near the boundary for solutions of elliptic partial differential equations satisfying general boundary conditions, I, Comm. Pure Appl. Math. 12 (1959) 623-727; II ibid. 17 (1964) 35-92.

[2] Anselone, P. M., Collectively Compact Operator Approximation Theory, London 1971.

[3] Atkinson, K., The numerical evaluation of the Cauchy transform on simple closed curves, SIAM J. Num Anal. 9 (1972) 284-299.

[4] Aubin, J. P., Approximation of Elliptic Boundary-Value Problems, Wiley 1972.

[5] Babuška, I., Numerical solution of boundary value problems by the perturbed variational principle, Univ. Maryland, Techn. Note BN-624, College Park Md. 1969.

[6] Babuška, I., A remark to the finite element method, Com.Math.U.Car. 12 (1971) 367-376.

[7] Babuška, I. and Aziz, A.K., Survey lectures on the mathematical foundations of the finite element method, in: The Math.Found.Finite El.Meth. Appl.Partial Diff. Equat. ed. by A. K. Aziz, Academic Press 1972, p. 3-359.

[8] Baker, G., Simplified proofs of error estimates of the least squares method for Dirichlet's problem, Math. Comp. 27 (1973) p. 229-235 .

[9] Ben Noble, Error analysis of colocation methods for solving Fredholm integral equations, "Topics in Num. Analysis", ed. by J.Miller, Academic Press 1972.

[10] Bramble, J. H. and Schatz, A. H., Least squares methods for 2mth order elliptic boundary-value problems, Math. Computation 25 (1971) p. 1-32.

[11] Bruhn, G. and Wendland, W., Über die näherungsweise Lösung von linearen Funktionalgleichungen; ISNM Vol. 7 (1967) Birkhäuser Basel S. 136-164.
[12] Dikanskii, A. S., Problems adjoint to elliptic pseudodifferential boundary value problems, Soviet Math. Doklady Vol. 12 (1971), p. 1520-1525.
[13] Dikanskii, A. S., Conjugate problems fo elliptic differential and pseudodifferential boundary value problems in a bounded domain, Math.USSR.Sb. 20 (1973) p.67-83.
[14] Fichera, G., Linear elliptic equations of higher order in two independent variables and singular integral equations, Proc. Conference on Part. Diff. Equations and Continuous Mechanics, Univ. Wisconsin Press, Madison, 1961.
[15] Gaier, D., Integralgleichungen erster Art und konforme Abbildung, Math. Z.to appear
[16] Gohberg, I. C. and Fel'dman, I. A., Convolution Equations and Projection Methods for their Solution, AMS, Translations of Math. Monog. Providence 1974.
[17] Haack, W. and Wendland, W., Lecture on Partial and Pfaffian Differential Equations, Pergamon Press 1972.
[18] Hildebrandt, St. and Wienholtz, E., Constructive Proofs of representation theorems in separable Hilbert space. Comm. Pure Appl. Math. 17 (1964) 369-373.
[19] Hsiao, G. C. and MacCamy, R. C., Solution of boundary value problems by integral equations of the first kind; SIAM Review 15 (1973) 687-705.
[20] Hsiao, G. C. and Wendland, W., A finite element method for some integral equations of the first kind, Journal Math. Anal. Appl., to appear.
[21] Kohn, J. J. and Nirenberg, L., An algebra of pseudodifferential operators, Comm. Pure Appl. Math. 18 (1965) 269-305.
[22] LeRoux, M.N., Equations intégrales pur le pour le problème du potential electrique dans le plan, C. R. Acad. Sc. Paris 278 (1974) A541.
[23] Le Roux, M.N., Résolution numérique du problème du potential dans le plan par une mèthode variationelle d'élements finis, These L'Université de Rennes 1974 Ser. A 38 No. 347.
[24] Le Roux, M.N., Method d'elements finis pur la resolution numérique de problèmes exterieurs en dimension deux, Rev.Franc. d'Aut.Inf. Rech. Operat., to appear.
[25] Lions, J. L. and Magénes, E., Non-Homogeneous Boundary-Value Problems, I, Spr. 1972.
[26] Michlin, S. G., Variationsmethoden der Mathematischen Physik, Akad. Berlin 1962.
[27] Mikhlin, S. G., Multidimensional Singular Integral and Integral Equations, Pergamon Press 1965.
[28] Nedelec, J. C., Methods d'elements finis courbes pour la resolution des surfaces de R^3, Rev. Franc. d'Aut. Inf. Rech. Op., to appear.
[29] Nedelec, J. C. and Planchard, J., Une methode variationelle d'elements finis pur la resolution numerique d'un problème exterieurs dans R^3. Rev. Franc. d'Aut. Inf. Rech. Operationelle (1973) R3.
[30] Nitsche, J., Umkehrsätze für Spline-Approximationen, Comp. Math. 21 (1969) 400-416.
[31] Nitsche, J., Über ein Variationsprinzip zur Lösung von Dirichlet-Problemen bei Verwendung von Teilräumen, die keinen Randbedingungen unterworfen sind, Abh. d. Hamb. Math. Sem. 36 (1971) 9-15.
[32] Ricci, P. E., Sui potenziali die semplice strato per le equazioni ellittiche di ordine superiore in due variabili, Rendiconti Mat. 7 (1974) 1-39.
[33] Schultz, M. H., Rayleigh-Ritz-Galerkin methods for multidimensional problems, SIAM J. Numer. Anal. 6 (1969) 523-538.
[34] Seeley, R., Topics in pseudo-differential operators in "Pseudo-Differential Operators", C.I.M.E., Edizioni Cremonese Roma 1969.
[35] Stephan, E., Differenzenapproximationen von Pseudo-Differentialoperatoren, Dissertation, Darmstadt 1975.
[36] Vainikko, G., On the question of convergence of Galerkin's method (Russian) Tartu Rükl. Ül. Toim. 177 (1965), 148-152.
[37] Vekua, I. N., Generalized Analytic Functions, Pergamon Press 1962.
[38] Višik, M. I., Elliptic equations in convolution in a bounded domain with applications, AMS-Transl., Series 2, 70 (1968) 257-266.
[39] Višik, M. I. and Eskin, G. I., Singular elliptic equations and systems of variable order, Soviet Math. Doklady 5 (1964) 615-619.
[40] Višik, M. I. and Eskin, G. I. Equations in convolutions in a bounded region, Russ. Math. Surveys Vol. 20 (1965) 85-151.

Boundary-Value Problems With Discontinuous Non-Linearities

C.A. STUART

§1. Introduction

We consider the boundary value problem

$$-u''(x) + cu(x) = \lambda f(u(x)) \quad \text{for} \quad 0 < x < 1 \qquad (1.1)$$
$$u(0) = u(1) = 0$$

where c is a given constant and $f: [0,\infty) \to \mathbb{R}$ is a given function. We seek a non-negative function u on [0,1] and a constant $\lambda \geq 0$ which together satisfy (1.1). This problem has been intensively studied and a wide variety of appealing results is available. Amann [1] has recently given an excellent survey. In almost all cases, however, only problems in which f is continuous on $[0,\infty)$ are considered. Here we allow f to have a discontinuity and we try to highlight some features of the problem which are peculiar to this situation.

For simplicity, we assume that f is continuous except possibly at 1. Indeed we suppose that

(H1) there are Lipschitz continuous functions $h: [0,1] \to \mathbb{R}$ and $k: [1,\infty) \to \mathbb{R}$ such that $f(p) = \begin{cases} h(p) & \text{for } 0 \leq p < 1 \\ k(p) & \text{for } 1 < p < \infty \end{cases}$

and $f(0) > 0$.

The value of f at 1 is not related to h and k and, as we shall see, it is unimportant except in some special circumstances. If $h(1) \neq k(1)$, then f is discontinuous at 1 and we must say what is meant by a solution of (1.1).

Definition 1 A solution of type I is a pair $(u,\lambda) \in C^1([0,1]) \times [0,\infty)$ such that $u(x) \geq 0$ for all $x \in [0,1]$,

$$u(0) = u(1) = 0,$$

u' is absolutely continuous on $[0,1]$ and

$$-u''(x) + cu(x) = \lambda f(u(x)) \quad \text{for almost all } x \in (0,1).$$

This is perhaps the most obvious definition. If f is continuous, it coincides with the usual concept of a classical solution. There is, however, a different and essentially broader definition of solution in which we consider f as "multi-valued". That is, we associate with a function f satisfying (H1) a set-valued map $\tilde{f}$ defined by

$\tilde{f}(p) = \{f(p)\}$ if $p \neq 1$ and

$\tilde{f}(1)$ is the closed interval with end points $h(1)$ and $k(1)$.

Definition 2 A solution of type II is a pair $(u,\lambda) \in C^1([0,1]) \times [0,\infty)$ such that $u(x) \geq 0$ for all $x \in [0,1]$,

$$u(0) = u(1) = 0,$$

u' is absolutely continuous on $[0,1]$ and

$$-u''(x) + cu(x) \in \lambda\tilde{f}(u(x)) \quad \text{for almost all } x \in (0,1).$$

If f is continuous, this definition again coincides with the usual concept of a classical solution. However, if f is discontinuous at 1, there are often solutions of type II which are not solutions of type I.

Because we are considering an autonomous ordinary differential equation, a complete description of the form of all possible solutions can be given. This exposes the essential difference between solutions of type I and type II; and is set out in section 2. In section 3, we consider some general properties of the set of all solutions. Again there is a striking difference between solutions of type I and type II. The results of this section hold true in much greater generality. The proofs (given in [5]) can be generalised without much difficulty to elliptic partial differential equations. In section 4, we consider how many solutions exist for a given value of λ.

Boundary value problems involving discontinuous non-linearities have been discussed in [2,3,4], where their study is motivated by Joule heating of a solid conductor which changes from one solid phase to another, at a temperature which we have normalised to unity. The results described in sections 2 and 3 are given in detail in [5] and those in section 4 in [6].

There is another kind of discontinuous non-linearity not considered here. That is where f is continuous on $(0,\infty)$ but $f(p) \to \infty$ as $p \to 0$. Results for problems of this type are given in [7,8,9,10].

§2. Properties of solutions

For $u \in C([0,1])$, let $\|u\| = \max\{|u(x)|: 0 \le x \le 1\}$ and let $I(u) = \{x: u(x) = 1\}$. Suppose that f satisfies (H1) and that (u,λ) is a solution of type I or II with $\lambda > 0$. If $\|u\| < 1$, then $I(u) = \emptyset$ and $u \in C^2([0,1])$. If $\|u\| > 1$ and $k(1) \neq h(1)$, then $I(u) \neq \emptyset$ and $u \notin C^2([0,1])$.

Theorem 2.1 Suppose that f satisfies (H1) and that (u,λ) is a solution of type I with $\|u\| \ge 1$ and $\lambda > 0$. Then $I(u) \neq \emptyset$ and, setting $t_0 = \inf I(u)$, we have that $u'(t_0) \ge 0$.

(a) If $\|u\| = 1$ then $u'(t_0) = 0$, $\lambda h(1) > c$ and $I(u) = [t_0, 1-t_0]$ where $t_0 = \frac{1}{2}$ provided that $\lambda f(1) \neq c$.

(b) If $\|u\| > 1$ and $u'(t_0) > 0$ then $t_0 < \frac{1}{2}$ and $I(u) = \{t_0, 1-t_0\}$.

(c) If $\|u\| > 1$, $u'(t_0) = 0$ and $\lambda f(1) \neq c$ then $\lambda h(1) > c$, $\lambda k(1) < c < \lambda k(\|u\|)/\|u\|$ and there exists a positive integer n such that $I(u) = \left\{t_0 + \frac{i(1-2t_0)}{n} : i = 0,1,\ldots,n\right\}$.

(d) If $\|u\| > 1$, $u'(t_0) = 0$ and $\lambda f(1) = c$ then $\lambda h(1) > c$, $\lambda k(1) < c < \lambda k(\|u\|)/\|u\|$ and $I(u) = \bigcup_{i=0}^{m} [a_i, b_i]$ where

$a_i \leq b_i < a_{i+1}$ for $i = 0,1,\ldots,m-1$,

$a_0 = t_0$, $b_m = 1-t_0$,

and $a_{i+1} - b_i = \dfrac{1 - 2t_0 - \sum_{j=0}^{m}(b_j - a_j)}{n}$ for some positive integer n.

<u>Theorem 2.2</u> Suppose that f satisfies (H1) and that (u,λ) is a solution of type II with $\|u\| \geq 1$ and $\lambda > 0$. Then $I(u) \neq \emptyset$ and, setting $t_0 = \inf I(u)$ we have that $u'(t_0) \geq 0$.

(a) If $\|u\| = 1$ then $u'(t_0) = 0$, $\lambda h(1) > c$ and $I(u) = [t_0, 1-t_0]$ where $t_0 = \frac{1}{2}$ provided that $c \notin \lambda\tilde{f}(1)$.

(b) If $\|u\| > 1$ and $u'(t_0) > 0$, then $t_0 < \frac{1}{2}$ and $I(u) = \{t_0, 1-t_0\}$.

(c) If $\|u\| > 1$ and $u'(t_0) = 0$ then $\lambda h(1) > c$, $\lambda k(1) < c < \lambda k(\|u\|)/\|u\|$ and $I(u)$ has the structure described in (d) of Theorem 2.1.

<u>Remarks</u> 1. If f is continuous on $[0,\infty)$ and (u,λ) is a solution of type I or II, then $u(x) = u(1-x)$ for all $x \in [0,1]$, $u'(\frac{1}{2}) = 0$ and only cases (a) and (b) can occur. In case (a), $I(u) = \{\frac{1}{2}\}$. Furthermore, u is monotonely increasing on $[0,\frac{1}{2}]$ and $\|u\| = u(\frac{1}{2})$.

2. If $k(1) \geq h(1)$ or if $k(p)/p$ is a non-increasing function of p, then the more complicated behaviour described in cases (c) and (d) cannot occur. Thus if $k(1) \geq h(1)$ every solution of type II is a solution of type I and has the additional properties described in Remark 1.

3. If $\lambda k(1) < c < \lambda h(1)$ then there can be solutions (u,λ) of type II for which I(u) contains non-trivial intervals and so has non-zero measure. If $\lambda f(1) \neq c$ and (u,λ) is a solution of type I, then I(u) contains only a finite number of points and so the value of f at 1 is not important since I(u) has zero measure.

4. Simple examples exhibiting all these features are given in [5].

§3. The principal component of solutions

Here we consider certain components of the set of all solutions. Let f satisfy (H1) and set,

$S(I) = \{(u,\lambda) \in C^1([0,1]) \times [0,\infty): (u,\lambda) \text{ is a solution of type I}\}$

and

$S(II) = \{(u,\lambda) \in C^1([0,1]) \times [0,\infty): (u,\lambda) \text{ is a solution of type II}\}$.

Since $f(0) > 0$, $(0,\lambda) \in S(I) \cup S(II)$ only if $\lambda = 0$. However $(0,0) \in S(I) \cap S(II)$ and, provided that $c \neq -\pi^2$, this is the only solution corresponding to $\lambda = 0$.

Regarding S(I) and S(II) as metric spaces with the metric of $C^1([0,1]) \times [0,\infty)$, we denote by $\mathcal{C}(I)$ and $\mathcal{C}(II)$ the connected components of S(I) and S(II) containing (0,0). They are called the principal components of solutions of type I and II. Since (1.1) does not involve u', the $C^1([0,1]) \times [0,\infty)$ topology on S(I) and S(II) is equivalent to the $C([0,1]) \times [0,\infty)$ topology.

If f is continuous on $[0,\infty)$, we have S(I) = S(II) and it is well-known that $\mathcal{C}$(I) (= $\mathcal{C}$(II)) is an unbounded subset of $C([0,1]) \times [0,\infty)$ provided that $c > -\pi^2$. This follows from the important technique introduced by Rabinowitz [10]. For discontinuous non-linearities, this result remains true for solutions of type II, but for solutions of type I the situation is more complicated.

<u>Theorem 3.1</u> Suppose that f satisfies (H1). Then $\mathcal{C}$(II) is an unbounded subset of $C([0,1]) \times [0,\infty)$ if and only if $c \geq -\pi^2$.

<u>Proof</u> Since $f(0) > 0$, there exists $\sigma \in (0,1)$ such that $f(p) \geq \frac{1}{2}f(0)$ for all $p \in [0,\sigma]$. Suppose now that $c < -\pi^2$ and that $(u,\lambda) \in S(II)$ with $\|u\| \leq \sigma$. Then $I(u) = \phi$ and $-u''(x) - \pi^2 u(x) + (c+\pi^2)u(x) = \lambda f(u(x))$ for all $x \in (0,1)$. Multiplying both sides by $\sin \pi x$ and integrating, we have

$$(c+\pi^2)\int_0^1 u(x)\sin \pi x \, dx = \lambda \int_0^1 f(u(x))\sin \pi x \, dx \geq \lambda f(0)/\pi.$$

Since $u(x)\sin \pi x \geq 0$ for all $x \in [0,1]$ and $c + \pi^2 < 0$, this implies that $\lambda = 0$ and consequently that $u \equiv 0$. Thus, if $c < -\pi^2$, $\mathcal{C}(II) = \{(0,0)\}$ and so is certainly bounded.

If $c = -\pi^2$, then $\{(\alpha\phi,0): \alpha \geq 0\} \subset \mathcal{C}(II)$ where $\phi(x) = \sin \pi x$. Hence, if $c = -\pi^2$, $\mathcal{C}$(II) is unbounded.

If $c > -\pi^2$ the unboundedness of $\mathcal{C}$(II) can be established by approximating f by a continuous function, using the result of Rabinowitz for this case and then passing to a limit. The details are given in [5].

<u>Corollary 3.2</u> Suppose that f satisfies (H1). Then $\mathscr{C}(I)$ is an unbounded subset of $C([0,1]) \times [0,\infty)$ provided that one of the following conditions is satisfied.

(i) $k(1) \geq h(1)$ and $c \geq -\pi^2$.

(ii) $h(1) > k(1) \geq 0$ and $-\pi^2 \leq c \leq 0$ with $k(1) > c$.

(iii) $h(1) < 0$ and $c \geq 0$.

<u>Proof</u> By Theorem 2.2, any one of the conditions (i) (ii) or (iii) implies that $S(II) \subset S(I)$ and so the result follows from Theorem 3.1.

However, if none of these conditions is satisfied, then $\mathscr{C}(I)$ can indeed be bounded.

<u>Theorem 3.3</u> Suppose that f satisfies (H1) and that

$$h(p) > \max\{0,k(1)\} \quad \text{for all } p \in [0,1].$$

Then there exists a constant $c \geq 0$ such that $\mathscr{C}(I)$ is a bounded subset of $C([0,1]) \times [0,\infty)$. If $k(1) < 0$ and $f(1) \neq 0$ we can choose $c = 0$.

This is proved in [5].

<u>Corollary 3.4</u> Suppose that f satisfies (H1) and that $h(p) \geq h(1) > 0$ for all $p \in [0,1]$. Then $\mathscr{C}(I)$ is an unbounded subset of $C([0,1]) \times [0,\infty)$ for every $c > -\pi^2$ if and only if $k(1) \geq h(1)$.

<u>Proof</u> This follows from Corollary 3.2 and Theorem 3.3.

Figure I shows how the boundedness of $\mathscr{C}(I)$ in Theorem 3.3 arises.

§4. The number of solutions for given λ

In special circumstances the number of solutions of (1.1) for given values of λ can be determined. In addition to (H1) we assume throughout this section that

(H2) $c = 0$, $f(p) > 0$ for all $p \geq 0$, $h \in C^2([0,1])$ and $k \in C^2([1,\infty))$.

Thus, as in Corollary 3.2 i) and ii), $S(I) = S(II)$ and $\mathcal{C}(I) = \mathcal{C}(II)$ is an unbounded subset of $C([0,1]) \times [0,\infty)$. Also, if $(u,\lambda) \in S(I)$, then $u(x) = u(1-x)$ for all $x \in [0,1]$ and u is monotone increasing on $[0,\frac{1}{2}]$. Thus (1.1) can be integrated to yield $\lambda^{\frac{1}{2}} = g(\|u\|)$

$$\text{where } g(\rho) = \sqrt{2}\int_0^\rho \{F(\rho) - F(\omega)\}^{-\frac{1}{2}} d\omega \text{ and } F(\omega) = \int_0^\omega f(s)ds \qquad (4.1)$$

for $\rho, \omega > 0$. Note that $g(\rho) < \infty$ since $f(\rho) > 0$. The relation (4.1) was used by Laetsch [11] to obtain some results on the multiplicity of solutions for fixed λ in the case where f is continuous on $[0,\infty)$. We find that, if f is discontinuous at 1 with $k(1) > h(1)$, then there are always some values of λ for which there is more than one solution of (1.1).

<u>Theorem 4.1</u> Suppose that conditions (H1) and (H2) are satisfied. Then, for each $\rho \in [0,\infty)$, there exists exactly one solution (u,λ) of type I such that $\|u\| = \rho$ and all the solutions lie on the principal component $\mathcal{C}(I)$. If f is asymptotically sublinear in the sense that $p^{-1}f(p) \to 0$ as $p \to \infty$, then for each $\lambda \in [0,\infty)$ there exists at least one solution (u,λ) of type I.

On the other hand, if f is asymptotically superlinear in the sense that $\liminf_{p\to\infty} p^{-1}f(p) \geq \alpha > 0$, then the set of λ for which (1.1) has a solution is a finite interval containing 0.

<u>Proof</u> The first assertions follow from (4.1).

Suppose that f is asymptotically sublinear and that

$$\{\lambda: \text{there exists } (u,\lambda) \in \mathcal{C}(I)\} \subset [0,M] \text{ for some } M < \infty.$$

Then there exists $\beta > 0$ such that $f(p) \leq \beta + {}^{2p}/_{M}$ for all $p \geq 0$. Thus if $(u,\lambda) \in \mathcal{C}(I)$, for $0 \leq x \leq \frac{1}{2}$, we have

$$0 \leq u'(x) \leq \lambda \int_x^{\frac{1}{2}} f(u(s))ds \leq \tfrac{1}{2}\lambda\{\beta + \tfrac{2}{M}\|u\|\}$$

and so

$$u(1) = \|u\| \leq \int_0^{\frac{1}{2}} \tfrac{1}{2}\lambda\{\beta + \tfrac{2}{M}\|u\|\}ds = \tfrac{1}{4}M\beta + \tfrac{1}{2}\|u\|.$$

Hence $\|u\| \leq \frac{1}{2}M\beta$ and we have shown that $\mathcal{C}(I)$ is a bounded subset of $C([0,1]) \times [0,\infty)$. This contradicts Corollary 3.2 and so we conclude that

$$\{\lambda: \text{there exists } (u,\lambda) \in \mathcal{C}(I)\} = [0,\infty).$$

Suppose now that $\liminf_{p\to\infty} p^{-1}f(p) \geq \alpha > 0$.

Then there exists $\gamma > 0$ such that $f(p) \geq \gamma p$ for all $p \geq 0$. Hence, if $(u,\lambda) \in \mathcal{C}(I)$, we have

$$-u''(x) - \pi^2 u(x) \geq (\lambda\gamma - \pi^2)u(x)$$

for $x \in (0,1) \setminus I(u)$. Hence multiplying by $\sin \pi x$ and integrating as in the proof of Theorem 3.1, we find that $\lambda\gamma - \pi^2 \leq 0$. Thus we have that

$\{\lambda: \text{there exists } (u,\lambda) \in \mathcal{C}(I)\} \subset [0, {}^{\pi^2}/\gamma]$.

<u>Theorem 4.2</u> Suppose that the conditions (H1) and (H2) are satisfied and that f is non-increasing on $[0,\infty)$. Then, for each $\lambda \in [0,\infty)$, there exists exactly one solution (u,λ) of type I and u depends continuously on λ in $C^1([0,1])$.

<u>Proof</u> Since f is bounded it follows from Theorem 4.1 that it is sufficient to prove that there is at most one solution for each $\lambda \geq 0$. But our hypothesis imply that g is differentiable and that $g'(\rho) > 0$ for all $\rho > 0$. Thus g is invertible. The details are given in [6].

<u>Theorem 4.3</u> Suppose that the conditions (H1) and (H2) are satisfied and that $k(1) > h(1)$. Then there exist $\lambda_2 > \lambda_1 > 0$ such that for each $\lambda \in (\lambda_1,\lambda_2)$ there are at least two solutions of type I.

<u>Proof</u> With these hypotheses we find that $g'(\rho) \to +\infty$ as $\rho \to 0+$ and $g'(\rho) \to -\infty$ as $\rho \to 1+$. Hence g has at least one maximum. The details are given in [6].

If more is assumed about the behaviour of f, the structure of S(I) can be determined. The following result is one example of this and the details of the proof are given in [6].

<u>Theorem 4.4</u> Suppose that the conditions (H1) and (H2) are satisfied. Assume, in addition, that

i) $f'(p) \leq 0$ and $\{f'(p)/f(p)^3\}' \leq 0$ for all $p \neq 1$.

ii) $k(1) > h(1)$ and $k'(1)/k(1)^3 \leq h'(1)/h(1)^3$.

iii) $1/f(0) > 1/h(1) - 1/k(1)$.

Then there exist $\lambda_2 > \lambda_1 > 0$ such that

for each $\lambda \in [0,\lambda_1) \cup (\lambda_2,\infty)$ there is exactly one solution of type I, for each $\lambda \in (\lambda_1,\lambda_2)$ there are exactly three solutions of type I and for $\lambda = \lambda_1$ or λ_2 there are exactly two solutions of type I.

Proof Again we study g and its turning points.

If h and k are increasing and concave and $k(1) > h(1)$ a similar result can be obtained, [6]. The conclusion of Theorem 4.4 is represented in Figure II.

Figure I

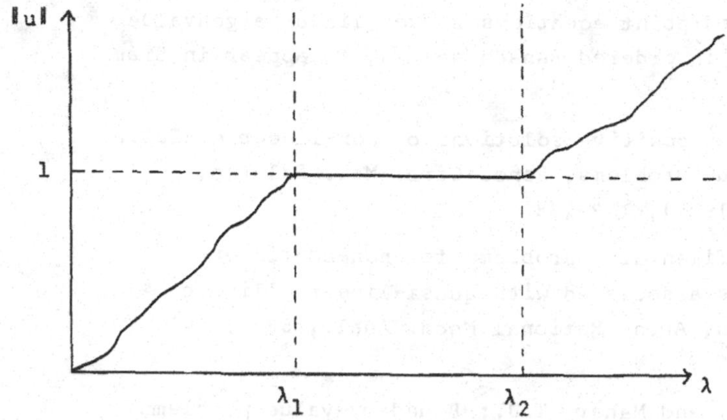

The solid curve represents the unbounded component $\mathcal{C}$(II) under the hypothesis of Theorem 3.3. For $\lambda \in (\lambda_1,\lambda_2)$ at most one of the solutions (u,λ) on $\mathcal{C}$(II) is a solution of type I since we must have $c = \lambda f(1)$ by Theorem 2.1 (a). Hence $\mathcal{C}(I) = \{(u,\lambda) \in \mathcal{C}(II): \lambda \le \lambda_1\}$ is bounded.

Figure II

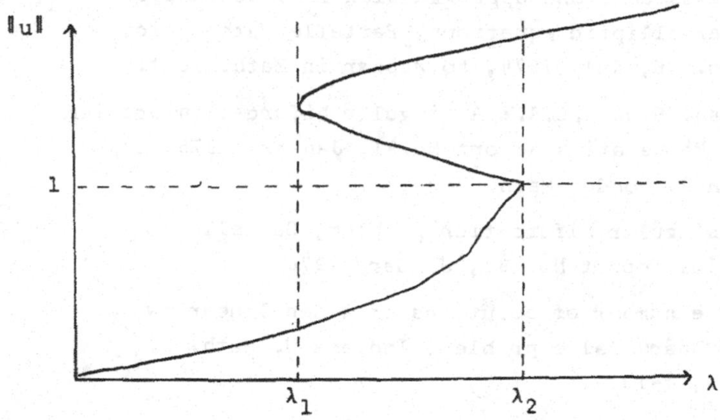

The solid curve represents the set of all solutions under the hypothesis of Theorem 4.4.

References

[1] Amann, H.: Fixed point equations and non-linear eigenvalue problems in ordered Banach spaces, to appear in Siam Review.

[2] Kuiper, H.J.: On positive solutions of non-linear elliptic eigenvalue problems, Rend. Circ. Mat. Palermo, (2) 20 (1971), 113-138.

[3] Kuiper, H.J.: Eigenvalue problems for non-continuous operators associated with quasi-linear elliptic equations, Arch. Rational Mech. Anal., 53 (1974), 178-186.

[4] Fleishman, B.A. and Mahar, T.J.: Boundary-value problems for non-linear differential equations with discontinuous non-linearities, Math. Balkanica, 3 (1973), 98-108.

[5] Stuart, C.A.: Differential equations with discontinuous non-linearities, Battelle Mathematics Report No.94, March 1975, to appear in Arch. Rational Mech. Anal.

[6] Stuart, C.A.: The number of solutions of boundary-value problems with discontinuous non-linearities, to appear.

[7] Stuart, C.A.: Concave solutions of singular non-linear differential equations, Math. Zeit., 136 (1974), 117-135.

[8] Stuart, C.A.: Existence and approximation of solutions of non-linear elliptic equations, Battelle Mathematics Report No. 86, July 1974, to appear in Math. Zeit.

[9] Nussbaum, R.D. and Stuart, C.A.: A singular bifurcation problem, Battelle Mathematics Report No.91, January 1975, to appear in J. London Math. Soc.

[10] Küpper, T.: A singular bifurcation problem, Battelle Mathematics Report No. 99, January 1976.

[11] Laetsch, T.: The number of solutions of a non-linear two point boundary value problem, Indiana U. Math. J., 20 (1970), 1-13.

ON AN ELLIPTIC SINGULAR PERTURBATION PROBLEM

A. van Harten

Here we consider the problem:

(1) $\varepsilon \Delta u - u_y = 0$ ε is a small parameter

(2) $u = \phi$ on S asymptotically: $\varepsilon \downarrow 0$.

This problem is considered on the domain

$$D = \{r < 1\} \subset \mathbb{R}^2;\ S = \bar{D}\backslash D = \{r = 1\}$$

r,θ denote polar coordinates.

fig. 1

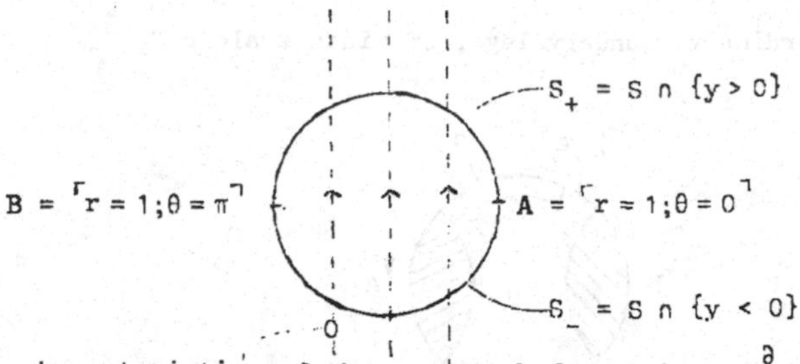

characteristics of the unperturbed operator $-\frac{\partial}{\partial y}$

Note that the problem is elliptic with a boundary condition of Dirichlet type. We suppose: $\phi \in C^\infty(S)$.

Physical background:

Parallel flow of a conducting fluid along a pipe with a circular cross-section in the presence of a uniform magnetic field perpendicular to the pipe.

$\varepsilon = M^{-1}$ with M = the Hartmann number.

ref. Roberts, 1967.

Survey of some literature relevant to this problem:

Levinson, 1950; Visik and Lyusternik, 1957

Eckhaus and de Jager, 1966:

these authors construct an approximation of order ε^N valid in a subdomain $D^* \subset \overline{D}$, $D^* = \overline{D} \setminus (V_A \cup V_B)$, where V_A, V_B are open ε-independent neighbourhoods of A, B, respectively. Their approximation has the structure U + G with:

U : the regular expansion in the interior of D^* and along S_-, corrected by

G : the ordinary boundary layer of width ε along S_+

fig.2.

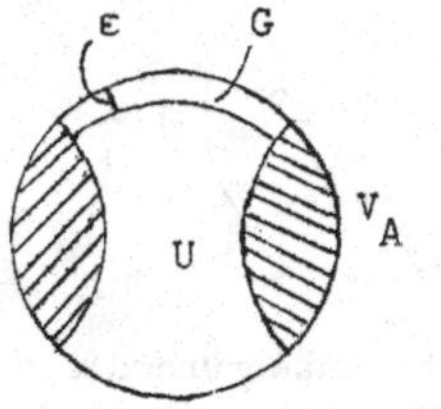

Grasman, 1971 :

this author starts to investigate the behaviour at A, B and for this purpose a double boundary layer structure is used

Y : the intermediate boundary layer of width $\varepsilon^{2/3} \times \varepsilon^{1/3}$

W : the internal boundary layer of width $\varepsilon \times \varepsilon$

fig.3.

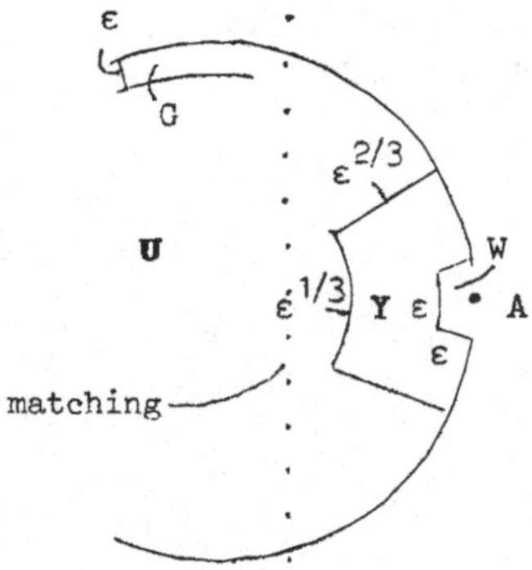

The local expansions at A, B are matched with the expansion $U + G$ in D^* in order to obtain uniformly valid results. Grasman only constructs terms upto the order ε.

van Harten, 1975:
in this work it is shown, that the double boundary layer structure at A, B used by Grasman is too complicated. In fact his internal boundary layers appear to be superfluous. As a consequence of this simpler structure an approximation, which is uniformly of order ε^m in $\overline{D}$ can be constructed (m arbitrarily large). Here below a discussion of this work will be given.

<u>OUTLINE</u>

1° The approximation in D^* : $U + G$ is constructed:

$$U = \sum_{n=0}^{N} \varepsilon^n \cdot U_n(x,y)$$

$$G = \sum_{n=0}^{N} \varepsilon^n \cdot G_n(\zeta,\theta) \text{ with } \zeta = \frac{1-r}{\varepsilon}.$$

2° At A the problem is investigated in the local coordinates:

$$\xi = \frac{1-r}{\varepsilon^{2/3}} \; ; \; \eta = \frac{\theta}{\varepsilon^{1/3}}$$

and we expand:

$$Y = \sum_{n=0}^{3N+3} \varepsilon^{n/3} \, Y_n(\xi,\eta)$$

NOTE: We prefer to call Y the parabolic boundary layer at A, which is a terminology different from the one used by Grasman. The adjective parabolic indicates the type of equations, which are found for the Y_n's.

3° A description of the matching between Y and U + G is given.

4° A function Z is composed, which is likely to be a uniform approximation in $\overline{D}$ and which satisfies:

$Z = \phi$ on S

5° It is verified, that Z is a formal approximation. So:

$|(\varepsilon\Delta - \partial/\partial y)Z|_o = O(\varepsilon^m)$

where $|\;\;|_o$ denotes the supnorm on D and m is a number dependent on N, which will be specified further on

6° It is proven, that:

$|u - Z|_o = O(\varepsilon^m)$.

1. CONSTRUCTION of U + G.

The regular expansion is found from the following recursive system:

$$(1.1)\qquad \frac{\partial U_n}{\partial y} = (1 - \delta_{n,0})\,\Delta U_{n-1}$$

$$(1.2)\qquad U_n(x, -\sqrt{1-x^2}) = \delta_{n,0}\,\phi_-(x)$$

Note, that U satisfies the b.c. on S_-. The solution of 1.1-2 is:

$$(1.3)\qquad U_n(x,y) = \delta_{n,0}\,\phi_-(x) + \int_{-\sqrt{1-x^2}}^{y} \Delta U_{n-1}(x,\tau)d\tau$$

U is singular at the points A and B (in general).

By rewriting the operator $\varepsilon\Delta - {}^{\partial}/\partial y$ in the coordinates θ, ζ, substituting U + G and recollecting equal powers of ε we find for the terms G_n of the ordinary boundary layer a recursive system of the following type:

(1.4) $$\left[\frac{\partial^2}{\partial\zeta^2} + \sin(\theta)\frac{\partial}{\partial\zeta}\right] G_n = (1-\delta_{n,0})\cdot g_n$$

(1.5) $$G_n(0,\theta) = \delta_{n,0}\,\phi(\theta) - U_n(\cos\theta, \sin\theta)$$

(1.6) $$\lim_{\zeta\to\infty} G_n(\zeta,\theta) = 0$$

Note that U + G satisfies the b.c. on S_+. By induction it is possible to show, that:

(1.7) $$G_n(\zeta,\theta) = \exp(-\zeta\sin\theta).\sum_{k=0}^{2n} p_k^{(n)}(\theta).\zeta^k$$

with $p_k^{(n)} \in C^\infty(0,\pi)$, but singular in $0,\pi$ (generally).

So outside V_A, V_B G decreases exponentially for $\zeta \to \infty$.

2. THE PARABOLIC BOUNDARY LAYER.

By rewriting the problem in the coordinates ξ,η, substituting Y and recollecting equal powers of ε we find a recursive system of the following type:

(2.1) $$\left[\frac{\partial^2}{\partial\xi^2} + \eta\frac{\partial}{\partial\xi} - \frac{\partial}{\partial\eta}\right] Y_n = h_n$$

(2.2) $$Y_n(0,\eta) = \frac{\phi^{(n)}(0)}{n!}\,\eta^n$$

The combination of a parabolic equation such as 2.1 with only a boundary condition such as 2.2 gives a rather unusual problem, for which existence, uniqueness and regularity properties of a solution are not at all clear.

However, the situation appears to be very nice.

Let E be the following class of functions:

(2.3) $u \in E \iff$

(i) $u, u_\xi, u_{\xi\xi}$ and u_η are continuous on $[0,\infty) \times (-\infty,\infty)$

(ii) $$\lim_{p\to\infty} \max_{\substack{\xi+\frac{1}{2}\eta^2=p \\ \xi\geq 0}} |u(\xi,\eta)\exp(-2\eta-p)| = 0$$

THEOREM

a. *There exists an ∞-differentiable solution* Y_n *of* 2.1-2 *for which:*

$$D_\xi^k D_\eta^l Y_n \in E \text{ for all } k \geq 0,\ l \geq 0.$$

b. *This solution* Y_n *is unique within* E.

Idea of the proof:

a. A function $J_{n,M,l}$ is constructed, such that the solution $Y_{n,M,l}$ of the problem 2.1-2 on the domain $\{\xi \geq 0;\ \eta \geq -M\}$ with the "initial" condition

$$Y_{n,M,l}\ (\xi,-M) = J_{n,M,l}(\xi)$$

fig.4

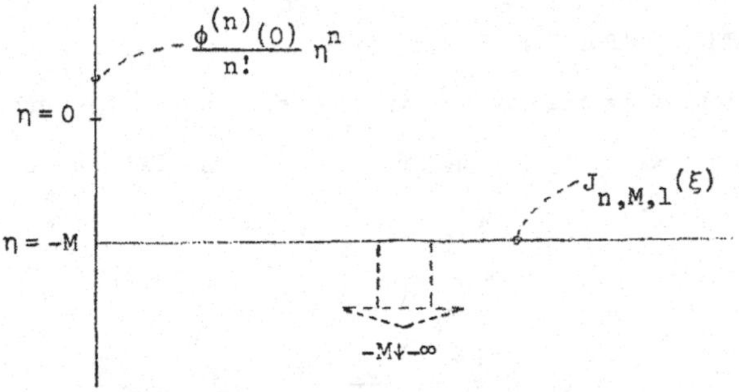

converges to Y_n for $M \to \infty$ in the following sense:

$$|(Y_n - Y_{n,M,l}) \exp(-\frac{\xi+2\eta+\frac{1}{2}\eta^2}{2})|_M^{2l,l} \to 0 \text{ for } M \to \infty$$

in which $|\ \ |_M^{2l,l}$ = the Höldernorm of order $\begin{cases} 2l \text{ in } \xi \\ l \text{ in } \eta \end{cases}$

on the domain $\{\xi \geq 0;\ \eta \geq -M\}$; l can be chosen arbitrarily large.

For the existence, uniqueness, regularity of $Y_{n,M,l}$ we refer to Ladyzenskaja, Solonnikov and Ural'ceva, 1967.

b. Uniqueness within E is shown by applying a Phragmen-Lindelöf transformation and using a maximum priciple for the transformed problem. □

3. ON THE MATCHING BETWEEN Y AND U + G

In order to investigate this matching we first reexpand U + G in the coordinates ξ,η. After that we will give a theorem in which the relation of this reexpansion with the behaviour of Y for $\xi,\eta \to \infty$ (that is $s = (2\xi + \eta^2)^{\frac{1}{2}} \to \infty$) is described.

fig.5.

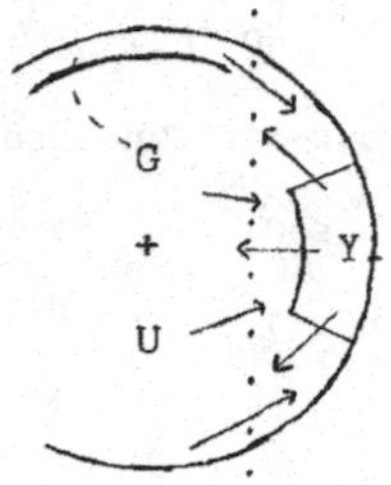

(i) We reexpand:

(3.1) $$\varepsilon^n U_n = \sum_{l=0} \varepsilon^{l/3}\, U_{n,l}(\xi,\eta)$$

(3.2) $$\varepsilon^n G_n = \sum_{l=0} \varepsilon^{l/3}\, G_{n,l}(\xi,\eta)$$

It is not difficult to find recursive systems for the $U_{n,l}$'s and $G_{n,l}$'s and to show that:

(3.3) $$U_{0,0} = \phi^{(0)}(0) \;;\; U_{n,0} = 0 \text{ for } n \geq 1$$

$$U_{n,l} = \sum_{i+j=l-3n} a^{n,l}_{i,j}\, s^i \eta^j$$

only finitely many $a^{n,l}_{i,j} \neq 0$

(3.4) $$G_{n,l} = 0 \text{ for } l \text{ even.}$$

$$G_{n,l} = \exp(-\xi\eta) \sum_{\substack{i-j=l-3n \\ j \geq 0}} b^{n,l}_{i,j}\, \eta^i \xi^j$$

only finitely many $b^{n,l}_{i,j} \neq 0$

The following estimates can be verified:

(3.5) $$\left|\varepsilon^n U_n - \sum_{l=0}^{k} \varepsilon^{l/3}\, U_{n,l}\right| \leq C(n,K).\varepsilon^{\frac{K+1}{3}}.\, s^{K+1-3n}$$

(3.6) $$|\varepsilon^n G_n - \sum_{l=0}^{K} \varepsilon^{l/3} G_{n,e}| \leq D(n,K).\varepsilon^{\frac{K+1}{3}} . s^{K+1-3n} . \exp(-\xi\eta)$$

for $n \geq 1$.

Now we consider the behaviour of Y_n for $s \to \infty$.

Define:

(3.7) $$Y_{n,l} = U_{l,n}$$

(3.8) $$\overline{Y}_{n,l} = G_{l,n}$$

(3.9) $$Y^*_{n,K} = \sum_{l=0}^{K} Y_{n,l} + H \sum_{l=0}^{K} \overline{Y}_{n,l}$$

with H an ∞ - differentiable out-off function

(3.10) $$H(\eta) = \begin{cases} 0 & \text{for } \eta \leq 0 \\ 1 & \text{for } \eta \geq 1 \end{cases}$$

THEOREM

There exist constants $M > 0$ *and* $B > 0$ *such that for* $s \geq M$:

(3.11) $$|(Y_n - Y^*_{n,K})\ (\xi,\eta)| \leq B\, s^{n-3K-3}$$

Idea of the proof:

It is verified that $v = Y_n - Y^*_{n,K}$ satisfies

(3.12) $$[\frac{\partial^2}{\partial\xi^2} + \eta \frac{\partial}{\partial\xi} - \frac{\partial}{\partial\eta}]v = r_{n,K}$$

with: $|r_{n,K}| \leq R(n,K)\ s^{n-3K-4}$ for $s \geq M$.

(3.13) $$v(0,\eta) = 0 \quad \text{for} \quad |\eta| \geq M .$$

Then by a maximum principle and the technique of barrierfunctions 3.11 is proven. □

4. COMPOSITION OF AN APPROXIMATION IN $\overline{D}$.

From now on the parabolic boundary layer at A is denoted by $Y^{(A)}$. Analogous to $Y^{(A)}$ we introduce at B : $Y^{(B)}$.

Further we introduce:

(4.1) $$M^{(.)} = \sum_{n=0}^{3N+3} \sum_{l=0}^{N} \varepsilon^{n/3} Y^{(.)}_{n,l} \; ; \; (.) = (A) \text{ or } (B)$$

$M^{(A)}$ consists of the matching terms of U and $Y^{(A)}$,

$M^{(B)}$ consists of the matching terms of U and $Y^{(B)}$.

(4.2) $$\overline{M}^{(.)} = \sum_{n=0}^{3N+3} \sum_{l=0}^{N} \varepsilon^{n/3} \overline{Y}^{(.)}_{n,l} \; ; \; (.) = (A) \text{ or } (B)$$

$\overline{M}^{(A)}$ consists of the matching terms of G and $Y^{(A)}$

$\overline{M}^{(B)}$ consists of the matching terms of G and $Y^{(B)}$

As our approximation in $\overline{D}$ we now take:

(4.3) $$Z = U + H_A H_B J.G + J_A \{Y^{(A)} - M^{(A)} - H_A \overline{M}^{(A)}\} + J_B \{Y^{(B)} - M^{(B)} - H_B \overline{M}^{(B)}\}$$

H_A, H_B, J, J_A and J_B are suitably chosen ∞-differentiable cut-off functions.

$$H_A = \begin{cases} 1 & \text{for } \theta \geq \varepsilon^{1/3} \\ 0 & \text{for } \theta \leq 0 \end{cases} \; ; \; H_B = \begin{cases} 1 & \text{for } \pi-\theta \geq \varepsilon^{1/3} \\ 0 & \text{for } \theta \geq \pi \end{cases}$$

$$J = \begin{cases} 1 & \text{for } r \geq \frac{3}{4} \\ 0 & \text{for } r \leq \frac{1}{4} \end{cases}$$

$$J_A = \begin{cases} 1 & \text{on } V_A \\ 0 & \text{on } \overline{D}\backslash V'_A \end{cases} \; ; \; J_B = \begin{cases} 1 & \text{on } V_B \\ 0 & \text{on } \overline{D}\backslash V'_B \end{cases}$$

V_A, V'_A open neighbourhoods of A ; $\overline{V}_A \subset V'_A$

V_B, V'_B open neighbourhoods of B ; $\overline{V}_B \subset V'_B$

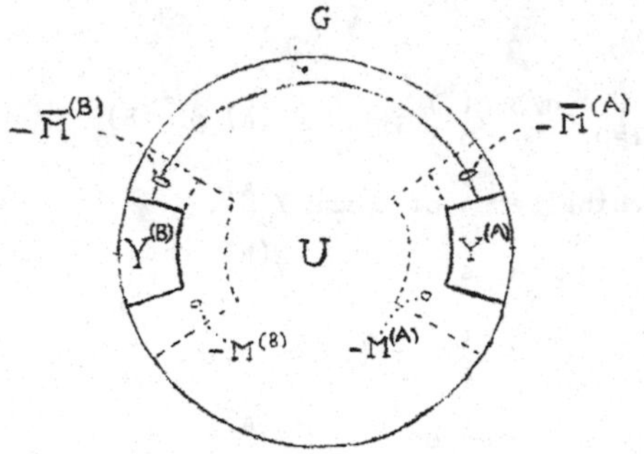

fig.6

It is easily verified, that:

(4.4) $\qquad Z = \phi$ on S

5. Z IS A FORMAL APPROXIMATION.

THEOREM.

(5.1) $$|(\varepsilon\Delta - \frac{\partial}{\partial y})Z|_o = O(\varepsilon^m)$$

with $m = \frac{N}{2} - \frac{1}{3}$.

Idea of the proof:

Direct calculation.

In $\overline{V}_A(\varepsilon^{1/6}) = \{s \le \varepsilon^{-1/6}\}$ we recombine the terms as follows:

$$(\varepsilon\Delta - \frac{\partial}{\partial y})Z = (\varepsilon\Delta - \frac{\partial}{\partial y})\ (U - M^{(A)}) +$$
$$(\varepsilon\Delta - \frac{\partial}{\partial y})\ H_A(G - \overline{M}^{(A)}) +$$
$$(\varepsilon\Delta - \frac{\partial}{\partial y})\ Y^{(A)}$$

We analogously proceed in $\overline{V}_B(\varepsilon^{1/6})$. Outside $\overline{V}_A(\varepsilon^{1/6}) \cup \overline{V}_B(\varepsilon^{1/6})$ we use the form of Z given in 4.3. □

6. Z IS A UNIFORM APPROXIMATION OF $O(\varepsilon^m)$

THEOREM

(6.1) $$|u - Z|_o = O(\varepsilon^m)$$

Idea of the proof:

From 4.2, 5.1 we obtain for $v = u - Z$:

(6.2) $|(\varepsilon\Delta - {}^{\partial}/\partial y)v|_{o} = O(\varepsilon^{m})$

(6.3) $v = 0$ on S

The proof is completed by using the maximum principle and the technique of barrier functions. □

REFERENCES.

ECKHAUS, W. and DE JAGER, E.M., 1966. Asymptotic solutions of singular perturbation problems for linear differential equations of elliptic type, Arch.Rat.Mech. and An., vol.23

GRASMAN, J., 1968. On the birth of boundary layers, Math. Centre tracts 36, Amsterdam, Holland.

VAN HARTEN, A., 1975. Singularly perturbed non-linear 2nd order elliptic boundary value problems, thesis, Univ. of Utrecht, Holland.

LADYZENSKAJA, O.A., SOLONNIKOV, V.A., URAL'CEVA, N.N., 1967. Linear and quasi-linear equations of parabolic type, Transl. of Math. Mon., Am.Math. Soc., vol.23

LEVINSON, N., 1950. The first boundary value problem, Annals of Math. Vol.51.

ROBERTS, P.H., 1967. Singularities of Hartmann layers, Proc. of the Royal Soc., ser.A, vol.300.

VISIK, M.I., LYUSTERNIK, L.A., 1957. Regular degeneration and boundary layer for linear differential equations with small parameter, Uspehi. Math. Nauk., vol.12 (1962, Am.Math.Soc.Transl., series 2, vol.20)

Über die Existenz von Lösungen der Differentialgleichung $u' = f(u)$ in einer abgeschlossenen Menge, wenn f eine k - Mengenkontraktion ist

Peter Volkmann

1. Einleitung. Martin /7/ beweist die Existenz von Lösungen der Differentialgleichung $u' = f(u)$ in einer abgeschlossenen Teilmenge M eines Banachschen Raumes E unter Bedingungen für f, die mit Hilfe des Kuratowskischen Nichtkompaktheitsmaßes formuliert worden sind [1]. Für den Spezialfall einer k-Mengenkontraktion wird in der vorliegenden Note gezeigt, daß ein solches Resultat auch ohne die in /7/ vorausgesetzte gleichmäßige Stetigkeit der Funktion f gilt [2]. Hilfsmittel beim Beweise ist, neben den Ergebnissen von Martin /6/, /7/ zur Konstruktion der Näherungslösungen, insbesondere die Formel für die Fortsetzung einer stetigen Funktion nach Dugundji /4/.

2. Ergebnis. Es sei E ein (reeller) Banachscher Raum mit Norm $|\ |$ und Nullelement θ. Ist B eine beschränkte Teilmenge von E, so bezeichne $\gamma(B)$ ihr Kuratowskisches Nichtkompaktheitsmaß, d.h. $\gamma(B)$ ist das Infimum aller Zahlen $\delta \geq 0$, so daß B durch endlich viele Mengen überdeckt werden kann, deren

---

[1]) Eine Übersicht über damit zusammenhängende Ergebnisse findet man sowohl in /7/ als auch in den Artikeln von Deimling /3/ und Lakshmikantham /5/.

[2]) Ein im Falle $M = E$ weitaus allgemeineres Ergebnis ist ein Existenzsatz von Pianigiani /8/.

Durchmesser höchstens δ beträgt. Genauso wird $\gamma(\Omega)$ definiert, wenn Ω eine beschränkte Teilmenge eines metrischen Raumes Ξ ist.

Satz. Es sei M eine abgeschlossene Teilmenge von E und U eine relativ offene Teilmenge von M. Die Funktion $f\colon U \to E$ sei eine k-Mengenkontraktion ($k \geq 0$), d.h. sie sei stetig, und für beschränktes $B \subseteq U$ gelte

$$\gamma(f(B)) \leq k\gamma(B). \tag{1}$$

Ferner sei die Bedingung

$$\varliminf_{h \to 0+} \frac{1}{h} \operatorname{dist}(M, x+hf(x)) = 0 \qquad (x \in U) \tag{2}$$

erfüllt [1]). Dann existiert zu jedem $u_0 \in U$ ein $\eta > 0$ und eine stetig differenzierbare Funktion $u\colon [0,\eta] \to U$, so daß

$$u(0) = u_0, \quad u'(t) = f(u(t)) \qquad (0 \leq t \leq \eta) \tag{3}$$

gilt.

3. Ein einfacher Reduktionsschritt. Es genügt, den Fall

$$U = M \tag{4}$$

und

$$K = \sup_{x \in M} |f(x)| < +\infty \tag{5}$$

zu behandeln. Zum Beweise dieser Tatsache sei $u_0 \in U$ fixiert. Man wähle $r > 0$, so daß die Menge $S(u_0;2r) \cap M$ [2]) in U enthalten und f auf ihr beschränkt ist. Man bilde für $x \in M$

$$f_1(x) = \begin{cases} f(x) & , \ |x-u_0| \leq r \\ (2 - \frac{1}{r}|x-u_0|)f(x), & r \leq |x-u_0| \leq 2r \\ \theta & , \ 2r \leq |x-u_0| \ . \end{cases}$$

---

1) $\operatorname{dist}(M,y) \equiv \inf_{p \in M} |p-y|$, allgemeiner $\operatorname{dist}(P,Q) \equiv \inf_{p \in P, q \in Q} |p-q|$ für $P,Q \subseteq E$.

2) $S(a;\rho) \equiv \{x \in E\colon |x-a| \leq \rho\}$.

Dann ist f_1: $M \to E$ eine beschränkte k-Mengenkontraktion, welche wegen (2) der Bedingung

$$\varliminf_{h\to 0+} \frac{1}{h} \operatorname{dist}(M,x+hf_1(x)) = 0 \qquad (x \in M)$$

genügt. Da f und f_1 auf $S(u_0;r) \cap M$ übereinstimmen, braucht (3) nur für f_1 an Stelle von f bewiesen zu werden, d.h. man kann (4), (5) voraussetzen.

4. Heranziehung des Dugundjischen Fortsetzungssatzes. Die Funktion f: $M \to E$ läßt sich nach Dugundji /4/ zu einer stetigen Funktion F: $E \to E$ fortsetzen, wobei wegen (5)

$$K = \sup_{x\in E} |F(x)| < +\infty \qquad (6)$$

erreicht werden kann. Dabei ist für $x \in E$ die Darstellung

$$F(x) = \sum_{\nu=1}^{n(x)} \lambda_\nu(x) f(a_\nu(x)) \qquad (7)$$

möglich (vgl. Deimling /2/, S.21, Formel (*)), wobei $a_\nu(x) \in M$, $\lambda_\nu(x) \geq 0$, $\sum_{\nu=1}^{n(x)} \lambda_\nu(x) = 1$ und (wie man leicht nachrechnet)

$$|x-a_\nu(x)| \leq 4 \operatorname{dist}(M,x) \qquad (8)$$

gilt. Es kann nun die Abschätzung

$$\gamma(F(B)) \leq 10k(\operatorname{diam} B + \operatorname{dist}(M,B)) \quad (B \subseteq E,\ \operatorname{diam} B < +\infty)\ ^{1)} \qquad (9)$$

gezeigt werden. Für eine beschränkte Teilmenge B von E sei nämlich

$$A = \{a_\nu(x) : x \in B,\ \nu = 1,2,\ldots,n(x)\}.$$

Dann ist nach (7)

$$F(B) \subseteq \operatorname{Konv} f(A),\ ^{2)}$$

und mit (1) folgt

$$\gamma(F(B)) \leq \gamma(\operatorname{Konv} f(A)) = \gamma(f(A)) \leq k\gamma(A) \leq k \operatorname{diam} A .$$

---

[1]) diam B bezeichnet den Durchmesser von B.

[2]) "Konv" bedeutet konvexe Hülle.

Daher genügt es, die Ungleichung

$$\operatorname{diam} A \leq 10(\operatorname{diam} B + \operatorname{dist}(M,B)) \qquad (10)$$

zu verifizieren. Zunächst ist für $x \in B$

$$\operatorname{dist}(M,x) \leq \operatorname{diam} B + \operatorname{dist}(M,B),$$

also folgt mit (8)

$$|x-a_\nu(x)| \leq 4(\operatorname{diam} B + \operatorname{dist}(M,B)) \qquad (x \in B).$$

Ist x_0 ein fest gewählter Punkt aus B, so folgt weiter

$$|x_0-a_\nu(x)| \leq |x_0-x| + |x-a_\nu(x)| \leq \operatorname{diam} B + |x-a_\nu(x)|$$
$$\leq 5(\operatorname{diam} B + \operatorname{dist}(M,B)),$$

also ist

$$A \subseteq S(x_0;5(\operatorname{diam} B + \operatorname{dist}(M,B))).$$

Das beweist (10).

5. Lemma. Es sei $\Phi = \{u_1,u_2,u_3,\ldots\}$ eine beschränkte Teilmenge des metrischen Raumes Ξ (mit Metrik ρ), ε_n eine Nullfolge reeller Zahlen und $0 \leq \alpha < 1$. Für den Operator $T: \Phi \to \Xi$ gelte

$$\gamma(T(\Omega)) \leq \alpha\gamma(\Omega) \qquad (\Omega \subseteq \Phi) \qquad (11)$$

und

$$\rho(u_n,Tu_n) \leq \varepsilon_n \qquad (n = 1,2,3,\ldots).$$

Dann besitzt die Folge u_n eine Cauchysche Teilfolge [1]).

Der Beweis hierzu sei nur kurz angedeutet: Ohne Beschränkung der Allgemeinheit kann Φ als unendliche Menge vorausgesetzt werden. Durch einen indirekten Schluß erhält man für jede unendliche Teilmenge $\Phi_1 \subseteq \Phi$ die Formel

$$\inf\{\gamma(\Omega): \Omega \subseteq \Phi_1,\ \Omega \text{ unendlich}\} = 0,$$

und hierdurch wird die Konstruktion der Cauchyschen Teilfolge mit Hilfe eines Diagonalverfahrens ermöglicht.

---

[1]) Allgemeiner ist $\gamma(\Phi) = 0$.

6. Näherungslösungen und Grenzübergang. Man wähle $\eta > 0$, so daß

$$\alpha = 10k\eta < 1 \tag{12}$$

ausfällt. Nach dem Vorgehen von Martin /6/, /7/ lassen sich für $n = 1,2,3,\ldots$ stetige Funktionen u_n: $[0,\eta] \to E$ (Näherungslösungen für (3)) finden, welche die folgenden Eigenschaften besitzen:

$$|u_n(t) - u_0 - \int_0^t F(u_n(s))ds| \leq \frac{1}{n} \qquad (0 \leq t \leq \eta), \tag{13}$$

$$\mathrm{dist}(M, u_n(t)) \leq \frac{1}{n} \qquad (0 \leq t \leq \eta), \tag{14}$$

$$|u_n(s) - u_n(t)| \leq (K+1)|s-t| \qquad (0 \leq s,t \leq \eta) \tag{15}$$

(K aus (6)).

Es sei nun $\Xi = C([0,\eta],E)$ der Raum der stetigen Funktionen mit der Maximumnorm $\|\ \|$, und $T: \Xi \to \Xi$ sei definiert durch

$$(Tu)(t) = u_0 + \int_0^t F(u(s))ds \qquad (u \in \Xi,\ \ 0 \leq t \leq \eta).$$

Dann besagt (13)

$$\|u_n - T(u_n)\| \leq \frac{1}{n} \qquad (n = 1,2,3,\ldots). \tag{16}$$

Wie unten gezeigt wird, gilt für

$$\Phi = \{u_1, u_2, u_3, \ldots\} \subseteq \Xi$$

die Formel (11). Da Φ auch beschränkt ist, läßt sich das Lemma aus Nr. 5 anwenden, und man erhält eine (in Ξ) konvergente Teilfolge u_{n_k} von u_n,

$$u_{n_k} \to u \in \Xi .$$

Mit (16) folgt $u = T(u)$, also ist u: $[0,\eta] \to E$ eine Lösung von (3), welche wegen (14) ganz in M verläuft.

7. Nachweis von (11) für die Menge Φ aus Nr. 6. Es sei

$$\Omega = \{u_{n_1}, u_{n_2}, u_{n_3}, \ldots\}$$

eine unendliche Teilmenge von Φ mit $\operatorname{diam}\Omega < \delta$. Zum Beweise von (11) genügt es,

$$\gamma(T(\Omega)) \leq \alpha\delta \tag{17}$$

zu zeigen. Wegen (15) (gleichgradige Stetigkeit der u_{n_k}) gibt es zunächst eine Einteilung

$$0 = t_0 < t_1 < \ldots < t_N = \eta$$

des Intervalls $[0,\eta]$ und Mengen $B_1, B_2, \ldots, B_N \subseteq E$ mit

$$\operatorname{diam} B_\nu < \delta \qquad (\nu = 1,2,\ldots,N), \tag{18}$$

so daß

$$u_{n_k}(t) \in B_\nu \qquad (t_{\nu-1} \leq t \leq t_\nu) \tag{19}$$

gilt. Wegen (14),(19) ist

$$\operatorname{dist}(M, B_\nu) = 0,$$

also folgt mit (9),(12),(18)

$$\gamma(F(B_\nu)) \leq \frac{\alpha}{\eta}\delta \qquad (\nu = 1,2,\ldots,N). \tag{20}$$

Nach Definition des Operators T ist wegen (19) für $0 \leq t \leq \eta$

$$(Tu_{n_k})(t) \in \overline{A(t)} \tag{21}$$

mit

$$A(t) = u_0 + (t_1 - t_0)\operatorname{Konv} F(B_1) + \ldots + (t_{\nu-1} - t_{\nu-2})\operatorname{Konv} F(B_{\nu-1}) + $$
$$+ (t - t_{\nu-1})\operatorname{Konv} F(B_\nu) \qquad (t_{\nu-1} \leq t \leq t_\nu)$$

(man betrachte Riemannsche Summen, wobei $t_0, t_1, \ldots, t_{\nu-1}$ auch als Einteilungspunkte vorkommen). Mit (20) ergibt sich hieraus

$$\gamma(\overline{A(t)}) = \gamma(A(t)) \leq \sum_{\nu=1}^{N} (t_\nu - t_{\nu-1})\gamma(\operatorname{Konv} F(B_\nu))$$
$$= \sum_{\nu=1}^{N} (t_\nu - t_{\nu-1})\gamma(F(B_\nu)) \leq \alpha\delta\,. \tag{22}$$

Da die Funktionen Tu_{n_k} noch gleichgradig stetig sind (F ist beschränkt), folgt nach einem Satze von Ambrosetti /1/ aus (21), (22)

$$\gamma(T(\Omega)) \leq \alpha\delta .$$

Das beweist (17).

Literatur

/1/ Ambrosetti, A.: Un teorema di esistenza per le equazioni differenziali negli spazi di Banach. Rend. Sem. mat. Univ. Padova 39, 349-361 (1967).

/2/ Deimling, K.: Nichtlineare Gleichungen und Abbildungsgrade. Berlin - Heidelberg - New York: Springer 1974.

/3/ ——— : On existence and uniqueness for Cauchy's problem in infinite dimensional Banach spaces. Proc. Colloquium Differential Equations, Keszthely, 2 - 6 September 1974 (erscheint demnächst).

/4/ Dugundji, J.: An extension of Tietze's theorem. Pacific J. Math. 1, 353-367 (1951).

/5/ Lakshmikantham, V.: Existence and comparison results for differential equations. Internat. Conf. Differential Equations, Univ. Southern California, 3 - 7 September 1974, edited by H.A. Antosiewicz, New York - San Francisco - London: Academic Press 1975, S. 459 - S. 473.

/6/ Martin, R.H., jr.: Differential equations on closed subsets of a Banach space. Trans. Amer. math. Soc. 179, 399-414 (1973).

/7/ —— : Approximation and existence of solutions to ordinary differential equations in Banach spaces. Funkcialaj Ekvac., Ser. internac. 16, 195-211 (1973).

/8/ Pianigiani, G.: Existence of solutions for ordinary differential equations in Banach spaces. Bull. Acad. Polon. Sci., Sér. Sci. math. astron. phys. 23, 853-857 (1975).

Bemerkungen zur Verwendung der Pfaffschen Formen bei der Definition der absoluten Temperatur nach Carathéodory

Johann Walter

Meinem Vater Hermann Walter zu seinem Geburtstag am 31.3.1976 gewidmet

1. Einleitung: Bei nicht zu niedrigen Temperaturen stellen Gasthermometer ein zufriedenstellendes Meßinstrument dar, weil die auf diese Weise definierte Temperaturskala im wesentlichen von der Natur der als Thermometersubstanz genommenen Edelgase unabhängig ist. Bei sehr niedrigen Temperaturen machen sich jedoch auch bei Edelgasen Abweichungen vom "idealen" Verhalten bemerkbar. Es stellt sich daher die Frage nach einer von der Wahl der Thermometersubstanz völlig freien Definition der Temperaturskala. Der Definition einer derartigen "absoluten" Skala- und zwar nach dem Vorbild von Carathéodory - ist die vorliegende Arbeit gewidmet. Die Verbesserungen gegenüber bereits vorliegenden vergleichbaren Darstellungen (vgl. etwa [1; p.161] , [2], [3; p.131], [6], [9], [10; p.26], [12; p.95], [13; p.64]) sind teils didaktischer Art (Vermeidung von ad-hoc-Methoden, Weglassung überflüssiger Beweiselemente, vgl. Bemerkung 1) teils inhaltlicher Art (Behebung einer von Cooper [4] bemerkten Schwierigkeit, vgl. § 5).

2. Thermodynamische Systeme: Thermodynamische Systeme sind Systeme, die eines thermischen Gleichgewichts fähig sind und deren sämtliche Teile miteinander in einem wärmeleitenden Kontakt stehen. Als ein Maß für den aus dem Alltagsleben entnommenen qualitativen Temperaturbegriff wird der Begriff der empirischen Temperatur eingeführt. Unter einer empirischen Temperatur versteht man eine an einem Thermometersystem beobachtbare Größe (z.B. bei einem idealen Gas das Produkt aus Druck und Volumen), die für alle Zustände des Thermometers, die miteinander thermisch

koexistieren können, denselben Wert hat. Hat diese Größe den Wert ϑ, so schreibt man dem Thermometer die Temperatur ϑ zu. Man schreibt einem beliebigen System Σ die Temperatur ϑ zu, wenn es mit dem Thermometer von der Temperatur ϑ thermisch koexistieren kann. Ein thermodynamisches System Σ wird außer durch seine empirische Temperatur durch eine geeignete Zahl von Variablen $x := (x_1, \cdots, x_n)$ charakterisiert. Alle weiteren Bestimmungsstücke von Σ können dann als Funktionen der $n+1$ Variablen (x, ϑ) dargestellt werden. Derartige Funktionen nennt man Zustandsfunktionen. Im folgenden denken wir uns die unabhängigen Variablen (x, ϑ) jedes thermodynamischen Systems ein für alle Mal festgelegt, wobei die einzige Einschränkung in der Wahl der Variablen darin besteht, daß wir jeweils die oben eingeführte empirische Temperatur ϑ als eine dieser Variablen nehmen wollen.

An thermodynamischen Systemen können Zustandsänderungen vorgenommen werden. Dabei versteht man unter einer Zustandsänderung einen in einem solchen System ablaufenden Prozeß, der mit einem Gleichgewichtszustand beginnt und endet. Sind zusätzlich auch alle dazwischen durchlaufenen Zustände Gleichgewichtszustände, so spricht man von einer quasistatischen Zustandsänderung. Derartige Zustandsänderungen können durch Kurven im (x, ϑ)-Raum, dem sogenannten Zustandsraum, dargestellt werden. Zustandsänderungen unter Ausschluß des Wärmekontaktes mit der Umgebung heißen adiabatisch. Bei adiabatischen Zustandsänderungen kann die Zufuhr oder Abfuhr von Energie nur auf rein mechanischem Wege geschehen. Im folgenden bezeichnen wir mit $A_z(x_2, \vartheta_2; x_1, \vartheta_1)$ die einem System bei einer Zustandsänderung z mit dem Anfangszustand (x_1, ϑ_1) und dem Endzustand (x_2, ϑ_2) zugeführte mechanische Arbeit.

<u>Verhalten von A bei adiabatischen Zustandsänderungen</u>:

<u>1. Hauptsatz</u>: Bei adiabatischen Zustandsänderungen hängt $A_z(x_2, \vartheta_2; x_1, \vartheta_1)$ nur vom Anfangs- und Endzustand ab.

Aus dem 1. Hauptsatz folgt die Existenz einer bis auf eine additive Konstante eindeutig bestimmten Zustandsfunktion $\mathcal{U}(x,\vartheta)$, der sog. inneren Energie, mit der Eigenschaft, daß für alle (x_1,ϑ_1), (x_2,ϑ_2) und alle adiabatischen Zustandsänderungen $\mathcal{Z}$ zwischen diesen beiden Zuständen

$$A_{\mathcal{Z}}(x_2,\vartheta_2;x_1,\vartheta_1) = \mathcal{U}(x_2,\vartheta_2) - \mathcal{U}(x_1,\vartheta_1).$$

Verhalten von A bei quasistatischen Zustandsänderungen:

Bei quasistatischen Zustandsänderungen, die ja nicht notwendig auch adiabatisch sein müssen, hängt zwar die zugeführte Arbeit bei gleichbleibendem Anfangs- und Endzustand noch von der Art des Prozesses ab, d.h. es gilt im allgemeinen

(1) $$A_{\mathcal{Z}}(x_2,\vartheta_2;x_1,\vartheta_1) \neq \mathcal{U}(x_2,\vartheta_2) - \mathcal{U}(x_1,\vartheta_1).$$

Da jedoch einer quasistatischen Zustandsänderung $\mathcal{Z}$ eine orientierte Kurve $\mathcal{L}$ im Zustandsraum entspricht kann man die zugeführte Arbeit als ein Kurvenintegral darstellen

$$A_{\mathcal{Z}}(x_2,\vartheta_2;x_1,\vartheta_1) = \int_{\mathcal{L}} K(x,\vartheta)\,dx$$

mit einem geeigneten n-tupel $K := (K_1,\dots,K_n)$ von Zustandsfunktionen. Entsprechend hat man bei einer infinitesimalen Zustandsänderung

$$A(x+dx,\vartheta+d\vartheta;x,\vartheta) = K(x,\vartheta)\,dx.$$

Dh. die bei einer infinitesimalen quasistatischen Zustandsänderung zugeführte Arbeit wird durch eine sogenannte Pfaffsche Form dargestellt. Diese Pfaffsche Form ist eine Funktion der $2n+1$ unabhängigen Variablen $\vartheta, x_1,\dots,x_n, dx_1,\dots,dx_n$ und wird im folgenden mit A bezeichnet.

Den in (1) zum Ausdruck kommenden Unterschied zwischen der zugeführten Arbeit und der Differenz der inneren Energie schreibt man der Tatsache zu, daß bei Aufhebung der Wärmeisolation dem System neben der mechanischen Arbeit eine hiervon verschiedene Form der Energie zugeführt wird. Man setzt

$$Q_{\mathcal{Z}}(x_2,\vartheta_2;x_1,\vartheta_1) = \mathcal{U}(x_2,\vartheta_2) - \mathcal{U}(x_1,\vartheta_1) - A_{\mathcal{Z}}(x_2,\vartheta_2;x_1,\vartheta_1)$$

und bezeichnet Q als die dem System beim quasistatischen Prozeß zugeführte Wärmemenge. Die bei einer infinitesimalen quasistatischen Zustandsänderung zugeführte Wärmemenge wird wieder durch eine Pfaffsche Form beschrieben

$$Q(x+dx,\vartheta+d\vartheta;x,\vartheta)=dU-Kdx.$$

Diese Pfaffsche Form wird mit Q bezeichnet.

3. Q besitzt einen integrierenden Nenner: 2. Hauptsatz (Carathéodory [3]): In beliebiger Nähe jedes Gleichgewichtszustandes (x_0,ϑ_0) eines thermodynamischen Systems liegen Gleichgewichtszustände dieses Systems, die von (x_0,ϑ_0) aus nicht durch eine adiabatische Zustandsänderung erreichbar sind.

Im folgenden benötigen wir nicht die volle Aussage des 2. Hauptsatzes (vgl. [5]) sondern nur das folgende

Korollar 1: In beliebiger Nähe jedes Zustandes (x_0,ϑ_0) liegen Zustände (x,ϑ), die von (x_0,ϑ_0) aus nicht durch quasistatisch-adiabatische Zustandsänderungen erreichbar sind.

Quasistatisch-adiabatische Zustandsänderungen sind nun einerseits reversibel und lassen sich andererseits interpretieren als Lösungskurven der Pfaffschen Differentialgleichung

(2) $$dU-Kdx=0.$$

Das Korollar 1 kann also auch folgendermaßen formuliert werden:

Korollar 2: In beliebiger Nähe jedes Zustandes (x_0,ϑ_0) liegen Zustände (x,ϑ) die mit (x_0,ϑ_0) nicht durch Lösungskurven der Pfaffschen Differentialgleichung (2) verbunden werden können.

Nach einem Satz von Carathéodory [3] (vgl. auch [1], [2], [7], [13]) folgt aus dem Korollar 2, daß für jedes thermodynamische System die entsprechende Pfaffsche Form Q integrabel ist, sich also in der Form $Q=\mu\, d\sigma$ darstellen läßt mit geeigneten Funktionen $\mu(x,\vartheta)$, $\sigma(x,\vartheta)$ μ heißt integrierender Nenner von Q.

4. Q besitzt sogar einen integrierenden Nenner der Form $t(\vartheta)$:

Wir betrachten spezielle thermodynamische Systeme $\Sigma^* := (\Sigma', \Sigma)$ mit den unabhängigen Variablen $(x', x, \vartheta) = (x'_1, \dots, x'_m, x_1, \dots, x_n, \vartheta)$, die sich aus einem als Eichsystem ausgezeichneten System Σ' mit den Variablen (x', ϑ) und einem beliebigen System Σ mit den Variablen (x, ϑ) durch thermische Kopplung bilden lassen. Offenbar hat der Zustandsraum von Σ^* nicht die Dimension $m+n+2$ (wie bei bloßer Juxtaposition von Σ' und Σ ohne Wärmekontakt! Allerdings läge dann auch kein thermodynamisches System im Sinn unserer Definition von Beginn des § 2 vor!) sondern nur noch die Dimension $m+n+1$. Es gilt $Q^* = Q' + Q$. Ferner existieren nach § 3 Funktionen $\mu^*(x', x, \vartheta)$, $\mu'(x', \vartheta)$, $\mu(x, \vartheta)$, $\sigma^*(x', x, \vartheta)$, $\sigma'(x', \vartheta)$, $\sigma(x, \vartheta)$ so, daß $\mu^* d\sigma^* = \mu' d\sigma' + \mu\, d\sigma$ bzw. nach Division durch μ^*

(3) $$d\sigma^* = \frac{\mu'}{\mu^*} d\sigma' + \frac{\mu}{\mu^*} d\sigma .$$

Nach der Kettenregel existiert nun eine Funktion $F(\xi', \xi)$ derart, daß

(4) $$\sigma^* = F(\sigma', \sigma), \quad \frac{\mu'}{\mu^*} = F_{\xi'}(\sigma', \sigma), \quad \frac{\mu}{\mu^*} = F_{\xi}(\sigma', \sigma).$$

Setzt man zur Abkürzung $\frac{F_\xi}{F_{\xi'}} = \phi$ so folgt aus (4)

(5) $$\mu(x, \vartheta) = \mu'(x', \vartheta)\, \phi(\sigma'(x', \vartheta), \sigma(x, \vartheta)).$$

Im folgenden nehmen wir der Einfachheit an $m = n = 1$. Der allgemeine Fall wird ganz analog behandelt. Sei nun

(6) $$x' = \varphi'(\xi', \tau), \quad x = \varphi(\xi, \tau), \quad \vartheta = \tau$$

die Umkehrtransformation der Transformation

(7) $$\xi' = \sigma'(x', \vartheta), \quad \xi = \sigma(x, \vartheta), \quad \tau = \vartheta.$$

Einsetzen von (6) in (5) ergibt

$$\mu(\varphi(\xi, \tau), \tau) = \mu'(\varphi'(\xi', \tau), \tau) \cdot \phi(\xi', \xi).$$

Ersetzung von ξ' durch eine Konstante a und Rücktransformation vermöge der letzten zwei Gleichungen von (7) ergibt

(8) $$\mu(x,\vartheta) = t(\vartheta)\cdot\psi(\sigma(x,\vartheta)),$$

wobei zur Abkürzung $t(\vartheta) := \mu'(\varphi'(a,\vartheta),\vartheta)$ und $\psi(\xi) := \phi(a,\xi)$ gesetzt wurde. Ist Ψ eine Stammfunktion von ψ , so kann man schließlich schreiben $Q = t(\vartheta)\cdot d\Psi(\sigma(x,\vartheta))$. Wir haben also gezeigt, daß die Funktion $t(\vartheta)$ für jedes System Σ ein integrierender Nenner der entsprechenden Pfaffschen Form Q ist.

<u>Bemerkung 1</u>: Die wichtigsten Merkmale der oben gegebenen Herleitung von $t(\vartheta)$ sind die folgenden: 1) Rückgriff auf die Kettenregel beim Übergang von (3) nach (4). Hierdurch lassen sich diverse ad-hoc-Methoden, wie sie in der Literatur an dieser Stelle benützt werden, auf ein wohlbekanntes Element der Analysis zurückführen. 2) Zur Gewinnung der Funktion $t(\vartheta)$ wird etwa bei Landé [9; p.292] (vergl. auch [1; p.172], [2], [6], [10; p.29], [13; p.79]) nicht von (5) sondern von den Quotienten $\frac{\mu'}{\mu^*}$ und $\frac{\mu}{\mu^*}$ ausgegangen. Diese Art der Darstellung ist geeignet, den Unterschied zwischen dem Eichsystem Σ' und dem beliebigen System Σ einzuebnen. Ferner ist man genötigt, noch einmal ausgiebig von partiellen Ableitungen Gebrauch zu machen. Die von uns gewählte Art der Darstellung findet man in [3; p.171], [12; p.96].

3) Landé begnügt sich nicht mit der Herleitung von (8), sondern beweist eine analoge Produktdarstellung auch für μ^* (ähnlich auch [1; p.173], [6], [10; p.30], [13; p. 80]). Dieser Nachweis ist jedoch überflüssig, da die Produktdarstellung (8) für beliebige Systeme, also insbesondere auch für alle Systeme der Form (Σ', Σ) gilt!

<u>5. Definition der absoluten Temperatur</u>: Die für die Identifikation von $t(\vartheta)$ mit der absoluten Temperatur noch fehlende Eigenschaft der strengen Monotonie würde nun sofort folgen, wenn man die Existenz eines idealen Gases annähme.[1] Diese Eigenschaft wird jedoch stillschweigend auch von denjenigen Autoren vorausgesetzt, die den Rück-

---

1) Im Fall eines Gasthermometers läßt sich leicht $t(\vartheta) \equiv \vartheta$ zeigen.

griff auf das ideale Gas vermeiden (vergl. [4], [8; p.332]). Sie läßt sich jedoch leicht zeigen. Sei nämlich $t(\vartheta)$ nicht streng monoton. Dann existiert ein ϑ^* mit $t'(\vartheta^*) = 0$. Sei weiter

$$(9) \qquad Q = dU + p\,dV = (U_V + p)\,dV + U_p\,dp$$

ein zweidimensionales System und $\vartheta = \varphi(V, p)$. Nach § 4 ist dann $t(\varphi(V, p))$ ein integrierender Nenner von Q, dh. es gilt [2]

$$(10) \qquad \left(\frac{U_V + p}{t(\varphi)}\right)_p = \left(\frac{U_p}{t(\varphi)}\right)_V .$$

Für $\varphi(V, p) = \vartheta^*$ ergibt dies nach Ausdifferenzieren einen Widerspruch.

Wir definieren schließlich die absolute Temperatur durch

$$T(\vartheta) := \frac{100}{t(\vartheta_1) - t(\vartheta_0)} \cdot t(\vartheta),$$

wobei ϑ_0 bzw. ϑ_1 die empirische Temperatur des Gefrier- bzw. Siedepunktes von Wasser bedeutet. [3]

Bemerkung 2: Zur effektiven Bestimmung der absoluten Temperatur benützt man die zu (10) äquivalente Differentialgleichung

$$(11) \qquad \frac{dt}{t} = \frac{d\varphi}{(U_V + p)\varphi_p - U_p \varphi_V},$$

wobei durch § 4 garantiert ist, daß die rechte Seite von (11) nur von φ abhängt. Dies kann aber auch leicht direkt hergeleitet werden und würde somit einen neuen Beweis für die Behauptung von § 4 liefern. Sei nämlich durch $\varphi_V\,dV_1 + \varphi_p\,dp_1 = 0$ und $(U_V + p)\,dV_2 + U_p\,dp_2 = 0$ ein zwischen den Temperaturen φ und $\varphi - d\varphi$ arbeitender [4] infinitesimaler Carnotscher Kreisprozeß gegeben. Die gewonnene Arbeit ist

---

2) Ist $\varphi(V, p)$ bereits die absolute Temperatur, so ist (10) eine bekannte Relation, vergl. [11; p.119, (36,17)].

3) $\frac{Q}{T}$ ist das Differential der bis auf eine additive Konstante eindeutig bestimmten Entropie.

4) $d\varphi = \varphi_V\,dV_2 + \varphi_p\,dp_2$.

$A = dV_1 dp_2 - dV_2 dp_1$ die umgesetzte Wärmemenge $Q = (u_v + p) dV_1 + u_p dp_1$. Der Wirkungsgrad $\frac{A}{Q}$ dieses Kreisprozesses stimmt, wie man leicht nachrechnet, mit der rechten Seite von (11) überein. Er ist aber bekanntlich eine Funktion der Temperatur allein.

Literatur

1. Born,M.: Ausgewählte Abhandlungen, Band 1. Göttingen: Vandenhoeck & Ruprecht 1963.
2. Buchdahl,H.A.: A formal treatment of the consequences of the second law of thermodynamics in Carathéodory's formulation. Zeitschr. f. Physik 152, 425-439 (1958).
3. Carathéodory,C.: Gesammelte mathematische Schriften, Band 2. München: C.H.Beck'sche Verlagsbuchhandlung 1955.
4. Cooper,J.L.B.: The foundations of thermodynamics. J.Math.Anal.Appl. 17, 172-1993 (1967).
5. Ehrenfest-Afanassjewa,T.: Zur Axiomatisierung des zweiten Hauptsatzes der Thermodynamik. Zeitschr.f.Physik 33, 933-946 (1925).
6. Eisenschitz,R.: The principle of Carathéodory. Science Progress 43, 246-260 (1955).
7. Falk,G., Jung,H.: Axiomatik der Thermodynamik. Handbuch der Physik, Band III/2, p. 118-175. Berlin-Göttingen-Heidelberg: Springer-Verlag 1959.
8. Jauch,J.M.: On a new foundation of equilibrium thermodynamics. Foundations of Physics 2, 327-332 (1972).
9. Landé,A.: Axiomatische Begründung der Thermodynamik durch Carathéodory. Handbuch der Physik, Band IX, p. 281-300. Berlin: Verlag von Julius Springer 1926.
10. Margenau,H., Murphy,G.M.: The mathematics of physics and chemistry. Princeton,N.J.: D. van Nostrand Company, Inc. 1956.

11. Päsler, M.: Phaenomenologische Thermodynamik. Berlin-New York: Walter de Gruyter 1975.

12. Planck, M.: Thermodynamik. Berlin: Walter de Gruyter & Co. 1954.

13. Wilson, A.H.: Thermodynamics and statistical mechanics. Cambridge: University Press 1966.

ON THE EIGENVALUES OF LINEAR AUTONOMOUS DIFFERENTIAL DELAY EQUATIONS

H. O. WALTHER

1. The stability of the zero solution of the equation

$$\dot{y}(t) = L(y_t), \tag{1}$$

with $L: C[-1,0] \to R$ linear and continuous with respect to the supremum-norm and with $y_t(a) = y(t+a)$, is determined by the distribution of the eigenvalues, i.e. of the complex solutions $\lambda = u + iv$ of

$$\lambda - L(\exp(\lambda\cdot)) = 0. \tag{2}$$

We consider the case $L \neq 0$, $L(\varphi) \leqslant 0$ for $\varphi \geqslant 0$, which includes the equations

$$\dot{y}(t) = -\alpha y(t) \tag{3}$$

and

$$\dot{y}(t) = -\alpha y(t-1) \tag{4}$$

with $\alpha > 0$. We may write $L(\varphi) = -\alpha\int_{-1}^{0} \varphi(a)ds(a)$ with $\alpha > 0$ and $s \in S := \{\sigma: [-1,0] \to R \,|\, \sigma(-1) = 0,\ \sigma \text{ increasing},\ \sigma(1) = 1\}$.

Equation (2) becomes

$$f(\lambda,\alpha,s) := \lambda + \alpha\int_{-1}^{0} \exp(\lambda a)ds(a) = 0. \tag{5}$$

The parameter α may serve as a measure of the power of the negative feedback in the system given by $\dot{y}(t) = -\alpha\int_{-1}^{0} y(t+a)ds(a)$ while the function s describes the hereditary dependence. For example, one might expect that for s concave the stability is in some way less than for s convex because the system takes longer to produce a sufficient reaction to perturbations of the equilibrium. In the extremal cases this conjecture is right in the following way.

Equation (3) corresponds to the minimal (convex) function in S, and we have asymptotic stability for all $\alpha > 0$. Equation (4) comes from the maximal (concave) function in S, and for every $\alpha > \pi/2$ there is at least one eigenvalue with $u > 0$, see the paper of Wright [6]. In addition, we have

Theorem 1: For every $s \in S$ and for every $\alpha < \pi/2$, every eigenvalue has negative real part.

Proof: [4].

We shall see how this behaviour of the minimal and maximal function in S carries over to two classes of smooth functions in S.

First, let us state some preliminary facts.

$$\lambda + \alpha\int_{-1}^{0} \exp(\lambda a)ds(a) = 0 \iff \left(u + \alpha\int_{-1}^{0} \exp(ua)\cos(va)ds(a) = 0 \ \wedge\ v + \alpha\int_{-1}^{0} \exp(ua)\sin(va)ds(a) = 0\right), \tag{6}$$

$$f(\lambda,\alpha,s) = 0 \iff f(\bar{\lambda},\alpha,s) = 0, \tag{7}$$

$$f(\lambda,\alpha,s) = 0 \wedge u \geqslant 0 \implies |\lambda| \leqslant \alpha. \tag{8}$$

Theorem 2 (Stability for all $\alpha > 0$): Let $s \in S \cap C^2[-1,0] \cap C^3(-1,0]$ and $s'(-1) = s''(-1) = 0$, $s''' \geqslant 0$, $s''' \neq 0$. Then for every

$\alpha > 0$, every eigenvalue has negative real part.

Sketch of proof: Integration by parts yields $\int_{-1}^{0}\cos(va)ds(a) > 0$ for all $v > 0$. Hence there are no eigenvalues on iR, by (6) and (7). Now the existence of an eigenvalue in $C^+ := R^+ + iR$ for certain α_o would imply the existence of an eigenvalue in C^+ for $\alpha = 1 < \pi/2$, by (8) and by the continuous dependence of the eigenvalues on α. But this contradicts Theorem 1.

Remark: Theorem 2 holds for $s:a \to (a+1)^\beta$, $\beta > 2$. The case $\beta = 2$ shows that Theorem 2 is optimal in a certain sense: $\tilde{s}:a \to (a+1)^2$ fulfills the hypotheses except of $\tilde{s}''' \neq 0$, and $f(2\pi ki,(2\pi k)^2/2,\tilde{s}) = 0$ for $k \in Z\setminus\{0\}$.

Theorem 3 (Instability): For $s \in S \cap \overset{1}{C}[-1,0]$ with $s(a) \geqslant a+1$, there are $\alpha > 0$ and λ with $u > 0$ and $f(\lambda,\alpha,s) = 0$.

Remark: The hypothesis in Theorem 3 can be replaced by "s concave".

Sketch of proof: Let $s \in S$. Define a mapping $F = \begin{pmatrix} F_1 \\ F_2 \end{pmatrix}: R^2 \times R^+ \to R^2$ by $F_1(u,v,\alpha) = \operatorname{Re} f(\lambda,\alpha,s)$, $F_2(u,v,\alpha) = \operatorname{Im} f(\lambda,\alpha,s)$. Suppose $iv \in iR^+$ is an eigenvalue for s and $\alpha > 0$. Then $F(0,v,\alpha) = 0$, and

$$d := \det\begin{pmatrix} \frac{\partial F_1}{\partial u} & \frac{\partial F_1}{\partial v} \\ \frac{\partial F_2}{\partial u} & \frac{\partial F_2}{\partial v} \end{pmatrix}(0,v,\alpha) \geqslant \alpha^2\left(\int_{-1}^{0} a\sin(va)ds(a)\right)^2.$$

For $d > 0$ there are neighbourhoods U of α and W of $(0,v)$ and a mapping $G:U \to W$ with $G(\alpha) = (0,v)$ and $F\circ(G,\mathrm{id}) = 0$ on U, hence $G_1(\alpha') + iG_2(\alpha')$ are eigenvalues for $\alpha' \in U$, and $G_1'(\alpha) > 0$ would imply the assertion. - We have $G_1(\alpha) = v/d \int_{-1}^{0} a\sin(va)ds(a)$. Therefore we only have to find an eigenvalue $iv \in iR^+$ with $\int_{-1}^{0} a\sin(va)ds(a)$ positive. - Let $s \in S \cap C^1[-1,0]$, $s(a) \geqslant a+1$. Then $\int_{-1}^{0}\cos(\pi a)ds(a) = 1 + \pi\int_{-1}^{0}\sin(\pi a)s(a)da \leqslant 1 + \pi\int_{-1}^{0}(a+1)\sin(\pi a)da = 0$, and the function $h:t \to \int_{-1}^{0}\cos(ta)ds(a)$ has a zero v in $(0,\pi]$. Obviously, $\int_{-1}^{0}\sin(va)ds(a) < 0$ and $\int_{-1}^{0} a\sin(va)ds(a) > 0$ and $f(iv,-v/\int_{-1}^{0}\sin(va)ds(a),s) = 0$.

2. For the simplest smooth function in S, $s(a) = a+1$, we can describe the location of all eigenvalues for all $\alpha > 0$.

Theorem 4: Let $s(a) = a+1$ for $-1 \leqslant a \leqslant 0$.

a) For every $\alpha > 0$, every eigenvalue lies in one of the strips $R + i(-2\pi,2\pi)$ and $R \pm i(2\pi k,2\pi k + 2\pi)$ with $k \in N$.

b) For every $\alpha > 0$ and every $k \in N$, there is exactly one eigenvalue $\lambda_k(\alpha)$ in $R + i(2\pi k, 2\pi k + 2\pi)$. We have

$$\lambda_k(\alpha)\begin{cases} \in R^- + i(2\pi k, 2\pi k + \pi) & \text{for } \alpha < \alpha_k := (2\pi k + \pi)^2/2 \\ = iv_k := i(2\pi k + \pi) & \text{for } \alpha = \alpha_k \\ \in R^+ + i(2\pi k + \pi, 2\pi k + 2\pi) & \text{for } \alpha > \alpha_k. \end{cases}$$

c) For every $\alpha > 0$, there are exactly two eigenvalues in $R + i(-2\pi, 2\pi)$. Let $\alpha^* := -2u^*\exp(u^*)$ with $u^* < 0$ and $2\exp(u^*) - 2 = u^*$. For $\alpha \leqslant \alpha^*$, both eigenvalues are real. If we denote them by $u_1(\alpha)$ and $u_2(\alpha)$ with $u_1(\alpha) \leqslant u_2(\alpha)$, then $u_1(\alpha) < u^* < u_2(\alpha) < 0$ for $\alpha < \alpha^*$, $u_1(\alpha) \to -\infty$ and $u_2(\alpha) \to 0$ for $\alpha \to 0$, and $u_1(\alpha^*) = u^* = u_2(\alpha^*)$.
For $\alpha^* < \alpha < \alpha_0 := \pi^2/2$, there is exactly one eigenvalue in $R^- + i(0,\pi)$, for $\alpha = \alpha_0$ $i\pi$ is an eigenvalue, and for $\alpha > \alpha_0$ there is exactly one eigenvalue in $R^+ + i(\pi, 2\pi)$.

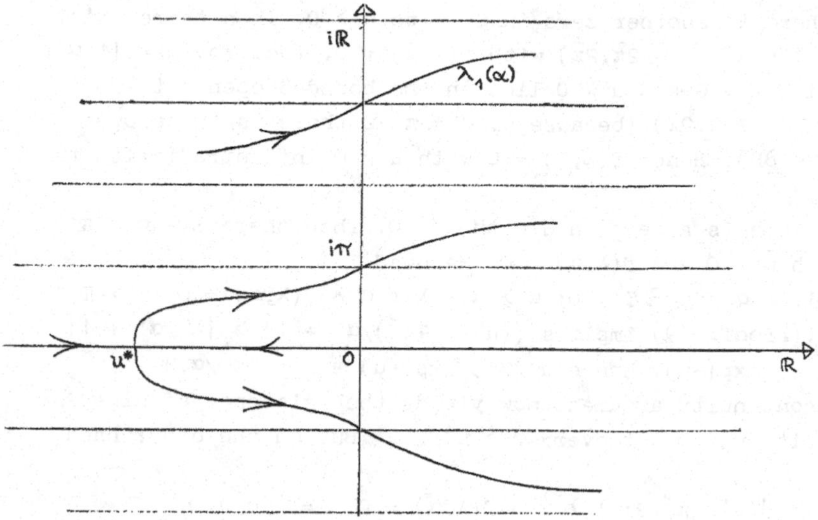

The arrows indicate the direction of increasing α.

Remarks: 1) We see: If λ is an eigenvalue with $u > 0$, then $|v| > \pi$. This exhibits one of the difficulties which arise if one tries to prove the existence of a nonconstant periodic solution of the nonlinear equation $\dot{x}(t) = -\alpha\int_{-1}^{0} x(t+a)da[1 + x(t)]$. - Even in the simpler case of $\dot{x}(t) = -\alpha x(t-1)[1 + x(t)]$ the existence of eigenvalues of the linearised equation with $u > 0$ and $0 < v < \pi$ is required, see the different proofs of Nussbaum [3], Grafton [2] and Chow [1].

2) A similar theorem concerning the equation $\lambda + \alpha\exp(-\lambda) = 0$ was proved by Wright [6]. He used elementary functions to derive his result. Our method is different:

Remarks on the proof of Theorem 4: Set $f(\lambda,\alpha) := f(\lambda,\alpha,\mathrm{id} + 1)$. We have

$$f(\lambda,\alpha) = 0 \iff \lambda \neq 0 \wedge (\lambda^2 + \alpha)\exp(\lambda) = \alpha. \tag{9}$$

From (9) we infer a) and

$\{(iv,\alpha) \in iR \times R^+ | f(iv,\alpha) = 0\} = \{((2\pi k + \pi)i, (2\pi k + \pi)^2/2) | k \in Z\}$.

To explain the method of our proof let us try to show that there are exactly two zeros of $f(\cdot,\pi^2/2)$ in $G := R + i(-2\pi,2\pi)$. We know that there are exactly two zeros in $G \cap iR$, namely $\pm i\pi$, and that $iv \in G$ and $f(iv,\alpha) = 0$ imply $v = \pm i\pi$, $\alpha = \pi^2/2$.

i) Suppose there is another zero in G, with $u > 0$. Then there exist $\alpha < \pi^2/2$ and $\lambda \in R^+ + i(-2\pi,2\pi)$ with $f(\lambda,\alpha) = 0$, too. For $\alpha' \in [1,\alpha]$, every zero with $\lambda \in G$ and $u > 0$ lies in the bounded open set $B := (0,\alpha+1) + i(-2\pi,2\pi)$ (because of $\alpha' \neq \pi^2/2$ there is no zero of $f(\cdot,\alpha')$ on $iR \cap \partial B$). Hence $f(\lambda,1) = 0$ with $u > 0$ in contradiction to Theorem 1.

ii) Suppose there is a zero in G with $u < 0$. Then there are $\alpha > \pi^2/2$ and $\lambda \in G$ with $u < 0$ and $f(\lambda,\alpha) = 0$. We need

Proposition 1: $\forall\, \alpha > 0\ \exists\, T < 0\colon \alpha' \geq \alpha \wedge \lambda \in G \wedge f(\lambda,\alpha) = 0 \Rightarrow T < u$.
(Proof: (9) implies $((u^2 + 4\pi^2)/\alpha + 1) \geq |\lambda^2/\alpha' + 1| \geq \exp(-u)$, hence $u^2/\alpha \geq \exp(-u) - 1 - 4\pi^2/\alpha$.)

As above, a continuity argument now yields the existence of eigenvalues in G with $u < 0$ for every $\alpha > \pi^2/2$. - But on the other hand we have

Proposition 2: $\exists\, \alpha^* > \pi^2/2\colon \lambda \in G \wedge f(\lambda,\alpha^*) = 0 \Rightarrow u > 0$.

3. There is another fact which expresses an increase of stability if the maximal (step-) function in S is replaced by a smaller, smooth function: The branches of eigenvalues in the right half-plane become bounded. Such a branch is a maximal connected subset of the set $P := \{\lambda \in C | u > 0 \wedge (\exists\, \alpha > 0\colon f(\lambda,\alpha,s) = 0\}$.

For $s(a) = 1$ on $(-1,0]$, there are unbounded branches: Choose $v \in (\pi/2,\pi)$, set $u_v := -v\cos(v)/\sin(v)$ and $\alpha_v := -u_v \exp(u_v)/\cos(v)$. Then $f(u_v + iv, \alpha_v, s) = 0$, and $\{u_v + iv \mid \pi/2 < v < \pi\}$ is an unbounded connected subset of P.

On the other hand, we have

<u>Theorem 5</u>: For $s \in S \cap C^3[-1,0]$ with $s'(-1) > 0$ and $s'(0) > 0$, every connected subset of P is bounded.

For $s: a \rightarrow a+1$, the proof is simple: From the preceding theorem we know that every connected subset Q of P has bounded imaginary part $\operatorname{Im} Q := \{\operatorname{Im} \lambda \mid \lambda \in Q\}$. For $\lambda \in Q$ and suitable $\alpha > 0$, (9) gives $\lambda^2/\alpha + 1 = \exp(-\lambda)$, $(u^2 - v^2)/\alpha + 1 = \exp(-u)\cos(v)$. For sequences λ_n, α_n with $\lambda_n \in Q$ and $u_n \rightarrow \infty$ we infer $1 \leqslant \lim_{n\rightarrow\infty} (u_n^2/\alpha_n + 1) = \lim_{n\rightarrow\infty} (v_n^2/\alpha_n + \exp(-u_n)\cos(v_n)) = 0$, contradiction.

4. The proofs of Theorems 2 - 5 can be found in [5].

References:

[1]Chow,C.N.: Existence of periodic solutions of autonomous functional differential equations. J.Differential Equations 15, 350 - 378 (1974).

[2]Grafton,R.B.: A periodicity theorem for autonomous functional differential equations. J.Differential Equations 6,87 - 109(1969).

[3]Nussbaum,R.D.: Periodic solutions of some nonlinear autonomous functional differential equations. Annali di Matematica Pura ed Applicata Vol. CI, 263 - 306 (1974).

[4]Walther,H.O.: Asymptotic stability for some functional differential equations. To appear in: Proceedings of the Royal Society of Edinburgh, March 1976.

[5]Walther,H.O.: On a transcendental equation in the stability analysis of a population growth model. To appear.

[6]Wright,E.M.: A non-linear differential-difference equation. J. Reine Angewandte Mathematik 194, 66 - 87 (1955).

An explicit St. Venant's principle in three-dimensional elasticity

Norbert Weck

o. Introduction: Consider a region $\Omega \subset \mathbb{R}^3$ which is supposed to be "long in x_1-direction".

St. Venant's principle for a boundary value problem

(1) $\quad Lu = f \quad$ in Ω ; $\quad Bu = g$ on $\partial\Omega$

can be roughly stated as follows: Suppose that

(2) $\quad f(x), \; g(x) = o \quad$ for $x_1 \geq o$

Then u is "small for large x_1".
A rigorous version of this statement uses the positive quadratic form (energy) which is (maybe) associated with (1):
St.Venant's principle: Let $U(s)$ be the energy contained in

$$\Omega(s) := \{x \in \Omega \mid x_1 > s\} .$$

Then

$$U(s) \leq U(o) \quad e^{-\Phi(s)}$$

Theorems of this kind have been proven by a number of authors.
Let us mention only TOUPIN [8] (three-dimensional elasticity), KNOWLES [5],[6] (two-dimensional elasticity, elliptic second order operators) and EDELSTEIN and KNOWLES [3], [4] (heat equation). In the following we want to treat again linear three-dimensional elasticity. Stress is laid on an explicit determination of the decay function $\Phi(s)$ i. e. we want to determine Φ from the geometry of Ω. This seems to be important in view of the origin of St. Venant's principle.

Notation: Latin indices range from 1 to 3; greek indices range from 2 to 3. We use summation convention, thus e. g.

$$u_i \, v_i := \sum_{i=1}^{3} u_i \, v_i$$

$$s_{i\alpha} \, t_{k\alpha} := \sum_{\alpha=2}^{3} s_{i\alpha} \, t_{k\alpha}$$

If s is a (simply or doubly) indexed quantitiy, then we denote by $\hat{s}$ the same quantity its indices being restricted to $\{2,3\}$. Thus for $x = (x_1,x_2,x_3) \in \mathbb{R}^3$

$$\hat{x} := (x_2,x_3)$$

Inner products are defined in the usual way. Thus e.g. it s,t are doubly indexed quantities then

$$(s,t) \;:=\; s_{ij}\, t_{ij} \qquad |s| \;:=\; (s,s)^{1/2}$$

$$(\hat{s},\hat{t}) \;:=\; s_{\alpha\beta}\, t_{\alpha\beta} \qquad |\hat{s}| \;:=\; (\hat{s},\hat{s})^{1/2}$$

Similarly if f, g are (indexed) functions and S is a region

$$(f,g)(S) \;:=\; \int_S (f,g)dx$$

$$||f||(S) \;:=\; (f,f)(S)^{1/2}$$

1. A Model problem: Consider Dirichlet's problem for $\Delta u = f$. Energy is defined by means of the Dirichlet integral

$$U \;:=\; (\nabla u, \nabla u)(\Omega)$$

For $s \geq o$ put

$$U(s) \;:=\; ||\nabla u||^2(\Omega(s))$$

Then

(3)
$$U'(s) = - ||\nabla u||^2(F(s))$$

$$F(s) \;:=\; \{x \in \Omega | x_1 = s\}$$

On the other hand by (2) and Green's formula $(\partial_i := \frac{\partial}{\partial x_i})$:

$$U(s) = - (\partial_1 u, u)(F(s))$$

(4)
$$U(s) \leq \frac{\alpha}{2} ||u||^2(F(s)) + \frac{1}{2\alpha} ||\partial_1 u||^2(F(s)) \qquad \alpha \in \mathbb{R}^+$$

If $\lambda_1(s)$ is the first Dirichlet eigenvalue of the twodimensional region $F(s)$ then (since $u|_{\partial F(s)} = o$)

$$\lambda_1(s) = \inf_{v|_{\partial F(s)} = o} \frac{||\hat{\nabla} v||^2(F(s))}{||v||^2(F(s))} \leq \frac{||\hat{\nabla} u||^2(F(s))}{||u||^2(F(s))}$$

$$||u||^2(F(s)) \leq \lambda_1^{-1}(s) \; ||\nabla u||^2(F(s))$$

We put $\alpha = \lambda_1^{1/2}(s)$ in (4) to obtain

$$U(s) \leq \frac{1}{2} \lambda_1^{-1/2}(s) ||\nabla u||^2(F(s))$$

Combining this with (2) we get $U'(s) + 2 \lambda_1^{1/2}(s) \; U(s) \leq o$

(5)
$$U(s) \leq U(o) \exp\left(- \int_o^s 2 \lambda_1^{1/2}(\sigma) \, d\sigma\right)$$

This is the desired decay property provided we can estimate λ_1 from below using onlythe geometry of Ω. But in the case of Dirichlet's problem this can be done easily using e. g. isoperimetric inequalities [7] or the monotonicity of eigenvalues [8] in case of more complicated problems.

2. The elasticity problem and some difficulties solved by TOUPIN:

Let c_{ijke} be given functions satisfying

(6) $$c_{ijkl} = c_{jikl} = c_{ijlk} = c_{klij}$$

(7) $$c_{ijkl}(x)\ s_{ij}\ s_{kl} \geq c_o\ s_{ij}\ s_{kl} \qquad c_o > o$$

for all symmetric matrices (s_{ij}).

(7') $$|(c_{ijkl}(x)s_{kl})|^2 \leq c_1|(s_{kl})|^2$$

We consider the following boundary value problem

(1') $$\begin{aligned} \partial_i(c_{ijkl}(x)\partial_k\ u_l(x)) &= f_j(x) & x \in \Omega \\ n_i(\xi)\ c_{ijkl}(\xi)\partial_k u_l(\xi) &= g_j(\xi) & \xi \in \partial\Omega \end{aligned}$$

We assume

(2') $$f_j(x),\ g_j(x) = o \qquad \text{for} \quad x_1 \geq o$$

and furthermore that the forces f and g are "equilibrated". Then one can find a solution u of (1'). We put

(8) $$U(s) := \int_{\Omega(s)} c_{ijkl}(x)\ \partial_i u_j(x)\ \partial_k u_l(x)\ dx$$

and find

(9) $$U'(s) = - \int_{F(s)} c_{ijkl}\ \partial_k u_j\ \partial_k u_l\ d\hat{x}$$

(10) $$U(s) = - \int_{F(s)} c_{1jkl}\ \partial_k u_l \cdot u_j\ d\hat{x}$$

Proceeding as in the model problem we would be faced with some difficulties:

<u>Difficulty 1:</u> u does not satisfy homogeneous boundary conditions since we are treating a Neumann type problem.

<u>Solution:</u> This is not a real difficulty since we get Neumann eigenvalues by Rayleigh quotients involving functions that need not satisfy boundary conditions.

<u>Difficulty 2:</u> The first eigenvalue is zero.

<u>Solution:</u> There are only finitely many eigenfunctions corresponding to $\lambda = o$ and these are known explicitly. Addition of them does not change $U(s)$. Therefore one may assume that u is orthogonal to them wherever this is needed.

<u>Difficulty 3:</u> On the two-dimensional region we can only estimate $\hat{u}$, not u_1.

<u>Solution:</u> F(s) is made "thick" by re-defining $U(s)$ to be

$$\tilde{U}(s) := \int_{\Omega} \chi_s(x_1)\, c_{ijkl}\, \partial_i u_j\, \partial_k u_l \, dx$$

with a cut-off function χ_s like

$$\chi_s(x_1) := \begin{cases} o & \text{for} \quad x_1 \le s \\ 1 & \text{for} \quad x_1 \ge s+1 \\ \text{linear} & \text{for} \quad s \le x_1 \le s+1 \end{cases}$$

Then

$$\tilde{U}'(s) = - \int_{\Omega(s,s+1)} c_{ijkl}\, \partial_i u_j\, \partial_k u_l \, dx$$

$$\Omega(s,s+1) := \{x \in \Omega \mid s < x_1 < s+1\}$$

The foregoing is a sketch of work done by TOUPIN [8]. If $s_c(1)$ is the first <u>non-zero</u> eigenvalue of $\Omega(s,s+1)$ he gets a St. Venant estimate of the following form

$$(11) \qquad U(s) \le U(o) \exp\left\{ -s_c(1)^{-1}(s-1) \right\}$$

Although this solves the two- and three-dimensional St. Venant problem in elasticity there still have been efforts to treat special cases [5] in order to make the decay functions more explicit. The problem in (11) is to obtain lower estimates for $s_c(1)$. This could be done (in principle) using the work of BRAMBLE and PAYNE [1]. Their estimates are however not too well adopted for treating regions like $\Omega(s,s+1)$. In particular they will become bad for $1 \to o$ (which is inevitable since $s_c(1) \to o$ then). On the other hand choosing 1 large will deteriorate (11) because of the term $s-1$. We shall try to remove this dilemma by treating $\hat{u}$ and u_1 separately. Another idea used is that $U(s)$ can also swallow L_2-norms of u.

<u>3. St. Venant's principle:</u> First we shall fix the geometric situation. For any $s \ge o$ we suppose that

$$F(s) = \{x \in \Omega \mid x_1 = s\}$$

is star-shaped with respect to (s,o,o) and thus in cylindrical coordinates can be written as

$$F(s) = \{(s, \rho\cos\phi, \rho\sin\phi) \mid o \le \phi \le 2\pi \text{ and } o \le \rho < r(s,\phi)\}$$

with r defined on $[o,X_1] \times [o,2\pi]$. We put

(12) $\quad r_1(s) \;:=\; \min_{\phi} r(s,\phi) \;\leq\; \max_{\phi} r(s,\phi) \;=:\; r_2(s)$

We assume that for any s there is a Lipschitz constant $L(s)$ such that

(13) $\quad |r(s,\phi) - r(x_1,\phi)| \;\leq\; L(s)\,(x_1-s) \quad \text{for } s \leq x_1 \leq \min(X_1,\, s+2r_1(s))$

Next define $\quad S(u) \;:=\; (S_{ij}(u)) \;:=\; \frac{1}{2}(\partial_i u_j + \partial_j u_i)$

From (6), (7) and (7')

(14) $\quad c_o|S(u)|^2 \leq c_{ijkl}\,\partial_i u_j \partial_k u_l \;\leq\; c_1|S(u)|^2$

The only solutions of $S(u) = o$ are linear combinations of

$$e^1 := (1,o,o) \qquad e^2 := (o,1,o) \qquad e^3 := (o,o,1)$$

$$d^1 := (o,x_3,-x_2) \qquad d^2 := (-x_3,o,x_1) \qquad d^3 := (x_2,-x_1,o)$$

We state some lemmas concerning $S(u)$.

Lemma 1 (cf. [1], p. 18): Let $B \subset \mathbb{R}^3$ be a ball with radius R. Then

$$\inf_{(w,e^i)(\partial B)=(w,d^j)(\partial B) = o} \frac{||Sw||^2(B)}{||w||^2(B)} = \frac{1}{2R}$$

The next lemma is concerned with a two-dimensional region F. To keep the analogy to $F(s)$ we denote vectors $x \in F$ by $\hat{x} = (x_2,x_3)$ and so on.

Lemma 2: $\mu(F) = \inf_{(\hat{u},\hat{e}^\alpha)(F) = (\hat{u},\hat{d}^1)(F) = o} \dfrac{||S(\hat{u})||^2(F)}{||\hat{u}||^2(F)} > o$

If F is "strongly" star-shaped with respect to the origin, i. e.

$$\inf_{\xi \in \partial F} |\xi|^{-1}(\xi,n(\xi)) - \frac{1}{2} =: H > o$$

and

$$o < r_1 := \min_{\xi\in\partial F} |\xi| \leq \max_{\xi \in \partial F} |\xi| =: r_2$$

then

$$\mu(F) \geq r_1^{-2} \left(\max\{4\mu^2 + 9H^{-1}\mu^5,\; 4+3(4+3H^{-1})\mu^5\}\right)^{-1}$$

Proof: Cf. [1]

We go back to three dimensions and assume that we can find $M = M(s) > o$ such that

i) $\quad B(s) := \{x \mid |x - (M+s,o,o)| < M\} \quad \Omega(s)$

ii) $\quad$ Every segment joining $(2M,o,o)$ and $(s,\hat{x}) \in F(s)$ is contained in $\Omega(s)$.

Under our assumptions a possible choice for M is

$$M(s) := \frac{1}{l} r_1(s) \qquad l := l(s) := L(s) + \sqrt{L^2(s)+1}$$

(If s is near X_1 - the end of - then we have to switch to $M(s) := \frac{1}{2}(X_1 - s)$.).

Lemma 3: In the geometrical situation described above we have

$$||v_1||^2(F(s)) \leq K_1(s)||S(v)||^2(\Omega(s)-B) + K_2(s)||v||^2(\partial B)$$

$$K_1(s) := M\left[4(1+\nu^2)^3 + (4+\nu^2)^2 (2 + \frac{\nu^2}{2})\right]$$

$$K_2(s) := 4(5 + 2\nu^2)$$

$$\nu := \nu(s) = \frac{r_2(s)}{M(s)} = \frac{r_2(s)}{r_1(s)} (L(s) + \sqrt{L^2(s) + 1})$$

A <u>proof</u> of this will be given in the appendix.

Now we define $U(s)$ by (8) and recall (9) and (10). By (10) we can split $U(s)$ in two parts:

(10') $$U_1(s) := - \int_{F(s)} c_{11kl}\ \partial_k u_l\ u_1\ d\hat{x}$$

(10'') $$U_2(s) := - \int_{F(s)} c_{1\alpha kl}\ \partial_k u_l\ u_\alpha\ d\hat{x}$$

Thus $$U(s) = U_1(s) + U_2(s)$$

<u>Lemma 4:</u> Adding e^i and d^j to u does not change $U(s)$. The same is true for U_1 and U_2 separately.

<u>Proof:</u> Since $S(e^i) = S(d^j) = o$ adding e^i or d^j does not change U. Adding e^1 does not change U_2, since u_1 does not occur there. So $U_1(s) = U(s) - U_2(s)$ is not changed either. Repeating this type of argument we find that U_1 and U_2 are <u>both</u> invariant.

During the following considerations s will be fixed. Therefore we shall often suppress the argument s. To estimate U_1 we replace u by $v := + \alpha_i e^i + \beta_j d^j + u$ choosing α_i, β_j such that

(15) $$(v, e^i)(\partial B) = (v, d^j)(\partial B) = o$$

Using lemmas 1 and 3 and (14) we find $(\varepsilon \in \mathbb{R}^+)$

$$|U_1(s)| = |\int_{F(s)} c_{11kl}\ \partial_k u_l\ v_1 d\hat{x}|$$

$$\leq \frac{\varepsilon}{2} ||c_{11kl}\ \partial_k u_l||^2(F) + \frac{1}{2\varepsilon} ||v_1||^2(F)$$

(16) $$|U_1(s)| \leq \frac{\varepsilon}{2}||c_{11kl}\ \partial_k u_l||^2(F) + \frac{1}{2\varepsilon \cdot c_o} \max\ \{K_1, 2MK_2\}\ U(s)$$

To estimate U_2 we choose α_i, β_j such that

(15') $\quad (v, e^{\alpha})(F) = (v, d^1)(F) = o$

$$|U_2(s)| = |\int_F c_{1\alpha k1} \partial_k u_1 \quad v_\alpha \; dx|$$

$$\leq \frac{1}{2}\mu(F)^{-1/2} \int_F \sum_\alpha c_{1\alpha k1} \partial_k u_1|^2 d\hat{x} + \frac{1}{2}\mu(F)^{1/2} ||\hat{v}||^2(F)$$

Using (15') and lemma 2 we find

(17) $$|U_2(s)| \leq \frac{1}{2} \mu(F)^{-1/2} \sum_\alpha \sum_i || c_{\alpha ik1} \partial_k u_1||^2(F)$$

choosing $\varepsilon := c_o^{-1} \max \{K_1, 2r_1 K_2\}$ and observing (9) we find

$$U(s) \leq \max \{\mu(F)^{-1/2}, \varepsilon\} \sum_{i,j} ||c_{ijk1} \partial_j u_1||^2(F)$$

$$\leq c_1 c_o^{-1} \max \{c_o \mu(F)^{-1/2}, K_1, 2r_1 K_2\} ||Su||^2(F)$$

$$\leq - c_1 c_o^{-2} \max \{\ldots\} U'(s)$$

Theorem 1: Under the assumptions made above we have

$$U(s) \leq U(o) \exp\left\{ -\int_o^s K(\sigma)\, d\sigma \right\}$$

$$K(s) := c_o^2 c_1^{-1} \max \{c_o^{-1} \mu(F(s))^{1/2}, K_1(s)^{-1}, \frac{1}{2} r_1^{-1} K_2^{-1}(s)\}$$

$$K_1(s) := \frac{r_1(s)}{1} 4(1+\nu^2)^3 + (4+\nu^2)^2(2+\frac{\nu^2}{2})$$

$$K_2(s) := 4(5+2\nu^2)$$

$$\nu := \nu(s) := \frac{r_2(s)}{r_1(s)} 1$$

$$1 := 1(s) := L(s) + \sqrt{L^2(s)+1}$$

(For s near X_1 this has to be modified. Cf. the remark before lemma 3.)

<u>4. Appendix (Proof of Lemma 3):</u> We consider the geometrical situation as described in the main text. We shall put $s = o$ und suppress the argument s. For $x = (o;\hat{x}) \varepsilon F$ we define two line segments $S(\hat{x})$ and $T(\hat{x})$ by means of a parametric representation ($\rho := |\hat{x}|$):

$$S(\hat{x}): \quad y(\sigma,\hat{x}) := (2M\sigma; (1-\sigma)\hat{x}) \quad o \leq \sigma \leq \Phi(\rho) := \frac{\rho^2}{4M^2+\rho^2}$$

$$T(\hat{x}): \quad z(\tau,\hat{x}) := (M\tau; (1-\tau)\hat{x}) \quad o \leq \tau \leq \psi(\rho) := 1 - \sqrt{\frac{M^2}{M^2+\rho^2}}$$

Their respective endpoints are $x \in F$ and $\eta := \eta(\hat{x}) := y(\Phi(s),\hat{x}) \in \partial B$ (for $S(\hat{x})$) and $x \in F$ and $\xi := \xi(\hat{x}) := z(\psi(\rho),\hat{x}) \in \partial B$ (for $T(\hat{x})$).

We consider two line integrals

$$\int_0^{\Phi(\rho)} \frac{d}{d\sigma}\, 2\, Mv_1(y(\sigma,\hat{x})) - x_\alpha v_\alpha(y(\sigma,\hat{x}))\, d\sigma$$

$$= 2Mv_1(\eta) - x_\alpha v_\alpha(\eta) - 2Mv_1(x) - x_\alpha v_\alpha(x)$$

$$\int_0^{\psi(\rho)} \frac{d}{d\tau}\, Mv_1(z(\tau,\hat{x})) - x_\alpha v_\alpha z(\tau,\hat{x})\, d\tau$$

$$= Mv_1(\xi) - x_\alpha v_\alpha(\xi) - Mv_1(x) + x_\alpha v_\alpha(x)$$

Subtracting these equations one from another we get

$$Mv_1(x) = -\int_0^{\Phi} \frac{d}{d\sigma}[\ldots]\, d\sigma + \int_0^{\psi} \frac{d}{d\tau}[\ldots]\, d\tau + (t,v(\xi)) - (s,v(\eta))$$

$$s := s(\hat{x}) := (-2M;\hat{x}) \quad ; \quad t := t(x) := (M;\hat{x})$$

We square this and integrate with respect to $\hat{x}$ using Schwarz' inequality and the following facts:

(16) $$|t(\hat{x})|^2 = M^2 + |\hat{x}|^2 \le M^2 + r_2^2 = M^2(1+\nu^2)$$

$$|s(\hat{x})|^2 = 4M^2 + |\hat{x}|^2 \le 4M^2 + r_2^2 = M^2(4+\nu^2)$$

i) As x ranges over F the segments $S(\hat{x}), T(\hat{x})$ fill subsets of $\Omega - B$. The corresponding volume elements are given by

$$2M(1-\sigma)^2\, d\sigma\, d\hat{x} \ge 2M(1-\Phi)^2 d\sigma d\hat{x} \ge 2M(1+\tfrac{1}{4}\nu^2)^{-1} d\sigma d\hat{x}$$

$$M(1-\tau)^2\, d\tau d\hat{x} \ge M(1-\psi)^2 d\tau d\hat{x} \ge M(1+\nu^2)^{-1} d\tau d\hat{x}$$

ii) $\eta(\hat{x})$ and $\xi(\hat{x})$ fill subsets of ∂B. The corresponding surface elements are given by

$$\lambda(\hat{x})d\hat{x} \; ; \; \mu(\hat{x})d\hat{x} \quad \text{with } \lambda(\hat{x}),\ \mu(\hat{x}) \ge 1 .$$

Taking all this into account we find

$$\frac{1}{4} M^2 \|v_1\|^2(F) \le \frac{1}{2M}(1+\frac{1}{4}\nu^2)\|\frac{d}{d\sigma}[\ldots]\|^2(\Omega-B)$$

$$+ \frac{1}{M}(1+\nu^2)\|\frac{d}{d\tau}[\ldots]\|^2(\Omega-B)$$

$$+ M^2(5+\nu^2)\|v\|^2(\partial B)$$

Now $$\frac{d}{d\tau}(Mv_1 - x_\alpha v_\alpha) = t_i(\hat{x})\, t_j(\hat{x})\, S_{ij}(v)$$

$$|\frac{d}{d\tau}(\ldots)|^2 \le |t|^4\, |S(v)|^2 \le (M^2+\nu^2)^2\, |S(v)|^2$$

Similarly

$$\left|\frac{d}{d\sigma}\ (\ldots)\right|^2 \leq |s|^4 |S(v)|^2 \leq (4M^2 + \nu^2)^2\ |S(v)|^2$$

Putting this into (17) yields lemma 3.

References:

[1] BRAMBLE, J. H. and PAYNE, L. E.: Some inequalities for vector functions with applications in elasticity.
Arch. Rat. Mech. Anal. 11 (1962), 16-26.

[2] COURANT, R. and HILBERT, D.: Methoden der mathematischen Physik I.
Heidelberger Taschenbücher, Band 30.
Springer Verlag Berlin - Heidelberg - New York 1968.

[3] EDELSTEIN, W. S.: A spatial decay estimate for the heat equation.
ZAMP 20 (1969), 900 - 905.

[4] KNOWLES, J. K.: On the spatial decay of solutions of the heat equation.
ZAMP 22 (1971).

[5] ————————: On Saint-Venant's principle in the two-dimensional linear theory of elasticity.
Arch. Rat. Mech. Anal. 21 (1966), 1 - 22.

[6] ————————: A Saint-Venant principle for a class of second-order-elliptic boundary value problems.
ZAMP 18 (1967), 473-490.

[7] POLYA, G.: Isoperimetric inequalities in mathematical physics.
Annals of Mathematical Studies 27- Princeton University Press, Princeton 1951.

[8] TOUPIN, R. A.: Saint-Venant's principle.
Arch. Rat. Mech. Anal. 18 (1965), 83-96.

TRACE CLASS METHODS FOR SCATTERING IN A HOMOGENEOUS ELECTRO-STATIC FIELD

Joachim Weidmann

Let T_o be the operator in $L_2(\mathbb{R}^m)$ which is defined by

$$D(T_o) = S = S(\mathbb{R}^m),$$
$$T_o f(x) = -\Delta f(x) + \varepsilon x_1 f(x) \quad \text{for} \quad f \in D(T_o)$$

with some $\varepsilon > 0$; S is the Schwartz space. By T we denote the uniquely determined self adjoint extension of T_o, $T = \overline{T_o}$. In a forthcoming paper [2] K.Veselić and the author prove the folowing results:

(A) Let V be a symmetric operator in $L_2(\mathbb{R}^m)$ which satisfies $D(V) \supset S$ and

$$\|Vf\|_2 \leq (1+r)^{-\Theta} \|Df\|_q$$

for $f \in S$ with $\{x \in \mathbb{R}^m : |x_1| < r\} \cap \text{supp } f = \emptyset$, where $\Theta > \frac{1}{2}$, $q \in [2,\infty]$ and D is a constant coefficient differential operator with respect to the variables $x_2, x_3, \ldots, x_m$ ($\|\cdot\|_q$ is the norm in $L_q(\mathbb{R}^m)$). Then for every selfadjoint extension H of $T_o + V$ the wave operators $W_\pm(H,T)$ exist.

(B) Let $m = 1$. If V is the operator of multiplication with a real valued measurable function v on $\mathbb{R}$, satisfying

$$|v(x)| \leq C(1+|x|)^{-\Theta}, \quad x \in \mathbb{R}$$

for some $\Theta > \frac{1}{2}$, then the wave operators $W_\pm(H,T)$ are complete. (With some additional assumptions it is proved that these wave operators are unitary).

The <u>proof</u> of (A) uses Cook's lemma and uniform estimates of $e^{-itT}f$, which are proved via Fourier transform. It should be noticed that (for $m \geq 3$) the Coulomb potential satisfies the condition of (A) in a somewhat modified form. The proof of (B) makes use of some special results on the spectral theory of second order ordinary differential operators.

In the present paper we shall prove the existence and completeness of the wave operators in the one-dimensional case by means of trace class methods (Birman-Kato-Rosenblum). It is surprising that, compared with (B), our assumption on V is much weaker for $x \to \infty$ while it is much stronger for $x \to -\infty$. A combination of both methods gives the strongest result which is known to us so far.

We first prove an auxiliary theorem.

<u>Theorem 1</u>. Let T be the selfadjoint Sturm-Liouville operator in $L_2(\mathbb{R})$ defined by

$$D(T) = \{f \in L_2(\mathbb{R}): f \text{ and } f' \text{ are locally absolutely continuous}, -f'' + \varepsilon x f \in L_2(\mathbb{R})\},$$

$$Tf(x) = -f''(x) + \varepsilon x f(x) \quad \text{for} \quad f \in D(T).$$

Let $q: \mathbb{R} \to \mathbb{R}$ be locally integrable with

$$\int_{n-1}^{n+1} |q(x)|\,dx \leq \begin{cases} C(1+n)^{-\delta} & \text{for } n \geq 0, \\ C(1+|n|)^{-2-\delta} & \text{for } n < 0 \end{cases}$$

for some $\delta > 0$. Then the operator

$$M: L_2(\mathbb{R}) \to L_2(\mathbb{R}), \quad f \mapsto |q(.)|^{1/2}(T+i)^{-1}f$$

is of Hilbert-Schmidt type.

It is clear that T is selfadjoint, since $lu = -u'' + \varepsilon x u$ is "limit point" at both end points. For the proof of Theorem 1 we need some preparations.

Let $\varphi \in C_o^\infty(\mathbb{R})$ with

$$\operatorname{supp} \varphi \subset [-\tfrac{1}{2}, \tfrac{1}{2}], \quad \varphi(x) \geq 0 \quad \text{and} \quad \int \varphi(x)\,dx = 1.$$

Then for

$$\varphi_n(x) = \int_{n-\frac{1}{2}}^{n+\frac{1}{2}} \varphi(x-y)\,dy \qquad (n \in \mathbb{Z},\ x \in \mathbb{R})$$

we have $\varphi_n \in C_o^\infty(\mathbb{R})$, $\operatorname{supp} \varphi_n \subset [n-1, n+1]$ and $\sum_{n \in \mathbb{Z}} \varphi_n(x) = 1$ for every $x \in \mathbb{R}$ (for every $x \in \mathbb{R}$ this sum consists of at most two terms).

<u>Lemma.</u> Let T be the operator of Theorem 1. There exists a $K > 0$ such that for every $f \in D(T)$

$$\|(\varphi_n f)'\| \leq \begin{cases} K(1+n)^{-1/2}(\|Tf\|+\|f\|) & \text{for } n \geq 0, \\ K(1+|n|)^{1/2}(\|Tf\|+\|f\|) & \text{for } n < 0. \end{cases}$$

Proof. For every real valued $\xi \in C_o^\infty(\mathbb{R})$ we have

$$\begin{aligned} \|\xi Tf\|^2 + \|\xi f\|^2 &\geq 2 \operatorname{Re} \langle \xi Tf, \xi f \rangle = 2 \operatorname{Re} \langle Tf, \xi^2 f \rangle \\ &= 2 \operatorname{Re} \{\langle f', (\xi^2 f)' \rangle + \varepsilon \langle x\xi f, \xi f \rangle\} \\ &= 2 \operatorname{Re} \{\|\xi f'\|^2 + 2 \langle \xi f', \xi' f \rangle + \varepsilon \langle x\xi f, \xi f \rangle\} \qquad (1) \\ &\geq 2 \{\|\xi f'\|^2 - \tfrac{1}{2}\|\xi f'\|^2 - 2\|\xi' f\|^2 + \varepsilon x_\xi \|\xi f\|^2\} \\ &= \|\xi f'\|^2 - 4\|\xi' f\|^2 + 2\varepsilon x_\xi \|\xi f\|^2, \end{aligned}$$

where $x_\xi = \inf\{\operatorname{supp} \xi\}$. This implies for $n \leq 1$

$$\|\varphi_n f'\| \leq C_1(\|Tf\| + (1+|n|)^{1/2}\|f\|)$$

and therefore

$$\|(\varphi_n f)'\| \leq \|\varphi_n' f\| + \|\varphi_n f'\| \leq C_2(1+|n|)^{1/2}(\|Tf\| + \|f\|).$$

We now consider $n > 1$. In this case we have

$$\begin{aligned} \|T(\varphi_n f)\|^2 &= \|-(\varphi_n f)'' + \varepsilon x \varphi_n f\|^2 \\ &= \|(\varphi_n f)''\|^2 - 2\varepsilon \operatorname{Re} \langle (\varphi_n f)'', x\varphi_n f \rangle + \varepsilon^2 \|x\varphi_n f\|^2 \\ &\geq 2\varepsilon \langle (\varphi_n f)', x(\varphi_n f)' \rangle \geq 2\varepsilon(n-1)\|(\varphi_n f)'\|^2, \end{aligned}$$

hence

$$\|(\varphi_n f)'\| \leq C_3 n^{-1/2} \|T(\varphi_n f)\|.$$

Together with $T(\varphi_n f) = \varphi_n Tf - 2\varphi_n' f' - \varphi_n'' f$ this implies

$$\begin{aligned} \|(\varphi_n f)'\| &\leq C_3 n^{-1/2}\{\|\varphi_n Tf\| + 2\|\varphi_n' f'\| + \|\varphi_n'' f\|\} \\ &\leq C_4 n^{-1/2}\{\|Tf\| + \|\varphi_n' f'\| + \|f\|\} \end{aligned}$$

Applying (1) with $\xi = \varphi_n'$ $(n > 1)$ we get

$$\|\varphi_n' f'\| \leq C_5(\|Tf\| + \|f\|),$$

hence

$$\|(\varphi_n f)'\| \leq C_6 n^{-1/2}(\|Tf\| + \|f\|).$$

This completes the proof of the lemma.

Proof of Theorem 1. In the first step of the proof we show that the operators

$$K_n : D(T) \to L_2(\mathbb{R}),\ f \mapsto |q|^{1/2}\varphi_n f$$

are of Hilbert-Schmidt type, where the corresponding Hilbert-Schmidt norms $|||K_n|||$ satisfy the inequalities

$$|||K_n|||^2 \le C_1(1+|n|)^{-1-\varepsilon} \qquad \text{for every } n \in \mathbb{Z}.$$

By means of the above lemma the operators

$$K_{n,1} : D(T) \to L_2(\mathbb{R}), f \mapsto (\varphi_n f)'$$

are bounded with

$$\|K_{n,1}\| \le \begin{cases} C_2(1+n)^{-1/2} & \text{for } n \ge 0, \\ C_2(1+|n|)^{1/2} & \text{for } n < 0. \end{cases}$$

The operators

$$K_{n,2} : R(K_{n,1}) \to L_2(\mathbb{R}), (\varphi_n f)' \mapsto |q|^{1/2}\varphi_n f$$

are of Hilbert-Schmidt type with the kernels

$$k_n(x,t) = \begin{cases} |q(x)|^{1/2} & \text{for } n-1 \le t \le x \le n+1, \\ 0 & \text{elsewhere;} \end{cases}$$

hence

$$|||K_{n,2}|||^2 \le 2 \int_{n-1}^{n+1} |q(x)|\,dx.$$

This implies the desired inequality for $K_n = K_{n,2}K_{n,1}$.

In the second step we shall now prove that the series $\sum_{n\in\mathbb{Z}} K_n$ converges with respect to the Hilbert-Schmidt norm of operators from $D(T)$ into $L_2(\mathbb{R})$. The sum of this series is obviously equal to the operator

$$K : D(T) \to L_2(\mathbb{R}), f \mapsto |q|^{1/2} f;$$

hence K is of Hilbert-Schmidt type.

Let $\{e_k : k \in \mathbb{N}\}$ be an orthonormal basis of $D(T)$ (with respect to the scalar product $\langle f,g\rangle_T = \langle f,g\rangle + \langle Tf,Tg\rangle$). Since $K_n e_k \perp K_m e_k$ for $|m-n| > 1$, we have for arbitrary $k \in \mathbb{N}$ and $n,m \in \mathbb{Z}$ with $n \le m$

$$\|\sum_{n\le l\le m} K_l e_k\|^2 = \sum_{\substack{n\le l,p\le m\\ |l-p|\le 1}} \langle K_l e_k, K_p e_k\rangle$$

$$\le \sum_{\substack{n\le l,p\le m\\ |l-p|\le 1}} \tfrac{1}{2}(\|K_l e_k\|^2 + \|K_p e_k\|^2) \le 3 \sum_{n\le l\le m} \|K_l e_k\|^2.$$

This implies

$$|||\sum_{n\le l\le m} K_l|||^2 \le 3 \sum_{n\le l\le m} |||K_l|||^2 .$$

Therefore the above estimate of $|||K_n|||$ implies the convergence of $\sum_{n\in\mathbb{Z}} K_n$ in the sense of the Hilbert-Schmidt norm.

Now it follows immediately that $M = |q|^{1/2}(T+i)^{-1}$ is of Hilbert-Schmidt type, since $(T+i)^{-1}$ is a bounded operator from $L_2(\mathbb{R})$ onto $D(T)$. This concludes the proof of Theorem 1.

We are now in the position to prove our main result.

<u>Theorem 2.</u> Let T be as in Theorem 1, $q: \mathbb{R} \to \mathbb{R}$ locally square integrable and

$$|q(x)| \le \begin{cases} C(1+|x|)^{-\delta} & \text{for } x \to \infty, \\ C(1+|x|)^{-2-\delta} & \text{for } x \to -\infty \end{cases}$$

for some $\delta > 0$ and $C \ge 0$. If V is the operator of multiplication by the function q, then $T + V$ is selfadjoint and the wave operators $W_\pm(T + V,T)$ exist and are complete.

<u>Proof</u>. It is well known that V is T-compact, hence $T + V$ is selfadjoint (V has T-bound 0 and V is T-small at infinity; cf. [3] Theorem 2.7). Therefore it suffices to show that $(T+V+i)^{-1} - (T+i)^{-1}$ is a trace class operator (cf. [1] Theorem X.4.12). From the second resolvent equation we get

$$(T+V+i)^{-1} - (T+i)^{-1} = -(T+V+i)^{-1}V(T+i)^{-1}$$

$$= -(T+V+i)^{-1}|q|^{1/2} \operatorname{sgn} q\, |q|^{1/2}(T+i)^{-1}.$$

From the above lemma we know that $|q|^{1/2}(T+i)^{-1}$ is of Hilbert-Schmidt type. Hence

$$|q|^{1/2}(T+V-i)^{-1} = |q|^{1/2}(T+i)^{-1}(T+i)(T+V-i)^{-1}$$

is also of Hilbert-Schmidt type, since $(T+i)(T+V-i)^{-1}$ is bounded. This implies that

$$\overline{(T+V+i)^{-1}|q|^{1/2}} = [|q|^{1/2}(T+V-i)^{-1}]^*$$

is of Hilbert-Schmidt type, and therefore $(T+V+i)^{-1} - (T+i)^{-1}$ is of trace class.

<u>Remarks.</u> A combination of the above result with the methods of [2] shows that the existence and completeness of the wave operators $W_\pm(T+V,T)$ holds, if q satisfies the condition

$$|q(x)| \leq \begin{cases} C(1+|x|)^{-\delta} & \text{for } x \to \infty, \\ C(1+|x|)^{-\frac{1}{2}-\delta} & \text{for } x \to -\infty. \end{cases}$$

In order to prove this we apply the method of [2] for the potential $q_1 = \chi_{(-\infty,0)}q$ and then the above method to the additional perturbation $q_2 = q-q_1$ (notice that the above method also applies for q_2 if T is replaced by $T_1 = T+V_1$, where V_1 is the multiplication by q_1). With the same additional assumption as in [2] the unitarity of the wave operators may be proved; we omit the details.

<u>References</u>

[1] T.Kato: Perturbation theory for linear operators. Berlin-Heidelberg-New York 1966.

[2] K.Veselič and J.Weidmann: Potential scattering in a homogeneous electro static field. To appear.

[3] J.Weidmann: Perturbations of selfadjoint operators in $L_2(G)$ with applications to differential operators. Proc. Conf. Ordinary and Partial Diff. Equ. Dundee/Scotland 1974, Lecture Notes in Mathematics Vol. 415, Berlin-Heidelberg-New York 1974.

Eigenvalue problems for free vibrations of rectangular elastic plates with random inhomogeneities

A. D. Wood and F. D. Zaman

1. Introduction

In recent years there has been increasing interest in the motion of elastic bodies whose material properties (density, elastic moduli, etc.) are not constant, but vary with position, perhaps in a random manner. This interest has been motivated by new materials currently in use. An approach to the problem of free vibration of such bodies has been made by Sobczyk [1] who determined the effect of random inhomogeneities in the Young's modulus E, and hence in the flexural rigidity D, of an elastic rectangular plate on its natural frequencies. Sobczyk assumed that the Poisson's ratio ν and density ρ remained constant throughout the plate. While certain real examples are covered by this analysis, there are materials, such as glass fibres in epoxy resin, where the considerable variation in ν and ρ cannot be ignored. We shall discuss composite plates in §3, but first we obtain a general mathematical theory for randomly inhomogeneous elastic plates where D, ν and ρ are random functions of position (x,y) and a statistical parameter γ (see Boyce [2]). We assume that the plate remains isotropic and the fundamental requirements of plate theory are satisfied. Our aim is to obtain an expression for the variance of the natural frequency Ω_{mn} about its mean $\Omega_{mn}^{(0)}$ which will be the natural frequency of a homogeneous comparison plate. We shall take simply supported boundary conditions so that we have expressions for $\Omega_{mn}^{(0)}$ and its associated normal mode in closed form. Under these conditions the equation of motion for free transverse vibrations is

$$\frac{\partial^2}{\partial x^2}\{D(\frac{\partial^2 w}{\partial x^2} + \nu\frac{\partial^2 w}{\partial y^2})\} + \frac{\partial^2}{\partial y^2}\{D(\frac{\partial^2 w}{\partial y^2} + \nu\frac{\partial^2 w}{\partial x^2})\} + 2\frac{\partial^2}{\partial x\partial y}\{D(1-\nu)\frac{\partial^2 w}{\partial x\partial y}\} + \rho h\frac{\partial^2 w}{\partial t^2} = 0 \tag{1}$$

where $w(x,y,t)$ is the transverse displacement and h the thickness of the plate, with boundary conditions

$$\begin{aligned} w(x,y,t) &= \frac{\partial^2 w}{\partial x^2}(x,y,t) = 0 \qquad \text{for } x = 0,a, \\ w(x,y,t) &= \frac{\partial^2 w}{\partial y^2}(x,y,t) = 0 \qquad \text{for } y = 0,b, \end{aligned} \tag{2}$$

We shall assume that the mean values $\langle D(x,y;\gamma)\rangle$, $\langle\nu(x,y;\gamma)\rangle$ and $\langle\rho(x,y;\gamma)\rangle$ are constants D_o, ν_o and ρ_o independent of position (x,y) (ergodic hypothesis). We take the random inhomogeneities to be small in the sense that

$$\begin{aligned} D(x,y;\gamma) &= D_o(1 + \varepsilon D_1(x,y;\gamma)) \\ \nu(x,y;\gamma) &= \nu_o(1 + \varepsilon\nu_1(x,y;\gamma)) \\ \rho(x,y;\gamma) &= \rho_o(1 + \varepsilon\rho_1(x,y;\gamma)) \end{aligned} \tag{3}$$

where ε is a small parameter, which may be thought of as the "amplitude" of the inhomogeneity. It follows that the mean values of these "errors" D_1, ν_1, ρ_1 vanish for all (x,y) and these functions are statistically homogeneous. We shall show in §2 that the mean value of the natural frequency is that of the homogeneous plate with material properties D_o, ν_o, ρ_o. Given these mean values and the correlation functions for D_1, ν_1, ρ_1 we obtain an approximation to the variance of the natural frequency.

2. The perturbed eigenvalue problem with random coefficients

For free vibrations we assume a solution of (1), (2) in the form

$$w(x,y,t) = \sum_{m,n=1}^{\infty} W_{mn}(x,y)e^{i\omega_{mn}t}$$

Substituting this in (1), using (3), gives

$$\frac{\partial^2}{\partial x^2}\{(1 + \varepsilon D_1)(\frac{\partial^2 W_{mn}}{\partial x^2} + \nu_o(1 + \varepsilon\nu_1)\frac{\partial^2 W_{mn}}{\partial y^2})\}$$

$$+ \frac{\partial^2}{\partial y^2}\{(1 + \varepsilon D_1)(\frac{\partial^2 W_{mn}}{\partial y^2} + \nu_o(1 + \varepsilon\nu_1)\frac{\partial^2 W_{mn}}{\partial x^2})\}$$

$$+ 2\frac{\partial^2}{\partial x\partial y}\{(1 + \varepsilon D_1)\{1 - \nu_o(1 + \varepsilon\nu_1)\}\frac{\partial^2 W_{mn}}{\partial x\partial y}\} \tag{4}$$

$$= \frac{\rho_o(1 + \varepsilon\rho_1)h\omega_{mn}^2}{D_o}W_{mn} \equiv \Omega_{mn}(1 + \varepsilon\rho_1(x,y;\gamma))W_{mn} \text{ say.}$$

It can easily be checked that $W_{mn}(x,y)$ satisfies the same boundary conditions (2) as $w(x,y,t)$. We now have an eigenvalue problem for the non-dimensional natural frequency Ω_{mn}. Given that ρ_1 is a continuous function of (x,y) and D_1 and ν_1 are twice continuously differentiable functions of (x,y), analytic perturbation theory is valid: see Kato [5] Ch. VII. Suppressing for the moment the dependence of Ω and W on m,n, the expansions

$$\Omega = \Omega^{(0)} + \varepsilon\Omega^{(1)} + \varepsilon^2\Omega^{(2)} + \ldots \tag{5}$$

$$W(x,y) = W^{(0)}(x,y) + \varepsilon W^{(1)}(x,y) + \varepsilon^2 W^{(2)}(x,y) + \ldots \tag{6}$$

are valid in a neighbourhood of $\varepsilon = 0$. Substituting (5) and (6) into (4) and (2) and comparing the terms in which ε does not appear gives the corresponding eigenvalue problem for the homogeneous plate with material constants D_o, ν_o, ρ_o. The solution of this problem is well-known (Warburton [6]) to be

$$W_{mn}^{(0)}(x,y) = A_{mn}\sin\frac{m\pi x}{a}\sin\frac{n\pi y}{b} \qquad m,n = 1,2,\ldots$$

$$\Omega_{mn}^{(0)} = (\frac{m^2\pi^2}{a^2} + \frac{n^2\pi^2}{b^2})^2 . \tag{7}$$

Now comparing the coefficients of ε gives the problem

$$\frac{\partial^2}{\partial x^2}\left[\frac{\partial^2 W^{(1)}}{\partial x^2} + \nu_o\frac{\partial^2 W^{(1)}}{\partial y^2} + \nu_o\nu_1\frac{\partial^2 W^{(0)}}{\partial y^2} + D_1\left(\frac{\partial^2 W^{(0)}}{\partial x^2} + \nu_o\frac{\partial^2 W^{(0)}}{\partial y^2}\right)\right]$$
$$+ \frac{\partial^2}{\partial y^2}\left[\frac{\partial^2 W^{(1)}}{\partial y^2} + \nu_o\frac{\partial^2 W^{(1)}}{\partial x^2} + \nu_o\nu_1\frac{\partial^2 W^{(0)}}{\partial x^2} + D_1\left(\frac{\partial^2 W^{(0)}}{\partial y^2} + \nu_o\frac{\partial^2 W^{(0)}}{\partial x^2}\right)\right]$$
$$+ 2\frac{\partial^2}{\partial x\partial y}\left[(1 - \nu_o)\frac{\partial^2 W^{(1)}}{\partial x\partial y} + (D_1(1 - \nu_o) - \nu_o\nu_1)\frac{\partial^2 W^{(0)}}{\partial x\partial y}\right]$$
$$= \Omega^{(0)}W^{(1)} + \Omega^{(1)}W^{(0)} + \rho_1\Omega^{(0)}W^{(0)} \qquad (8)$$

with the same boundary conditions (2). Continuing in this way we may obtain eigenvalue problems for the higher coefficients in the series in ε. We shall be content with the first approximation $\Omega = \Omega^{(0)} + \varepsilon\Omega^{(1)}$.

$\Omega^{(1)}$ may be obtained from (8) by the method given in Titchmarsh [7] Ch. 19. We obtain

$$M\Omega^{(1)} = \int_0^a\int_0^b\left\{\frac{\partial^2}{\partial x^2}\left[(D_1\nu_o + \nu_o\nu_1)\frac{\partial^2 W^{(0)}}{\partial y^2} + D_1\frac{\partial^2 W^{(0)}}{\partial x^2}\right]W^{(0)}\right.$$
$$+ \frac{\partial^2}{\partial y^2}\left[(D_1\nu_o + \nu_o\nu_1)\frac{\partial^2 W^{(0)}}{\partial x^2} + D_1\frac{\partial^2 W^{(0)}}{\partial y^2}\right]W^{(0)}$$
$$\left.+ 2\frac{\partial^2}{\partial x\partial y}\left[\{D_1(1 - \nu_o) - \nu_o\nu_1\}\frac{\partial^2 W^{(0)}}{\partial x\partial y}\right]W^{(0)} - \Omega^0\rho_1 W^{(0)^2}\right\}dxdy$$

where $M = \int_0^a\int_0^b W^{(0)^2}\,dxdy$. (9)

Substituting (7) in (9) and integrating by parts gives, on using (2),

$$\Omega^{(1)}_{mn} = P_{mn}\int_0^a\int_0^b D_1\sin^2\frac{m\pi x}{a}\sin^2\frac{n\pi y}{b}\,dxdy$$
$$+ Q_{mn}(1 - \nu_o)\int_0^a\int_0^b D_1\cos^2\frac{m\pi x}{a}\cos^2\frac{n\pi y}{b}\,dxdy$$
$$+ 2Q_{mn}\nu_o\int_0^a\int_0^b \nu_1\left\{\sin^2\frac{m\pi x}{a}\sin^2\frac{n\pi y}{b} - \cos^2\frac{m\pi x}{a}\cos^2\frac{n\pi y}{b}\right\}dxdy$$
$$- R_{mn}\int_0^a\int_0^b \rho_1\sin^2\frac{m\pi x}{a}\sin^2\frac{n\pi y}{b}\,dxdy \qquad (10)$$

where

$$P_{mn} = \frac{4\pi^4}{ab}\left(\frac{m^4}{a^4} + \frac{n^4}{b^4} + 2\nu_o \frac{m^2n^2}{a^2b^2}\right), \quad Q_{mn} = \frac{8m^2n^2\pi^4}{a^3b^3}, \quad R_{mn} = 4ab\left(\frac{m^2\pi^2}{a^2} + \frac{n^2\pi^2}{b^2}\right)^2. \tag{11}$$

Neglecting terms of order >2 in ε, the variance of Ω_{mn} is given by

$$\varepsilon^2 \langle \Omega_{mn}^{(1)^2} \rangle . \tag{12}$$

The right hand side of (10) is of form $\sum_{i=1}^{4} \int_0^a\int_0^b p_i(x,y,\gamma)\phi_i(x,y)dxdy$

where $p_1(x,y,\gamma) = D_1(x,y:\gamma)$, $\phi_1(x,y) = P_{mn} \sin^2 \frac{m\pi x}{a} \sin^2 \frac{n\pi y}{b}$, etc..

This is a weakly correlated random process of the type described in [2], page 26. Hence the mean square of $\Omega_{mn}^{(1)}$ is given by

$$\langle \Omega_{mn}^{(1)} \rangle = \sum_{i=1}^{4} \sum_{j=1}^{4} \int_0^a\int_0^b\int_0^a\int_0^b K_{ij}(x_1,y_1;x_2,y_2)\phi_i(x_1,y_1)\phi_j(x_2,y_2)dx_1dy_1dx_2dy_2 \tag{13}$$

where $K_{ij}(x_1,y_1,x_2,y_2)$ is the correlation function for $p_i(x_1,y_1;\gamma)$ and $p_j(x_2,y_2;\gamma)$.

3. The correlation functions for special types of composite plates

There are well-documented experimental methods for the determination of the correlation functions in (13): see Corson [3] or Miller [4]. Consider a composite plate which is macroscopically isotropic, for instance glass spheres in an epoxy resin: at this stage our analysis cannot cope with anisotropic plates. Let $P_{ij}(x_1,y_1;x_2,y_2)$ be the probability that (x_1,y_1) lies in phase i and (x_2, y_2) in phase j of the material. These probabilities have been determined by Corson for aluminium-lead plates by electron miscroscopy. Following Corson we take the correlation between the values of D_1, say, at two different points

(x_1,y_1) and $x_2,y_2)$ in a 2-phase composite to be

$$K_{11}(x_1,y_1;x_2,y_2) = D_{1,1}^2 P_{11}(x_1,y_1;x_2,y_2) + 2D_{1,1}D_{1,2}P_{12}(x_1,y_1;x_2,y_2) + D_{1,2}^2 P_{22}(x_1,y_1;x_2,y_2) \tag{14}$$

where $D_{1,i}$ is the value of D_1 in phase i, i = 1,2. There are similar expressions for the other correlations. The results of Corson suggest taking

$$\begin{aligned} P_{11}(x_1,y_1;x_2,y_2) &= c^2 + c(1-c)\{(1+c)\,P_1(r) - cP_2(r)\} \\ P_{12}(x_1,y_1;x_2,y_2) &= c(1-c)\{1 - cP_1(r) - (1-c)P_2(r)\} \\ P_{22}(x_1,y_1;x_2,y_2) &= (1-c)^2 + c(1-c)\{(2-c)P_1(r) - (1-c)P_2(r)\} \end{aligned} \tag{15}$$

where r is the distance between (x_1,y_1) and (x_2,y_2), c is the volume concentration of phase 1 and $P_i(r) = \exp[-k_i r^{n_i}]$, i = 1,2 where the constants k_i, n_i are determined experimentally and depend on the geometry of the material. Substituting (15) in (14) and then in (13) we may evaluate the integrals to obtain an expression for $\langle\eta_{mn}^{(1)^2}\rangle$.

In this paper we have taken D_o, ν_o and ρ_o as data. The problem of predicting D_o and ν_o from the values of E, ν and ρ in the separate phases is extremely difficult.

Both authors wish to acknowledge helpful discussions with Norman Laws, Professor of Theoretical Mechanics at Cranfield.

References

1. K. Sobczyk, Journal of Sound and Vibration (1974) 22 (1), 33-39.

2. W. E. Boyce, article in "Probabilistic Methods in Applied Mathematics" (editor A. T. Bharucha-Reid), London, Academic Press (1968).

3. P. B. Corson, Journal of Applied Physics (1974) Vol. 45, No. 7, 3159-3170.

4. M. N. Miller, Journal of Mathematical Physics (1969) Vol. 10, No. 11, 1988-2019.

5. T. Kato, "Perturbation Theory of Linear Operators", Berlin, Springer Verlag (1966).

6. G. B. Warburton, Proc. Inst. Mech. Engineers (London) (1954), 168, 371-384.

7. E. C. Titchmarsh, "Eigenfunction Expansions", Part II, Oxford (1958).

Limit Point Conditions for Powers

Anton Zettl

We consider linear ordinary real symmetric differential expressions

(1) $$My = \sum_{i=0}^{n} (-1)^i (p_i y^{(i)})^{(i)}$$

on the interval $[0,\infty)$. The coefficient functions p_i are assumed to be real valued with $p_n(t) > 0$ for $t \geq 0$. Since we wish to take powers $M^2, M^3, \ldots$ we assume, for simplicity, that $p_i \in C^\infty$ for each $i = 0,\ldots,n$.

The deficiency index $d(M)$ can be defined as the number of linearly independent solutions of

(2) $$My = \lambda y$$

which are in $L^2(0,\infty)$. Here λ is a non-real complex number. It was shown by Glazman [9] that the integer $d(M)$ is independent of the non-real complex number λ, satisfies the inequality

(3) $$n \leq d(M) \leq 2n$$

and every integer in this range is realized as $d(M)$ for some M of type (1).

Powers of M can be formed in the natural way: $M^2y = M(My),\ldots,M^{n+1}y = m(M^n y)$. These powers M^k are again symmetric expressions of type (1) - see [1]. Thus the coefficients p_i determine not only the deficiency index of $d(M)$ but a whole sequence $d(M)$, $d(M^2)$, $d(M^3),\ldots$ of deficiency indices.

Which sequences of positive integers are realizable as deficiency sequences?

Clearly every term must satisfy (3), but are there any other restrictions? In the Proceedings of the 1974 Dundee Conference, Springer-Verlag, "Lecture Notes in Mathematics" no. 415, pp. 293-301 we presented some new results which placed further restrictions on the possible such sequences. In particular it was shown there that

$$(4) \qquad d(M^k) \geq kd(M).$$

By comparing the deficiency indices of different powers with each other rather than just with $d(M)$ R. M. Kauffman improved (for $k > 2$) (4):

<u>Theorem 1</u>. (Kauffman [13]). The sequence $d(M^2) - d(M)$, $d(M^3) - d(M^2)$. $d(M^4) - d(M^3)$,... is non-decreasing and bounded above by the order of M.

T. T. Read [16] constructed examples to show that all sequences not specifically ruled out by Theorem 1 actually occur. Thus we know precisely which sequence of positive integers are realizable as deficiency index sequences $d(M)$, $d(M^2)$, $d(M^3)$,... .

We say that the expression M in (1) is in the limit-point case or simply limit-point if the deficiency index is minimal i.e. $d(M) = n$.

In this paper we survey the results which have been obtained very recently, most of them are not in print at the time of this writing, dealing with the question: Under what conditions do all the terms of the sequence $d(M)$, $d(M^2)$, $d(M^3)$,... take on a given set of values allowed by Theorem 1? In particular, when do all terms of this sequence take on their maximum possible values? their minimum values? The first question is relatively easy to answer.

Theorem 2. (Zettl [19]). Let M be given by (1). Then $d(M) = 2n$ if and only if $d(M^k) = 2kn$ for any $k = 2, 3, 4, \ldots$.

To get some answers to the second question we look for sufficient conditions on the coefficients such that $d(M^k) = kd(M)$. In particular we examine some of the well known sufficient conditions for the limit-point case and ask if they are also sufficient for (some or all) powers of M to be in the limit-point case.

Consider second order expressions first. Let

$$My = -(py')' + qy \tag{5}$$

on $[0,\infty)$ with p, q real valued C^∞ functions. Perhaps the best known limit-point condition is that of Levinson [15]:

(6) There exists a positive differentiable function Q and positive constants k_1, k_2 such that

(a) $q(t) \geq -k_1 Q(t)$ for $t \geq 0$

(b) $p(t)Q'^2(t)Q^{-3}(t) \leq k_3$ for $t \geq 0$

(c) $\int_0^\infty \{p(t)Q(t)\}^{-1/2} dt = \infty$.

We can now state our first general result. This is due to W. D. Evans and A. Zettl. Some special cases of this result were previously obtained by Everitt and Giertz [6,7], Read [17], and Kumar [14].

Theorem 3. (Evans-Zettl [3]). Let M be given by (5) with the coefficients p, q satisfying Levinson's condition (6). Then $d(M^k) = k$ for $k = 1,2,3,\ldots$.

Consequently, by a result of Kauffman [12], P(M) is limit point for any real polynomial P(x).

As a special case we have

Corollary 4. Suppose there exist positive numbers k_1, k_2 and a number $\alpha \leq 2$ such that

(7) $$0 < p(t) \leq k_1 t^{\alpha}, \quad q(t) \geq -k_2 t^{2-\alpha}.$$

Then M^k is limit point for any $k = 1,2,3,\ldots$. In particular $d(M^k) = k$ if

(8) $$p(t) = 1, \text{ for } t \geq 0 \text{ and } q(t) \geq -k_2 t^2, \; k_2 > 0, \; t \geq 0.$$

Conditions (8) are known to be best possible in the sense that if $p(t) = 1$ and $q(t) = t^{2+\varepsilon}$ for $\varepsilon > 0$ then $d(M) = 2$, see [1].

It is interesting and perhaps surprising to note that, although conditions (7) are best possible as far as powers of t are concerned, they nevertheless can be made to yield more information. To put it another way, all the possible sequences of deficiency indices $d(M)$, $D(M^2)$, $d(M^3),\ldots$ with $d(M) = 1$ which are not minminal (i.e. $d(M^k)$ is not k for all k) must be generated by an expression (5) which is in the limit-point case but whose coefficients p and q do not satisfy the Levinson condition (6).

In the higher order case perhaps the best known limit-point condition on the coefficients is that of Hinton [16]. It almost reduces to the Levinson condition in the second order case.

(9) Suppose there exists a positive C^n function R and positive constants k_1, k_2 such that

i) $|p_i| R^{4i}/p_n \leq k_1$, $i = 1,2,\ldots,n-1$

ii) $-p_o R^{4n}/p_n \leq k_2$,

iii) $R'R$ and $R^2 p_n'/p_n$ are bounded,

iv) $(R^{4n-2}/p_n)^{(j)} = O(R^{4n-2-2j}/p_n)$, $j = 1,2,\ldots,n$,

v) $(R^{4n}/p_n)^{(j)} = O(R^{4n-2j}/p_n)$, $j = 1,2,\ldots,n$,

vi) $\int_0^\infty (R^{4n-2}/p_n) = \infty$.

Theorem 5. (Evans-Zettl [4]). Let M be given by (1). If the coefficients p_i satisfy (9), then M^k is in the limit-point case for $k = 1,2,3,\ldots$. Furthermore $P(M)$ is limit-point for any real polynomial $P(x)$ by Kauffman's result [12].

Taking $R(t) = t^{(\alpha-1)/(4n-2)}$ we get

Corollary 6. Let $p_n(t) = t^\alpha$, $\alpha \leq 2n$, $|p_i| = O(t^{\gamma_i})$ where $\gamma_i = [4i + \alpha(4n-4i-2)]/(4n-2)$, $i = 1,\ldots,n-1$ and $p_0(t) \geq -Kt^{(4n-2\alpha)/(4n-2)}$ for $t \geq 0$ and some $k > 0$. Then M^k is limit-point for any $k = 1,2,3,\ldots$ and moreover $P(M)$ is limit-point for any polynomial $P(x)$.

Applying Corollary 5 to the special case

(10) $$My = (-1)^n y^{(2n)} + qy$$

we have

Corollary 7. Let M be given by (10) and suppose

(11) $$q(t) \geq -Kt^{4n/(4n-2)}, \text{ for some } K > 0, \ t \geq 0.$$

then M^k is limit-point for $k = 1,2,3,\ldots$, and in fact $P(M)$ is limit-point for any polynomial $P(x)$.

For $n = 2$ the power $4/3$ in (11) is known to be best possible [5].

The next result extends some of the "interval type" limit point conditions, first considered by P. Hartman [8], to powers.

Theorem 8. (Evans-Zettl [3]). Let M be given by (1). Suppose there exists a sequence of intervals $[a_r, b_r]$ such that $a_r \to \infty$ as $r \to \infty$ and $b_r - a_r \geq \delta > 0$ for all r. If there exist positive constants k_1, k_2 such that on these intervals

i, $p_n = 1$ and $|p_i| \leq k_1$, $i = 1,2,\ldots,n-1$,

ii, $p_0 \geq -k_2$,

then M^k (and hence any polynomial in M) is limit-point, $k = 1,2,3,\ldots$.

An interesting example of an expression satisfying the conditions of theorem 7 is

$$My = (-1)^n y^{(2n)} + (t^\alpha \sin t)y, \text{ any } \alpha.$$

The proofs of Theorem 2 and 4, but not 7, are based on the concept of partial separation.

Definition. Let M be given by (1) and let k be a positive integer. Then M^k is partially separated if $f \in L^2(0,\infty)$, $f^{(2nk-1)}$ absolutely continuous on compact subintervals of $[0,\infty)$ and $M^k f \in L^2(0,\infty)$ together imply that $M^r f \in L^2(0,\infty)$ for every $r = 1,2,\ldots,k-1$.

Theorem 9. (Zettl [19]). Let M be given by (1) and let k be a positive integer. Then $d(M^k) = kd(M)$ if and only if M^k is partially separated.

The next result is the main lemma in the proofs of Theorems 2 and 4 (but not 7). It is stated here since it may be of independent interest.

Theorem 10. (Evans-Zettl [3,4]). Let M be given by (1). If $n = 1$ and the coefficients satisfy Levinson's condition (6) or $n \geq 2$ and the coefficients satisfy Hinton's condition (9), then M^k is partially separated for any $k = 1,2,3,\ldots$.

The proofs of these theorems are too long to be given here. The reader is referred to the indicated references.

In [8] Everitt and Giertz showed that conditions (8) are best possible - as far as powers of t are concerned - not only for M to be limit-point but also for M^2 to be limit-point. They did this by constructing, for any given $\varepsilon > 0$, a function q satisfying

$$q(t) \geq -Kt^{2+\varepsilon}, \quad K > 0$$

such that $My = -y'' + qy$ is limit-point but $d(M^2) = 3$. A shorter and simpler construction has recently been obtained which also yields a slight improvement in this result.

Theorem 11. (Eastham-Zettl [2]). For any $\varepsilon > 0$ and any $K > 0$ there exits a function q satisfying

$$q(t) \geq -Kt^2(\log t)^{2+\varepsilon} \tag{12}$$

such that $My = -y'' + qy$ is limit-point but M^2 is not i.e. $d(M^2) = 3$. ($d(M^2)$ can't be 4 by Theorem 2 and Levinson's theorem.)

Conditions (12) can be improved, by the same construction, to

$$q(t) \geq -Kt^2(\log t)^2(\log\log t)^2\ldots(\log\log\ldots\log t)^{2+\varepsilon}. \qquad (13)$$

For second order expressions M with $d(M) = 1$ and $d(M^2) = 3$ - such as the ones obtained with the above mentioned construction - we know from Theorem 1 that $d(M^k) = 2k - 1$ for $k \geq 2$. We also know, by Theorem 9, that for these M's no power M^k is partially separated.

References

1. Dunford, N. and Schwartz, J. T., "Linear operators," part II, Interscience, New York, 1963.

2. Eastham, M. S. P. and Zettl, Anton, "Second-order differential expressions whose squares are limit - 3." Proc. Royal Soc. Edinburgh, series A, (to appear).

3. Evans, W. D. and Zettl, Anton, "Levinson's limit-point criterion and powers," submitted for publication.

4. ____________________, "On the deficiency indices of powers of real $2n^{th}$ order symmetric differential expressions," J. London Math. Soc. (to appear).

5. Everitt, W. N., "On the limit-point classification of fourth-order differential equations," J. London Math. Soc. 44 (1969), 273-281.

6. Everitt, W. N. and Giertz, M., "On some properties of the powers of a formally self-adjoint differential expression." Proc. London Math. Soc. (3), 24, 149-170 (1972).

7. ____________________, "On the deficiency indices of powers of formally symmetric differential expressions." Lecture Notes in Mathematics No. 448. Berlin-Heidelberg-New York: Springer-Verlag 1975.

8. ____________________, "A critical class of examples concerning the integrable-square classification of ordinary differential expressions." Proc. Royal Soc. Edinburgh, Sec. A (to appear).

9. Glazman, I.M., "On the theory of singular differential operators," Uspehi, Mat. Nauk. (N.S.) 5, No. 6 (40), 102-135 (1950). (Russian). Amer. Math. Soc. Transl. no. 96 (1953).

10. Hartman, P., "The number of L^2 solution of $x'' + q(t)x- 0$", Amer. J. Math., 73 (1951), 635-645.

11. Hinton, D. G., "Limit-point criteria for differential equations," Can. J. Math. 24 2 (1972), 293-305.

12. Kauffman, R. M., "Polynomials and the limit-point condition," Trans. Amer. Math. Soc.

13. __________, "A rule relating the deficiency index of L^j to that of L^k," Proc. Royal Soc. Edinburgh, series A, (to appear).

14. Kumar, K. V., "A criterion for a formally symmetric fourth order differential expression to be in the limit-2 case at ∞." J. London Math. Soc. (2) 8, 134-138 (1974).

15. Levinson, N., "Criteria for the limit-point case for second order linear differential operators." Casopis Pest. Mat. Fys. 74, 17-20 (1949).

16. Read, T. T., "Sequences of deficiency indices, " Proc. Royal Soc. Edinburgh, (to appear).

17. __________, "Limit-point criteria for polynomials in a non-oscillatory expression."

18. Zettl, Anton, "Deficiency indices of polynomials in symmetric differential expressions II," Proc. Royal Soc. Edinburgh, series A, (to appear).

19. ___________, "Deficiency indices of polynomials in symmetric differential expressions II," Proc. Royal Soc. Edinburgh, series A, (to appear).

Addendum to "Limit Point Conditions for Powers"

by

W. D. Evans and Anton Zettl

The hypotheses of Theorem 3 in the above report can be weakened to include the very recent "interval type" limit point criteria of T. T. Read (see also the report by Evans [2] in these Proceedings). Let

$$My = -(py')' + qy$$

on $[0,\infty)$ with p and q real C^∞ functions and $p > 0$.

Theorem. Let $q = q_1 + q_2$. Suppose there exists a nonnegative function w such that, for some positive constants K_1, K_2, K_3, and a

i) $pw'^2 \leq K_1$ a.e.

ii) $\int_a^\infty p^{-1/2} w = \infty$

iii) $-q_1 w^2 \leq K_2$

iv) $p^{-1/2} w | \int_a^x q_2 \leq K_3$.

Then all powers of M and, more generally, all ploynomials in M are in the limit-point case.

The conditions on q_1 and q_2 can be described by saying that q_1 satisfies Levinson's condition (relative to p) and q_2 can be highly oscillatory with large amplitude but is such that its integral over $[a,x]$ does not grow too fast as a function of x. To illustrate the types of coefficients q allowed by the Theorem we mention some examples. The first two are from [3], the third from [1].

1. $My = -(ty')' - (t+te^t \sin e^t)y$

and

2. $My = -y'' -[t + t^3(\sin t)^4 + t^5 \sin t^6]y.$

In a recent paper [1] Atkinson, Eastham and McLeod studied the limit-point, limit-circle classification of

3. $My = -y'' \pm (t^\alpha \sin t^\beta)y, \alpha > 0, \beta > 0.$

Letting $q_1 = 0$ and $q_2 = \pm t^\alpha \sin t^\beta$ a simple computation shows that the hypothesis of the Theorem are satisfied for $\alpha \leq \beta$. Hence for $\alpha \leq \beta$ all powers of M given by 3 are limit-point.

REFERENCES

1. F. V. Atkinson, M. S. P. Eastham, J. B. McLeod, "The limit-point, limit-circle nature of rapidly oscillating potentials," Proc. Royal Soc. Edinburgh, ser. A (to appear).

2. W. D. Evans, "On limit-point and Dirichlet type results for second order differential expressions," these Proceedings.

3. T. T. Read, "A limit-point criterion for $-(py')' + qy$," these Proceedings.

Vol. 399: Functional Analysis and its Applications. Proceedings 1973. Edited by H. G. Garnir, K. R. Unni and J. H. Williamson. II, 584 pages. 1974.

Vol. 400: A Crash Course on Kleinian Groups. Proceedings 1974. Edited by L. Bers and I. Kra. VII, 130 pages. 1974.

Vol. 401: M. F. Atiyah, Elliptic Operators and Compact Groups. V, 93 pages. 1974.

Vol. 402: M. Waldschmidt, Nombres Transcendants. VIII, 277 pages. 1974.

Vol. 403: Combinatorial Mathematics. Proceedings 1972. Edited by D. A. Holton. VIII, 148 pages. 1974.

Vol. 404: Théorie du Potentiel et Analyse Harmonique. Edité par J. Faraut. V, 245 pages. 1974.

Vol. 405: K. J. Devlin and H. Johnsbråten, The Souslin Problem. VIII, 132 pages. 1974.

Vol. 406: Graphs and Combinatorics. Proceedings 1973. Edited by R. A. Bari and F. Harary. VIII, 355 pages. 1974.

Vol. 407: P. Berthelot, Cohomologie Cristalline des Schémas de Caracteristique p > o. II, 604 pages. 1974.

Vol. 408: J. Wermer, Potential Theory. VIII, 146 pages. 1974.

Vol. 409: Fonctions de Plusieurs Variables Complexes, Séminaire François Norguet 1970–1973. XIII, 612 pages. 1974.

Vol. 410: Séminaire Pierre Lelong (Analyse) Année 1972–1973. VI, 181 pages. 1974.

Vol. 411: Hypergraph Seminar. Ohio State University, 1972. Edited by C. Berge and D. Ray-Chaudhuri. IX, 287 pages. 1974.

Vol. 412: Classification of Algebraic Varieties and Compact Complex Manifolds. Proceedings 1974. Edited by H. Popp. V, 333 pages. 1974.

Vol. 413: M. Bruneau, Variation Totale d'une Fonction. XIV, 332 pages. 1974.

Vol. 414: T. Kambayashi, M. Miyanishi and M. Takeuchi, Unipotent Algebraic Groups. VI, 165 pages. 1974.

Vol. 415: Ordinary and Partial Differential Equations. Proceedings 1974. XVII, 447 pages. 1974.

Vol. 416: M. E. Taylor, Pseudo Differential Operators. IV, 155 pages. 1974.

Vol. 417: H. H. Keller, Differential Calculus in Locally Convex Spaces. XVI, 131 pages. 1974.

Vol. 418: Localization in Group Theory and Homotopy Theory and Related Topics. Battelle Seattle 1974 Seminar. Edited by P. J. Hilton. VI, 172 pages 1974.

Vol. 419: Topics in Analysis. Proceedings 1970. Edited by O. E. Lehto, I. S. Louhivaara, and R. H. Nevanlinna. XIII, 392 pages. 1974.

Vol. 420: Category Seminar. Proceedings 1972/73. Edited by G. M. Kelly. VI, 375 pages. 1974.

Vol. 421: V. Poénaru, Groupes Discrets. VI, 216 pages. 1974.

Vol. 422: J.-M. Lemaire, Algèbres Connexes et Homologie des Espaces de Lacets. XIV, 133 pages. 1974.

Vol. 423: S. S. Abhyankar and A. M. Sathaye, Geometric Theory of Algebraic Space Curves. XIV, 302 pages. 1974.

Vol. 424: L. Weiss and J. Wolfowitz, Maximum Probability Estimators and Related Topics. V, 106 pages. 1974.

Vol. 425: P. R. Chernoff and J. E. Marsden, Properties of Infinite Dimensional Hamiltonian Systems. IV, 160 pages. 1974.

Vol. 426: M. L. Silverstein, Symmetric Markov Processes. X, 287 pages. 1974.

Vol. 427: H. Omori, Infinite Dimensional Lie Transformation Groups. XII, 149 pages. 1974.

Vol. 428: Algebraic and Geometrical Methods in Topology, Proceedings 1973. Edited by L. F. McAuley. XI, 280 pages. 1974.

Vol. 429: L. Cohn, Analytic Theory of the Harish-Chandra C-Function. III, 154 pages. 1974.

Vol. 430: Constructive and Computational Methods for Differential and Integral Equations. Proceedings 1974. Edited by D. L. Colton and R. P. Gilbert. VII, 476 pages. 1974.

Vol. 431: Séminaire Bourbaki – vol. 1973/74. Exposés 436–452. IV, 347 pages. 1975.

Vol. 432: R. P. Pflug, Holomorphiegebiete, pseudokonvexe Gebiete und das Levi-Problem. VI, 210 Seiten. 1975.

Vol. 433: W. G. Faris, Self-Adjoint Operators. VII, 115 pages. 1975.

Vol. 434: P. Brenner, V. Thomée, and L. B. Wahlbin, Besov Spaces and Applications to Difference Methods for Initial Value Problems. II, 154 pages. 1975.

Vol. 435: C. F. Dunkl and D. E. Ramirez, Representations of Commutative Semitopological Semigroups. VI, 181 pages. 1975.

Vol. 436: L. Auslander and R. Tolimieri, Abelian Harmonic Analysis, Theta Functions and Function Algebras on a Nilmanifold. V, 99 pages. 1975.

Vol. 437: D. W. Masser, Elliptic Functions and Transcendence. XIV, 143 pages. 1975.

Vol. 438: Geometric Topology. Proceedings 1974. Edited by L. C. Glaser and T. B. Rushing. X, 459 pages. 1975.

Vol. 439: K. Ueno, Classification Theory of Algebraic Varieties and Compact Complex Spaces. XIX, 278 pages. 1975

Vol. 440: R. K. Getoor, Markov Processes: Ray Processes and Right Processes. V, 118 pages. 1975.

Vol. 441: N. Jacobson, PI-Algebras. An Introduction. V, 115 pages. 1975.

Vol. 442: C. H. Wilcox, Scattering Theory for the d'Alembert Equation in Exterior Domains. III, 184 pages. 1975.

Vol. 443: M. Lazard, Commutative Formal Groups. II, 236 pages. 1975.

Vol. 444: F. van Oystaeyen, Prime Spectra in Non-Commutative Algebra. V, 128 pages. 1975.

Vol. 445: Model Theory and Topoi. Edited by F. W. Lawvere, C. Maurer, and G. C. Wraith. III, 354 pages. 1975.

Vol. 446: Partial Differential Equations and Related Topics. Proceedings 1974. Edited by J. A. Goldstein. IV, 389 pages. 1975.

Vol. 447: S. Toledo, Tableau Systems for First Order Number Theory and Certain Higher Order Theories. III, 339 pages. 1975.

Vol. 448: Spectral Theory and Differential Equations. Proceedings 1974. Edited by W. N. Everitt. XII, 321 pages. 1975.

Vol. 449: Hyperfunctions and Theoretical Physics. Proceedings 1973. Edited by F. Pham. IV, 218 pages. 1975.

Vol. 450: Algebra and Logic. Proceedings 1974. Edited by J. N. Crossley. VIII, 307 pages. 1975.

Vol. 451: Probabilistic Methods in Differential Equations. Proceedings 1974. Edited by M. A. Pinsky. VII, 162 pages. 1975.

Vol. 452: Combinatorial Mathematics III. Proceedings 1974. Edited by Anne Penfold Street and W. D. Wallis. IX, 233 pages. 1975.

Vol. 453: Logic Colloquium. Symposium on Logic Held at Boston, 1972–73. Edited by R. Parikh. IV, 251 pages. 1975.

Vol. 454: J. Hirschfeld and W. H. Wheeler, Forcing, Arithmetic, Division Rings. VII, 266 pages. 1975.

Vol. 455: H. Kraft, Kommutative algebraische Gruppen und Ringe. III, 163 Seiten. 1975.

Vol. 456: R. M. Fossum, P. A. Griffith, and I. Reiten, Trivial Extensions of Abelian Categories. Homological Algebra of Trivial Extensions of Abelian Categories with Applications to Ring Theory. XI, 122 pages. 1975.

Vol. 457: Fractional Calculus and Its Applications. Proceedings 1974. Edited by B. Ross. VI, 381 pages. 1975.

Vol. 458: P. Walters, Ergodic Theory – Introductory Lectures. VI, 198 pages. 1975.

Vol. 459: Fourier Integral Operators and Partial Differential Equations. Proceedings 1974. Edited by J. Chazarain. VI, 372 pages. 1975.

Vol. 460: O. Loos, Jordan Pairs. XVI, 218 pages. 1975.

Vol. 461: Computational Mechanics. Proceedings 1974. Edited by J. T. Oden. VII, 328 pages. 1975.

Vol. 462: P. Gérardin, Construction de Séries Discrètes p-adiques. »Sur les séries discrètes non ramifiées des groupes réductifs déployés p-adiques«. III, 180 pages. 1975.

Vol. 463: H.-H. Kuo, Gaussian Measures in Banach Spaces. VI, 224 pages. 1975.

Vol. 464: C. Rockland, Hypoellipticity and Eigenvalue Asymptotics. III, 171 pages. 1975.

Vol. 465: Séminaire de Probabilités IX. Proceedings 1973/74. Edité par P. A. Meyer. IV, 589 pages. 1975.

Vol. 466: Non-Commutative Harmonic Analysis. Proceedings 1974. Edited by J. Carmona, J. Dixmier and M. Vergne. VI, 231 pages. 1975.

Vol. 467: M. R. Essén, The Cos $\pi\lambda$ Theorem. With a paper by Christer Borell. VII, 112 pages. 1975.

Vol. 468: Dynamical Systems – Warwick 1974. Proceedings 1973/74. Edited by A. Manning. X, 405 pages. 1975.

Vol. 469: E. Binz, Continuous Convergence on C(X). IX, 140 pages. 1975.

Vol. 470: R. Bowen, Equilibrium States and the Ergodic Theory of Anosov Diffeomorphisms. III, 108 pages. 1975.

Vol. 471: R. S. Hamilton, Harmonic Maps of Manifolds with Boundary. III, 168 pages. 1975.

Vol. 472: Probability-Winter School. Proceedings 1975. Edited by Z. Ciesielski, K. Urbanik, and W. A. Woyczyński. VI, 283 pages. 1975.

Vol. 473: D. Burghelea, R. Lashof, and M. Rothenberg, Groups of Automorphisms of Manifolds. (with an appendix by E. Pedersen) VII, 156 pages. 1975.

Vol. 474: Séminaire Pierre Lelong (Analyse) Année 1973/74. Edité par P. Lelong. VI, 182 pages. 1975.

Vol. 475: Répartition Modulo 1. Actes du Colloque de Marseille-Luminy, 4 au 7 Juin 1974. Edité par G. Rauzy. V, 258 pages. 1975. 1975.

Vol. 476: Modular Functions of One Variable IV. Proceedings 1972. Edited by B. J. Birch and W. Kuyk. V, 151 pages. 1975.

Vol. 477: Optimization and Optimal Control. Proceedings 1974. Edited by R. Bulirsch, W. Oettli, and J. Stoer. VII, 294 pages. 1975.

Vol. 478: G. Schober, Univalent Functions – Selected Topics. V, 200 pages. 1975.

Vol. 479: S. D. Fisher and J. W. Jerome, Minimum Norm Extremals in Function Spaces. With Applications to Classical and Modern Analysis. VIII, 209 pages. 1975.

Vol. 480: X. M. Fernique, J. P. Conze et J. Gani, Ecole d'Eté de Probabilités de Saint-Flour IV-1974. Edité par P.-L. Hennequin. XI, 293 pages. 1975.

Vol. 481: M. de Guzmán, Differentiation of Integrals in R^n. XII, 226 pages. 1975.

Vol. 482: Fonctions de Plusieurs Variables Complexes II. Séminaire François Norguet 1974–1975. IX, 367 pages. 1975.

Vol. 483: R. D. M. Accola, Riemann Surfaces, Theta Functions, and Abelian Automorphisms Groups. III, 105 pages. 1975.

Vol. 484: Differential Topology and Geometry. Proceedings 1974. Edited by G. P. Joubert, R. P. Moussu, and R. H. Roussarie. IX, 287 pages. 1975.

Vol. 485: J. Diestel, Geometry of Banach Spaces – Selected Topics. XI, 282 pages. 1975.

Vol. 486: S. Stratila and D. Voiculescu, Representations of AF-Algebras and of the Group U (∞). IX, 169 pages. 1975.

Vol. 487: H. M. Reimann und T. Rychener, Funktionen beschränkter mittlerer Oszillation. VI, 141 Seiten. 1975.

Vol. 488: Representations of Algebras, Ottawa 1974. Proceedings 1974. Edited by V. Dlab and P. Gabriel. XII, 378 pages. 1975.

Vol. 489: J. Bair and R. Fourneau, Etude Géométrique des Espaces Vectoriels. Une Introduction. VII, 185 pages. 1975.

Vol. 490: The Geometry of Metric and Linear Spaces. Proceedings 1974. Edited by L. M. Kelly. X, 244 pages. 1975.

Vol. 491: K. A. Broughan, Invariants for Real-Generated Uniform Topological and Algebraic Categories. X, 197 pages. 1975.

Vol. 492: Infinitary Logic: In Memoriam Carol Karp. Edited by D. W. Kueker. VI, 206 pages. 1975.

Vol. 493: F. W. Kamber and P. Tondeur, Foliated Bundles and Characteristic Classes. XIII, 208 pages. 1975.

Vol. 494: A Cornea and G. Licea. Order and Potential Resolvent Families of Kernels. IV, 154 pages. 1975.

Vol. 495: A. Kerber, Representations of Permutation Groups II. V, 175 pages. 1975.

Vol. 496: L. H. Hodgkin and V. P. Snaith, Topics in K-Theory. Two Independent Contributions. III, 294 pages. 1975.

Vol. 497: Analyse Harmonique sur les Groupes de Lie. Proceedings 1973–75. Edité par P. Eymard et al. VI, 710 pages. 1975.

Vol. 498: Model Theory and Algebra. A Memorial Tribute to Abraham Robinson. Edited by D. H. Saracino and V. B. Weispfenning. X, 463 pages. 1975.

Vol. 499: Logic Conference, Kiel 1974. Proceedings. Edited by G. H. Müller, A. Oberschelp, and K. Potthoff. V, 651 pages 1975.

Vol. 500: Proof Theory Symposion, Kiel 1974. Proceedings. Edited by J. Diller and G. H. Müller. VIII, 383 pages. 1975.

Vol. 501: Spline Functions, Karlsruhe 1975. Proceedings. Edited by K. Böhmer, G. Meinardus, and W. Schempp. VI, 421 pages. 1976.

Vol. 502: János Galambos, Representations of Real Numbers by Infinite Series. VI, 146 pages. 1976.

Vol. 503: Applications of Methods of Functional Analysis to Problems in Mechanics. Proceedings 1975. Edited by P. Germain and B. Nayroles. XIX, 531 pages. 1976.

Vol. 504: S. Lang and H. F. Trotter, Frobenius Distributions in GL_2-Extensions. III, 274 pages. 1976.

Vol. 505: Advances in Complex Function Theory. Proceedings 1973/74. Edited by W. E. Kirwan and L. Zalcman. VIII, 203 pages. 1976.

Vol. 506: Numerical Analysis, Dundee 1975. Proceedings. Edited by G. A. Watson. X, 201 pages. 1976.

Vol. 507: M. C. Reed, Abstract Non-Linear Wave Equations. VI, 128 pages. 1976.

Vol. 508: E. Seneta, Regularly Varying Functions. V, 112 pages. 1976.

Vol. 509: D. E. Blair, Contact Manifolds in Riemannian Geometry. VI, 146 pages. 1976.

Vol. 510: V. Poènaru, Singularités C^∞ en Présence de Symétrie. V, 174 pages. 1976.

Vol. 511: Séminaire de Probabilités X. Proceedings 1974/75. Edité par P. A. Meyer. VI, 593 pages. 1976.

Vol. 512: Spaces of Analytic Functions, Kristiansand, Norway 1975. Proceedings. Edited by O. B. Bekken B. K. Øksendal, and A. Stray. VIII, 204 pages. 1976.

Vol. 513: R. B. Warfield, Jr. Nilpotent Groups. VIII, 115 pages. 1976.

Vol. 514: Séminaire Bourbaki vol. 1974/75. Exposés 453 – 470. IV, 276 pages. 1976.

Vol. 515: Bäcklund Transformations. Nashville, Tennessee 1974 Proceedings. Edited by R. M. Miura. VIII, 295 pages. 1976.